Perspectives on Astronomy

About the Authors

Mike Seeds has been a Professor of Physics and Astronomy at Franklin and Marshall College in Lancaster, Pennsylvania, since 1970. In 1989 he received F&M College's Lindback Award for Distinguished Teaching. Mike's love for the history of astronomy led him to create upper-level courses on "Archaeoastronomy" and "Changing Concepts of the Universe." His research interests focus on variable stars and the automation of astronomical telescopes. Mike is the author of *Horizons: Exploring the Universe,* Tenth Edition (2008); *Astronomy: The Solar System and Beyond,* Fifth Edition (2007); and *Foundations of Astronomy,* Ninth Edition (2006), all published by Brooks/Cole. He was Senior Consultant for creation of the 20-episode telecourse accompanying his book *Horizons: Exploring the Universe.*

Dana Backman taught in the physics and astronomy department at Franklin and Marshall College in Lancaster, Pennsylvania, from 1991 until 2003. He invented and taught a course titled "Life in the Universe" in F&M's interdisciplinary Foundations program. Dana now teaches introductory astronomy, astrobiology, and cosmology courses in Stanford University's Continuing Studies Program. His research interests focus on infrared observations of planet formation, models of debris disks around nearby stars, and evolution of the solar system's Kuiper Belt. Dana is with the SETI Institute in Mountain View, California, in charge of the education and public outreach program for SOFIA (Stratospheric Observatory for Infrared Astronomy) at NASA's Ames Research Center.

Perspectives on Astronomy

Michael A. Seeds

Joseph R. Grundy Observatory,
Franklin and Marshall College

Dana E. Backman

SOFIA (Stratospheric Observatory for
Infrared Astronomy)
SETI Institute & NASA Ames Research
Center

THOMSON

BROOKS/COLE

Australia • Brazil • Canada • Mexico • Singapore • Spain
United Kingdom • United States

THOMSON

★

BROOKS/COLE

Perspectives on Astronomy
Michael A. Seeds and Dana E. Backman

Astronomy Editor: *Chris Hall*
Development Editor: *Alyssa White*
Editorial Assistant: *Shawn Vasquez*
Assistant Editor: *Sylvia Krick*
Technology Project Manager: *Samuel Subity*
Publisher/Executive Editor: *David Harris*
Marketing Manager: *Mark Santee*
Marketing Assistant: *Melissa Wong*
Managing Marketing Communications Manager: *Bryan Vann*
Content Project Manager, Editorial Production: *Teri Hyde*
Creative Director: *Rob Hugel*
Executive Art Director: *Vernon Boes*
Executive Print Buyer: *Karen Hunt*
Production Service: *Carol O'Connell, Graphic World Publishing Services*

Text Designer: *Linda Beaupré*
Photo Researcher: *Kathleen Olson*
Copy Editor: *Margaret Pinette*
Illustrator: *Precision Graphics*
Cover Designer: *Irene Morris*
Cover Image: Perspective view of Coprates Chasma and Catena—looking north (large background image): Copyright ESA/DLR/FU Berlin (G. Neukum). *New Insight:* Montage of planets, NASA/JPL. *Galaxies:* False-colored Spitzer M31 Andromeda photo, NASA/JPL-Caltech/Harvard-Smithsonian CfA/NOAO. *Water Volcanoes:* Enceladus ice volcanoes, copyright Michael Carroll.
Cover and Text Printer: *Courier Corporation/Kendallville*
Compositor: *Graphic World Inc.*

Library of Congress Control Number: 2006908077

ISBN-13: 978-0-495-11352-2

ISBN-10: 0-495-11352-2

Thomson Higher Education
10 Davis Drive
Belmont, CA 94002-3098
USA

For more information about our products, contact us at:
Thomson Learning Academic Resource Center
1-800-423-0563

For permission to use material from this text or product, submit a request online at **http://www.thomsonrights.com**.
Any additional questions about permissions can be submitted by e-mail to **thomsonrights@thomson.com**.

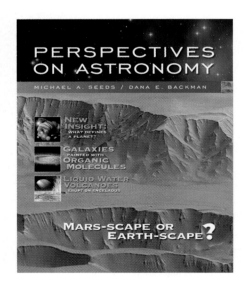

About the Cover:
Mars-scape or Earth-scape? Although this canyon wall could be mistaken for part of the Arizona desert, it is actually on Mars. This is a perspective view of Coprates Chasma and Cantena—looking north. Coprates Chasma is a major trough in the Valles Marineris canyon system. Earthly deserts and Mars look similar because they are governed by similar geological processes. By using comparative planetology, astronomers can match features seen on Mars with similar features found on Earth and unravel some of the mysteries of the red planet.

To All Our Students

Part 1: Exploring the Sky

CHAPTER 1 THE SCALE OF THE COSMOS: FROM SOLAR SYSTEM TO GALAXY TO UNIVERSE 2
CHAPTER 2 USER'S GUIDE TO THE SKY: PATTERNS AND CYCLES 10
CHAPTER 3 CIRCLES AROUND THE EARTH BECOME ORBITS AROUND THE SUN 32
CHAPTER 4 ASTRONOMICAL TELESCOPES AND INSTRUMENTS: EXTENDING HUMANITY'S VISION 56

Part 2: The Stars

CHAPTER 5 SUN LIGHT AND SUN ATOMS 76
CHAPTER 6 THE FAMILY OF STARS 102
CHAPTER 7 THE STRUCTURE AND FORMATION OF STARS 124
CHAPTER 8 THE DEATHS OF STARS 146

Part 3: Galaxies and Cosmology

CHAPTER 9 THE MILKY WAY GALAXY 174
CHAPTER 10 GALAXIES 196
CHAPTER 11 COSMOLOGY IN THE 21ST CENTURY 220

Part 4: The Solar System

CHAPTER 12 THE ORIGIN OF THE SOLAR SYSTEM 240
CHAPTER 13 COMPARATIVE PLANETOLOGY OF THE TERRESTRIAL PLANETS 264
CHAPTER 14 COMPARATIVE PLANETOLOGY OF THE OUTER PLANETS 290

Part 5: Life

CHAPTER 15 LIFE ON OTHER WORLDS 312

Contents

Part 1: Exploring the Sky

Chapter 1 | The Scale of the Cosmos: From Solar System to Galaxy to Universe 2

Chapter 2 | User's Guide to the Sky: Patterns and Cycles 10

2-1   THE STARS 12

2-2   THE SKY AND ITS MOTIONS 14

2-3   THE CYCLE OF THE SUN 18

2-4   THE CYCLES OF THE MOON 22

2-5   ECLIPSES 23

Chapter 3 | Circles Around the Earth Become Orbits Around the Sun 32

3-1   ASTRONOMY BEFORE COPERNICUS 34

3-2   NICOLAUS COPERNICUS 34

3-3   TYCHO BRAHE, JOHANNES KEPLER, AND PLANETARY MOTION 40

3-4   GALILEO GALILEI 43

3-5   ISAAC NEWTON, GRAVITY, AND ORBITS 46

Chapter 4 | Astronomical Telescopes and Instruments: Extending Humanity's Vision 56

4-1   RADIATION: INFORMATION FROM SPACE 58

4-2   TELESCOPES 60

4-3   OBSERVATORIES ON EARTH–OPTICAL AND RADIO 64

4-4   ASTRONOMICAL INSTRUMENTS AND TECHNIQUES 67

4-5   AIRBORNE AND SPACE OBSERVATORIES 69

Windows on Science

2-1   Frameworks for Thinking about Nature: Scientific Models 15

2-2   Astrology and Pseudoscience: Misusing the Rules of Science 22

3-1   Creating New Ways to See Nature: Scientific Revolutions 39

3-2   Hypothesis, Theory, and Law: Levels of Confidence 42

3-3   Universal Gravitation 48

3-4   Testing a Theory by Prediction 53

4-1   Resolving Power and the Accuracy of a Measurement 63

Concept Art Portfolios

The Sky Around Us 16–17

The Cycle of the Seasons 20–21

The Phases of the Moon 24–25

The Ancient Universe 36–37

Orbiting Earth 50–51

Modern Astronomical Telescopes 65

The Great Observatories in Space 72–73

Part 2: The Stars

Chapter 5 | Sun Light and Sun Atoms 76

5-1   THE SUN–BASIC CHARACTERISTICS   78

5-2   THE ORIGIN OF SUNLIGHT   78

5-3   THE SUN'S SURFACE   81

5-4   LIGHT, MATTER, AND MOTION   81

5-5   THE SUN'S ATMOSPHERE   85

5-6   SOLAR ACTIVITY   91

Chapter 6 | The Family of Stars   102

6-1   STAR DISTANCES   104

6-2   APPARENT BRIGHTNESS, INTRINSIC BRIGHTNESS,
      AND LUMINOSITY   105

6-3   STAR TEMPERATURES   107

6-4   STAR SIZES   109

6-5   MASSES OF STARS–BINARY STARS   113

6-6   TYPICAL STARS   117

Chapter 7 | The Structure and Formation
          of Stars   124

7-1   STELLAR STRUCTURE   126

7-2   NUCLEAR FUSION IN THE SUN AND STARS   129

7-3   MAIN-SEQUENCE STARS   132

7-4   THE BIRTH OF STARS   134

Chapter 8 | The Deaths of Stars   146

8-1   GIANT STARS   148

8-2   THE DEATHS OF LOWER-MAIN-SEQUENCE STARS   150

8-3   THE EVOLUTION OF BINARY SYSTEMS   155

8-4   THE DEATHS OF MASSIVE STARS   156

8-5   NEUTRON STARS   161

8-6   BLACK HOLES   166

Windows on Science

5-1   Confirmation, Confidence, and Consolidation   98

6-1   Learning about Nature through Chains
      of Inference   115

7-1   Quantitative Thinking with Mathematical
      Models   128

8-1   The Impossibility of Proof in Science   165

Concept Art Portfolios

Atomic Spectra   86–87

Sunspots and the Sunspot Cycle   92–93

Magnetic Solar Phenomena   96–97

The Family of Stars   120–121

Three Kinds of Nebulae   136–137

Star Formation in the Orion Nebula   142–143

Star Cluster H–R Diagrams   152–153

Celestial Profile 1 | The Sun   79

Part 3: Galaxies and Cosmology

Chapter 9 | The Milky Way Galaxy 174

9-1 THE DISCOVERY OF THE GALAXY 176

9-2 SPIRAL ARMS AND STAR FORMATION 183

9-3 THE ORIGIN AND HISTORY OF THE MILKY WAY 187

9-4 THE NUCLEUS 190

Chapter 10 | Galaxies 196

10-1 THE FAMILY OF GALAXIES 198

10-2 MEASURING THE PROPERTIES OF GALAXIES 198

10-3 THE EVOLUTION OF GALAXIES 205

10-4 ACTIVE GALAXIES AND QUASARS 210

Chapter 11 | Cosmology in the 21st Century 220

11-1 INTRODUCTION TO THE UNIVERSE 222

11-2 THE BIG BANG THEORY 224

11-3 SPACE AND TIME 229

11-4 21ST-CENTURY COSMOLOGY 232

Windows on Science

9-1 The Calibration of One Parameter to Find Another 178

10-1 Selection Effects in Science 199

10-2 It Wouldn't Stand Up in Court: Statistical Evidence 211

Concept Art Portfolios

Sagittarius A* 192–193

Galaxy Classification 200–201

Interacting Galaxies 208–209

Part 4: The Solar System

Chapter 12 | The Origin of the Solar System 240

12-1 THE GREAT CHAIN OF ORIGINS 242

12-2 A SURVEY OF THE PLANETS 243

12-3 SPACE DEBRIS: PLANET BUILDING BLOCKS 244

12-4 THE STORY OF PLANET FORMATION 249

12-5 PLANETS ORBITING OTHER STARS 258

Chapter 13 | Comparative Planetology of the Terrestrial Planets 264

13-1 A TRAVEL GUIDE TO THE TERRESTRIAL PLANETS 266

13-2 EARTH: THE ACTIVE PLANET 267

13-3 THE MOON 273

13-4 MERCURY 277

13-5 VENUS 277

13-6 MARS 281

Chapter 14 | Comparative Planetology of the Outer Planets 290

14-1 A TRAVEL GUIDE TO THE OUTER PLANETS 292

14-2 JUPITER 293

14-3 SATURN 297

14-4 URANUS 303

14-5 NEPTUNE 306

14-6 PLUTO: PLANET NO MORE 308

Windows on Science

12-1 Reconstructing the Past from Evidence and Hypothesis 251

12-2 Scientists: Courteous Skeptics 261

13-1 The Present Is the Key to the Past 287

14-1 Who Pays for Science? 302

Concept Art Portfolios

Terrestrial and Jovian Planets 246–247

The Active Earth 270–271

The Ice Rings of Saturn 300–301

Celestial Profile 2 | Earth 269

Celestial Profile 3 | The Moon and Mercury 273

Celestial Profile 4 | Venus compared with Earth 278

Celestial Profile 5 | Mars compared with Earth 282

Celestial Profile 6 | Jupiter and Saturn 295

Celestial Profile 7 | Uranus and Neptune 305

Part 5: Life

Chapter 15 | Life on Other Worlds 312

15-1 THE NATURE OF LIFE 314

15-2 THE ORIGIN OF LIFE 318

15-3 COMMUNICATION WITH DISTANT CIVILIZATIONS 325

Windows on Science

15-1 The Nature of Scientific Explanation 314

15-2 UFOs and Space Aliens 326

Concept Art Portfolios

DNA: The Code of Life 316–317

AFTERWORD 333

APPENDIX A: UNITS AND ASTRONOMICAL DATA 335

APPENDIX B: OBSERVING THE SKY 341

GLOSSARY 343

INDEX 351

A Note to the Student

From Mike and Dana

We are excited that you are taking an astronomy course because we believe knowing astronomy will give you a new perspective on your life. People who know astronomy not only see the stars and the moon in a different way, but they see everything around them with a deeper insight. Once you know how stars can explode and how planets can form and how galaxies can collide, you will see Earth in a new way. We are excited that you are about to get an astronomer's perspective on your world and on yourself.

You will find *Perspectives on Astronomy* to be a brief but complete survey of all of astronomy at the introductory college level for students with no college science or mathematics background. While this text will introduce you to the fundamentals of astronomy and to the latest scientific research, it remains focused on helping you learn concepts rather than memorize details. We want to build on your curiosity about the sky and the discoveries you have seen covered in the news by focusing on key questions you may have already considered. Questions such as: Why isn't Pluto a planet? How do we know there was a big bang? What's it like to fall into a black hole?

To help you find yourself in this very large universe, we have arranged the topics in this book in a modern order. You will study the night sky, the sun, the stars, the galaxies, and the universe as a whole. Then, when you have an astronomer's understanding of the entire cosmos, you will return to Earth to see how the sun and our solar system formed and evolved and how Earth took shape. This order will give you a new insight into how you fit into the universe.

The planets of our solar system are amazingly diverse, so we have taken a modern view of them using comparative planetology. That is, rather than discuss each planet alone and memorize a list of facts, you can use comparative planetology to compare the planets with each other and identify important concepts. Once you understand the principles, the collection of individual planets in our solar system will make sense as a family of relatives that share certain important properties.

To give you perspective on your role in the universe, we have focused on helping you answer two very important questions:

What are we?
How do we know?

By the question *What are we?* we mean how do we fit into the universe and its history? The atoms you are made of had their first birthday in the big bang when the universe began, but those atoms were cooked and remade inside stars and now they are inside you. Where will they be in a billion years? Astronomy is the only course on campus that can tell you that story, and it is a story that everyone should know.

By the question *How do we know?* we mean how does science work? How do scientists understand nature? For instance, how can anyone know there was a big bang? In today's world, you need to think carefully about the things so-called experts say. You should demand explanations. Scientists have a special way of knowing based on evidence. Scientific knowledge isn't just opinion or policy or marketing or public relations. It is the human race's best understanding of nature. To understand the world around you, you need to understand how science works. Woven throughout this text, you will find boxes called *Windows on Science*. These boxes will help you answer the question *How do we know?* by showing you how scientists use the methods of science to know what the universe is like.

Innovative Chapter Organization

- To introduce you to astronomical objects and distances, Chapter 1 will take you on a quick cosmic zoom from Earth out past planets, stars, and galaxies and finally to the clouds of galaxies that fill the universe. This first glimpse of the cosmos will give you perspective as you study each kind of astronomical object and process.

- Chapter 2, *Users Guide to the Sky: Patterns and Cycles,* will introduce you to the appearance and motion of the sky in all its beauty and diversity. It is the rest of the universe as seen from our moving planet.

- Understanding how astronomers use spectroscopes to analyze light is critical to your study of astronomy. Chapter 5 uses the sun as a case study. You will discover how astronomers can look at the sun's spectrum and learn about temperatures, compositions, motions, and even magnetic fields.

- Deep space is filled with galaxies each containing billions of stars, and you will study them in Chapter 10. Some galaxies are colliding and interacting, and others are erupting in tremendous explosions. Comparing such different galaxies will reveal important principles, and you will learn some of nature's biggest and most violent secrets.

- Cosmology, the study of the universe as a whole, is a truly mind-twisting subject, but you will find that Chapter 11 leads you from things you know, step by step, to new discoveries and finally to the most recent and most exciting research on the origin of the universe.

- Asteroids, comets, and meteors are exciting because they provide clues to how the solar system formed. Furthermore, asteroids and comets must hit Earth now and then, and a major impact could have driven the dinosaurs to extinction. Will an impact end human civilization on Earth? You will learn about these peculiar objects in Chapter 12 when you consider the formation of the solar system.

- The newest research has uncovered astonishing insights into the formation of the outer planets and the migration of their orbits. As you study the outer planets you will see why Pluto is no longer considered one of the classical planets and why it and its dwarf-planet cousins reveal new secrets in the history of our solar system.

Key Features

- **Guideposts** on the opening page of each chapter help you see the organization of the book. Cast as a series of essential questions that are answered within the text narrative, the Guideposts connect the chapter with the preceding and following chapters to provide an overall organizational guide. The essential questions from the Guidepost are revisited in the summaries at the end of each chapter.

- **Concept Art Portfolios** are "visual essays" of key concepts presented in a collage format. They provide an opportunity for you to create your own understanding. Flow through the material is clarified with numerical markers linked to guiding questions in the text.

- **Windows on Science** boxes provide a parallel commentary on how all of science works—on how scientists know about the natural world. For example, different Windows discuss the use of statistical evidence, the difference between a hypothesis and a theory, and the construction of scientific models.

- **Building Scientific Arguments** at the end of many text sections are questions designed to help you review and synthesize concepts from the section. A short answer follows to show how scientists construct scientific arguments that lead to a conclusion based on observations, evidence, theories, and natural laws. A further question then gives you a chance to construct your own argument on a related issue.

- **Celestial Profiles** use comparative planetology to show how objects in our solar system directly compare and contrast planets with each other. This is the way planetary scientists understand planets, not as isolated unrelated bodies, but as siblings with noticeable differences but many characteristics and a family history in common.

- **End-of-Chapter Review Questions** are designed to help you review and test your understanding of the material.

- **Learning-to-Look Questions** give you a chance to interpret astronomical images and build your own understanding.

- **End-of-Chapter Discussion Questions** go beyond the text and invite you to think critically and creatively about scientific questions. You can ponder these questions yourself or discuss them in class.

- **ThomsonNOW.** Take charge of your learning with the first assessment-centered learning aid for astronomy. Access ThomsonNOW via the web with the access code card bound into this book and begin to maximize your study time with a host of interactive tutorials and quizzes that help you focus on what you need to learn to master astronomy.

- **Virtual Astronomy Laboratories.** This set of 20 online labs is free with a pass code included with every new copy of this textbook. The labs cover topics from helioseismology to dark matter and allow you to submit your results electronically to your instructor or print them out to hand in notes which labs correlate to that chapter.

- **TheSky Student Edition CD-ROM.** With this CD-ROM, a personal computer becomes a powerful personal planetarium. Loaded with data on 118,000 stars and 13,000 deep-sky objects with images, it allows you to view the universe at any point in time from 4000 years ago to 8000 years in the future, to see the sky in motion, to view constellations, to print star charts, and much more.

- **Starry Night™ CD-ROM.** Packaged on request with the textbook at no additional charge, Starry Night is an incredibly powerful planetarium software package, yet is easy enough to make learning about the night sky accessible to everyone, from the backyard stargazer to the serious student. Containing data on 1 million stars as well as 110 deep-space objects such as galaxies, star clusters, and nebulae, it allows the user to view the sky from an exact location on Earth or from elsewhere within 20,000 light-years of our solar system. The user also has the option to travel through 200,000 years, from the remote past to the distant future. Numerous user-friendly features make it easy to identify constellations or jump to your location of choice. The user even has the option to create personal QuickTime movies of the experience.

Acknowledgments

Through the years we have both had the guidance of a great many people who care about astronomy and teaching, including the students and teachers who have responded so enthusiastically to Mike's books. Their comments and suggestions have been very helpful in shaping this new book's first edition.

Certain unique diagrams in Chapters 7 and 8 are based on figures Mike designed for an article on stellar evolution that appeared in *Astronomy* (February 1979).

We are happy to acknowledge the use of images and data from a number of sources. Many observatories, research institutes, laboratories, and individual astronomers listed on the credits page have supplied figures and diagrams for this book. We would like to thank them here for their generosity. In preparing materials for this book we also used NASA's SkyView facility located at Goddard Space Flight Center, plus atlas images and mosaics from 2MASS, a joint project of the University of Massachusetts and IPAC/Caltech, funded by NASA and the NSF. A number of solar images are used by the courtesy of the SOHO consortium, a project of international cooperation between the European Space Agency (ESA) and NASA.

It is always a pleasure to work with the Thomson Learning team at Brooks/Cole. Special thanks go to all of the people at Brooks/Cole who have contributed to this project including Teri Hyde, Hal Humphrey, Mark Santee, Samuel Subity, Colin Blake, and Shawn Vasquez. We have enjoyed working with Margaret Pinette, and appreciate her understanding and goodwill. We would like to thank Carol O'Connell of Graphic World for her help in keeping everything on track. We would especially like to thank our editor, Chris Hall, and our developmental editor, Alyssa White, for their help, patience, and guidance throughout this project.

Most of all we would like to thank our families and friends for putting up with "the books." They know all too well that textbooks are made of time.

Mike Seeds
mseeds@fandm.edu
Dana Backman
dbackman@sofia.usra.edu

Reviewer Acknowledgments

David Alexander, Rice University

Hyder Ali, College of the Canyons

Celso Batalha, Evergreen Valley College

Bernard Bates, University of Puget Sound

Jon U. Bell, Indian River Community College

David Bennum, University of Nevada, Reno

John L. Berkley, SUNY Fredonia

Thomas J. Berlin, Alderson-Broaddus College

S. Leslie Blatt, Clark University

Jeff R. Bodart, Chipola College

Erik Bodegom, Portland State University

Vincent J. Bonina, Johns Hopkins University

Christian Y. Cardall, University of Tennessee, Knoxville

Stan Celestian, Glendale Community College

Dr. Kristi D. Concannon, King's College, Wilkes-Barre, PA

Thomas Corwin, The University of North Carolina at Charlotte

Christopher J. Crow, IPFW

Gregory L. Dolise, Harrisburg Area Community College

Ryan E. Droste, Trident Technical College

Patrick Durrell, Youngstown State University

Robert A. Egler, North Carolina State University

Gregg Erickson, Sullivan County Community College

Tim Farris, Volunteer State Community College

Steven Federman, University of Toledo

Charles Figura, Wartburg College

Kent D. Fisher, Columbus State Community College

Giovanni Fossati, Rice University

Michael Frey, Cypress College

Robert H. Gowdy, Virginia Commonwealth University

Robert M. Hagood, Washtenaw Community College

G. Barry Hillard, Baldwin-Wallace College

Michael D. Inglis, Suffolk County Community College

Craig M. Jensen, Northern Virginia Community College

Jeremy King, Clemson University

Cindy Klevickis, James Madison University

Patrick L. Koehn, Eastern Michigan University

Ichishiro Konno, University of Texas at San Antonio

Lois Breur Krause, Clemson University

Kevin M. Lee, University of Nebraska

John Patrick Lestrade, Mississippi State University

Rudi Paul Lindner, The University of Michigan, Ann Arbor

Robert C. Mania, Jr., Kentucky State University

Mark McConnell, University of New Hampshire

Rick McDaniel, Henderson State University

Linda McDonald, North Park University

Mark R. McGovern, Antelope Valley College

Kenneth W. McLaughlin, Loras College

Rahul Mehta, University of Central Arkansas

Barry R. Metz, Delaware County Community College

Scott Miller, Pennsylvania State University

Joseph Mills, San Jacinto College

Jeff Molony, University of LaVerne

Edward Murphy, University of Virginia

Max S. Nam, Montgomery College Takoma Park/Silver Spring

Paul Nienaber, Saint Mary's University of Minnesota

Douglas A. O'Handley, Santa Clara University

Christine O'Leary, Wallace State Community College

Jacqueline Omland, Northern State University

Nicola Orsini, Marshall University

Larry Ovall, Yakima Valley Community College

Slawomir Piatek, New Jersey Institute of Technology

Dale Pleticha, Gordon College

Jerry Polson, Southeaster Oklahoma State University

Ed Rosenthal, Paradise Valley Community College

Kenneth S. Rumstay, Valdosta State University

Earl F. Skelton, Georgetown University

Joseph L. Spradley, Wheaton College

C. Michael Stinson, Southside Virginia Community College

Jack W. Sulentic, University of Alabama

Glenn P. Tiede, Bowling Green State University

Rusty Towell, Abilene Christian University

Arun C. Venkatachar, Ohio University–Chillicothe

James R. Wade, St. Clair County Community College

Robert L. Warasila, Suffolk County Community College

Alex Webster, Antelope Valley College

Richard Williamon, Emory University

Perspectives on Astronomy

1 | The Scale of the Cosmos: From Solar System to Galaxy to Universe

The longest journey begins with a single step.

CONFUCIUS

Y OU ARE ABOUT to go on a voyage to the limits of the known universe, traveling outward, away from your home on Earth, past the moon and the sun and the other planets of our solar system, past the stars you see in the night sky, and beyond billions more stars that can be seen only with the aid of telescopes. You will visit the most distant galaxies—great globes and whirlpools of stars—and continue on, carried only by experience and imagination, seeking to understand the structure of the universe. | Astronomy is more than the study of planets, stars, and galaxies—it is the study of the whole universe in which you live. Although humanity is confined to a small planet circling an average star, the study of astronomy can take you beyond these boundaries and help you not only see where you are but also understand what you are.

| Continued on page 4 |

Guidepost

Getting Started

You are already an expert on astronomy. You have enjoyed sunsets and moonrises, admired the stars, and may know a few constellations. You have probably read about Mars rovers and the Hubble Space Telescope. That is more than most Earthlings know about astronomy. Still, you owe it to yourself to understand where you are. You should know what it means to live on a planet that whirls around a star sailing through one galaxy in a universe full of galaxies.

This Chapter

Here you will take a "cosmic zoom," a quick trip from objects you know up to the largest things in the universe. This fast survey will answer a few essential questions:

Where is Earth in relation to the sun, the planets, the stars, and the galaxies?

How do astronomers describe distances?

Which of these objects are big relative to the others, and which are small?

Are there other worlds like Earth?

This chapter will give you a sense of the scale of the universe that will carry you to new discoveries.

Looking Ahead

It is easy to learn a few facts, but it is the relationships among facts that are important. This chapter will give you the sense of scale that you need to understand where you are in the universe. The remaining chapters in this book will fill in the details, give evidence, describe theories, and illustrate the wonderful intricacy and beauty of the universe. That journey begins here.

ThomsonNOW The ThomsonNOW icon throughout the text indicates an opportunity for you to test yourself on key concepts and to explore animations and interactions on the web at: **www.thomsonedu.com**

Your imagination is the key to discovery; it will be your scientific space-and-time machine transporting you across the universe and into the past and future. Go back in time to watch the formation of the sun and Earth, the birth of the first stars, and ultimately the creation of the universe. Then, rush into the future to see what will happen when the sun dies and Earth withers.

Although you will discover a beginning to the universe, you will not find an edge or an end in space. No matter how far you voyage, you will not run into a wall. In a later chapter you will discover evidence that the universe may be infinite; that is, it may extend in all directions without limit.

Astronomy will introduce you to sizes, distances, and times far beyond your usual experience on Earth. Your task in this chapter is to grasp the meaning of these unfamiliar sizes, distances, and times. The solution lies in a single word: scale. In this chapter, you will compare objects of different sizes in order to comprehend the scale of the universe.

You should begin with something familiar, like the size of yourself and your surroundings. ■ Figure 1-1 shows a region about 16 meters (52 feet) across occupied by a human being, a sidewalk, and a few trees—all objects with sizes you can relate to.

■ **Figure 1-3**

(NASA, Landsat infrared photograph)

Each picture in the following sequence shows you a frame or **field of view** within the universe that is 100 times wider than the preceding picture.

In ■ Figure 1-2, your field of view has increased in size by a factor of 100, and you can now see an area about 1.6 kilometers (1 mile) in diameter. Notice the arrow that points to the scene shown in the previous figure. Now you can see an entire college campus and the surrounding streets and houses. This is still the world you know and can relate to the scale of your body.

■ Figure 1-3 has a span of 160 kilometers (100 miles). Take a look at the picture and notice the red color. This is an infrared photograph, in which healthy green leaves and crops show up as red. Your eyes are sensitive to only a narrow range of colors. As you explore the universe, you will need to learn to use a wide range of "colors," from X rays to radio waves, revealing sights invisible to your unaided eyes.

The college campus is now invisible, and the patches of gray are towns and cities, including Wilmington, Delaware, visible in the lower right corner. At this scale, you can see natural features of Earth's surface: the Allegheny Mountains of southern Pennsylvania crossing the picture at the upper left and the Susquehanna River flowing southeast into Chesapeake Bay. What look like white bumps are actually puffs of clouds. Mountains and valleys are only temporary features on Earth that are slowly but constantly changing. As you explore the universe, you will come to see that it is also always evolving.

■ **Figure 1-2**

This box ■ represents the relative size of the previous figure. (USGS)

■ **Figure 1-1**

(Michael A. Seeds)

In the next step of the journey, you will see the entire planet Earth (■ Figure 1-4), which is about 13,000 kilometers (8000 miles) in diameter. This picture shows most of the daylight side of the planet. However, the blurriness at the extreme right is the

sunset line. The rotation of Earth on its axis each 24 hours carries you eastward, and as you cross the sunset line into darkness you say that the sun has set. At the scale of this figure, the atmosphere on which your life depends is thinner than a strand of thread.

Enlarge your field of view again by a factor of 100, and you see a region 1,600,000 kilometers (1 million miles) wide (■ Figure 1-5). Earth is the small blue dot in the center, and the moon, with a diameter only about one-fourth that of Earth, is an even smaller dot along its orbit. If you've had a high-mileage car, it may have made the equivalent of a trip to the moon, which has an average distance from Earth of 380,000 kilometers (240,000 miles). These numbers are so large that it is inconvenient to write them out. Astronomy is the science of big numbers, and you will use numbers much larger than these to describe the universe.

Rather than writing out numbers such as those in the previous paragraph, it is more convenient to write them in **scientific notation.** This is nothing more than a simple way to write numbers without writing a great many zeros. For example, in scientific notation, you can write 380,000 as 3.8×10^5.

When you once again enlarge your view by a factor of 100 (■ Figure 1-6), Earth, its moon, and the moon's orbit all lie in the small red box at lower left. This figure has a diameter of 1.6×10^8 kilometers. Now you see the sun and two other planets that are part of our solar system. The **solar system** consists of

the sun, its family of planets, and some smaller bodies, such as moons, asteroids, and comets.

Like Earth, Venus and Mercury are **planets,** small, nonluminous bodies that shine by reflecting sunlight. Venus is about the size of Earth, and Mercury is a bit larger than Earth's moon. In this figure they are both too small to be seen as anything but tiny dots. The sun is a **star,** a self-luminous ball of hot gas that generates its own energy. The sun is about 110 times larger in diameter than Earth (inset), but it too is nothing more than a dot in this view. Earth orbits the sun once a year.

Another way astronomers deal with large numbers is to define new units. The average distance from Earth to the sun is called the **Astronomical Unit (AU),** a distance of 1.5×10^8 kilometers (93 million miles). Using that unit you can then say, for example, that the average distance from Venus to the sun is about 0.7 AU.

Your first field of view was only about 16 meters (52 feet) in width. After just six steps, each enlarging by a factor of 100, you now see the entire solar system, all the major

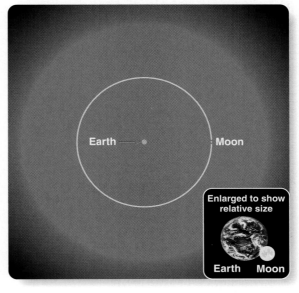

Earth —— Moon

Enlarged to show relative size

Earth   Moon

Venus

Sun

1 AU

Mercury

Earth

Enlarged to show relative size

Earth

Sun

plants, and their slightly elliptical orbits (■ Figure 1-7). Your view now is 1 trillion (10^{12}) times wider than in the first figure. The details of the previous figure are lost in the red square at the center of this figure. You see only the brighter, more widely separated objects as you back away. The sun, Mercury, Venus, and Earth are so close together that you cannot separate them at this scale.

Mars, the next outward planet, lies only 1.5 AU from the sun. In contrast, Jupiter, Saturn, Uranus, and Neptune are so far from the sun that they are easy to find in this figure. Light from the sun reaches Earth in only 8 minutes, but it takes over 4 hours to reach Neptune. Pluto orbits mostly outside Neptune's orbit, but it is no longer considered a major planet.

When you again enlarge your view by a factor of 100, the solar system becomes invisibly small (■ Figure 1-8). The sun is only a point of light, and all the planets and their slightly elliptical orbits are now crowded into the small red square at the center. The planets are too small and reflect too little light to be visible so near the brilliance of the sun.

Nor are any stars visible except for the sun. The sun

is a fairly typical star, a bit larger than average, and it is located in a fairly normal neighborhood in the universe. Although there are many billions of stars like the sun, none is close enough to be visible in this figure. The stars are separated by average distances

■ Figure 1-9

about 30 times larger than this view, which has a diameter of 11,000 AU. It is difficult to grasp the isolation of the stars. If the sun were represented by a golf ball in New York City, the nearest star would be another golf ball in Chicago.

In ■ Figure 1-9, your view has expanded to a diameter a bit over 1 million AU. The sun is at the center, and you see a few of the nearest stars. These stars are so distant that it is not reasonable to give their distances in AU. Astronomers have defined a new larger unit of distance, the light-year. One **light-year (ly)** is the distance that light travels in 1 year, roughly 10^{13} km or 63,000 AU. The diameter of your view in Figure 1-9 is 17 ly. The nearest star to the sun, Proxima Centauri, is 4.2 ly from Earth. In other words, light from Proxima Centauri takes 4.2 years to reach Earth.

Although these stars are roughly the same size as the sun, they are so far away that you cannot see them as anything but points of light. Even with the largest telescopes on Earth, you still see only points of light when you look at stars, and any planets that might circle those stars are much too small and faint to be visible. In this figure, the sizes of the dots represent not the sizes of the stars but their brightness. This is the custom in astronomical diagrams, and it is also how starlight is recorded. Bright stars make larger spots in a photograph or electronic picture than faint stars. The size of a star image in a photograph tells you not how big the star really is but only how bright it looks.

In ■ Figure 1-10, you expand your field of view by another factor of 100, and the sun and its neighboring stars van-

■ Figure 1-8

■ Figure 1-7

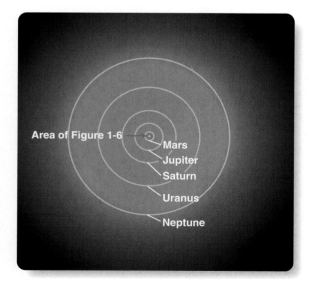

ish into the background of thousands of stars. This figure is 1700 ly in diameter. Of course, no one has ever journeyed thousands of light-years from Earth to look back and photograph the sun's neighborhood, so this is a representative picture from Earth of a part of the sky that can be used as a

■ Figure 1-10

reasonable simulation. The sun is faint enough that it would not be easily located in a picture at this scale.

Some things that are invisible in this figure are actually critically important. You do not see the thin gas that fills the spaces between the stars. Although those clouds of gas are thinner than the best vacuum produced in laboratories on Earth, it is those clouds that give birth to new stars. The sun formed from such a cloud about 5 billion years ago. You will see more star formation in the next figure.

If you expand your view again by a factor of 100, you see our galaxy (■ Figure 1-11). A **galaxy** is a great cloud of stars, gas, and dust bound together by the combined gravity of all the matter. In the night sky, you see our galaxy from the inside as a great, cloudy band of stars ringing the sky as the **Milky Way,** and our galaxy is called the **Milky Way Galaxy.** Of course, no one has photographed our galaxy, so this figure actually shows a galaxy that astronomers have evidence is similar to our own. Our sun would be invisible in such a picture, but if you could see it, you would find it about two-thirds of the way from the center to the edge. Our galaxy contains over 100 billion stars, and, like many others, has graceful **spiral arms** winding outward through the disk. You will discover in a later chapter that stars are born in great clouds of gas and dust as they pass through the spiral arms.

The visible disk of our galaxy is roughly 80,000 ly in diameter. Only a century ago, astronomers thought it was the entire

universe—an island universe of stars in an otherwise empty vastness. Now the Milky Way Galaxy is known to be not unique; it is a typical galaxy in many respects, although larger than most. In fact, ours is only one of many billions of galaxies scattered throughout the universe.

As you expand your field of view by another factor of 100, our galaxy appears as a tiny luminous speck surrounded by other specks (■ Figure 1-12). This figure includes a region 17 million ly in diameter, and each of the dots represents a galaxy. Notice that our galaxy is part of a cluster of a few dozen galaxies. You will find that galaxies are commonly grouped together in clusters. Some of these galaxies have beautiful spiral patterns like our own galaxy, but others do not. One of the questions you will investigate in a later chapter is what produces these differences among the galaxies.

Figure ■ 1-13 represents a view with a diameter of 1.7 billion light years by combining observations with theoretical calculations. The figure shows clusters of galaxies connected in a vast network. Clusters are grouped into

■ Figure 1-11

(©Anglo-Australian Telescope Board)

■ Figure 1-12

Milky Way Galaxy →

■ Figure 1-13

(Detail from galaxy map from M. Seldner, B. L. Siebers, E. J. Groth, and P. J. E. Peebles, *Astronomical Journal 82* [1977].)

superclusters—clusters of clusters—and the superclusters are linked to form long filaments and walls outlining voids that seem nearly empty of galaxies. These filaments and walls appear to be the largest structures in the universe.

Were you to expand your view frame one more time, you would probably see a uniform fog of filaments and voids. When you puzzle over the origin of these structures, you are at the frontier of human knowledge. The sequence of figures ends here because it has reached the limits of the best telescopes. Humanity's view does not extend as far as the region that would be covered by a figure 100 times larger than Figure 1-13.

A problem in studying astronomy is keeping a proper sense of scale. Remember that each of the billions of galaxies contains billions of stars. Many of those stars probably have families of planets like our solar system, and on some of those billions of planets liquid-water oceans and protective atmospheres may have sheltered the spark of life. It is possible that some other planets are inhabited by intelligent creatures who share your curiosity, wonder at the scale of the cosmos, and are looking back at you when you gaze into the heavens.

Study and Review

Summary

Your goal in this chapter was to get an overview of the scale of astronomical objects. To do so, you journeyed outward from a college campus scene by expanding your view frame by steps of factors of 100. Only 12 such steps took you to the largest known structures in the universe.

Where is Earth in relation to the sun, the planets, the stars, and the galaxies?
■ You live on planet Earth, which orbits our star, the sun, once a year.

■ As Earth rotates once a day, you see the sun rise and set.

■ The other major planets in our solar system, Mercury, Venus, Mars, Jupiter, Saturn, Uranus, and Neptune, orbit our sun in ellipses that are nearly circular.

■ Our sun is just one out of the billions of stars that fill our home galaxy, the Milky Way, which is a spiral galaxy.

■ The Milky Way Galaxy is just one of billions of galaxies arranged in great clusters, clouds, walls, and filaments that fill the universe.

How do astronomers describe distances?
■ Astronomers use scientific notation for very large or very small numbers.

■ Astronomers have also invented new units of measure such as the Astronomical Unit (AU) and the light-year (ly). One AU is the average distance from Earth to the sun. Mars, for example, orbits 1.5 AU from the sun. The light-year (ly) is the distance light can travel in one year. The nearest star is 4.2 ly from the sun.

Which objects are big relative to the others, and which are small?
■ The moon is only about one-fourth the diameter of Earth, but the sun is about 110 times larger than Earth—a typical size for a star.

■ Galaxies contain many billions of stars. You live in the Milky Way Galaxy, which is about 80,000 ly in diameter and contains over 100 billion stars.

■ The largest things in the universe are the walls and long filaments containing many clusters of galaxies.

Are there other worlds like Earth?
■ Planets as small as Earth are difficult to detect around other stars. Many stars may have families of planets like our solar system, and some of those billions of planets may resemble Earth.

New Terms

field of view   (p. 4)

scientific notation   (p. 5)

solar system   (p. 5)

planet   (p. 5)

star   (p. 5)

Astronomical Unit (AU)   (p. 5)

light-year (ly)   (p. 6)

galaxy   (p. 7)

Milky Way   (p. 7)

Milky Way Galaxy   (p. 7)

spiral arm   (p. 7)

supercluster   (p. 8)

Review Questions

ThomsonNOW™ Assess your understanding of this chapter's topics with additional quizzing and animations at **www.thomsonedu.com**

1. What is the largest dimension you have personal knowledge of? Have you run a mile? Hiked 10 miles? Run a marathon? Driven across the country?

2. In Figure 1-4, the division between daylight and darkness is at the right on the globe of Earth. How do you know this is the sunset line and not the sunrise line?

3. What is the difference between our solar system, our galaxy, and the universe?

4. Look at Figure 1-6. How can you tell that Mercury follows an elliptical orbit? Can you detect the elliptical shape of any other orbits in this figure or the next?

5. Which is the outermost major planet in our solar system?

6. Why are light-years more convenient than miles, kilometers, or astronomical units for measuring certain distances?

7. Why is it difficult to detect planets orbiting other stars?

8. What does the size of the star image in a photograph tell us?

9. What is the difference between the Milky Way and the Milky Way Galaxy?

10. What are the largest known structures in the universe?

11. Of the objects listed here, which would be contained inside the object shown at right? Which would contain the object? Stars, planets, galaxy clusters, supercluster filaments, spiral arms.

(Bill Schoening/ NOAO/AURA/ NSF)

2 | User's Guide to the Sky: Patterns and Cycles

The Southern Cross I saw every

night abeam. The sun every

morning came up astern;

every evening it went down

ahead. I wished for no other

compass to guide me, for

these were true.

CAPTAIN JOSHUA SLOCUM,
SAILING ALONE AROUND THE WORLD

THE NIGHT SKY is the rest of the universe as seen from our planet. When you look up at the stars, you look out through a layer of air only about 100 kilometers deep. Beyond that, space is nearly empty, with the planets of our solar system several AU away and the far more distant stars scattered many light-years apart. You can begin your understanding of the natural laws that govern the universe by carefully noting what the universe looks like and how it behaves. As you read this chapter, keep in mind that you live on a planet, a moving platform. Earth rotates on its axis once a day, so from our viewpoint sky objects appear to rotate around us each day. For example, the sun rises in the east and sets in the west, and also so do the stars. The sun, the moon, planets, stars, and galaxies all have an apparent daily motion that is not real but is caused by a real motion of Earth.

Guidepost

Looking Back

As you learn about astronomy, you learn about yourself. The "cosmic zoom" in the previous chapter showed you and Earth in relation to other objects in the universe. Even in that quick preview, as you learned about stars and galaxies, you were also learning about your own place in the universe.

This Chapter

Now it is time to return to Earth and look carefully at the sky as it appears to your unaided eyes. You have noticed stars and may know a few constellations and star names, but now you can understand those terms more completely. The daily, monthly, and yearly cycles of the sky have direct influences on your life and also produce dramatic sights. As you explore the sky's patterns and cycles you will find answers to these questions:

How do astronomers refer to stars?

How can you compare the brightness of stars?

How does the sky appear to move as Earth rotates?

What causes the seasons?

Why does the moon go through phases?

What is a solar eclipse?

What is a lunar eclipse?

This chapter will give you a tour of the sky and vocabulary to describe it. This chapter also illustrates a powerful way to analyze certain kinds of problems. If a process repeats, you can understand it by searching for regular cycles. The apparent motions of the sun and moon become simple and elegant when you see them as cycles.

Looking Ahead

The remaining chapters of this book will fill in details and help you understand objects in the sky. Appearances can be deceiving; although it seems as if you and Earth don't move and are at the center of the universe, Chapter 3 will show how humans finally came to understand that we live on a moving planet. In fact, you will discover that modern science was born as the result of people trying to understand fully the appearance of the sky and its cycles.

ThomsonNOW The ThomsonNOW icon throughout the text indicates an opportunity for you to test yourself on key concepts and to explore animations and interactions on the web at: **www.thomsonedu.com**

2-1 The Stars

ON A DARK NIGHT FAR FROM CITY LIGHTS, you can see a few thousand stars in the sky. Your observations can be summarized by naming individual stars and groups of stars and by specifying their relative brightness.

Constellations

All around the world, ancient cultures celebrated heroes, gods, and mythical beasts by naming groups of stars called **constellations** (■ Figures 2-1a and 2-1b). You should not be surprised that the star patterns do not look like the creatures they are named after any more than Columbus, Ohio, looks like Christopher Columbus.

The constellations named within western culture originated in ancient civilizations of Mesopotamia, Babylon, Egypt, and Greece beginning as much as 5000 years ago. Of those ancient constellations, 48 are still in use. In those former times, a constellation was simply a loose grouping of bright stars, and many of the fainter stars were not included in any constellation. Regions of the southern sky not visible to ancient astronomers living at northern latitudes also were not identified with any constellations. Constellation boundaries, when they were defined at all, were only approximate (Figure 2-1a), so a star like Alpheratz could be thought of both as part of Pegasus and also part of Andromeda.

In recent centuries, astronomers have added 40 modern constellations to fill gaps, and in 1928 the International Astronomical Union (IAU) established a total of 88 official constellations with clearly defined permanent boundaries (Figure 2-1b) that together cover the entire sky. Thus a constellation now represents not a group of stars but a section of the sky—a viewing direction—and any star within the region belongs only to that one constellation.

In addition to the 88 official constellations, the sky contains a number of less formally defined groupings called **asterisms.** The Big Dipper, for example, is an asterism you probably recognize that is part of the constellation Ursa Major (the Great Bear). Another asterism is the Great Square of Pegasus (Figure 2-1b) that includes three stars from Pegasus and the previously mentioned star Alpheratz, now considered to be part of Andromeda only.

Although constellations and asterisms are named as if they were real groupings, most are made up of stars that are not physically associated with one another. Some stars may be many times farther away than others in the same constellation and moving through space in different directions. The only thing they have in common is that they lie in approximately the same direction as viewed from Earth (■ Figure 2-2).

■ Figure 2-1

(a) In antiquity, constellation boundaries were poorly defined, as shown on this map by the curving dotted lines that separate Pegasus from Andromeda. (From Duncan Bradford, *Wonders of the Heavens*, Boston: John B. Russell, 1837.) (b) Modern constellation boundaries are precisely defined by international agreement.

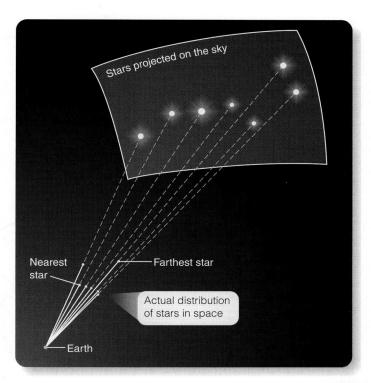

■ Figure 2-2

The stars you see in the Big Dipper—an asterism consisting of the brighter stars of the constellation Ursa Major, the Great Bear—are not at the same distance from Earth. You see the stars in a group in the sky because they lie in the same general direction as seen from Earth, not because they are all actually near each other. The sizes of the star dots in the star chart represent the apparent brightness of the stars.

The Names of the Stars

In addition to naming groups of stars, ancient astronomers named the brighter stars, and modern astronomers still use many of those names. The names of the constellations are in Latin or Greek, the languages of science in Medieval and Renaissance Europe. Most individual star names derive from ancient Arabic, much altered over centuries. For example, the name of Betelgeuse, the bright red star in Orion, comes from the Arabic phrase *yad aljawza,* meaning "armpit of Jawza [Orion]." Aldebaran, the bright red eye of Taurus the bull, comes from the Arabic *al-dabar an,* meaning "the follower."

Another way to identify stars is to assign Greek letters to the bright stars in a constellation in approximate order of brightness. Thus the brightest star is usually designated alpha (α), the second brightest beta (β), and so on. For many constellations, the letters follow the order of brightness, but some constellations are exceptions (■ Figure 2-3). A Greek-letter star name also includes the possessive form of the constellation name;

for example, the brightest star in the constellation Canis Major is alpha Canis Majoris. This name identifies the star and the constellation and gives a clue to the relative brightness of the star. Compare this with the ancient individual name for that star, Sirius, which tells you nothing about its location or brightness.

The Brightness of Stars

Astronomers measure the brightness of stars using the **magnitude scale.** The ancient astronomers divided the stars into six brightness groups. The brightest were called first-magnitude stars. The scale continued downward to sixth-magnitude stars, the faintest visible to the human eye. Thus, the *larger* the magnitude number, the *fainter* the star. This makes sense if you think of bright stars as first-class stars and the faintest stars visible as sixth-class stars.

The Greek astronomer Hipparchus (190–120 BC) is believed to have compiled the first star catalog, and there is evidence he used the magnitude system in that catalog. About 300 years later (around AD 140) the Egyptian-Greek astronomer Claudius Ptolemy definitely used the magnitude system in his own catalog, and successive generations of astronomers have continued to

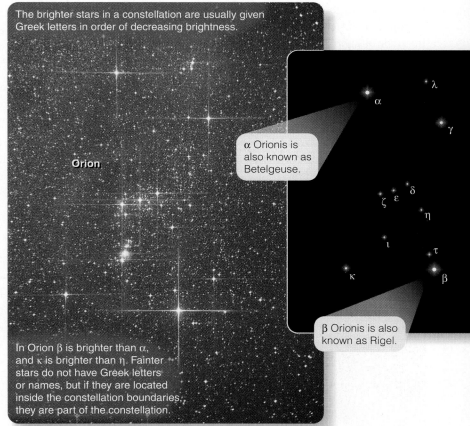

■ Figure 2-3

Stars in a constellation can be identified by names derived from Arabic or by Greek letters plus the name of the constellation. The spikes on the star images in the photograph were produced by the optics in the camera. (William Hartmann)

use the system. Star brightnesses expressed in this system are known as **apparent visual magnitudes (m_V),** describing how the stars look to human eyes observing from Earth.

Brightness is quite subjective, depending on both the physiology of eyes and the psychology of perception. To be scientifically accurate you should refer to **flux**—a measure of the light energy from a star that hits one square meter in one second. With modern scientific instruments astronomers can measure the flux of starlight with high precision and then use a simple mathematical relationship that relates light flux to apparent visual magnitude. Instead of saying that the star known by the charming name Chort (Theta Leonis) is about third magnitude, you can say its magnitude is 3.34. Thus, precise modern measurements of the brightness of stars are still connected to observations of apparent visual magnitude that go back to the time of Hipparchus.

Limitations of the apparent visual magnitude system have motivated astronomers to supplement it in various ways: (1) Some stars are so bright the magnitude scale must extend into negative numbers (■ Figure 2-4). On this scale, Sirius, the brightest star in the sky, has a magnitude of -1.47. (2) With a telescope you can find stars much fainter than the limit for your unaided eyes. Thus the magnitude system has also been extended to numbers larger than sixth magnitude to include faint stars (Figure 2-4). (3) Although some stars emit large amounts of infrared or ultraviolet light, those types of radiation (discussed further in Chapters 4 and 5) are invisible to human eyes. The subscript "V" in m_V is a reminder that you are counting only light that is visible. Other magnitude systems have been invented to express the brightness of invisible light arriving at Earth from the stars. (4) An apparent magnitude tells only how bright the star is as seen from Earth but doesn't tell anything about a star's true power output because the star's distance is not included. You can describe the true power output of stars with another magnitude system that will be described in Chapter 6.

■ **Figure 2-4**

The scale of apparent visual magnitudes extends into negative numbers to represent the brightest objects and to positive numbers larger than 6 to represent objects fainter than the human eye can see.

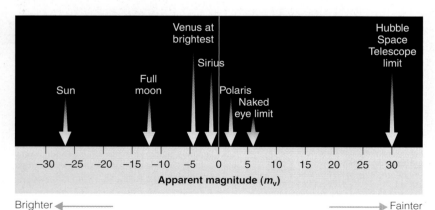

Brighter ◄———————— ————————► Fainter

THE SKY ABOVE YOU seems to be a great blue dome in the daytime and a sparkling ceiling at night. Learning to understand the sky requires that you first recall the perspectives of people who observed the sky thousands of years ago.

The Celestial Sphere

Ancient astronomers believed the sky was a great sphere surrounding Earth with the stars stuck on the inside like thumbtacks in a ceiling. Modern astronomers know that the stars are scattered through space at different distances, but it is still convenient and useful in some contexts for you to think of the sky as a great sphere enclosing Earth with stars all at one distance. The **celestial sphere** is an example of a **scientific model,** a common feature of scientific thought (**Window on Science 2-1**). You can use the celestial sphere as a convenient model of the sky. Notice that a scientific model does not have to be true to be useful. You will learn about many scientific models in the chapters that follow.

As you study **The Sky around Us** on pages 16–17, notice three important points:

1 Sky objects appear to rotate westward around Earth each day, but that is a consequence of the eastward rotation of Earth. This produces day and night.

2 Astronomers measure distances across the sky as angles expressed in units of degrees and subdivisions of degrees called arc minutes and arc seconds.

3 What you can see of the sky depends on where you are on Earth. For example, Australians see many constellations and asterisms invisible from North America, but they never see the Big Dipper.

Precession

In addition to the daily motion of the sky, Earth's rotation adds a second motion to the sky that can be detected only over centuries.

More than 2000 years ago, Hipparchus compared a few of his star positions with those made by other astronomers nearly two centuries before him and realized that the celestial poles and equator were slowly moving relative to the stars. Later astronomers understood that this apparent motion is caused by a special motion of Earth called **precession.**

If you have ever played with a toy top or gyroscope, you may recall that the axis of such a rapidly spinning object sweeps around relatively slowly in a circle (■ Figure 2-5a). The weight of the top tends to make it tip, and this combines with its rapid rotation to make its axis sweep around slowly in precession motion.

Frameworks for Thinking about Nature: Scientific Models

In everyday language, you can use the word *model* in various ways—fashion model, model airplane, model student—but scientists use the word in a very specific way. A scientific model is a carefully devised mental conception of how something works, a framework that helps scientists think about some aspect of nature. For example, astronomers use the celestial sphere as a way to think about the motions of the sky, sun, moon, and stars.

Although a scientific model is a mental conception, it can take many forms. Some models are imprecise—the psychologist's model of how the human mind processes visual information into images, for instance. But other models are so specific that they can be expressed as a set of mathematical equations. For example, an astronomer might use a set of equations to describe in detail how gas falls into a black hole. You could refer to such a calculation as a model. Of course, you could use metal and plastic to build a celestial globe, but the thing you build wouldn't really

be the model any more than the equations are a model. A scientific model is a mental conception, an idea that helps you think about nature.

On the other hand, a model is not meant to be a statement of truth. The celestial sphere is not real; you know the stars are scattered through space at various distances. Nevertheless you can imagine a celestial sphere and use it to help you think about the sky. A scientific model does not have to be true to be useful. Chemists, for example, think about the atoms in molecules by visualizing them as spheres joined together by sticks. This model of a molecule is not really correct, but it is a helpful way to think about molecules; it gives chemists a framework within which to organize their ideas.

Because scientific models are not meant to be totally correct, you must always remember the assumptions on which they are based. If you begin to think a model is true, it can mislead you instead of helping you. The celestial sphere, for instance, can help you think

The ancient celestial sphere is a useful model of the sky.

about the sky, but you must remember that it is only a mental crutch. The universe is much larger and much more interesting than this ancient scientific model of the heavens.

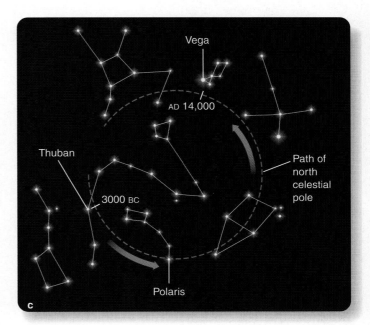

■ Figure 2-5

Precession. (a) A spinning top precesses in a slow circular motion around the perpendicular to the floor because its weight tends to make it fall over. (b) Earth precesses around the perpendicular to its orbit because the gravity of the sun and moon tend to twist it "upright." (c) Precession causes the north celestial pole to drift among the stars, completing a circle in 26,000 years.

The Sky Around Us

1 The eastward rotation of Earth causes the sun, moon, and stars to move westward in the sky as if the **celestial sphere** were rotating westward around Earth. From any location on Earth you see only half of the celestial sphere, the half above the **horizon.** The **zenith** marks the top of the sky above your head, and the **nadir** marks the bottom of the sky directly under your feet. The drawing at right shows the view for an observer in North America. An observer in South America would have a dramatically different horizon, zenith, and nadir.

The apparent pivot points are the **north celestial pole** and the **south celestial pole** located directly above Earth's north and south poles. Halfway between the celestial poles lies the **celestial equator.** Earth's rotation defines the directions you use every day. The **north point** and **south point** are the points on the horizon closest to the celestial poles. The **east point** and the **west point** lie halfway between the north and south points. The celestial equator always touches the horizon at the east and west points.

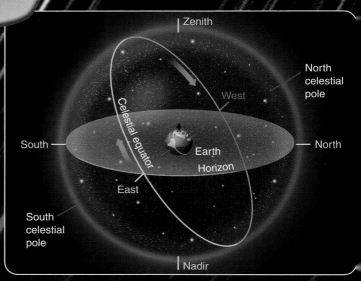

ThomsonNOW™ Sign in at www.thomsonedu.com and go to ThomsonNOW to see Active Figure "Celestial Sphere." Notice how each location on Earth has its unique horizon.

AURA/NOAO/NSF

Looking north

Looking east

Looking south

1a This time exposure of about 30 minutes shows stars as streaks, called star trails, rising behind an observatory dome. The camera was facing northeast to take this photo. The motion you see in the sky depends on which direction you look, as shown at right. Looking north, you see the star Polaris, the North Star, located near the north celestial pole. As the sky appears to rotate westward, Polaris hardly moves, but other stars circle the celestial pole. Looking south from a location in North America, you can see stars circling the south celestial pole, which is invisible below

ThomsonNOW™ Sign in at www.thomsonedu.com and go to ThomsonNOW to see Active Figure "Rotation

Astronomers measure distance across the sky as angles.

Angular distance

2 Astronomers might say, "The star was only 2 degrees from the moon." Of course, the stars are much farther away than the moon, but when you think of the celestial sphere, you can measure distances *on the sky* as **angular distances** in degrees, minutes of arc, and seconds of arc. A **minute of arc** is 1/60th of a degree, and a **second of arc** is 1/60th of a minute of arc. Then the **angular diameter** of an object is the angular distance from one edge to the other. The sun and moon are each about half a degree in diameter, and the bowl of the Big Dipper is about 10° wide.

3 What you see in the sky depends on your latitude as shown at right. Imagine that you begin a journey in the ice and snow at Earth's North Pole with the north celestial pole directly overhead. As you walk southward, the celestial pole moves toward the horizon, and you can see further into the southern sky. The angular distance from the horizon to the north celestial pole always equals your latitude—the basis for celestial navigation. As you cross Earth's equator, the celestial equator would pass through your zenith, and the north celestial pole would sink below your northern horizon.

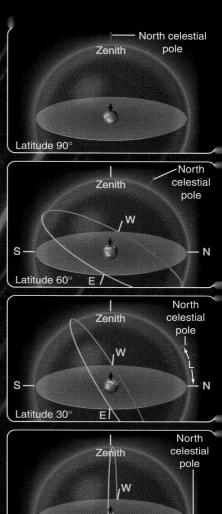

North celestial pole

Zenith

Latitude 90°

North celestial pole

Zenith

S — N

Latitude 60° E

North celestial pole

Zenith

W

L

S — N

Latitude 30° E

North celestial pole

Zenith

W

S — N

Latitude 0° E

South celestial pole

Zenith

W

S — N

Latitude –30° E

A few circumpolar constellations

Cassiopeia

Cepheus

Perseus

Rotation of sky

Rotation of sky

Polaris

Ursa Minor

Ursa Major

3a **Circumpolar constellations** are those that never rise or set. From mid-northern latitudes, as shown at left, you see a number of familiar constellations circling Polaris and never dipping below the horizon. As the sky rotates, the pointer stars at the front of the Big Dipper always point toward Polaris. Circumpolar constellations near the south celestial pole never rise as seen from mid-northern latitudes. From a high latitude such as Norway, you would have more circumpolar constellations, and from Quito, Ecuador, located on Earth's equator, you would have no circumpolar constellations at all.

Sign in at www.thomsonedu.com and go to ThomsonNOW to see Active Figure "Constellations from Different Latitudes."

ThomsonNOW

Earth spins like a giant top, but it does not spin upright relative to its orbit around the sun. You can say either that Earth's axis is tipped 23.5 degrees from vertical or that Earth's equator is tipped 23.5 degrees relative to its orbit; the two statements are equivalent. Earth's large mass and rapid rotation keep its axis of rotation pointed toward a spot near the star Polaris (alpha Ursa Minoris), and its axis direction would not move if Earth were a perfect sphere. However, Earth has a slight bulge around its middle, and the gravity of the sun and moon both pull on this bulge, tending to twist Earth's axis upright relative to its orbit. The combination of these forces and Earth's rotation causes Earth's axis to precess in a slow circular sweep, taking about 26,000 years for one cycle (Figure 2-5b).

Because the celestial poles and equator are defined by Earth's rotational axis, precession moves these reference marks. You would notice no change at all from night to night or year to year, but precise measurements reveal their slow apparent motion. Over centuries, precession has dramatic effects. Egyptian records show that 4800 years ago the north celestial pole was near the star Thuban (alpha Draconis). The pole is now approaching Polaris

and will be closest to it in about the year 2100. In another 12,000 years the pole will have moved to the apparent vicinity of the very bright star Vega (alpha Lyrae). Figure 2-5c shows the apparent path followed by the north celestial pole over thousands of years.

2-3 The Cycle of the Sun

THE ENGLISH LANGUAGE defines **rotation** as the turning of a body on its axis. **Revolution** means the motion of a body around a point outside the body. Earth rotates on its axis, and that produces day and night. Earth also revolves around the sun, and that produces the yearly cycle.

The Annual Motion of the Sun

Even in the daytime the sky is actually filled with stars, but the glare of sunlight fills our atmosphere with scattered light, and you can see only the brilliant blue sky. If the sun were fainter and you could see the stars in the daytime, you would notice that the sun

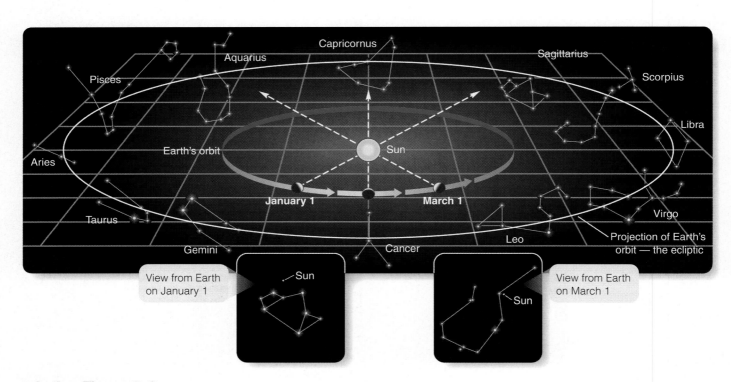

■ Active Figure 2-6

The motion of Earth around the sun makes the sun appear to move against the background of the stars. The circular orbit of Earth is thus projected on the sky as the ecliptic, the circular path of the sun during the year as seen from Earth. If you could see the stars in the daytime, you would notice the sun crossing in front of the distant constellations as Earth moves along its orbit.

ThomsonNOW™ Sign in at www.ThomsonNOW.com and go to ThomsonNow to see the Active Figure called "Constellations in Different Seasons."

appears to be moving slowly eastward relative to the background of the distant stars. This apparent motion is caused by the real orbital motion of Earth around the sun. In January you would see the sun in front of the constellation Sagittarius (■ Figure 2-6). By March you would see the sun in front of Aquarius. Note that your angle of view in Figure 2-6 makes the Earth's orbit seem very elliptical when it is really almost a perfect circle.

Through the year, the sun moves eastward among the stars following a line called the **ecliptic,** the apparent path of the sun among the stars. Recall the concept of the celestial sphere; if the sky were a great screen, the ecliptic would be the shadow cast by Earth's orbit. In other words, you can call the ecliptic the projection of Earth's orbit on the celestial sphere. Earth circles the sun in 365.26 days, and consequently the sun appears to go once around the sky in the same period. You don't notice this motion because you cannot see the stars in the daytime, but the apparent motion of the sun caused by a real motion of Earth has an important consequence that you do notice—the seasons.

The Seasons

The seasons are caused by the revolution of Earth around the sun combined with a simple fact you have already encountered: Earth's equator is tipped 23.5 degrees relative to its orbit. As you study **The Cycle of the Seasons** on pages 20–21, notice two important principles:

1 The seasons are *not* caused by variation in the distance between Earth and the sun. Earth's orbit is nearly circular, so it is always about the same distance from the sun.

2 The seasons *are* caused by changes in the amount of solar energy that Earth's northern and southern hemispheres receive at different times of the year, resulting from the tip of the Earth's equator and axis relative to its orbit.

The seasons are so important as a cycle of growth and harvest that cultures around the world have attached great significance to the ecliptic. It marks the center line of the **zodiac** ("circle of animals") and the motion of the sun, moon and the five visible planets (Mercury, Venus, Mars, Jupiter, and Saturn) are the basis of the ancient superstition of astrology (Window on Science 2-2). The signs of the zodiac are no longer important in astronomy. You can look for the planets along the ecliptic appearing like bright stars. Mars looks quite orange in color. Because Venus and Mercury orbit inside Earth's orbit, they never get far from the sun and are visible in the west after sunset or in the east before sunrise (■ Figure 2-7). Venus can be very bright, but Mercury is difficult to see near the horizon. By tradition, any planet in the sunset sky is called an **evening star,** and any planet in the dawn sky is called a **morning star.** Perhaps the most beau-

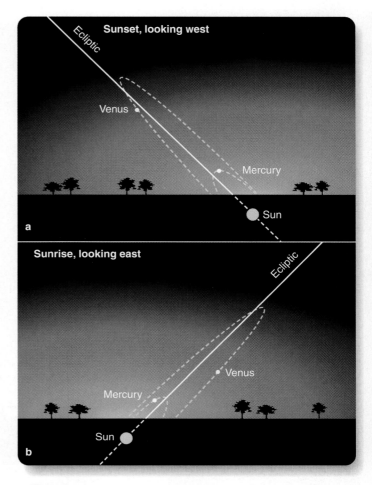

■ **Figure 2-7**

Mercury and Venus follow orbits that keep them near the sun, and they are visible only soon after sunset or just before sunrise. Venus takes 584 days to move from morning sky to evening sky and back again, but Mercury zips around in only 116 days.

tiful is Venus, which can become as bright as magnitude −4.7. As Venus moves around its orbit, it can dominate the western sky each evening for many weeks, but eventually its orbit appears to carry it back toward the sun as seen from Earth, and it is lost in the haze near the horizon. A few weeks later you can see Venus reappear in the dawn sky as a brilliant morning star. Months later it will switch back to being an evening star again.

ThomsonNOW™ Sign in at www.ThomsonNOW.com and go to ThomsonNOW to see Astronomy Exercises "Sunrise through the Seasons" and "The Seasons."

Building Scientific Arguments

How does the elliptical shape of Earth's orbit affect the seasons?

In the critical analysis of an idea it can be helpful to exaggerate the importance of a single factor. Doing so not only

The Cycle of the Seasons

1 You can use the celestial sphere to help you think about the seasons. The celestial equator is the projection of Earth's equator on the sky, and the ecliptic is the projection of Earth's orbit on the sky. Because Earth is tipped in its orbit, the ecliptic and equator are inclined to each other by 23.5° as shown at right. As the sun moves eastward around the sky, it spends half the year in the southern half of the sky and half of the year in the northern half. That causes the seasons.

The sun crosses the celestial equator going northward at the point called the **vernal equinox**. The sun is at its farthest north at the point called the **summer solstice**. It crosses the celestial equator going southward at the **autumnal equinox** and reaches its most southern point at the **winter solstice**. These labels refer to seasons in the northern hemisphere.

1a The seasons are defined by the dates when the sun crosses these four points, as shown in the table at the right. *Equinox* comes from the word for "equal"; the day of an equinox has equal amounts of daylight and darkness. *Solstice* comes from the words meaning "sun" and "stationary." *Vernal* comes from the word for "green." The "green" equinox marks the beginning of spring in the northern hemisphere.

Event	Date*	Season
Vernal equinox	March 20	Spring begins
Summer solstice	June 22	Summer begins
Autumnal equinox	September 22	Autumn begins
Winter solstice	December 22	Winter begins

* Give or take a day due to leap year and other factors.

1b On the day of the summer solstice in late June, Earth's northern hemisphere is inclined toward the sun, and sunlight shines almost straight down at northern latitudes. At southern latitudes, sunlight strikes the ground at an angle and spreads out. North America has warm weather, and South America has cool weather.

Earth's axis of rotation points toward Polaris, and, like a top, the spinning Earth holds its axis fixed as it orbits the sun. On one side of the sun, Earth's northern hemisphere leans toward the sun; on the other side of its orbit, it leans away. However, the direction of the axis of rotation does not change.

North celestial pole

Celestial equator

Autumnal equinox

Winter solstice

Ecliptic

Summer solstice

23.5°

Vernal equinox

South celestial pole

23.5°

To Polaris

40° N latitude

Equator

40° S latitude

Sunlight nearly direct on northern latitudes

To sun →

Sunlight spread out on southern latitudes

Earth at summer solstice

Summer solstice light

1c Light striking the ground at a steep angle spreads out less than light striking the ground at a shallow angle. Light from the summer-solstice sun strikes northern latitudes from nearly overhead and is concentrated.

Winter solstice light

Light from the winter-solstice sun strikes northern latitudes at a much steeper angle and spreads out. The same amount of energy is spread over a larger area, so the ground receives less energy from the winter sun.

2 The two causes of the seasons are shown at right for someone in the northern hemisphere. First, the noon summer sun is higher in the sky and the winter sun is lower, as shown by the longer winter shadows. Thus winter sunlight is more spread out. Second, the summer sun rises in the northeast and sets in the northwest, spending more than 12 hours in the sky. The winter sun rises in the southeast and sets in the southwest, spending less than 12 hours in the sky. Both of these effects mean that northern latitudes receive more energy from the summer sun, and summer days are warmer than winter days.

At summer solstice

At winter solstice

Sunlight spread out on northern latitudes →

23.5°

To Polaris

40° N latitude

← To sun

Equator

Sunlight nearly direct on southern latitudes →

40° S latitude

1d On the day of the winter solstice in late December Earth's northern hemisphere is inclined away from the sun, and sunlight strikes the ground at an angle and spreads out. At southern latitudes, sunlight shines almost straight down and does not spread out. North America has cool weather and South America has warm weather.

Earth's orbit is only very slightly elliptical. About January 3, Earth is at **perihelion,** its closest point to the sun, when it is only 1.7 percent closer than average. About July 5, Earth is at **aphelion,** its most distant point from the sun, when it is only 1.7 percent farther than average. This small variation does not significantly affect the seasons.

Astrology and Pseudoscience: Misusing the Rules of Science

Astronomers have a low opinion of astrology not only because it is groundless but because it pretends to be a science. It is a pseudoscience, from the prefix *pseudo,* meaning "false." There are many examples of pseudoscience, and it is illuminating to consider the difference between pseudoscience and science.

A pseudoscience is a set of beliefs that appear to be based on scientific ideas but that fail to obey the most basic rules of science. For example, some years ago a fad arose in which people placed objects under pyramids made of paper, plastic, and so on. The claim was that a pyramidal shape would focus cosmic forces on anything inside and so preserve fruit, sharpen razor blades, and do other miraculous things. Many books promoted this idea, but simple experiments showed that the pyramidal shape and mysterious cosmic forces were not involved. A cover in any shape would protect a piece of fruit from airborne spores and allow it to dry without rotting. Likewise, any shape would allow oxidation, thus improving the cutting edge of a razor blade. In short, experimental evidence contradicted the claim. Nevertheless, supporters of the theory declined to abandon or revise their claims. Refusal to change beliefs in re-

sponse to facts is characteristic of a pseudoscience.

Another characteristic of a pseudoscience is that it appeals to our needs and desires. Thus, some pseudoscientific claims are self-fulfilling. For example, some people bought pyramidal tents to put over their beds and thus improve their rest. While there is no logical mechanism by which such a tent could affect a sleeper, people slept more soundly because they wanted and expected the claim to be true. Many pseudoscientific claims involve medical cures, ranging from copper or magnetic bracelets to crystals that focus spiritual power to astonishingly expensive and illegal treatments for cancer. Logic is a stranger to pseudoscience, but human fears and needs are not.

Astrology is a pseudoscience. Over the centuries, astrology has been tested repeatedly, and no correlation between the movement of the heavens and human personalities or fates has been found. But it survives, and its supporters disregard any evidence that it doesn't work. Like all pseudosciences, astrology is not open to revision in the face of contradictory evidence. Furthermore, astrology fulfills our human need to believe that there is order and meaning to our lives. It may comfort you to believe that your sweetheart has rejected you

Astrology may be the oldest pseudo- (meaning, false-) science.

because of the motions of the sun, moon, and planets along the zodiac rather than to admit that you behaved badly on your last date. Comfort aside, astrology is a poor basis for life decisions.

Human nature and human needs probably ensure that pseudoscientific beliefs will continue to plague humans like emotional viruses propagating from person to person. If you recognize them for what they are, you can more easily guide your life by rational principles and not assign the stars blame for your failures and credit for your successes.

---

reveals the effect of that factor but can also reveal the inner workings of the process itself. Earth's orbit is nearly circular, so its distance from the sun varies by only about 3.3 percent from nearest to farthest. You might try to imagine what life on Earth would be like if Earth's orbit were more elliptical than it really is. Earth is at perihelion—closest to the sun— about January 3. If Earth's orbit were more elliptical, winters in the northern hemisphere would be warmer than they are now; and the weather in the southern hemisphere, normally warm in January, would be extremely hot. Six months later, about July 5, Earth reaches aphelion, its farthest point from the sun. If Earth had a more elliptical orbit, the northern hemisphere would experience a much cooler summer than it now does, and in the southern hemisphere the cool season would be bitterly cold. You can see that the net effect of the elliptical shape of Earth's orbit is to make northern seasons milder and southern seasons more extreme, but the effect is

hardly noticeable. Now use exaggeration to analyze the true cause of the seasons. **Try to imagine what the seasons would be like if Earth's equator were inclined more than the present 23.5 degrees or less than the present 23.5 degrees, relative to the plane of its orbit.**

■   ■   ■

2-4 The Cycles of the Moon

THE MOON ORBITS EASTWARD around Earth once a month. Starting this evening, look for the moon in the sky. If it is a cloudy night or if the moon is in the wrong part of its orbit, you may not see it; but keep trying on successive evenings, and within a week or two you will see the moon. Then, watch for the moon

■ Figure 2-8

In this sequence of pictures spanning about 10 nights, you see the same face of the moon, the same mountains, craters, and plains, but the changing direction of sunlight produces the lunar phases. (UCO/Lick Observatory)

on following evenings, and you will see it move along its orbit around Earth and cycling through its phases as it has done for billions of years.

The Motion of the Moon

If you watch the moon night after night, you will notice two things about its motion. *First,* you will see it moving relative to the background of stars; *second,* you will notice that the markings on its face don't change. These two observations will help you understand the motion of the moon and the origin of the moon's phases.

The moon moves rapidly among the constellations. If you watch the moon for just an hour, you can see it move eastward against the background of stars by slightly more than its own apparent diameter. Each night when you look at the moon, you will see it is roughly half the width of a zodiac constellation (about 13 degrees) to the east of its location the night before. This movement is the result of the motion of the moon along its orbit around Earth.

The Cycle of Moon Phases

The changing shape of the illuminated part of the moon as it orbits Earth is one of the most easily observed phenomena in astronomy. You have surely seen the full moon rising dramatically or a thin crescent moon hanging in the evening sky. Study **The Phases of the Moon** on pages 24–25 and notice three important points:

1 The moon always keeps the same side facing Earth, and you never see the far side of the moon. "The man in the moon" (some cultures see "the rabbit in the moon" instead) is produced by familiar features on the moon's near side.

2 The changing shape of the moon as it passes through its cycle of phases is produced by sunlight illuminating different parts of the side of the moon you can see. You always see the same side of the moon looking down on you, but the shifting shadows make the "man in the moon" change moods as the moon cycles through its phases (see also ■ Figure 2-8).

3 The orbital period of the moon around the Earth is not the same as the length of a moon phase cycle.

ThomsonNOW™ Sign in at www.ThomsonNOW.com and go to ThomsonNOW to see Astronomy Exercises "Phases of the Moon" and "Moon Calendar."

2-5 Eclipses

ECLIPSES ARE DUE to a seemingly complicated combination of apparent motions of the sun and moon, yet they are actually easy to predict once all the cycles are understood. Eclipses are also among the most spectacular of nature's sights you might witness.

Solar Eclipses

From planet Earth you can see a phenomenon that is not visible from most planets. It happens that the sun is 400 times larger than our moon and, on the average, 390 times farther away, so the sun and moon have nearly equal apparent diameters. Thus, the moon is just about the right size to cover the bright disk of the sun and cause a **solar eclipse.** In a solar eclipse, it is the sun that is being hidden (eclipsed) and the moon that is "in the way."

A shadow consists of two parts. The **umbra** is the region of total shadow. For example, if you were in the umbra of the moon's shadow, you would see no portion of the sun. The umbra of the moon's shadow usually just barely reaches Earth's surface and covers a relatively small circular zone (■ Figure 2-9a). Standing in that

The Phases of the Moon

1 As the moon orbits Earth, it rotates to keep the same side facing Earth as shown at right. Consequently you always see the same features on the moon, and you never see the far side of the moon. A mountain on the moon that points at Earth will always point at Earth as the moon revolves and rotates.

(Not to scale)

2 As seen at left, sunlight always illuminates half of the moon. Because you see different amounts of this sunlit side, you see the moon cycle through phases. At the phase called "new moon," sunlight illuminates the far side of the moon, and the side you see is in darkness. At new moon you see no moon at all. At full moon, the side you see is fully lit, and the far side is in darkness. How much you see depends on where the moon is in its orbit.

Notice that there is no such thing as the "dark side of the moon." All parts of the moon experience day and night in a month-long cycle.

In the diagram at the left, you see that the new moon is close to the sun in the sky, and the full moon is opposite the sun. The time of day depends on the observer's location on Earth.

First quarter

Waxing gibbous

Waxing crescent

Sunset

North Pole

Midnight — Noon

New

Earth's rotation

Sunrise

Full

← Sunlight

Waning crescent

Waning gibbous

Third quarter

2a The first two weeks of the cycle of the moon are shown below by its position at sunset on 14 successive evenings. As the moon grows fatter from new to full, it is said to wax.

Gibbous comes from the Latin word for humpbacked.

The first quarter moon is one week through its 4-week cycle.

Waxing gibbous

Waxing crescent

The full moon is two weeks through its 4-week cycle.

New moon is invisible near the sun

THE SKY AT SUNSET

Full moon rises at sunset

East

South

West

3 The moon orbits eastward around Earth in 27.32 days. This is how long the moon takes to circle the sky once and return to the same position among the stars.

A complete cycle of lunar phases takes 29.53 days.

Although you think of the lunar cycle as being about 4 weeks long, it is actually 1.53 days longer than 4 weeks. The calendar divides the year into 30-day periods called months (literally "moonths") in recognition of the 29.53-day synodic cycle of the moon.

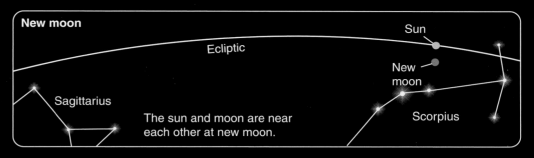

New moon

Ecliptic

Sun

New moon

Sagittarius

Scorpius

The sun and moon are near each other at new moon.

One orbital period after new moon

Ecliptic

Sun

Sagittarius

Moon

Scorpius

One sidereal period after new moon, the moon has returned to the same place among the stars, but the sun has moved on along the ecliptic.

One phase period after new moon

Ecliptic

Sun

New moon

Sagittarius

Scorpius

One synodic period after new moon, the moon has caught up with the sun and is again at new moon.

You can use the diagram on the opposite page to determine when the moon rises and sets at different phases.

TIMES OF MOONRISE AND MOONSET

Phase	Moonrise	Moonset
New	Dawn	Sunset
First quarter	Noon	Midnight
Full	Sunset	Dawn
Third quarter	Midnight	Noon

2b The last two weeks of the cycle of the moon are shown below by its position at sunrise on 14 successive mornings. As the moon shrinks from full to new, it is said to wane.

The first quarter moon is 3 weeks through its 4-week cycle.

Waning crescent

Waning gibbous

New moon is invisible near the sun

THE SKY AT SUNRISE

Full moon sets at sunrise

East

South

West

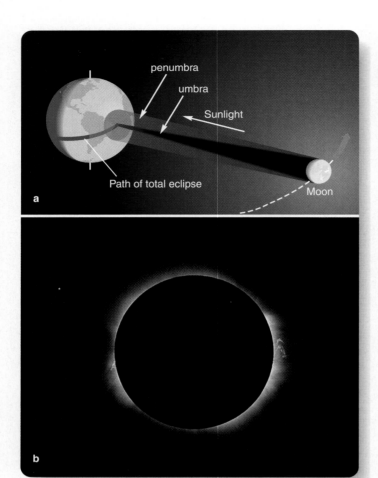

A Total Solar Eclipse

The moon moving from the right just begins to cross in front of the sun.

The disk of the moon gradually covers the disk of the sun.

Sunlight begins to dim as more of the sun's disk is covered.

During totality, faint features of the atmosphere are often visible.

A longer-exposure photograph during totality shows the extended solar atmosphere.

■ **Figure 2-9**

(a) The umbral shadow of the moon sweeps over a narrow strip of Earth during a solar eclipse. (b) From a location inside the umbral shadow, you would see the moon cover the bright surface of the sun in a total solar eclipse. Note faint features of the sun's atmosphere around the edge of the moon's disk. (Daniel Good)

■ **Figure 2-10**

This sequence of photos shows the first half of a total solar eclipse. (Daniel Good)

umbral zone, you would be in total shadow, unable to see any part of the sun's surface (Figure 2-9b). That is called a total eclipse. If you moved into the **penumbra,** however, you would be in partial shadow but could also see part of the sun peeking around the edge of the moon. This is called a partial eclipse. Of course, if you are outside the penumbra, you see no eclipse at all.

Because of the orbital motion of the moon and the rotation of Earth, the moon's shadow sweeps rapidly across Earth in a long, narrow path of totality. If you want to see a total solar eclipse you must be in the path of totality (Figure 2-9a). When the umbra of the moon's shadow sweeps over you, you see one of the most dramatic sights in the sky—the totally eclipsed sun.

The eclipse begins as the moon slowly crosses in front of the sun (■ Figure 2-10). It takes about an hour for the moon to cover the solar disk, but as the last sliver of sun disappears darkness arrives in a few seconds. Automatic streetlights come on, drivers of cars turn on their headlights, and birds go to roost. The sky usually becomes so dark you can even see the brighter stars. The darkness lasts only a few minutes because the umbra is never more than 270 km (170 miles) in diameter on the surface of

Sunlight

Pinhole

Image of partially
eclipsed sun

■ **Figure 2-11**

A safe way to view the partial phases of a solar eclipse. Use a pinhole in a card to project an image of the sun on a second card. The greater the distance between the cards, the larger (and fainter) the image will be.

Earth and sweeps across the landscape at over 1600 km/hr (1000 mph). The period of totality lasts on average only 2 or 3 minutes and never more than 7.5 minutes. During totality you can see subtle features of the sun's atmosphere such as red flamelike projections (Figure 2-9b) that are visible only during these moments when the brilliant disk of the sun is completely covered by the moon. (The sun's atmosphere will be discussed in more detail in Chapter 5.) As soon as part of the sun's disk reappears, the fainter features vanish in the glare, and the period of totality is over. The moon moves on in its orbit and in an hour the sun is completely visible again.

Sometimes when the moon crosses in front of the sun it is too small to fully cover the sun, and then you would witness an **annular eclipse.** That is a solar eclipse in which an annulus (meaning "ring") of the sun's disk is visible around the disk of the moon. The eclipse never becomes total; it never quite gets dark, and you can't see the faint features of the solar atmosphere. Annular eclipses occur because the moon follows a slightly elliptical orbit around Earth. If the moon is in the farther part of its orbit during totality, its apparent diameter will be less than the apparent diameter of the sun, and thus you see an annular eclipse. Furthermore, Earth's orbit is slightly elliptical, so the Earth-to-sun distance varies slightly, and so does the apparent diameter of the solar disk, contributing to the effect of the moon's varying apparent size.

If you plan to observe a solar eclipse, remember that the sun is bright enough to burn your eyes and cause permanent damage if you look at it directly. This is true whether there is an eclipse

or not. Solar eclipses can be misleading, tempting you to look at the sun in spite of its brilliance and thus risking your eyesight.

During the few minutes of totality the brilliant disk of the sun is hidden, and it is safe to look at the eclipse, but the partial eclipse phases and annular eclipses can be dangerous. See ■ Figure 2-11 for a safe way to observe the partially eclipsed sun. ■ Table 2-1 will allow you to determine when some upcoming solar eclipses will be visible from your location.

Lunar Eclipses

Occasionally you can see the moon darken and turn copper-red in a **lunar eclipse.** A lunar eclipse occurs at full moon when the moon moves through the shadow of Earth. Because the moon shines only by reflected sunlight, you see the moon gradually darken as it enters the shadow. If you were on the moon and in the umbra of Earth's shadow, you would see no portion of the sun. If you moved into the penumbra, however, you would be in partial shadow and would see part of the sun peeking around the edge of Earth so the sunlight would be dimmed but not extinguished. In a lunar eclipse, it is the moon that is being hidden in the Earth's shadow and Earth that is "in the way" of the sunlight.

■ **Table 2-1 I Total and Annular Eclipses of the Sun, 2007–2010**

Date	Total/ Annular (T/A)	Time* of Mid-eclipse (GMT)	Maximum Length of Total or Annular Phase (Min:Sec)	Area of Visibility
2008 Feb. 7	A	4h	2:14	S. Pacific, Antarctica
2008 Aug. 1	T	10h	2:28	Canada, Arctic, Siberia
2009 Jan. 26	A	8h	7:56	S. Atlantic, Indian Ocean
2009 July 22	T	3h	6:40	Asia, Pacific
2010 Jan. 15	A	7h	11:10	Africa, Indian Ocean
2010 July 11	T	20h	5:20	Pacific, S. America

There are no total or annular eclipses of the sun in 2007. The next major total solar eclipse visible from the United States will occur on August 21, 2017.

*Times are Greenwich Mean Time. Subtract 5 hours for Eastern Standard Time, 6 hours for Central Standard Time, 7 hours for Mountain Standard Time, and 8 hours for Pacific Standard Time.

During a total lunar eclipse, the moon takes a number of hours to move through Earth's shadow.

A cross section of Earth's shadow shows the umbra and penumbra.

Sunlight scattered from Earth's atmosphere bathes the totally eclipsed moon in a coppery glow.

Motion of moon

Orbit of moon

To sun

Umbra

Penumbra

(Not to scale)

a

b

■ **Figure 2-12**

During a total lunar eclipse the moon passes through Earth's shadow, as shown in this multiple-exposure photograph. A longer exposure was used to record the moon while it was totally eclipsed. The moon's path appears curved in the photo because of photographic effects. (© 1982 Dr. Jack B. Marling)

If the orbit of the moon carries it through the umbra of Earth's shadow, you see a total lunar eclipse (■ Figure 2-12a). As you watch the moon in the sky, it first moves into the penumbra and dims slightly; the deeper it moves into the penumbra, the more it dims. In about an hour, the moon reaches the umbra, and you see the umbral shadow darken part of the moon. It takes about an hour for the moon to enter the umbra completely and become totally eclipsed. The period of total eclipse may last as long as 1 hour 45 minutes, though the timing of the eclipse depends on where the moon crosses the shadow.

When the moon is totally eclipsed, it does not disappear completely. Although it receives no direct sunlight, the moon in the umbra does receive some sunlight refracted (bent) through Earth's atmosphere. If you were on the moon during totality, you would not see any part of the sun because it would be entirely

■ **Figure 2-13**

Umbral shadows of Earth and the moon. Notice how easy it is, because of the tilt of the moon's orbit relative to the ecliptic, for the shadows to miss their mark at full moon and at new moon and thus fail to produce eclipses. That is, it is easy for the moon to reach full phase and not enter Earth's shadow. It is also easy for the moon to reach new phase and not cast its shadow on Earth. (The diameters of Earth and the moon are exaggerated by a factor of 2 for clarity.)

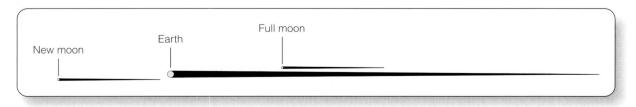

Full moon

Earth

New moon

■ Table 2-2 | Total and Partial Eclipses of the Moon, 2007–2010

Date	Time of Mid-eclipse (GMT)	Length of Totality (Min)	Length of Eclipse (Hr:Min)
2007 Mar. 3	23:22	74	3:40
2007 Aug. 28	10:38	90	3:32
2008 Feb. 21	3:27	50	3:24
2008 Aug. 16	21:11	Partial	3:08
2009 Dec. 31	19:24	Partial	1:00
2010 June 26	11:40	Partial	2:42
2010 Dec. 21	8:18	72	3:28

*Times are Greenwich Mean Time. Subtract 5 hours for Eastern Standard Time, 6 hours for Central Standard Time, 7 hours for Mountain Standard Time, and 8 hours for Pacific Standard Time. From your time zone, lunar eclipses that occur between sunset and sunrise will be visible, and those at midnight will have the moon highest in the sky.

hidden behind Earth. However, you would be able to see Earth's atmosphere illuminated from behind by the sun. The red glow from this ring consisting of all the Earth's simultaneous sunsets and sunrises illuminates the moon during totality and makes it glow coppery red, as shown in Figure 2-12b.

If the moon passes a bit too far north or south of the center of Earth's shadow, it may only partially enter the umbra, and you see a partial lunar eclipse. The part of the moon that remains outside the umbra in the penumbra receives some direct sunlight, and the glare is usually great enough to prevent your seeing the faint coppery glow of the part of the moon in the umbra.

Lunar eclipses always occur at full moon but not at every full moon. The moon's orbit is tipped about 5 degrees to the ecliptic, so most full moons cross the sky north or south of Earth's shadow and there is no lunar eclipse that month (■ Figure 2-13). For the same reason, solar eclipses always occur at new moon but not at every new moon. The orientation of the moon's orbit in space varies slowly and as a result solar and lunar eclipses repeat in a pattern called the **Saros cycle** lasting 18 years 11⅓ days. Ancient peoples who understood the Saros cycle could predict eclipses without understanding what the sun and moon really were.

Although there are usually no more than one or two lunar eclipses each year, it is not difficult to see one. You need only be on the dark side of Earth when the moon passes through Earth's shadow. That is, the eclipse must occur between sunset and sunrise at your location to be visible. ■ Table 2-2 will allow you to determine when some upcoming lunar eclipses will be visible from your location.

Building Scientific Arguments

What would solar eclipses be like if the moon's orbit were not tipped to the plane of Earth's orbit?

One way to analyze a complex process is to change one thing and imagine what would happen. In this case, you can review your understanding of real eclipses by trying to imagine a geometry that does not exist. If the moon's orbit were not inclined, then the moon would follow the ecliptic, and it would cross in front of the sun at every new moon. There would be a solar eclipse every month at new moon and a lunar eclipse every month at full moon. You could repeat that analysis for lunar eclipses, but it might be more interesting to change a different factor. **What would lunar eclipses look like if Earth had no atmosphere?**

■   ■   ■

Summary

2-1 | The Stars

How do astronomers refer to stars?

- Astronomers divide the sky into 88 constellations. The constellations originated in Greek and Middle Eastern mythology. Even the modern constellations, added to fill in some spaces between the ancient figures and in the southern part of the sky, have Latin names.

- The names of individual bright stars usually come from ancient Arabic. Modern astronomers often refer to bright stars also by a name composed of the constellation name plus a Greek letter assigned according to the star's brightness within the constellation.

How can you compare the brightness of stars?

- The magnitude system is the astronomer's brightness scale. First-magnitude stars are brighter than second-magnitude stars, which are brighter than third-magnitude stars, and so on.

- The magnitude you see when you look at a star in the sky is its apparent visual magnitude, which does not take into account its distance from Earth.

2-2 | The Sky and Its Motions

How does the sky appear to move as Earth rotates?

- The celestial sphere is a model of the sky, carrying the celestial objects around Earth. Because Earth rotates eastward, the celestial sphere appears to rotate westward on its axis. The northern and southern celestial poles are the pivots on which the sky appears to rotate.

- The celestial equator, an imaginary line around the sky above Earth's equator, divides the sky in half.

- What you see of the celestial sphere depends on your latitude. Much of the southern hemisphere of the sky is not visible from northern latitudes. To see that part of the sky, you would have to travel southward over Earth's surface.

- The angular distance from the horizon to the north celestial pole always equals your latitude. This is an important basis for celestial navigation.

- The gravitational forces of the moon and sun act on the spinning Earth and cause it to precess like a top. Earth's axis of rotation sweeps around in a circular motion with a period of 26,000 years, and consequently the celestial poles and celestial equator move slowly against the background of the stars.

2-3 | The Cycle of the Sun

What causes the seasons?

- Because Earth orbits the sun, the sun appears to move eastward along the ecliptic through the constellations of the zodiac in a yearly cycle.

- Because the ecliptic is tipped relative to the celestial equator, the sun spends half the year in the northern celestial hemisphere and half in the southern celestial hemisphere.

- In the summer, the sun is above the horizon longer and shines more directly down on the ground. Both effects cause warmer weather.

2-4 | The Cycles of the Moon

Why does the moon go through phases?

- Because you see the moon by reflected sunlight, its shape appears to change as it orbits Earth.

- The moon's illumination grows from new moon to first quarter to full moon and then shrinks from full moon to third quarter to new moon.

- A complete cycle of lunar phases takes 29.53 days.

2-5 Eclipses

What is a solar eclipse?

- A total solar eclipse occurs if a new moon passes directly between the sun and Earth and the moon's shadow sweeps over the part of the Earth's surface where you are located. If you are outside the path of totality but the penumbra sweeps over your location, you see a partial eclipse.

- During a total eclipse, the bright disk of the sun is covered, and you can see fainter features of the surrounding solar atmosphere.

- If the moon is in the farther part of its orbit during a solar eclipse, it does not appear large enough to cover the solar disk and you see only an annular eclipse.

What is a lunar eclipse?

- If a full moon passes through Earth's shadow while you are on the night side of Earth, you see the moon darken in a lunar eclipse. If the moon only grazes the shadow, the eclipse is partial and not total.

- The totally eclipsed moon looks copper-red because of sunlight scattered through Earth's atmosphere.

New Terms

constellation (p. 12)	rotation (p. 18)
asterism (p. 12)	revolution (p. 18)
magnitude scale (p. 13)	ecliptic (p. 19)
apparent visual magnitude (m_v) (p. 14)	zodiac (p. 19)
flux (p. 14)	evening star (p. 19)
celestial sphere (p. 14)	morning star (p. 19)
scientific model (p. 14)	vernal equinox (p. 20)
precession (p. 14)	summer solstice (p. 20)
horizon (p. 16)	autumnal equinox (p. 20)
zenith (p. 16)	winter solstice (p. 20)
nadir (p. 16)	perihelion (p. 21)
north and south celestial poles (p. 16)	aphelion (p. 21)
celestial equator (p. 16)	solar eclipse (p. 23)
north, south, east, and west points (p. 16)	umbra (p. 23)
angular distance (p. 17)	penumbra (p. 26)
arc minute (p. 17)	annular eclipse (p. 27)
arc second (p. 17)	lunar eclipse (p. 27)
angular diameter (p. 17)	Saros cycle (p. 29)
circumpolar constellation (p. 17)	

Review Questions

Assess your understanding of this chapter's topics with additional quizzing and animations at
www.thomsonedu.com

1. Why have astronomers added constellations to the sky in the past few hundred years?

2. Do people from other cultures on Earth see the same stars you see? The same constellations and asterisms?

3. How does the Greek-letter designation of a star give you a clue to its brightness?

4. What does the word *apparent* mean in *apparent visual magnitude?*

5. In what ways is the celestial sphere a scientific model?

6. Where would you go on Earth if you wanted to be able to see both the north celestial pole and the south celestial pole at the same time?

7. Where would you go on Earth to see a celestial pole at your zenith?

8. Explain how to make a simple astronomical observation that would determine your latitude.

9. Why does the number of circumpolar constellations depend on the latitude of the observer?

10. How could you detect Earth's precession by examining star charts from ancient Egypt?

11. What is the difference between the daily and annual motions of the sun?

12. What would our seasons be like if Earth's equator were tipped 35 degrees instead of 23.5 degrees to its orbit? What would they be like if Earth's equator were parallel to its orbit?

13. Why are the seasons reversed in the southern hemisphere relative to the northern hemisphere?

14. Why does the eccentricity of Earth's orbit make the contrast between summer and winter in the northern hemisphere different from the contrast in the southern hemisphere?

15. Do the phases of the moon look the same from every place on Earth, or is the moon full at different times as seen from different locations?

16. What phase would you see Earth in if you were on the moon when the moon is full? When the moon is at first quarter?

17. Why have many people seen a total lunar eclipse, while relatively few have seen a total solar eclipse?

18. Why isn't there an eclipse at every new moon and at every full moon?

19. Why is the moon red during a total lunar eclipse?

Discussion Questions

1. All cultures on Earth have named constellations. Why do you suppose this was such a common practice?

2. If you were lost at sea, you could find your approximate latitude by measuring the altitude of Polaris. But Polaris isn't exactly at the celestial pole. What else would you need to know to measure your latitude more accurately?

3. Do planets orbiting other stars have ecliptics? Could they have seasons?

4. Why would it be difficult to see faint features in the solar atmosphere if you were on the moon during a total lunar eclipse?

Circles Around the Earth Become Orbits Around the Sun

*The passions of astronomy
are no less profound
because they are not noisy.*

JOHN STEINBECK
The Short Reign of Pippin IV

f you are concerned about the environment, you owe a debt to the 16th century Polish astronomer, Nicolaus Copernicus. He proposed that Earth is not the center of the universe, but just one of the planets that circle the sun. His theory made Earth part of the cosmos and led to the modern understanding of humanity's place among all the creatures of Earth. ▌ As you read about Copernicus and his theory, you will see astronomers struggling with two related problems—the place of the Earth and the motion of the planets. That struggle led Galileo before the Inquisition and it led Isaac Newton to discover gravity. ▌ As you read about the birth of modern astronomy notice that it is also the story of the invention of science as a way to know about the world we live in. Before Copernicus the world seemed filled with mysterious influences;

▌ Continued on page 34 ▌

Aristotle, Copernicus, Galileo, and many other astronomers tried to understand the place of Earth in the universe but could never fully succeed until Isaac Newton understood gravity and orbital motion. This fanciful drawing shows Newton warily eyeing a small satellite orbiting his head.

Guidepost

Looking Back

The previous chapters gave you a view of the sky. You are now familiar with what you see there, and you also can imagine how the sun, moon, and planets move through space. Now you are ready to understand the first worldwide revolution in human thought: the realization that we live on a moving planet.

This Chapter

By the 16th century, many astronomers were uncomfortable with the ancient model that Earth sits at the center of a universe of crystalline spheres. In this chapter, you will discover how Nicolaus Copernicus changed the model, how Galileo Galilei changed the rules of debate, and how Isaac Newton invented physics and thereby changed humanity's concept of nature. Here you will find answers to these questions:

How did people in ancient civilizations describe Earth's place in the universe?

How did Copernicus revise that ancient model?

Why was the Copernican model gradually accepted?

How did Tycho Brahe and Johannes Kepler contribute to the Copernican revolution?

Why was Galileo condemned by the Inquisition?

How did Isaac Newton change humanity's view of nature?

This chapter is not just about the origin of modern astronomy. As they struggled to understand Earth and the heavens, the astronomers of the Renaissance invented science—a new way of understanding nature.

Looking Ahead

This is a book of science, and every chapter that follows will use the ways of thinking that were invented when Copernicus tried to repair the ancient model that put Earth at the center of the universe. As you think about planets, stars, and galaxies through the rest of this book, you will be using the intellectual methods that Galileo, Newton, and many others struggled to devise.

after Newton, scientists understood that the world is described by natural laws that are open to human study. The mysteries of nature are mysteries because they are unknown, not because they are unknowable.

(3-1) Astronomy before Copernicus

To understand why Copernicus's model was so important, you first need to backtrack to ancient Greece and meet the two great authorities of ancient astronomy, the brilliant philosopher Aristotle and a later follower of Aristotle's principles, Claudius Ptolemy (pronounced TAHL-eh-mee; the initial P is silent).

First, though, you should remember the point made in Chapter 1 that the terms *solar system, galaxy,* and *universe* have very different meanings. You know now that our solar system, consisting of Earth, the sun, Earth's sibling planets, their moons, and so on, is your very local neighborhood, much smaller than the Milky Way Galaxy, which in turn is tiny compared with the observable universe. However, from ancient times up through Copernicus's day, it was thought that the whole universe, everything that exists, did not extend much beyond the farthest planet of our solar system. Asking whether Earth or the sun is the center of the solar system was then the same question as asking whether Earth or the sun is the center of the universe. As you read this chapter, but only this chapter, you can pretend to have the old-fashioned view in which "solar system" and "universe" meant much the same thing.

Aristotle's Universe

Philosophers of the ancient world attempted to deduce truth about the universe by reasoning from **first principles.** A first principle was something that seemed obviously true to everyone and supposedly needed no further examination. That may strike you as peculiar; modern thinkers tend to observe how things work and then from that evidence make principles and conclusions that can always be reexamined. Before the Renaissance, however, reasoning from evidence (what you might call "scientific thinking") was not widespread.

Study ■ **The Ancient Universe** on pages 36-37 and notice three important ideas:

1 Ancient philosophers and astronomers accepted without question—as first principles—that heavenly objects must move on circular paths at constant speeds, and that the Earth is motionless at the center of the universe. Although a few ancient writers mentioned the possibility that Earth might move, most of them did so in order to point out how that idea is "obviously" wrong.

2 As viewed by you from Earth, the planets seem to follow complicated paths in the sky, including episodes of "back-

ward" motion that are difficult to explain in terms of motion on circular paths at constant speeds.

3 Finally, you can see how Ptolemy created an elaborate geometrical and mathematical model to explain details of the observed motions of the planets while assuming Earth is motionless at the center of the universe.

Aristotle lived in Greece from 384 to 322 BC. He believed as a first principle that the heavens were perfect. Because the sphere and circle were considered the only perfect geometrical figures, Aristotle also believed that all motion in the perfect heavens must be caused by the rotation of spheres carrying objects around in uniform circular motion. Aristotle's writings became so famous that he was known throughout the Middle Ages as "*The* Philosopher," and the geocentric universe of nested spheres that he devised dominated astronomy. His opinions on the nature of Earth and the sky were widely accepted for almost 2000 years.

Claudius Ptolemy, a mathematician who lived roughly 500 years after Aristotle, believed in the basic ideas of Aristotle's universe but was interested in practical rather than philosophical questions. For Ptolemy, first principles took second place to accuracy. He set about making an accurate mathematical description of the motions of the planets. Ptolemy weakened the first principles of Aristotle by moving Earth a little off-center in the model and inventing a way to slightly vary the planets' speeds. This made his model (published around AD 140) a better match to the observed motions.

Aristotle's universe, as embodied in the mathematics of Ptolemy's model, dominated ancient astronomy. At first the Ptolemaic model predicted positions of the planets with fair accuracy; but as centuries passed, errors accumulated, and Islamic and later European astronomers had to update the model, adjusting the sizes and locations of the circles and changing the rates of motion.

ThomsonNOW Log into ThomsonNOW and select this chapter to see Astronomy Exercise "Parallax."

(3-2) Nicolaus Copernicus

Nicolaus Copernicus (originally, Mikolaj Kopernik, ■ Figure 3-1) was born in 1473 in what is now Poland. At the time of his birth—and throughout his life—astronomy was based on Ptolemy's model of Aristotle's universe. In spite of many revisions, the Ptolemaic model was still a poor predictor of planet positions, but because of the authority of Aristotle it was the officially accepted model. Moreover, because in Aristotle's philosophy the most perfect region was in the heavens and the most imperfect region was at Earth's center, the classical **geocentric** universe model matched the commonly held Christian view of the geometry of heaven and hell. Anyone who criticized Aristotle's model of the universe was thereby also challenging belief in the locations of heaven and hell, risking at least criticism and perhaps a serious charge of heresy with a possible death penalty.

Nicolaus Copernicus (1473-1543) pursued a lifetime career in the Church, but he was also a talented mathematician and astronomer. His work triggered a revolution in human thought. This stamp was issued in 1973 to mark the 500th anniversary of his birth.

Copernicus's Model

Copernicus was associated with the Roman Catholic Church throughout his life. His uncle, by whom he was raised and educated, was a bishop. After studying medicine and Church law in some of the major universities in Europe, Copernicus became a Church employee, serving as secretary and personal physician to his powerful uncle for 15 years and moving into quarters in the cathedral after the uncle died. Because of this connection to the Church and his fear of persecution, he hesitated to publish his revolutionary ideas that challenged the Ptolemaic model and the geometry of heaven and hell.

What were these revolutionary ideas? Copernicus believed that the sun and not Earth was the center of the universe and that Earth rotated on its axis and revolved around the sun. Copernicus apparently began doubting Ptolemy's geocentric model during his college days. A **heliocentric** model (sun-centered, from the Greek word for sun, *helios*) had been discussed occasionally before Copernicus's time, but Copernicus was the first person to produce a detailed model with substantial justifying arguments. Sometime before 1514, Copernicus wrote a short pamphlet summarizing his model and distributed it in handwritten form while he worked on his book.

De Revolutionibus

Copernicus's book *De Revolutionibus Orbium Coelestium (About the Revolutions of Celestial Spheres)* was essentially finished by about 1530. He hesitated to publish, although other astronomers, and even church officials concerned about reform of the calendar, knew about his work, sought his advice, and looked forward to the book's publication. In 1542 Copernicus finally sent the manuscript for *De Revolutionibus* off to be printed. He died in 1543 before the printing was completed.

The most important idea in the book was placing the sun at the center of the universe. That single innovation had an impressive consequence—the retrograde motion of the planets was immediately explained in a straightforward way without the epicycles that Ptolemy used.

In Copernicus's model, Earth moves faster along its orbit than the planets that lie farther from the sun. Consequently, Earth periodically overtakes and passes these planets. Imagine that you are a runner on a track moving along an inside lane. Runners well ahead of you appear to be moving forward relative to background scenery. As you overtake and pass slower runners in outside lanes, they fall behind, seeming to move backward for a few moments relative to the scenery. The same thing happens as Earth passes a planet such as Mars. Although Mars moves steadily along its orbit, as seen from Earth it seems to slow to a stop and move westward (retrograde) relative to the background stars as Earth passes it (■ Figure 3-2).

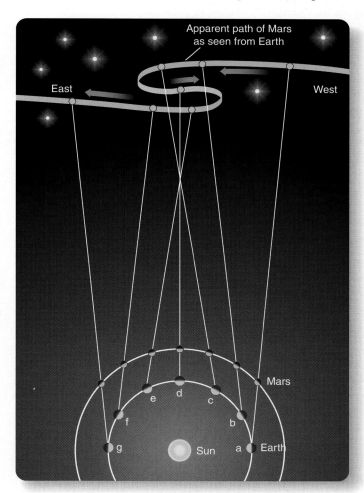

The Copernican explanation of retrograde motion. As Earth overtakes Mars (a-c), Mars appears to slow its eastward motion. As Earth passes Mars (d), Mars appears to move westward. As Earth draws ahead of Mars (e-g), Mars resumes its eastward motion against the background stars. The positions of Earth and Mars are shown at equal intervals of one month.

The Ancient Universe

1 For 2000 years, the minds of astronomers were shackled by two ideas. Greek philosophers argued that the heavens were perfect. Because the only perfect geometrical shape is a circle and the only perfect motion is uniform motion (motion at a constant speed), Aristotle believed that all motion in the heavens must be made up of combinations of circles turning at uniform rates. This was called **uniform circular motion**.

Aristotle also argued that Earth was imperfect and lay at the center of the universe. That is, he argued for a geocentric universe. He devised a model universe with 55 spheres turning at different rates and at different angles to carry the seven known planets (the moon, Mercury, Venus, the sun, Mars, Jupiter, and Saturn) across the sky.

Aristotle was known as the greatest philosopher in the ancient world, and the respect that later scholars had for his ideas caused them to expect the universe to be geocentric and its components to show uniform circular motion. This blinded them to other possibilities. See model at right by Peter Apian.

From *Cosmographica* by Peter Apian (1539).

Seen by left eye   Seen by right eye

1a Ancient astronomers believed that Earth did not move because they saw no **parallax,** the apparent motion of an object because of the motion of the observer. To demonstrate parallax, close one eye and cover a distant object with your thumb held at arm's length. Switch eyes, and your thumb appears to shift position as shown at left. If Earth moves, ancient astronomers reasoned, you should see the sky from different locations at different times of the year, and you should see parallax distorting the shapes of the constellations. They saw no parallax, so they concluded Earth could not move. Actually, the parallax of the stars is too small to see with the unaided eye.

2 Planetary motion was a big problem for ancient astronomers. In fact, the word *planet* comes from the Greek word for "wanderer," referring to the eastward motion of the planets against the background of the fixed stars. The planets did not, however, move at a constant rate, and they could occasionally stop and move westward for a few months before resuming their eastward motion. This backward motion is called **retrograde motion**.

Jan. 29, 2008

East

Nov. 15, 2007

Position of Mars at 5-day intervals

Dec. 12, 2005

Oct. 3, 2005

Ecliptic

Taurus

West

Betelgeuse

Every 2.14 years, Mars passes through a retrograde loop. Two successive loops are shown here. Each loop occurs further east along the ecliptic and has its own shape.

Orion

2a Simple uniform circular motion centered on Earth could not explain retrograde motion, so ancient astronomers combined uniformly rotating circles much like gears in a machine to try to reproduce the motion of the planets.

Rigel

3   Uniformly rotating circles were key elements of ancient astronomy. Claudius Ptolemy created a mathematical model of Aristotle's universe in which each planet moved on a small circle called an **epicycle** that slid around a larger circle. By adjusting the size and rate of rotation of the circles, he could approximate the retrograde motions of the planets. See illustration at right.

To adjust the speed of the planet, Ptolemy supposed that Earth was slightly off center and that the center of the epicycle moved such that it appeared to move at a constant rate as seen from an arbitrary point, not Earth.

To further adjust his model, Ptolemy added small epicycles riding on top of larger epicycles (not shown here), producing a highly complex model.

3a   Ptolemy's great book *Mathematical Syntaxis* (c. AD 140) contained the details of his model. Islamic astronomers preserved and studied the book through the Middle Ages, and they called it *Al Magisti* (The Greatest). When the book was found and translated from Arabic to Latin in the 12th century, it became known as *Almagest*.

3b   The Ptolemaic model of the universe shown below was geocentric and based on uniform circular motion. Note that Mercury and Venus were treated differently from the rest of the planets. The centers of the epicycles of Mercury and Venus had to remain on the Earth–Sun line as the sun circled Earth through the year.

Many smaller epicycles are not shown here. Some versions contained nearly 100 epicycles as generations of astronomers tried to fine-tune the model to better reproduce the motion of the planets.

Notice that this modern illustration shows rings around Saturn and sunlight illuminating the globes of the planets, features that could not be observed before the invention of the telescope.

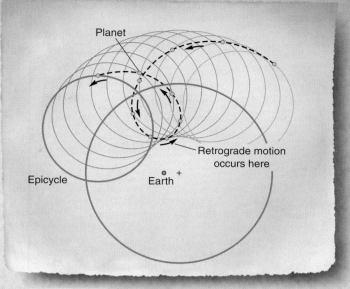

ThomsonNOW Sign in at www.thomsonedu.com and go to ThomsonNOW to see "Epicycles." Notice how the counter-clockwise rotation of the epicycle produces retrograde motion.

Because the planets' orbits do not lie in precisely the same plane, a planet does not resume its eastward motion in precisely the same path it followed earlier. Instead, it describes a loop with a shape depending on the angle between the two orbital planes.

Copernicus's model was simple and straightforward compared with the multiple off-center circles of the Ptolemaic model. However, *De Revolutionibus* failed to disprove the geocentric model immediately for one critical reason—the Copernican model could not predict the positions of the planets any more accurately than the Ptolemaic model could.

Although Copernicus proposed a revolutionary idea in making the solar system heliocentric, he was a classically trained astronomer with great respect for the old concept of uniform circular motion. Copernicus objected to Ptolemy's schemes for moving Earth slightly off-center and varying the speeds of planet motions. That seemed arbitrary and ugly to Copernicus, so he returned to a strong but incorrect belief in uniform circular motion. Therefore, even though his model put the sun correctly at the center of the solar system, it could not accurately predict the positions of the planets as seen from Earth. Copernicus even had to reintroduce small epicycles to match minor variations in the motions of the sun, moon, and planets. Astronomers today recognize those variations as due to the planets' real motions in elliptical orbits (discussed in Section 3-3).

■ **Figure 3-3**

(a) The Copernican universe, as reproduced in his book *De Revolutionibus*. Earth and all the known planets revolve in separate circular orbits about the sun *(Sol)* at the center. The outermost sphere carries the immobile stars of the celestial sphere. Notice the orbit of the moon around Earth *(Terra)*. (Yerkes Observatory) (b) The model is simple not only in the arrangement of the planets but also in their motions. Orbital velocities (blue arrows) decrease from that of Mercury, the fastest, to that of Saturn, the slowest. Compare the elegance of this model with the complexity of the Ptolemaic model as shown on page 37.

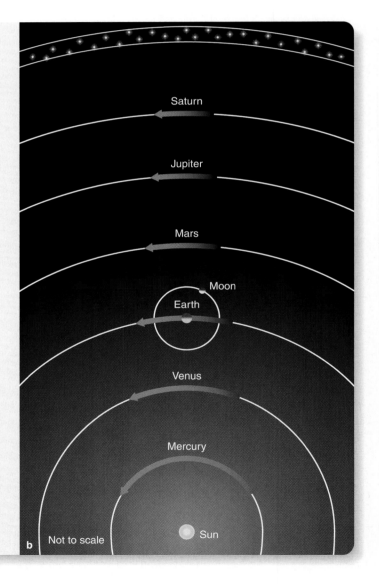

Creating New Ways to See Nature: Scientific Revolutions

The Copernican Revolution is often cited as the perfect example of a scientific revolution. Over a few decades, astronomers abandoned a way of thinking about the universe that was almost 2000 years old and adopted a new set of ideas and assumptions. The American philosopher of science Thomas Kuhn has referred to a commonly accepted set of scientific ideas and assumptions as a scientific **paradigm.** Thus, the pre-Copernican astronomers had a geocentric paradigm that included the perfection of the heavens, uniform circular motion, and a motionless Earth. That paradigm survived for many centuries, until a new generation of astronomers was able to overthrow the old paradigm and establish a new paradigm that included a heliocentric solar system with a moving Earth.

A scientific paradigm is powerful because it shapes perceptions. It determines what are judged to be important questions and what is judged to be significant evidence. Thus, it is often difficult to recognize how paradigms limit what you can understand. For example, the geocentric paradigm contained problems that seem obvious to a modern mind, but because astronomers before Copernicus lived and worked inside that paradigm they had difficulty seeing the problems. Overthrowing an outdated paradigm is not easy because you must learn to see nature in an entirely new way. Galileo, Kepler, and Newton saw nature from a new paradigm that would have been almost incomprehensible to astronomers of earlier centuries.

You can find examples of scientific revolutions in many fields, including biology, geology, genetics, and psychology. These scientific revolutions have been difficult and controversial because they have involved the overthrow of accepted paradigms, but that is why scientific revolutions are exciting. They give you not just a new idea or a new theory but an entirely new insight into how nature works—a new way of seeing the world.

The ancients believed the stars were attached to a starry sphere. (NOAO and Nigel Sharp)

You should notice the difference between the Copernican model and the Copernican hypothesis. The Copernican *model* is inaccurate. It includes uniform circular motion and thus does not precisely describe the motions of the planets. But the Copernican *hypothesis* that the solar system is heliocentric is correct—the planets do circle the sun, not Earth. Why that hypothesis gradually won acceptance is a question historians still debate. There are probably a number of reasons, including the revolutionary spirit of the times, but the most important factor may be the simplicity of the idea. For one thing, placing the sun at the center of the universe produced a symmetry among the motions of the planets that is elegant, pleasing to the eye and mind (■ Figure 3-3). In the Ptolemaic model, Mercury and Venus had to be treated differently from the rest of the planets: Their epicycles had to remain centered on the Earth-sun line. In Copernicus's model, all of the planets were treated the same. They all followed orbits that circled the sun at the center.

Although astronomers throughout Europe read and admired *De Revolutionibus* and found Copernicus's astronomical observations and mathematics to have great value, few astronomers believed, at first, that the sun actually was the center of the solar system and that Earth moved. How the Copernican hypothesis was gradually recognized as correct has been called the Copernican Revolution, because it was not just the adoption of a new idea but a to-

tal change in the way astronomers and the rest of humanity thought about the place of the Earth (see ■ Window on Science 3-1).

The most important consequence of the Copernican hypothesis was not what it said about the sun but what it said about Earth. By placing the sun at the center, Copernicus made Earth move along an orbit like the other planets. By making Earth a planet, Copernicus revolutionized humanity's view of its place in the universe and triggered a controversy that would eventually bring the astronomer Galileo Galilei before the Inquisition, a controversy over the nature of scientific and religious truths that continues even today.

Building Scientific Arguments

Why do we say that the Copernican hypothesis was correct but the model was wrong?

The Copernican hypothesis was that the sun and not Earth was the center of the universe. Given the lack of knowledge of the Renaissance astronomers about distant stars and galaxies, that hypothesis was correct. The sun is at the center of our solar system. That hypothesis produces a simpler description of the universe (with fewer moving parts) that explains retrograde motion in a straightforward way. The Copernican model, however, included not only the heliocentric hypothesis but also uniform circular motion. The model is wrong because the planets don't really follow circular orbits at constant speeds. Thus, the Copernican model could not make accurate predictions of planet positions even though it correctly placed the sun at the center of the universe.

The Copernican hypothesis won converts because it is simple and can easily explain retrograde motion. **How does its explanation of retrograde motion work, and how is it simpler than the Ptolemaic explanation?**

■ ■ ■

(3-3) Tycho Brahe, Johannes Kepler, and Planetary Motion

AS ASTRONOMERS STRUGGLED TO UNDERSTAND the place of the Earth, they also faced the problem of planetary motion. How exactly do the planets move? That problem was solved by a nobleman who built a fabulous observatory and a poor commoner with a talent for mathematics.

Tycho Brahe

The Danish nobleman Tycho Brahe is remembered in part for wearing false noses to hide a dueling scar from his college days (■ Figure 3-4). He was reportedly very proud of his noble station so his disfigurement probably did little to improve his lordly disposition.

In 1572, astronomers were startled to see a new star (now called Tycho's supernova) appear in the sky. Aristotle had argued that the heavens were perfect and therefore unchanging, so astronomers concluded that the new star had to be nearer than the moon. Tycho measured the new star's lack of parallax and showed it had to be far beyond the moon and was something changing in the supposedly unchanging starry sphere.

When Tycho wrote a book about his discovery, the king of Denmark honored him with a generous income and the gift of an island, Hveen, where Tycho built a fabulous observatory. Tycho lived before the invention of the telescopes, so his observatory was equipped with wonderful instruments for measuring the positions of the sun, moon, and planets using the naked eye and peep sights. For 20 years, Tycho and his assistants measured the positions of the stars and planets.

After the death of the Danish king, Tycho moved to Prague where he became the Imperial Mathematician to the Holy Roman Emperor Rudolph II. Tycho hired a few assistants including a German school teacher named Johannes Kepler. Just before Tycho died in 1601, he asked Rudolph II to make Kepler Impe-

rial Mathematician. Thus the newcomer, Kepler, became Tycho's replacement (though at one-sixth Tycho's salary).

Johannes Kepler

No one could have been more different from Tycho Brahe than was Johannes Kepler (■ Figure 3-5). He was born in 1571, the oldest of six children in a poor family living in what is now southwest Germany. His father was principally employed as a mercenary soldier, fighting for whomever paid enough, and eventually disappeared. Kepler's mother was apparently an unpleasant and unpopular woman. She was accused of witchcraft in her later years, and Kepler defended her (successfully) in a trial that dragged on for three years. Kepler himself had poor health, even as a child, so it is surprising that he did well in school, winning promotion to a Latin school and eventually a scholarship to the university at Tübingen, where he studied to become a Lutheran pastor.

While still a college student, Kepler had become a believer in the Copernican hypothesis. During his last year of study, Kepler accepted a teaching job in the town of Graz, in what is now Austria, which allowed him to continue his studies in mathematics and astronomy. By 1596, the same year Tycho arrived in Prague, Kepler had learned enough to publish a book called *The Forerunner of Dissertations on the Universe, Containing the Mystery of the Universe*. The book, like nearly all scientific works of

Johannes Kepler (1571-1630) derived three laws of planetary motion from Tycho Brahe's observations of the positions of the planets. This Romanian stamp commemorates the 400th anniversary of Kepler's birth. Ironically, it contains an astronomical error—the horns of the crescent moon should be inclined upward from the horizon.

him to Prague in 1600, Kepler went eagerly, ready to work with the famous astronomer. Tycho's sudden death in 1601 left Kepler in a position to use Tycho's extensive records of observations to analyze the motions of the planets.

Kepler began by studying the motion of Mars, trying to deduce from the observations how the planet actually moved. By 1606, he had solved the mystery: The orbit of Mars is an ellipse, not a circle. Thus, he abandoned the ancient belief in the circular motion of the planets. But the mystery was even more complex. The planets do not move at uniform speeds along their elliptical orbits. Kepler recognized that they move faster when close to the sun and slower when farther away. Thus Kepler abandoned both uniform motion and circular motion and thereby finally solved the problem of planetary motion. Later, he discovered that the radii of the planets' orbits are related to the planets' orbital periods. Kepler published his results in 1609 and 1619 in books called, respectively, *Astronomia Nova (New Astronomy)* and *Harmonice Mundi (The Harmony of the World)*.

that age, was written in Latin and is now known as *Mysterium Cosmographicum.*

By modern standards, the book contains almost nothing of value. It begins with a long appreciation of Copernicus's model and then goes on to mystical speculation on the reason for the spacing of the planets' orbits. The second half has one virtue—as Kepler tried to understand planet orbits, he demonstrated that he was a talented mathematician and that he had become well versed in astronomy. He sent copies to Tycho and to Galileo, who both recognized Kepler's talent in spite of the mystical parts of the book.

Life was unsettled for Kepler in Graz because of the persecution of Protestants in that region, so when Tycho Brahe invited

Kepler's Three Laws of Planetary Motion

Although Kepler dabbled in the philosophical arguments of the day, he was a mathematician, and his triumph was the solution of the problem of the motion of the planets. The key to his solution was the ellipse. An **ellipse** is a figure drawn around two points, called the *foci,* in such a way that the distance from one focus to any point on the ellipse and back to the other focus equals a constant. This makes it easy to draw ellipses with two thumbtacks and a loop of string. Press the thumbtacks into a board, loop the string about the tacks, and place a pencil in the loop. If you keep the string taut as you move the pencil, it traces out an ellipse (■ Figure 3-6a).

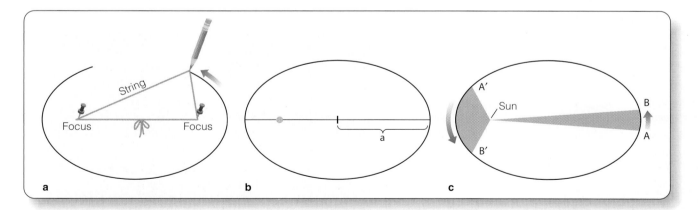

■ **Figure 3-6**

The geometry of elliptical orbits. (a) Drawing an ellipse with two tacks and a loop of string. (b) The semi-major axis, *a*, is half of the longest diameter. (c) Kepler's second law is demonstrated by a planet that moves from *A* to *B* in one month and from *A'* to *B'* in the same amount of time. The two blue segments have the same area.

Hypothesis, Theory, and Law: Levels of Confidence

Even scientists misuse the words *hypothesis, theory,* and *law.* You must try to distinguish these terms from one another because they are key elements in science.

A **hypothesis** is a single assertion or conjecture that can be tested. It could be true or false. "All Texans love chili" is a hypothesis. To know whether it is true or false, you need to test it against reality by making observations or performing experiments. Copernicus asserted that the universe was heliocentric; his assertion was a hypothesis subject to testing.

In Chapter 2, you saw that a model (Window on Science 2-1) is a description of some natural phenomenon; it can't be right or wrong. A model is not a conjecture of truth but merely a convenient way to think about a natural phenomenon. Consequently, a model such as the celestial sphere is not a hypothesis. Copernicus used his hypothesis to build a model, but they are not the same thing.

A **theory** is a system of rules and principles that can be applied to a wide variety of circumstances. A theory may have begun as one or more hypotheses, but it has been tested, expanded, and generalized. Many textbooks refer to the "Copernican theory," but some historians argue that it was not complete and had not been tested enough to be a theory. It is probably better to call it the Copernican hypothesis.

A **natural law** is a theory that has been refined, tested, and confirmed so often that scientists have great confidence in it. Laws are the most fundamental principles of scientific knowledge. Kepler's laws are good examples.

Confidence is the key to understanding these terms. Scientists have more confidence in a theory than in a hypothesis and great confidence in a natural law. Nevertheless, scientists are not always consistent about these words. For example, Einstein's theory of relativity is much more accurate than Newton's laws of gravity and motion, but traditionally scientists refer to "Einstein's theories" and "Newton's laws." Darwin's theory of evolution has been tested many times, and scientists have great confidence in it, but no one refers to Darwin's law. These distinctions are subtle and sometimes depend more on custom than on levels of confidence.

The geometry of an ellipse is described by two simple numbers. The **semi-major axis, *a*,** is half of the longest diameter (Figure 3-6b). The **eccentricity, *e*,** of an ellipse is half the distance between the foci divided by the semi-major axis. The eccentricity of an ellipse tells you its shape: If e is nearly equal to one, the ellipse is very elongated; if e is close to zero, the ellipse is more circular. To draw a circle with the string and tacks shown in Figure 3-6a, you would move the two thumbtacks together, which shows that a circle is the same as an ellipse with eccentricity equal to zero. As you move the thumbtacks farther apart, the ellipse becomes flatter, and the value of its eccentricity moves closer to 1.

Kepler used ellipses to describe the motion of the planets in three fundamental rules that have been tested and confirmed so many times that astronomers now refer to them as "natural laws" (▪ Window on Science 3-2). They are commonly called Kepler's laws of planetary motion (▪ Table 3-1).

Kepler's first law states that the orbits of the planets around the sun are ellipses with the sun at one focus. Thanks to the precision of Tycho's observations and the sophistication of Kepler's mathematics, Kepler was able to recognize the elliptical shape of the orbits even though they are nearly circu-

▪ Figure 3-7

The orbits of the planets are nearly circular. You can measure the horizontal and vertical diameters of this orbit to detect its elliptical shape.

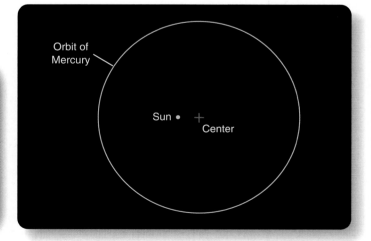

Orbit of Mercury

Sun • +
Center

▪ Table 3-1 | Kepler's Laws of Planetary Motion

  I. The orbits of the planets are ellipses with the sun at one focus.
 II. A line from a planet to the sun sweeps over equal areas in equal intervals of time.
III. A planet's orbital period squared is proportional to its average distance from the sun cubed:

$$P_y^2 = a_{AU}^3$$

lar. Of the planets known to Kepler, Mercury has the most elliptical orbit, but even it deviates only slightly from a circle (■ Figure 3-7).

Kepler's second law states that a line from the planet to the sun sweeps over equal areas in equal intervals of time. This means that when the planet is closer to the sun and the line connecting it to the sun is shorter, the planet moves more rapidly to sweep over the same area that is swept over when the planet is farther from the sun. Thus the planet in Figure 3-6c would move from point *A'* to point *B'* in one month, sweeping over the area shown. But when the planet is farther from the sun, one month's motion would be shorter, from *A* to *B*. The time that a planet takes to travel around the sun once is its orbital period, *P*, and its average distance from the sun equals the semi-major axis of its orbit, a. Kepler's third law tells us that these two quantities, orbital period and semi-major axis, are related: Orbital period squared is proportional to the semi-major axis cubed. For example, Jupiter's average distance from the sun (which equals the semi-major axis of its orbit) is 5.2 AU. The semi-major axis cubed would be about 140.6, so the period must be the square root of 140.6, roughly 11.8 years.

It is important to notice that Kepler's three laws are **empirical.** That is, they describe a phenomenon without explaining why it occurs. Kepler derived them from Tycho's extensive observations without referring to any first principles, fundamental assumptions, or theory. In fact, Kepler never knew what held the planets in their orbits or why they continued to move around the sun in the ways he discovered.

Building Scientific Arguments

What principles of classical astronomy were barriers that Kepler had to overcome to understand planetary motion? Critical analysis is more than careful thinking. You can think carefully about what Kepler did; but, until you analyze the paradigm (the network of assumptions and principles) within which he worked, you cannot understand how impressive his achievement was. As you have learned, Kepler lived at a time when every astronomer believed in uniform circular motion. Before Kepler could adopt elliptical orbits and variable planet speeds, he had to step outside the classical paradigm and see nature in a new way. Only then could he recognize how the planets really moved.

This might make you wonder what false paradigms surround you and prevent you from understanding nature better. Kepler was never confused by believing in the geocentric model. From his days in college, he accepted that the sun was the center of the universe. **Of course, his first law partially violates an interpretation of the heliocentric model. How?**

■   ■   ■

GALILEO GALILEI (■ Figure 3-8) was born in the Italian city of Pisa in 1564 and studied medicine at the university there. His true love, however, was mathematics, and he eventually became professor of mathematics at the university at Padua, where he remained for 18 years. During this time, Galileo seems to have adopted the Copernican model, although he admitted in a 1597 letter to Kepler that at that time he did not support that model publicly, fearing criticism.

Most people know two "facts" about Galileo, and both are wrong. Galileo did not invent the telescope, and he was not condemned by the Inquisition for believing that Earth moved around the sun. As you learn about Galileo, you will discover that what was on trial were not just his opinions about the place of the Earth but also the methods of science itself.

■ **Figure 3-8**

Galileo Galilei (1564-1642), remembered as the great defender of Copernicus's model, also made important discoveries in the physics of motion. Although he did not invent it, Galileo will always be remembered as associated with the telescope, as on these postage stamps. By depending on direct observation instead of first principles from philosophy and theology, Galileo led the way to the invention of modern astronomy and modern science as a way to know about the natural world.

Telescopic Observations

It was the telescope that drove Galileo to publicly defend the heliocentric model. Galileo did not invent the telescope. It was apparently invented around 1608 by lens makers in Holland. Galileo, hearing descriptions in the fall of 1609, was able to build working telescopes in his workshop. Galileo was also not the first person to look at the sky through a telescope, but he was the first person to observe the sky carefully and apply his observations to the main theoretical problem of the day—the place of Earth.

What Galileo saw through his telescopes was so amazing he rushed a small book into print, *Sidereus Nuncius (The Starry Messenger)*. In that book he reported two major discoveries about the solar system. First, the moon was not perfect. It had mountains and valleys on its surface, and Galileo used the shadows to calculate the height of the mountains. Aristotle's philosophy held that the moon was perfect, but Galileo showed that it was not only imperfect but was even a world like Earth. Second, Galileo's telescope revealed four new "planets" circling Jupiter, planets that we know today as the Galilean moons of Jupiter (■ Figure 3-9).

The moons of Jupiter supported the Copernican over the Ptolemaic model. Critics of Copernicus had said Earth could not move because the moon would be left behind; but Jupiter moved and kept its satellites. Galileo's discovery suggested that Earth, too, could move and keep its moon. Also, Aristotle's philosophy included the belief that all heavenly motion was centered on Earth. Galileo showed that Jupiter's moons revolve around Jupiter, so there could be centers of motion other than Earth. Later, after the *Messenger* was published, Galileo noticed that Jupiter's innermost moon had the shortest orbital period and the moons further from Jupiter had proportionally longer periods. In this way, Jupiter's moons made up a harmonious system ruled by Jupiter, just as the planets in the Copernican universe were a harmonious system ruled by the sun. This similarity didn't constitute proof, but Galileo saw it as an indication that the solar system could be sun centered and not Earth centered.

In the years of further exploration with his telescope, Galileo made additional fundamental discoveries. When he observed Venus, Galileo saw that it was going through phases like those of the moon. In the Ptolemaic model, Venus moves around an epicycle centered on a line between Earth and the sun. If that were true, it would always be seen as a crescent (■ Figure 3-10a). But Galileo saw Venus go through a complete set of phases, including full and gibbous, which proved that it did indeed revolve around the sun (Figure 3-10b).

Sidereus Nuncius was popular and made Galileo famous. In 1611, Galileo visited Rome and was treated with great respect. He had friendly discussions with the powerful Cardinal Barberini, but because he was outspoken, forceful, and some-

■ Figure 3-9

(a) On the night of January 7, 1610, Galileo saw three small "stars" near the bright disk of Jupiter and sketched them in his notebook. On subsequent nights (excepting January 9, which was cloudy), he saw that the stars were actually four moons orbiting Jupiter. (b) This photograph, taken through a modern telescope, shows the overexposed disk of Jupiter and three of the four Galilean moons. (Grundy Observatory)

times tactless he offended other important people who questioned his telescopic discoveries. Some critics said he was wrong, and others said he was lying. Some refused to look through a telescope lest it mislead them, and others looked and claimed to see nothing (hardly surprising given the awkwardness of those first telescopes) (■ Figure 3-11). When Galileo visited Rome again in 1616, Cardinal Bellarmine interviewed him privately and ordered him to cease public debate about models of the universe, an order Galileo appears to have mostly followed.

The Inquisition (formally named the Congregation of the Holy Office) banned books relevant to the Copernican hypothesis, although *De Revolutionibus* itself was only suspended pending revision because it was recognized as useful for its predictions of planet positions. Everyone who owned a copy of the book was required to cross out certain statements and add handwritten corrections stating that Earth's motion and the central location of the sun were only theories and not facts, a situation that you will recognize as recurring today in connection with textbooks discussing biological evolution.

(a) If Venus moved in an epicycle centered on the Earth-sun line, it would always appear as a crescent. (b) Galileo's telescope showed that Venus goes through a full set of phases, proving that it must orbit the sun.

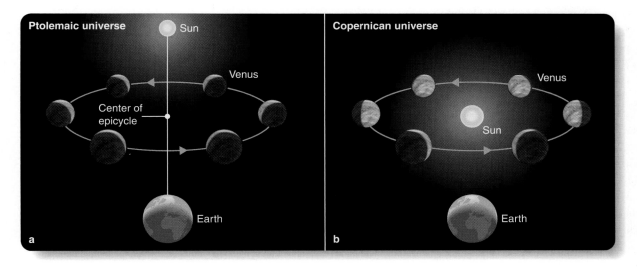

Dialogo and Trial

In 1623 Galileo's friend Cardinal Barberini became pope, taking the name Urban VIII. Galileo went to Rome in an attempt to have the 1616 order to cease debate lifted. Although that attempt was unsuccessful, Galileo began to write a massive defense of Copernicus's model, completing it in 1629. After some delay Galileo's book was approved by both the local censor in Florence and the head censor of the Vatican in Rome. It was printed in 1632.

Called *Dialogo Dei Due Massimi Sistemi (Dialogue Concerning the Two Chief World Systems)*, it confronts the ancient astronomy of Aristotle and Ptolemy with the Copernican model. Galileo wrote the book as a debate among three friends. Salviati is a swift-tongued defender of Copernicus; Sagredo is intelligent but largely uninformed; Simplicio is a dim-witted defender of Ptolemy. The book was a clear defense of Copernicus, and, either intentionally or unintentionally, Galileo exposed the pope's authority to ridicule. Urban VIII was fond of arguing that, as God was omnipotent, God could construct the universe in any form while making it appear to humans to have a different form, and thus its true nature could not be deduced by mere observation. Galileo placed the pope's argument in the mouth of Simplicio. The pope took offense and ordered Galileo to face the Inquisition.

Galileo was interrogated by the Inquisition and threatened with torture. He must have thought of Giordano Bruno, a monk who was tried, condemned, and burned at the stake in Rome in 1600 for, among other offenses, supporting Copernicus. However, Galileo's trial did not center on his belief in Copernicus's model. *Dialogo* had been approved by two censors. Rather, the trial centered on a record of the meeting in 1616 between Galileo and Cardinal Bellarmine that included the statement that Galileo was "not to hold, teach, or defend in any way" the principles of Copernicus. Many historians believe that this document, which was signed neither by Galileo nor by Bellarmine nor by a legal secretary, was a forgery, or perhaps a draft that was never used. By this time Bellarmine was dead and could not testify about the meeting or the document.

■ **Figure 3-11**

Galileo's telescopic discoveries generated intense interest and controversy. Some critics refused to look through a telescope lest it deceive them. (Yerkes Observatory)

The Inquisition condemned Galileo not primarily for heresy but for disobeying the orders given him in 1616. In 1633, at the age of 70, kneeling before the Inquisition, Galileo read a recantation admitting his errors. Tradition has it that as he rose he whispered, *"E pur si muove"* ("Still it moves"), referring to Earth. Although he was sentenced to life imprisonment, he was actually confined at his villa for the next 10 years, perhaps through the intervention of the pope. He died there in 1642, 99 years after the death of Copernicus.

Two Ways to Understand the World

Galileo was tried and condemned on a charge you might call a technicality. Why then is his trial so important that historians have studied it for almost four centuries? Why have some of the world's greatest authors, including Bertolt Brecht, written about Galileo's trial? Why in 1979 did Pope John Paul II create a commission to reexamine the case against Galileo?

To understand the trial, you must recognize that it was the result of a conflict between two ways of understanding the universe. Since the Middle Ages, scholars had taught that the only path to true understanding was through religious faith. St. Augustine (AD 354-430) wrote *"Credo ut intelligame,"* which can be translated as, "Believe in order to understand." Galileo and other scientists of the Renaissance, however, used their own observations to try to understand the universe; and, when their observations contradicted Scripture, they assumed their observations of reality were correct. Galileo paraphrased Cardinal Baronius in saying, "The Bible tells us how to go to heaven, not how the heavens go." The significance of Galileo's trial is about the birth of modern science as a new way to understand the universe.

Building Scientific Arguments

How were Galileo's observations of the moons of Jupiter evidence against the Ptolemaic model?

Galileo presented his arguments in the form of evidence and conclusions, and the moons of Jupiter were key evidence. Astronomers supporting Ptolemy's model argued that Earth could not move or it would lose its moon; but although even in the Ptolemaic universe Jupiter moved, the telescope showed that Jupiter has moons and manages to keep them. Perhaps Earth could also move and not leave its moon behind. Furthermore, moons circling Jupiter did not fit the classical belief that all motion was centered on Earth. Obviously there could be other centers of motion. Finally, the orbital periods of the moon were related to their distance from Jupiter, just as the orbital periods of the planets were, in the Copernican model, related to their distance from the sun. This similarity suggested that Jupiter and its moons constitute a system that is a miniature version of the sun and the planets.

Of all of Galileo's telescopic observations, the moons of Jupiter caused the most debate, but the craters on the moon

and the phases of Venus were also critical evidence. **How did they argue against the Ptolemaic model?**

■ ■ ■

(3-5) Isaac Newton, Gravity, and Orbits

THE BIRTH OF MODERN ASTRONOMY and of modern science dates from the 144 years between the publication of Copernicus's book in 1543 and Newton's book in 1687. The Renaissance is commonly taken to be the period between 1350 and 1600, so the 144 years of this story lie at the climax of the European reawakening of learning in all fields (■ Figure 3-12).

The problem of the place of Earth was resolved by the Copernican Revolution, but the problem of planetary motion was only partly solved by Kepler's laws. For the last 10 years of his life, Galileo studied the nature of motion, especially the accelerated motion of falling bodies. Although he made some important progress, he was not able to relate his discoveries about motion to the heavens. That final step was taken by Isaac Newton.

Isaac Newton

Galileo died in January 1642. Some 11 months later, on Christmas day 1642, Isaac Newton (■ Figure 3-13) was born in the English village of Woolsthorpe. Newton's life represented the flowering of the seeds planted by the previous four astronomers in this story, Copernicus, Tycho, Kepler, and Galileo.

Newton was a quiet child from a farming family, but his work at school was so impressive that his uncle financed his education at Trinity College, where he studied mathematics and physics. In 1665, plague swept through England, and the colleges were closed. During 1665 and 1666, Newton spent his time back home in Woolsthorpe, thinking and studying. It was during these years that he made most of his scientific discoveries. Among other things, he studied optics, developed three laws of motion, probed the nature of gravity, and invented differential calculus. The publication of his work in his book *Principia* in 1687 placed the fields of physics and astronomy on a new firm base.

It is beyond the scope of this book to analyze all of Newton's work, but his laws of motion and gravity had an important impact on the future of astronomy. From his study of the work of Galileo, Kepler, and others, Newton extracted three laws that relate the motion of a body to the forces acting on it (■ Table 3-2). These laws made it possible to predict exactly how a body would move if the forces were known.

When Newton thought carefully about motion, he realized that some force must pull the moon toward Earth's center. If

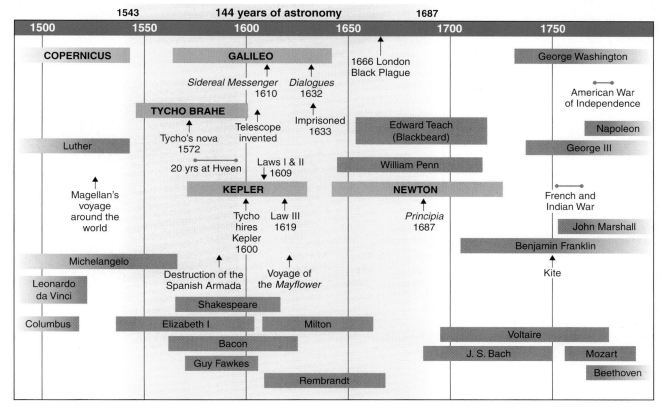

Figure 3-12

The 144 years between the publication of Copernicus's book *De Revolutionibus* in 1543 and Newton's book *Principia* in 1687 saw the birth of modern scientific astronomy.

there were no such force altering the moon's motion, it would continue moving in a straight line and leave Earth forever. It can circle Earth only if Earth attracts it. Newton's insight was to recognize that the force that holds the moon in its orbit is the same as the force that makes apples fall from trees.

Newtonian gravitation is sometimes called universal mutual gravitation (■ Window on Science 3-3). Newton's third law points out that forces occur in pairs. If one body attracts another, the second body must also attract the first. Thus, gravitation must be mutual. Furthermore, gravity must be universal. That is, all objects that contain mass must attract all other masses in the universe. The **mass** of an object is a measure of the amount of matter in the object, usually expressed in kilograms. Mass is not the same as **weight.** An object's weight is the force that Earth's gravity exerts on the object. Thus an object in space far from Earth might have no weight, but it would contain the same amount of matter and would thus have the same mass that it has on Earth.

Newton also realized that the distance between the objects is important. In other words, the gravitational force between two bodies depends not only on the masses of

Figure 3-13

Isaac Newton (1642-1727) worked from the discoveries of Galileo and Kepler to study motion and gravitation. He and some of his discoveries were honored on this English one-pound note.

Universal Gravitation

Isn't it weird that Isaac Newton is said to have "discovered" gravity in the late 17th century—as if people didn't have gravity before that, as if people in the 16th century floated around holding onto tree branches? Newton's accomplishment was that he realized that gravity was universal. The same force that makes an apple fall holds the moon in its orbit and also guides the planets around the sun and shapes other motions in the universe. Newton's discovery made gravity the master of the universe and changed the way people thought about nature.

Gravity, Newton said, is a characteristic of matter, and any object made of matter exerts gravity on other objects. Earth, the moon, the sun, the planets, and the stars have their gravitational fields, and they move under the combined influence of each others' gravitational tugs. Before Newton, most people thought the heavens were governed by a different set of rules; the heavens were separate from humanity and un-Earthly. Universal gravitation linked everything together and made humans part of the universe. The Copernican Revolution made Earth into a planet, and the Newtonian Revolution unified the heavens and Earth.

The universe is filled with gravitational forces, and our planet, our star, and our galaxy move through space under the combined influence of every other object made of matter. The universe is a swirling waltz of matter dancing to the music of gravity.

Gravity's tidal effect distorts two galaxies as they whirl past each other. (NASA, ESA and STScI)

■ Table 3-2 I Newton's Three Laws of Motion

I. A body continues at rest or in uniform motion in a straight line unless acted upon by some force.
II. A body's change of motion is proportional to the force acting on it and is in the direction of the force.
III. When one body exerts a force on a second body, the second body exerts an equal and opposite force back on the first body.

---

the bodies but also on the distance between them. He recognized that the force of gravity decreases as the square of the distance between the objects increases. Specifically, if the distance from, say, Earth to the moon were doubled, the gravitational force between them would decrease by a factor of 2^2, or 4. If the distance were tripled, the force would decrease by a factor of 3^2, or 9. This relationship is known as the **inverse square relation.** Newton guessed that gravity works by an inverse square relation because he had already discovered that light behaves this way (to be discussed in more detail in Chapter 6).

To summarize, the force of gravity attracting two objects to each other equals a constant times the product of their masses divided by the square of the distance between the objects. Gravity is universal: Your mass affects the planet Neptune and the galaxy M31, and every other object in the universe, and their masses af-

fect you—although not much, because they are so far away and your mass is relatively very small.

Orbital Motion

Newton's laws of motion and gravitation make it possible for you to understand why and how the moon orbits Earth and the planets orbit the sun and to discover why Kepler's laws work.

To understand how an object can orbit another object, you need to see orbital motion as Newton did. Objects in orbit are falling. Study ■ Orbiting Earth on pages 50-51 and notice three important ideas:

1 An object orbiting Earth is actually falling (being accelerated) toward Earth's center. An object in a stable orbit continuously misses Earth because of its orbital velocity.

2 Objects orbiting each other actually revolve around their mutual center of mass.

3 Notice the difference between closed orbits and open orbits. If you want to leave Earth never to return, you must give your spaceship a high enough velocity so it will follow an open orbit.

When the captain of a spaceship says to the pilot, "Put us into a standard orbit," the ship's computers must quickly calculate the velocity needed to achieve a circular orbit. That circular velocity depends only on the mass of the planet and the distance from the center of the planet. Once the engines fire and the ship reaches circular velocity, the engines can shut down. The ship is in orbit and will fall around the planet forever so long as it is above the atmosphere's friction. No further effort is needed to maintain orbit, thanks to the laws Newton discovered.

ThomsonNOW Log into ThomsonNOW and select this chapter to see Astronomy Exercises "Falling Bodies," "Orbital Motion," and "Escape Velocity."

Tides: Gravity in Action

Newton understood that gravity is mutual—Earth attracts the moon, and the moon attracts Earth—and that means the moon's gravity can explain the ocean tides. But Newton also realized that gravitation is universal, and that means there is much more to tides than just Earth's oceans.

Tides are caused by small gravitational forces. For example, Earth's gravity attracts your body downward with a force equal to your weight. The moon is less massive and more distant, so the moon attracts your body with a force equal to roughly 0.0003 percent of your weight. You don't notice that tiny force, but Earth's oceans, being much larger, respond more dramatically.

The side of Earth facing the moon is about 6400 km (4000 mi) closer to the moon than is the center of Earth. Consequently, the moon's gravity, tiny though it is at the distance of Earth, is just a bit stronger when it acts on the near side of Earth than on the center. It pulls on the oceans on the near side of Earth a bit more strongly than on Earth's center, and the oceans respond by flowing into a bulge of water on the side of Earth facing the moon. There is also a bulge on the side away from the moon, which develops because the moon pulls more strongly on Earth's center than on the far side. Thus the moon pulls Earth away from the oceans, which flow into a bulge on the far side as shown at the top of ■ Figure 3-14.

You might wonder: If Earth and the moon accelerate toward each other, why don't they smash together? The answer, of course, is that they are orbiting around their common center of gravity. The ocean tides are caused by the accelerations Earth and the oceans feel as they move around that center of gravity. The rocky bulk of Earth also responds to these tidal forces. Although you don't notice, Earth flexes, and the mountains and plains rise and fall by a few centimeters in response to the moon's gravitational acceleration.

You can see dramatic evidence of this effect if you watch the ocean shore for a few hours. Though Earth rotates on its axis, the tidal bulges remain fixed with respect to the moon. As the turning Earth carries you and your beach into a tidal bulge, the ocean water deepens, and the tide crawls up the sand. Later, when Earth's rotation carries you out of the bulge, the ocean becomes shallower, and the tide falls. Because there are two bulges on opposite sides of Earth, the tides should rise and fall twice a day on an ideal coast.

In reality, the tidal cycle at any given location can be quite complex because of the latitude of the site, shape of the shore, winds, and so on. Tides in the Bay of Fundy (New Brunswick, Canada), for example, occur twice a day and can exceed 13 meters (40 feet), whereas the northern coast of the Gulf of Mexico has only one tidal cycle a day of roughly 30 centimeters (1 foot).

Gravity is universal, so the sun, too, produces tides on Earth. The sun is much more massive than the moon, but it is also much farther from Earth. Consequently, tides on Earth caused by the sun are less than half those caused by the moon. At new moon and at full moon, the moon and sun produce tidal bulges that add together and produce extreme tidal changes; high tide is very high, and low tide is very low. Such tides are called **spring tides,** even though they occur at every new and full moon and not just in the spring. **Neap tides** occur at first- and third-quarter moons, when the moon and sun pull at right angles to each other, as shown in Figure 3-14. Then the tides caused by the sun reduce the tides caused by the moon, and the rise and fall of the ocean is less extreme than usual.

Galileo tried to understand tides, but not until Newton described gravity could astronomers analyze tidal forces and recognize their surprising effects. For example, the friction of the ocean waters with the ocean beds slows Earth's rotation and makes the length of a day grow by 0.0023 seconds per century. Fossils of marine animals and tidal sediments confirm that only 900 million years ago Earth's day was 18 hours long. In addition, Earth's gravitation exerts tidal forces on the moon, and although there are no bodies of water on the moon, friction within the flexing rock has slowed the moon's rotation to the point that it now always keeps the same face toward Earth.

Tidal forces can also affect orbital motion. Friction with the ocean beds drags the tidal bulges eastward out of a direct Earth-moon line. These tidal bulges contain a large amount of mass, and their gravitational field pulls the moon forward in its orbit, as shown at the bottom of Figure 3-14. As a result, the moon's orbit is growing larger by about 4 cm a year, an effect that astronomers can measure by bouncing laser beams off reflectors left on the lunar surface by the Apollo astronauts.

Newton's gravitation is much more than just the force that makes apples fall. In later chapters, you will see how tides can pull gas away from stars, rip galaxies apart, and melt the interiors of small moons orbiting near massive planets. Tidal forces produce some of the most surprising and dramatic processes in the universe.

Newton's Universe

Newton's insight gave the world a new conception of nature. His laws of motion and gravity were *general* laws that described the motions of all bodies under the action of external forces. In addition, the laws were *predictive* because they made possible spe-

Orbiting Earth

1 You can understand orbital motion by thinking of a cannonball falling around Earth in a circular path. Imagine a cannon on a high mountain aimed horizontally as shown at right. A little gunpowder gives the cannonball a low velocity, and it doesn't travel very far before falling to Earth. More gunpowder gives the cannonball a higher velocity, and it travels farther. With enough gunpowder, the cannonball travels so fast it never strikes the ground. Earth's gravity pulls it toward Earth's center, but Earth's surface curves away from it at the same rate it falls. It is in orbit. The velocity needed to stay in a circular orbit is called the **circular velocity.** Just above Earth's atmosphere, circular velocity is 7790 m/s or about 17,400 miles per hour, and the orbital period is about 90 minutes.

A satellite above Earth's atmosphere feels no friction and will fall around Earth indefinitely.

North Pole

Earth satellites eventually fall back to Earth if they orbit too low and experience friction with the upper atmosphere.

ThomsonNOW™

Sign in at www.thomsonedu.com and go to ThomsonNOW to see "Newton's Cannon" and fire your own version of Newton's cannon.

1a A **geosynchronous satellite** orbits eastward with the rotation of Earth and remains above a fixed spot — ideal for communications and weather satellites.

A Geosynchronous Satellite

At a distance of 42,250 km (26,260 miles) from Earth's center, a satellite orbits with a period of 24 hours.

The satellite orbits eastward, and Earth rotates eastward under the moving satellite.

The satellite remains fixed above a spot on Earth's equator.

1b According to Newton's first law of motion, the moon should follow a straight line and leave Earth forever. Because it follows a curve, Newton knew that some force must continuously accelerate it toward Earth — gravity. Each second the moon moves 1020 m (3350 ft) eastward and falls about 1.6 mm (about 1/16 inch) toward Earth. The combination of these motions produces the moon's curved orbit. The moon is falling but nevertheless stays in its orbit.

Straight line motion of the moon

Motion toward Earth

Curved path of moon's orbit

Earth

1c Astronauts in orbit around Earth feel weightless, but they are not "beyond Earth's gravity," to use a term from old science fiction movies. Like the moon, the astronauts are accelerated toward Earth by Earth's gravity, but they travel fast enough along their orbits that they continually "miss the Earth." They are literally falling around Earth. Inside or outside a spacecraft, astronauts feel weightless because they and their spacecraft are falling at the same rate. Rather than saying they are weightless, you should more accurately say they are in free fall.

NASA

2 To be precise you should not say that an object orbits Earth. Rather the two objects orbit each other. Gravitation is mutual, and if Earth pulls on the moon, the moon pulls on Earth. The two bodies revolve around their common **center of mass,** the balance point of the system.

2a Two bodies of different mass balance at the center of mass, which is located closer to the more massive object. As the two objects orbit each other, they revolve around their common center of mass as shown at right. The center of mass of the Earth–moon system lies only 4708 km (2926 miles) from the center of Earth — inside the Earth. As the moon orbits the center of mass on one side, the Earth swings around the center of mass on the opposite side.

Center of mass

3 **Closed orbits** return the orbiting object to its starting point. The moon and artificial satellites orbit Earth in closed orbits. Below, the cannonball could follow an elliptical or a circular closed orbit. If the cannonball travels as fast as **escape velocity,** the velocity needed to leave a body, it will enter an open orbit. An **open orbit** does not return the cannonball to Earth. It will escape.

ThomsonNOW Sign in at www.thomsonedu.com and go to ThomsonNOW to see "Center of Mass." Change the mass ratio to move the center of mass.

Hyberbola

A cannonball with a velocity greater than escape velocity will follow a hyperbola and escape from Earth.

Parabola

A cannonball with escape velocity will follow a parabola and escape.

North Pole

Ellipse

Circle

3a As described by Kepler's Second Law, an object in an elliptical orbit has its lowest velocity when it is farthest from Earth (apogee), and its highest velocity when it is closest to Earth (perigee). Perigee must be above Earth's atmosphere, or friction will rob the satellite of energy and it will eventually fall back to Earth.

cific calculations of predictions that could be tested by observation. For example, Newton's laws of motion can be used to derive Kepler's third law from the law of gravity. Newton's discoveries remade astronomy into an analytical science in which astronomers could measure the positions and motions of celestial bodies, calculate the gravitational forces acting on them, and predict their future motion (■ Window on Science 3-4).

One of the most often used and least often stated principles of science is *cause and effect*. You could argue that Newton's second law of motion was the first clear statement of that principle.

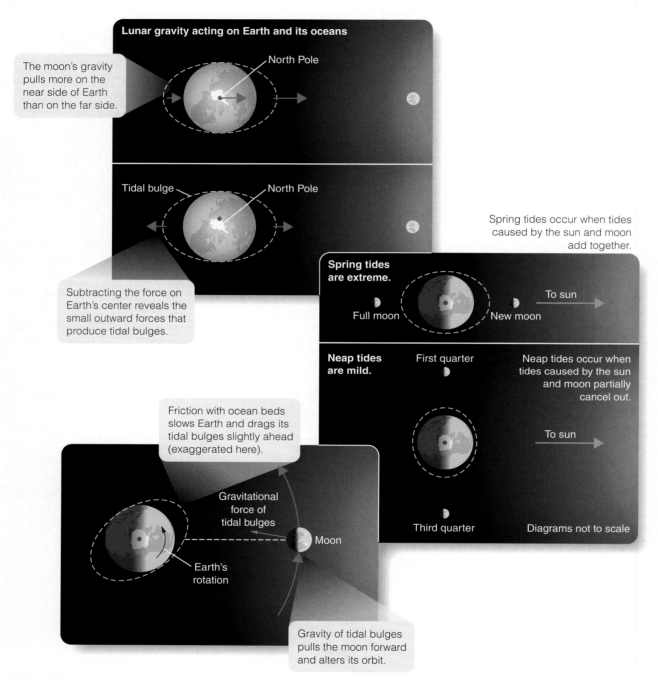

Lunar gravity acting on Earth and its oceans

The moon's gravity pulls more on the near side of Earth than on the far side.

North Pole

Tidal bulge · North Pole

Subtracting the force on Earth's center reveals the small outward forces that produce tidal bulges.

Spring tides occur when tides caused by the sun and moon add together.

Spring tides are extreme.

Full moon · New moon · To sun

Neap tides are mild. · First quarter · Neap tides occur when tides caused by the sun and moon partially cancel out.

To sun

Third quarter · Diagrams not to scale

Friction with ocean beds slows Earth and drags its tidal bulges slightly ahead (exaggerated here).

Gravitational force of tidal bulges · Moon

Earth's rotation

Gravity of tidal bulges pulls the moon forward and alters its orbit.

■ Figure 3-14

Tides are produced by small differences in the gravitational force exerted on different parts of an object. The side of Earth nearest the moon feels a larger force than the side farthest away. Relative to Earth's center, small forces are left over, and they cause the tides. Both the moon and the sun produce tides on Earth. Tides can alter both an object's rotation and orbital motion.

Testing a Theory by Prediction

When you read about any science, you should notice that scientific theories face in two directions. They look back into the past and explain previously observed phenomena. For example, Newton's laws of motion and gravity explained how the planets moved. But theories also face forward in that they enable you to make predictions about what you should find as you explore further. Thus, Newton's laws allowed astronomers to calculate the orbits of comets, predict their return, and eventually understand their origin.

Scientific predictions are important in two ways. First, if a theory leads to a prediction and scientists later discover the prediction was true, the theory is confirmed, and scientists gain confidence that it is a true description of nature. But predictions are important in science in a second way. Using an existing theory to make a prediction may lead you into an unexplored avenue of knowledge. Thus, the first theories of genetics made predictions that confirmed the genetic theory of inheritance, but those predictions also created a new understanding of how living creatures evolve.

As you read about any scientific theory, think about both what it can explain and what it can predict.

The motion of comets can be predicted using Newton's laws. (Nigel Sharp/NOAO/AURA/NSF)

Ancient philosophers such as Aristotle believed that objects moved because of innate tendencies; for example, objects made of earth or water had a natural tendency to move toward Earth at the center of the universe. In contrast, Newton's second law says if an object changes its motion by an acceleration, then it must have been acted on by a force. The principle of cause and effect gives scientists confidence that every effect has a cause. Hearing loss in certain laboratory rats or explosions on certain stars are both effects that scientists believe must have causes. If the universe were not rational, then you could never expect to discover causes. Newton's second law of motion was arguably the first explicit statement that the behavior of the universe is rational and depends on causes.

Building Scientific Arguments

What do the words *universal* and *mutual* mean in the term *universal mutual gravitation*?
Newton realized that the force that makes an apple accelerate downward is the same as the force that accelerates the moon and holds it in its orbit. You can learn more by thinking about Newton's third law of motion, which says that forces always occur in pairs. If Earth attracts the moon, then the moon must attract Earth. That is, gravitation is *mutual* between any two objects.

Furthermore, if Earth's gravity attracts the apple and the moon, then it must attract the sun, and the third law tells you that the sun must attract Earth. But if the sun attracts Earth, then it must also attract the other planets and even distant stars, which, in turn, must attract the sun and each other. Step by step, Newton's third law of motion leads you to conclude that gravitation must apply to all masses in the universe. That is, gravitation must be *universal*.

Newton's explanation of orbital motion was powerful because he explained it in terms of common processes known on Earth's surface. **For example, why is it correct to say that the moon is falling?**

■　　■　　■

Summary

3-1 | Astronomy before Copernicus

How did people in ancient civilizations describe Earth's place in the universe?

■ Ancient philosophers accepted as a first principle that Earth was the unmoving center of the universe. Another first principle was that the heavens were perfect; and, because the circle was the only perfect geometrical form, objects in the heavens must move in uniform (constant speed) circular motion.

■ The geocentric (Earth-centered) universe became part of the teachings of the great philosopher Aristotle, who argued that the sun, moon, and stars were carried around Earth on rotating crystalline spheres.

■ About AD 140, Ptolemy gave mathematical form to Aristotle's model. Ptolemy kept the geocentric (Earth-centered) and uniform circular motion principles, but he added off-center circles and variable speeds to better predict the motions of the planets.

3-2 | Nicolaus Copernicus

How did Copernicus revise that ancient model?

■ Copernicus devised a heliocentric (sun-centered) model. He preserved the principle of uniform circular motion, but he argued that Earth rotates on its axis and circles the sun once a year. His theory was controversial because it contradicted Church teaching.

■ Copernicus published his theory in his book *De Revolutionibus* in 1543, the year he died.

Why was the Copernican model gradually accepted?

■ Because Copernicus kept uniform circular motion, his model did not predict the motions of the planets well, but it did offer a simple explanation of retrograde motion of planets without using large epicycles (circles on circles). He did have to include small epicycles to account for some observed planetary motions.

■ The Copernican model was also more simple and straightforward. Venus and Mercury were treated the same as all the other planets, and the velocity of each planet was related to its distance from the sun.

3-3 | Tycho Brahe, Johannes Kepler, and Planetary Motion

How did Tycho Brahe and Johannes Kepler contribute to the Copernican Revolution?

■ Tycho's great contribution was to compile the most precise and detailed naked-eye observations of the planets and stars ever made, observations that were later analyzed by Kepler.

■ Kepler inherited Tycho's books of observations in 1601 and used them to uncover three laws of planetary motion. He found that the planets follow ellipses with the sun at one focus, that they move faster when near the sun, and that a planet's orbital period squared is proportional to its orbital radius cubed.

3-4 | Galileo Galilei

Why was Galileo condemned by the Inquisition?

■ Galileo used the newly invented telescope to observe the heavens, and he recognized the significance of what he saw there. His discoveries of the phases of Venus, the satellites of Jupiter, the mountains of the moon, and other phenomena helped undermine the Ptolemaic model.

■ Galileo based his analysis on observational evidence. In 1633, he was condemned before the Inquisition for refusing to obey an order to halt his defense of Copernicus's model.

3-5 | Isaac Newton, Gravity, and Orbital Motion

How did Isaac Newton change humanity's view of nature?

■ Newton used the work of Kepler and Galileo to discover three laws of motion and the law of gravity. These laws made it possible to understand such phenomena as orbital motion and the tides.

■ Newton's laws gave scientists a unified way to think about nature. Every effect has a cause, and science is the search for those causes.

■ The 144 years from Copernicus's book *De Revolutionibus* to Newton's book *Principia* marked the beginning of modern science. From that time on, science depended on evidence to test theories and relied on the methods of mathematical analysis demonstrated by Kepler and Newton.

New Terms

first principle (p. 34)	eccentricity, *e* (p. 42)
geocentric (p. 34)	empirical (p. 43)
heliocentric (p. 35)	mass (p. 47)
uniform circular motion (p. 36)	weight (p. 47)
parallax (p. 36)	inverse square relation (p. 48)
retrograde motion (p. 36)	spring tide (p. 49)
epicycle (p. 37)	neap tide (p. 49)
paradigm (p. 39)	circular velocity (p. 50)
ellipse (p. 41)	geosynchronous satellite (p. 50)
hypothesis (p. 42)	center of mass (p. 51)
theory (p. 42)	closed orbit (p. 51)
natural law (p. 42)	escape velocity (p. 51)
semi-major axis, *a* (p. 42)	open orbit (p. 51)

Review Questions

ThomsonNOW™ Assess your understanding of this chapter's topics with additional quizzing and animations at **www.thomsonedu.com**

1. Why did classical Greek astronomers believe that the heavens were made up of perfect crystalline spheres moving at constant speeds?

2. How did the Ptolemaic model explain retrograde motion of the planets?

3. In what ways were the models of Ptolemy and Copernicus similar?

4. Why did the Copernican hypothesis gradually win acceptance?

5. Why is it difficult for scientists to replace an old paradigm with a new paradigm?

6. Why did Tycho Brahe expect the new star of 1572 to show parallax? Why was the lack of parallax evidence against the Ptolemaic model?

7. Explain how Kepler's laws contradict uniform circular motion.

8. What are the differences between a hypothesis, a theory, and a law?

9. Review Galileo's telescope discoveries and explain why they supported the Copernican model and contradicted the Ptolemaic model.

10. If you lived on Mars, which planets would describe retrograde loops? Which would be always seen near the sun? Which would never be visible as crescent phases?

11. Galileo was condemned by the Inquisition, but Kepler, also a supporter of Copernicus, was not. Why not?

12. Why did Newton conclude that gravitation had to be universal?

13. Explain why you might describe the orbital motion of the moon with the statement, "The moon is falling."

Discussion Questions

1. Although ancient astronomers didn't use evidence as modern scientists do, they did observe the world around them. What observations led them to conclude that Earth didn't move?

2. Historian of science Thomas Kuhn has said that *De Revolutionibus* was a revolution-making book but not a revolutionary book. In what ways did it remain within the bounds of ancient astronomical theory?

3. How does the modern controversy over creationism and evolution reflect two ways of knowing about the physical world?

4 | Astronomical Telescopes and Instruments: Extending Humanity's Vision

LIGHT IS A TREASURE that links us to the sky. An astronomer's quest is to gather as much light as possible from the moon, sun, planets, stars, and galaxies in order to extract information about their natures. Telescopes, which gather and focus light for analysis, can help us do that. Nearly all the interesting objects in the sky are very faint sources of light, so large telescopes are built to collect the greatest amount of light possible (■ Figure 4-1). This chapter's discussion of astronomical research concentrates on large telescopes and the special instruments and techniques used to analyze light. | To gather light that is visible to your unaided eye, a normal telescope would work fine. However, visible light is only one type of radiation arriving here from distant objects. Astronomers can extract information from other forms of radiation by using other

| Continued on page 58 |

Guidepost

Looking Back

In the early chapters of this book, you looked at the sky the way ancient astronomers did, with the unaided eye. In the last chapter, you had a glimpse through Galileo's telescope that revealed wonderful things about the moon, Jupiter, and Venus. Now you can study the telescopes, instruments, and techniques of the modern astronomer.

This Chapter

Begin by understanding that these telescopes and instruments gather and focus light and other related forms of radiation. For that reason, the first of the central questions addressed in this chapter is about light:

What is light?

How do telescopes work? What are their capabilities and limitations?

How are observatories built, and how are good locations chosen for them?

What kinds of instruments and techniques do astronomers use to record and analyze light?

Why do astronomers sometimes use X-ray, ultraviolet, and infrared telescopes, and why must these types of telescopes operate in the upper atmosphere or in orbit?

As you answer these questions, you will discover how light in all its forms brings a wealth of information about the rest of the universe.

Looking Ahead

Astronomy is almost entirely an observational science. Astronomers cannot visit distant galaxies and far-off worlds, so they must observe celestial objects using astronomical telescopes. Eleven chapters remain in this book, and every one will discuss information gathered by telescopes. The chapter immediately after this one will show how observations of the sun and other stars can be understood with the help of experiments in laboratories on Earth that study the interactions of light and matter.

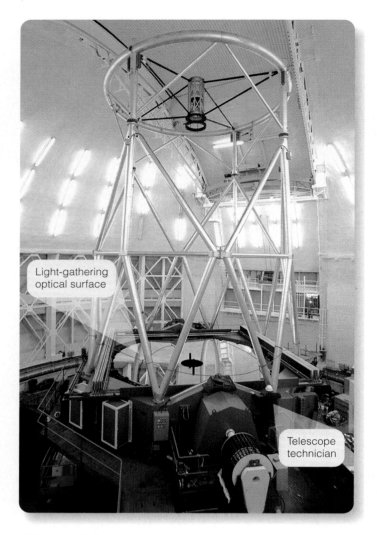

Light-gathering optical surface

Telescope technician

■ Figure 4-1

Astronomical telescopes are often very large to gather large amounts of starlight. The northern Gemini telescope stands over 19 m (60 ft) high when pointed straight up. Its optical mirror is 8.1 m (26.5 ft) in diameter—larger than some classrooms. The dome of this telescope is shown at the left of the photo on the opening page of this chapter. Note the telescope technician in the white hard hat for scale. (NOAO/AURA/NSF)

types of telescopes. Radio telescopes, for example, give an entirely different view of the sky. Some of these specialized telescopes can be used from Earth's surface, but others must go high in Earth's atmosphere or even above it.

4-1 Radiation: Information from Space

ASTRONOMERS NO LONGER STUDY THE SKY by mapping constellations or charting the phases of the moon. Modern astronomers analyze light using sophisticated instruments and techniques to investigate the compositions, motions, internal processes, and evolution of celestial objects. To understand this, you must learn about the nature of light.

Light as a Wave and as a Particle

Have you ever noticed the colors in a soap bubble? If so, then you have seen one effect of light behaving as a wave. When that same light enters the light meter on a camera, it behaves as a particle. How light behaves depends on how you treat it—light has both wavelike and particlelike properties. Sound is another type of wave that you have already experienced. Sound waves are an air pressure disturbance that travels through the air from source to ear. Sound requires a solid, liquid, or gas medium to carry it; so, for example, in space outside a spacecraft, there can be no sound. In contrast, light is composed of a combination of electric and magnetic waves that can travel through empty space. Unlike sound, light waves do not require a medium and thus can travel through a vacuum.

Because light is made up of both electric and magnetic fields, it is referred to as **electromagnetic radiation.** Visible light is only one form of electromagnetic radiation. Electromagnetic radiation is a wave phenomenon—that is, it is associated with a periodically repeating disturbance (a wave) that carries energy. Imagine waves in water: If you disturb a pool of water, waves spread across the surface. Now imagine placing a ruler parallel to the travel direction of the wave. The distance between peaks is the **wavelength.** The changing electric and magnetic fields of electromagnetic waves travel through space at about 300,000 kilometers per second (186,000 miles per second). That is commonly referred to as the speed of light, but it is the speed of all electromagnetic radiation.

It may seem odd to use the word *radiation* when talking about light, but radiation really refers to anything that spreads outward from a source. Light radiates from a source, so you can correctly refer to light as a form of radiation. Confusingly, radiation also refers to high-energy particles emitted from radioactive atoms.

The Electromagnetic Spectrum

The electromagnetic spectrum is simply the types of electromagnetic radiation arranged in order of increasing wavelength. Rainbows are spectra (plural) of visible light. The colors of visible light have different wavelengths: Red has the longest wavelength and violet the shortest, as shown in the visible spectrum at the top of ■ Figure 4-2.

The average wavelength of visible light is about 0.0005 mm. Fifty light waves would fit end-to-end across the thickness of a sheet of paper. It is too awkward to measure such short distances in millimeters, so physicists and astronomers describe the wavelength of light using either the unit of **nanometer (nm),** one-billionth of a meter (10^{-9} m), or the **Ångstrom (Å),** named after the Swedish astronomer Anders Ångström, equal to 10^{-10} m or 0.1 nm. The wavelength of visible light ranges between about

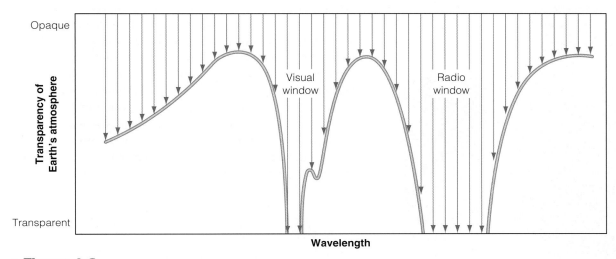

Figure 4-2

The spectrum of visible light, extending from red to violet, is only part of the electromagnetic spectrum. Most radiation is absorbed in Earth's atmosphere, and only radiation in the visual window and the radio window can reach Earth's surface.

400 nm and 700 nm, or, equivalently, 4000 Å and 7000 Å. Infrared astronomers often refer to wavelengths using units of microns (10^{-6} m), while radio astronomers use millimeters, centimeters, or meters. Figure 4-2 shows how the visible spectrum makes up only a small part of the electromagnetic spectrum.

Beyond the red end of the visible range lies **infrared (IR)** radiation, with wavelengths ranging from 700 nm to about 1 mm. Your eyes are not sensitive to this radiation, but your skin senses it as heat. A heat lamp is nothing more than a bulb that gives off large amounts of infrared radiation. ■ Figure 4-3 is an artist's conception of the English astronomer William Herschel measuring infrared radiation—and thus, discovering that there is such a thing as invisible light.

Radio waves have even longer wavelengths than IR radiation. The radio radiation used for AM radio transmissions has wavelengths of a few hundred meters, while FM, television, and also military, governmental, and amateur radio transmissions have wavelengths from a few tens of centimeters to a few tens of meters. Microwave transmissions, used for radar and long-distance

telephone communications, have wavelengths from about 1 millimeter to a few centimeters.

Look back at the electromagnetic spectrum in Figure 4-2 and notice that electromagnetic waves with wavelengths shorter than violet light are called **ultraviolet (UV)**. Shorter-wavelength electromagnetic waves than UV are called **X rays**, and the shortest are **gamma rays**.

The distinction between these wavelength ranges is mostly arbitrary—they are simply convenient human-invented labels. For example, the longest-wavelength infrared radiation and the shortest-wavelength microwaves are the same. Similarly, very short-wavelength ultraviolet light can be considered to be X rays. Nonetheless, it is all electromagnetic radiation, and you could say we are "making light" of it all, because all these types of radiation are the same phenomenon as light: Some types your eyes can see, some types your eyes can't see.

Although light behaves as a wave, under certain conditions it also behaves as a particle. A particle of light is called a **photon**, and you can think of a photon as a minimum-sized bundle of electromagnetic waves. The amount of energy a photon carries depends on its wavelength. Shorter-wavelength photons carry more energy, and longer-wavelength photons carry less energy. A photon of visible light carries a small amount of energy, but an X-ray photon carries much more energy and a radio photon carries much less.

Astronomers are interested in electromagnetic radiation because it carries almost all available clues to the nature of planets, stars, and other celestial objects. Earth's atmosphere is opaque to most electromagnetic radiation, as shown by the graph at the bottom of Figure 4-2. Gamma rays, X rays, and some radio waves are absorbed high in Earth's atmosphere, and a layer of ozone (O_3) at an altitude of about 30 km absorbs almost all UV radiation. Water vapor in the lower atmosphere absorbs long-wavelength IR radiation. Only visible light, some short-wavelength infrared radiation, and some radio waves reach Earth's surface through

what are called **atmospheric windows.** To study the sky from Earth's surface, you must look out through one of these "windows" in the electromagnetic spectrum.

ThomsonNOW™ **Log into ThomsonNOW** and select this chapter to see Astronomy Exercise "The Electromagnetic Spectrum."

Building Scientific Arguments

What could you see if your eyes were sensitive only to X rays?
Sometimes the critical analysis of an idea is easier if you try to imagine a totally new situation. In this case, you might at first expect to be able to see through walls, but remember that your eyes detect only light that already exists. There are almost no X rays bouncing around at Earth's surface, so if you had X ray eyes, you would be in the dark and would be unable to see anything. Even when you looked up at the sky, you would see nothing, because Earth's atmosphere is not transparent to X rays. X-ray vision would not help Superman see through a wall unless there was some source of X rays on the other side.

But suppose your eyes were sensitive only to radio waves. Would you be in the dark?

■ ■ ■

(4-2) Telescopes

ASTRONOMERS BUILD OPTICAL TELESCOPES to gather light and focus it into sharp images. This requires careful optical and mechanical designs, and it leads astronomers to build very large telescopes. To understand that, you need to learn the terminology of telescopes, starting with the different types of telescopes and why some are better than others.

Two Kinds of Telescopes

Astronomical telescopes focus light into an image in one of two ways. Either: (1) a lens bends (refracts) the light as it passes through the glass and brings it to a focus to form an image, or (2) a mirror—a curved piece of glass with a reflective surface—forms an image by bouncing light. Because there are two ways to focus light, there are two types of astronomical telescopes. **Refracting telescopes** use a lens to gather and focus the light, whereas **reflecting telescopes** use a mirror (■ Figure 4-4).

The main lens in a refracting telescope is called the **primary lens,** and the main mirror in a reflecting telescope is called the **primary mirror.** Both kinds of telescopes form a small, inverted image that is difficult to observe directly, so a lens called the **eyepiece** is used to magnify the image and make it convenient to view. The **focal length** is the distance from a lens or mirror to

■ **Figure 4-3**

Depiction of William Herschel in 1800 discovering that sunlight contains radiation detectable by thermometers but not by human eyes, which he named infrared ("beyond red") light. Note the third thermometer outside the sunlight beam serving as an experimental control so that changes in the lab air temperature would not be confused with measurements of changing radiation intensity. (© CCI Archives/SPL/Photo Researchers, Inc.)

■ Active Figure 4-4

(a) A refracting telescope uses a primary lens to focus starlight into an image that is magnified by a lens called an eyepiece. The primary lens has a long focal length, and the eyepiece has a short focal length. (b) A reflecting telescope uses a primary mirror to focus the light by reflection. A small secondary mirror reflects the starlight back down through a hole in the middle of the primary mirror to the eyepiece. Note that light is reflected from the mirror surfaces and does not enter the glass.

ThomsonNOW Log into ThomsonNOW and select this chapter to see the Active Figure called "Refractors and Reflectors."

the image it forms of a distant light source such as a star. Creating the proper optical shape to produce a good focus is an expensive process. The surfaces of lenses and mirrors must be shaped and polished to have no irregularities larger than the wavelength of light. Creating the optics for a large telescope can take months or years; involve huge, precision machinery; and employ several expert optical engineers and scientists.

Refracting telescopes have serious disadvantages. Most importantly, they suffer from an optical distortion that limits their usefulness. When light is refracted through glass, shorter wavelengths bend more than longer wavelengths. As a result, you see a color blur around every image. This color separation is called **chromatic aberration** and it can be only partially corrected. Another disadvantage is that the glass in primary lenses must be pure and flawless because the light passes all the way through it. For that same reason the weight of the

lens can be supported only around its outer edge. In contrast, light reflects from the front surface of a reflecting telescope's primary mirror but does not pass through it, so reflecting telescopes have no chromatic aberration. Also, mirrors are less expensive to make than similarly sized lenses and the weight of telescope mirrors can be supported easily. For these reasons, every large astronomical telescope built since 1900 has been a reflecting telescope.

ThomsonNOW Log into ThomsonNOW and select this chapter to see Astronomy Exercises "Lenses—Focal Length" and "Telescopes: Primary Lens and Eyepiece."

Optical telescopes gather visible light, but astronomers also build **radio telescopes** to gather radio radiation. Radio waves from celestial objects, like visible light waves, penetrate Earth's atmosphere and reach the ground. You can see in ■ Figure 4-5 how the dish reflector of a typical radio telescope focuses the radio waves so their intensity can be measured. Because radio wavelengths are so long, the disk reflector does not have to be as perfectly smooth as the mirror of a reflecting optical telescope.

The Powers of a Telescope

Astronomers struggle to build large telescopes because a telescope can help human eyes in three important ways—these are called the three powers of a telescope. The two most important of these three powers depend on the diameter of the telescope.

Most celestial objects of interest to astronomers are faint, so you need a telescope that can gather large amounts of light to pro-

■ Figure 4-5

In most radio telescopes, a dish reflector concentrates the radio signal on the antenna. The signal is then amplified and recorded. For all but the shortest radio waves, wire mesh is an adequate reflector. (Seth Shostak/SETI Institute)

duce a bright image. **Light-gathering power** refers to the ability of a telescope to collect light. Catching light in a telescope is like catching rain in a bucket—the bigger the bucket, the more rain it catches. The light-gathering power is proportional to the *area* of the primary mirror, that is, proportional to the square of the primary's diameter. A telescope with a diameter of 2 meters has four times (4×) the light-gathering power of a 1-meter telescope. That is why astronomers use large telescopes and why telescopes are ranked by their diameters. One reason radio astronomers build big radio dishes is to collect enough radio photons, which have low energies, and concentrate them for measurement.

The second power of telescopes, **resolving power,** refers to the ability of the telescope to reveal fine detail. One consequence of the wavelike nature of light is that there is an inevitable small blurring called a **diffraction fringe** around every point of light in the image, and you cannot see any detail smaller than the fringe (■ Figure 4-6). Astronomers can't eliminate diffraction fringes, but the fringes are smaller in larger telescopes, and that means they have better resolving power and can reveal finer detail. For example, a 2-meter telescope has diffraction fringes ½ as large, and thus 2× better resolving power, than a 1-meter telescope. The size of the diffraction fringes also depends on wavelength, and at the long wavelengths of radio waves, the fringes are large and the resolving power is poor. That's another reason radio telescopes need to be larger than optical telescopes.

One way to improve resolving power is to connect two or more telescopes in an **interferometer,** which has a resolving power equal to that of a telescope as large as the maximum separation between the individual telescopes (■ Figure 4-7). The first interferometers were built by radio astronomers connecting radio dishes kilometers apart. Modern technology has allowed astronomers to connect optical telescopes to form interferometers with very high resolution.

Aside from diffraction fringes, two other factors—optical quality and atmospheric conditions—limit resolving power. A

Beams combined to produce final image   Precision optical paths in tunnels

■ **Figure 4-7**

In an astronomical interferometer, smaller telescopes can combine their light through specially designed optical tunnels to simulate a larger telescope with a resolution set by the separation of the smaller telescopes.

telescope must contain high-quality optics to achieve its full potential resolving power. Even a large telescope shows little detail if its optical surfaces have imperfections. In addition, when you look through a telescope, you look through miles of turbulence in Earth's atmosphere, which makes images dance and blur, a condition that astronomers call **seeing** (■ Figure 4-8). A related phenomenon is the twinkling of a star. The twinkles are caused by turbulence in Earth's atmosphere, and a star near the horizon, where you look through more air, will twinkle more than a star overhead. On a night when the atmosphere is unsteady, the stars twinkle, the images are blurred, and the seeing is bad. Even with good seeing, the detail visible through a large telescope is limited, not just by its diffraction fringes but by the steadiness of the air through which the observer must look. A telescope per-

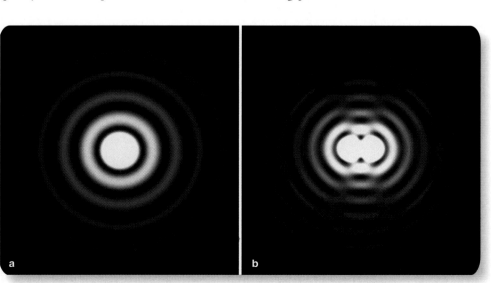

■ **Figure 4-6**

(a) Stars are so far away that their images are points, but the wave nature of light surrounds each star image with diffraction fringes, much magnified in this computer model. (b) Two stars close to each other have overlapping diffraction fringes and become impossible to detect separately. (Computer model by M. A. Seeds)

Resolving Power and the Accuracy of a Measurement

Have you ever seen a movie in which the hero magnifies a newspaper photo and reads some tiny detail? It isn't really possible, because newspaper photos are made up of tiny dots of ink, and no detail smaller than a single dot will be visible no matter how much you magnify the photo. In fact, all images are made up of elements of some sort. On a computer screen, the picture elements are called pixels. That means there is a limit to the amount of detail you can see in the image. In an image formed by a telescope on Earth's surface, the size of the picture element is set by the blurring called seeing. It would be foolish to try to resolve (detect) any detail smaller than this limit.

This limitation is true of all measurements in science. A zoologist might be trying to measure the length of a live snake, or a soci-ologist might be trying to measure the attitudes of people toward drunk driving, but both face limits to the resolution of their measurements. The zoologist might specify that the snake was 43.28932 cm long, and the sociologist might say that 98.2491 percent of people oppose drunk driving, but a critic might point out that it isn't possible to make these measurements that accurately. The resolution of the techniques does not justify the accuracy implied.

Science is based on measurement, and whenever you make a measurement you should ask yourself how accurate that measurement can be. The accuracy of the measurement is limited by the resolution of the measurement technique, just as the amount of detail in a photograph is limited by the resolution of the photo.

A high-resolution image of Mars reveals details such as mountains, craters, and the southern polar cap. (NASA/STScI)

Visual-wavelength image

■ Active Figure 4-8

The left half of this photograph of a galaxy is from an image recorded on a night of poor seeing. Small details are blurred. The right half of the figure is from an image recorded on a night when Earth's atmosphere above the telescope was steady and the seeing was better. Much more detail is visible under good seeing conditions. (William Keel)

ThomsonNOW™ Log into ThomsonNOW and select this chapter to see the Active Figure called "Resolution and Telescopes."

forms best on a high mountaintop where the air is thin and steady, but even at good sites atmospheric turbulence spreads star images into blobs 0.5 to 1 arc seconds in diameter. That situation can be improved by a difficult and expensive technique called **adaptive optics,** in which rapid computer calculations adjust the telescope optics and partly compensate for seeing distortions.

This limitation on the amount of information in an image is related to the limitation on the accuracy of a measurement. All measurements have some built-in uncertainty (■ **Window on Science 4-1**), and scientists must learn to work within those limitations.

The third, and, you may be surprised to learn, the least important power of a telescope is **magnifying power,** the ability to make an image large. The magnifying power of a telescope equals the focal length of the primary mirror or lens divided by the focal length of the eyepiece. For example, a telescope with a primary mirror that has a focal length of 700 mm and an eyepiece with a focal length of 14 mm has a magnifying power of 50×. Higher magnifying power does not necessarily show you more detail because the amount of detail you can see in practice is limited by a combination of the seeing conditions plus the telescope's resolving power and optical quality.

A telescope's primary function is to gather light and thus make faint things appear brighter, so the light-gathering power is the most important power and the diameter of the telescope is its most important characteristic. Light-gathering power and resolving power are fundamental properties of a telescope that cannot be altered, whereas magnifying power can be changed simply by changing the eyepiece.

(4-3) Observatories on Earth—Optical and Radio

MOST MAJOR OBSERVATORIES ARE LOCATED FAR FROM BIG CITIES and usually on high mountains. Optical astronomers avoid cities because **light pollution,** the brightening of the night sky by light scattered from artificial outdoor lighting, can make it impossible to see faint objects. In fact, many residents of cities are unfamiliar with the beauty of the night sky because they can see only the brightest stars. Radio astronomers face a problem of radio interference analogous to light pollution. Weak radio signals from the cosmos are easily drowned out by human radio interference—everything from automobiles with faulty ignition systems to poorly designed transmitters in communication. To avoid that, radio astronomers locate their telescopes as far from civilization as possible. Hidden deep in mountain valleys, they are able to listen to the sky protected from human-made radio noise.

As you learned previously, astronomers prefer to place optical telescopes on mountains because the air there is thin and more transparent, but, most important, they carefully select mountains where the airflow is usually not turbulent so the seeing is good. Building an observatory on top of a high mountain far from civilization is difficult and expensive, but the dark sky and good seeing make it worth the effort (■ Figure 4-9).

Study ■ **Modern Astronomical Telescopes** on page 65, and you will notice two important points:

1 Research telescopes must focus their light to positions at which cameras and other instruments can be placed.

2 Small telescopes can use other focal arrangements that would be inconvenient in larger telescopes.

Telescopes located on the surface of Earth, whether optical or radio, must move continuously to stay pointed at a celestial object as Earth turns on its axis. This is called **sidereal tracking** ("sidereal" refers to the stars).

The days when astronomers worked beside their telescopes through long, dark, cold nights are nearly gone. The complexity and sophistication of telescopes require a battery of computers, and almost all research telescopes are run from warm control rooms (■ Figure 4-10).

An international collaboration of astronomers built the Gemini telescopes with 8.1-meter thin mirrors (■ Figure 4-11a). One is located in the northern hemisphere and one in the southern hemisphere to allow observations of the entire sky. The European Southern Observatory built the Very Large Telescope (VLT) high in the remote Andes of northern Chile (Figure 4-11b). The VLT consists of four telescopes with computer-controlled mirrors 8.2 m in diameter and only 17.5 cm (6.9 in.) thick. The four telescopes can work singly or can combine their light to work as one large telescope. Italian, American, and German astronomers are building the Large Binocular Telescope, which carries a pair of 8.4-m mirrors on a single mounting, as shown in Figure 4-11c. Around the world astronomers are drawing plans for large telescopes, including truly gigantic instruments 30 m and even 100 m in diameter.

The largest radio dish in the world is 300 m (1000 ft) in diameter, built into a mountain valley in Arecibo, Puerto Rico (■ Figure 4-12a). Although this telescope can look only nearly overhead, the operators can change its aim slightly by moving the antenna above the dish and waiting for Earth's rotation to point the telescope in the proper direction. This telescope's ability to detect weak radio sources, coupled with its good resolving power, makes it one of the most important radio observatories in the world.

■ **Figure 4-9**

The domes of four giant telescopes are visible at upper left at Paranal Observatory, built by the European Southern Observatory at an altitude of 2635 m (8640 ft) atop the mountain Cerro Paranal in the Atacama desert of northern Chile. Located 120 km (75 mi) from the nearest city, the mountaintop was chosen for its steady air and isolation from lights. It is believed to be the driest area on Earth. (ESO)

Modern Astronomical Telescopes

1 Large astronomical telescopes must gather light and guide it to locations where it can be recorded with cameras or other instruments.

In larger telescopes the light can be focused to a **prime focus** position high in the telescope tube as shown at the right. Although it is a good place to image faint objects, the prime focus is inconvenient for large instruments. A **secondary mirror** can reflect the light through a hole in the primary mirror to a **Cassegrain focus.** This focal arrangement may be the most common form of astronomical telescope.

Secondary mirror

With the secondary mirror removed, the light converges at the prime focus. In large telescopes, astronomers can ride inside the prime-focus cage, although most observations are now made by instruments connected to computers in a separate control room.

The mirrors in astronomical telescopes must be ground and polished to precise shape to form the sharpest possible images. The mirrors must be carefully supported to prevent them from sagging under their own weight as the telescope moves around the sky, and many modern telescopes use computer controlled thrusters to maintain the mirror shape.

The Cassegrain focus is convenient and has room for large instruments.

2 Smaller telescopes can use other focal arrangements. The **Newtonian focus** that Isaac Newton used in his first reflecting telescope is awkward for large telescopes as shown at right, but is common for small telescopes.

Newtonian focus

2a Many small telescopes such as the one on your left use a **Schmidt-Cassegrain focus.** A thin correcting plate improves the image but is too slightly curved to introduce serious chromatic aberration.

Thin correcting lens

Schmidt-Cassegrain telescope

1a Shown below, the 4-meter Mayall Telescope at Kitt Peak National Observatory in Arizona can be used at either the prime focus or the Cassegrain focus. Note the human figure at lower right.

Prime focus cage

Secondary mirror

Primary mirror (inside)

Cassegrain focus

Astronomer

AURA/NOAO/NSF

■ Figure 4-10

In the control room of the 4-meter telescope at Kitt Peak National Observatory, the telescope operator at left manages the operation and safety of the telescope. The astronomer at right operates the instruments, records data, and makes decisions on the observing program. Astronomers work through the night using the computers that control the telescope and its instruments. (NOAO/AURA/NSF)

The Very Large Array (VLA) consists of 27 dish antennas spread in a Y-shape across the New Mexico desert (Figure 4-12b). In combination as an interferometer, they have the resolving power of a radio telescope up to 36 km (22 mi) in diameter. Another large radio interferometer, the Very Long Baseline Array (VLBA), consists of matched radio dishes spread from Hawaii to the Virgin Islands and has an effective diameter almost as large as Earth.

Building Scientific Arguments

Why do astronomers build optical observatories at the tops of mountains?

Astronomers have joked that the hardest part of building a new optical observatory is constructing the road to the top of the mountain. It certainly isn't easy to build a large, delicate telescope at the top of a high mountain, but it is worth the effort. A telescope on top of a high mountain is above the thickest part of Earth's atmosphere. There is less

air to dim the light, and there is less water vapor to absorb infrared radiation. Even more important, the thin air on a mountaintop causes less image disturbance, so the seeing is better. A large telescope with modern, high-quality optics is capable of detecting very small details. It is the seeing that limits the detail that astronomers can detect. It really is worth the trouble to build telescopes atop high mountains where the air is thin, dry, and steady.

Astronomers not only build telescopes on mountaintops; they also build gigantic telescopes many meters in diameter. **What are the advantages of building such giant telescopes?**

ThomsonNOW **Log into ThomsonNOW** and select this chapter to see Astronomy Exercises "Telescopes and Resolution I," "Telescopes and Resolution II," "Particulates," "Heat," and "Light Pollution."

■ ■ ■

■ Figure 4-11

Astronomers have begun building multiple telescopes to solve different problems. (a) Two Gemini telescopes have been built, one in Hawaii and one in Chile, to observe the entire sky. (NOAO/AURA/NSF) (b) The Large Binocular Telescope (LBT) carries two 8.4-m mirrors that combine their light. (Large Binocular Telescope Project and European Industrial Engineering) (c) The four telescopes of the VLT are housed in separate domes, but they can combine to work as a single very large telescope, an interferometer. (ESO)

The LBT has the light-gathering power of a 11.8-m telescope and the resolving power of a 22.8-m telescope.

The mirrors in the VLT telescopes are 8.2 m in diameter.

The two Gemini telescopes have 8.1-m mirrors with active optics.

■ Figure 4-12

(a) The 300-m (1000-ft) radio telescope in Arecibo, Puerto Rico, hangs from cables over a mountain valley. The Arecibo Observatory is part of the National Astronomy and Ionosphere Center, operated by Cornell University and the National Science Foundation. (David Parker/SPL/Photo Researchers, Inc.) (b) The Very Large Array uses 27 radio dishes that can be moved to different positions along a Y-shaped set of tracks across the New Mexico desert. They are shown here in the most compact arrangement. Signals from the dishes are combined as an interferometer to create very-high-resolution radio maps of celestial objects. (NRAO/AUI/NSF)

(4-4) Astronomical Instruments and Techniques

JUST LOOKING THROUGH A TELESCOPE DOESN'T TELL YOU MUCH. To learn about planets, stars, and galaxies, you must be able to analyze the light the telescope gathers. Special instruments attached to the telescope make that possible.

Imaging Systems and Photometers

The **photographic plate** was the first image-recording device used with telescopes. Brightness of objects imaged on a photographic plate can be measured with a lot of hard work, yielding only moderate precision. Astronomers also build **photometers,** sensitive light meters to measure the brightness of individual objects very precisely. Most modern astronomers use **charge-coupled devices (CCDs)** as both image-recording devices and photometers. A CCD is a specialized computer chip containing as many as a million or more microscopic light detectors

arranged in an array about the size of a postage stamp. These **array detectors** can be used like a small photographic plate, but they have dramatic advantages over both photometers and photographic plates. CCDs can detect both bright and faint objects in a single exposure and are much more sensitive than a photographic plate. CCD images are **digitized,** meaning, converted to numerical data, and can be read directly into a computer memory for later analysis. Although CCDs for astronomy are extremely sensitive and therefore expensive, less sophisticated CCDs are now used in commercial video and digital cameras. Infrared astronomers also use array detectors similar in operation to optical CCDs. At some other wavelengths, photometers are still used for measuring brightness of celestial objects.

The digital data representing an image from a CCD or other array detector are easy to manipulate to bring out details that would not otherwise be visible. For example, astronomical images are often reproduced as negatives, with the sky white and the stars dark. This makes the faint parts of the image easier to see (■ Figure 4-13). Astronomers also manipulate images to produce **false-color images** in which the colors represent different levels of intensity and are not related to the true colors of the object. For example, because humans can't see radio waves, astronomers must convert them into something perceptible. One way is to measure the strength of the radio signal at various places in the sky and draw a map in which contours mark areas of uniform radio intensity. Compare such a map to a seating diagram for a baseball stadium in which the contours mark areas in which the seats have the same price (■ Figure 4-14a). Contour maps are very common in radio astronomy and are often reproduced using false colors (Figure 4-14b).

Spectrographs

To analyze light in detail, you need to spread the light out according to wavelength into a spectrum, a task performed by a **spectrograph.** You can understand how this works by reproducing an experiment performed by Isaac Newton in 1666. Boring a hole in his window shutter, Newton admitted a thin beam of sunlight into his darkened bedroom. When he placed a prism in the beam, the sunlight spread into a beautiful spectrum on the far wall. From this, Newton concluded that white light was made of a mixture of all the colors.

Newton didn't think in terms of wavelength, but you can use that modern concept to see that the light passing through the prism is bent at an angle that depends on the wavelength. Violet (short-wavelength) light bends most, and red (long-wavelength) light least. Thus, the white light entering the prism is spread into what is called a **spectrum** (■ Figure 4-15). A typical prism spectrograph contains more than one prism to spread the spectrum wider, plus lenses to guide the light into the prism and to focus the light onto a photographic plate. Most modern spectrographs use a grating in place of a prism. A **grating** is a piece of glass with thousands of microscopic parallel lines scribed onto its surface. Different wavelengths of light reflect from the grating at slightly different angles, so white light is spread into a spectrum and can

be recorded, often by a CCD camera. Recording the spectrum of a faint star or galaxy can require a long time exposure, so astronomers have developed multiobject spectrographs that can record the spectra of as many as 100 objects simultaneously. Multiobject spectrographs automated by computers have made large surveys of many thousands of stars or galaxies possible.

Because astronomers understand how light interacts with matter, a spectrum carries a tremendous amount of information (as you will see in more detail in Chapter 5). That makes a spectrograph the astronomer's most powerful instrument. Astronomers are likely to remark, "We don't know anything about an object until we get a spectrum," and that is only a slight exaggeration.

Building Scientific Arguments

What is the difference between light going through a lens and light passing through a prism?

A refracting telescope producing chromatic aberration and a prism dispersing light into a spectrum are two examples of the same thing, but one is bad and one is good. When light passes through the curved surfaces of a lens, different wavelengths are bent by slightly different amounts, and the different colors of light come to focus at different focal lengths. This produces the color fringes in an image called chromatic aberration, and that's bad. But the surfaces of a prism are made to be precisely flat, so all of the light enters the prism at the same angle, and any given wavelength is bent by the same amount wherever it meets the prism. Thus, white light is dispersed into a spectrum. You could call the dispersion of light by a prism "controlled chromatic aberration," and that's good.

CCDs have been very good for astronomy, and they are now widely used. **Explain why they are more useful than photographic plates.**

Galaxy NGC 891 in true color. It is edge-on and contains thick dust clouds.

Visual-wavelength image

In these negative images of NGC 891, the sky is white and the stars are black.

■ Figure 4-13

Astronomical images can be manipulated in many ways to bring out details. The photo of the disk galaxy at upper left is dark, and the details of the dust clouds in the disk of the galaxy do not show well. The two negative images of the galaxy have been processed to show the dust clouds more clearly. (C. Howk, B. Savage, N. A. Sharp NOAO/WIYN/NSF)

Visual-wavelength negative images

Seat prices in a
baseball stadium
Red most expensive
Violet least expensive

Radio intensity map
Red strongest
Violet weakest

a

b

■ **Figure 4-14**

(a) A contour map of a baseball stadium shows regions of similar admission prices. The most expensive seats are those behind home plate. (b) A false color radio map of Tycho's supernova remnant, the expanding shell of gas produced by the explosion of a star in 1572. The radio contour map has been color-coded to show intensity. (NRAO)

4-5 Airborne and Space Observatories

YOU HAVE LEARNED ABOUT THE OBSERVATIONS that ground-based telescopes can make through the two atmospheric windows in the visible and radio parts of the electromagnetic spectrum. Most of the rest of the electromagnetic spectrum—infrared, ultraviolet, X-ray, and gamma-ray radiation—never reaches Earth's surface. To observe at these wavelengths, telescopes must fly above the atmosphere in high-flying aircraft, rockets, balloons, and satellites. The only exceptions are observations that can be made in what are called the near-infrared and the near-ultraviolet, almost the same wavelengths as visible light.

The Ends of the Visual Spectrum

Astronomers can observe from the ground in the near-infrared just beyond the red end of the visible spectrum because some of this radiation leaks through the atmosphere in narrow, partially open atmospheric windows ranging in wavelength from 1200 nm to about 30,000 nm. Infrared astronomers usually describe wave-

lengths in micrometers or microns (10^{-6} m), so they refer to this wavelength range as 1.2 to 30 microns. In this range, most of the radiation is absorbed by water vapor, carbon dioxide, or ozone molecules in Earth's atmosphere, so it is an advantage to place telescopes on the highest mountains where the air is especially thin and dry. For example a number of important infrared telescopes are located on the summit of Mauna Kea in Hawaii at an altitude of 4205 m (13,800 ft) (■ Figure 4-16).

The far-infrared range, which includes wavelengths longer than 30 micrometers, can tell us about planets, comets, forming stars, and other cool objects, but these wavelengths are absorbed high in the atmosphere. To observe in the far-infrared, telescopes must venture to high altitudes. Remotely operated infrared telescopes suspended under balloons have reached altitudes as high as 41 km (25 mi). For many years, the NASA Kuiper Airborne Observatory (KAO) carried a 91-cm infrared telescope and crews of astronomers to altitudes of over 12 km (40,000 ft) to get above 99 percent or more of the water vapor in Earth's atmosphere. Now retired from service, the KAO will soon be replaced by the Stratospheric Observatory for Infrared Astronomy (SOFIA), a Boeing 747-SP aircraft that will carry a 2.5-m (100-in.) telescope to the fringes of the atmosphere (■ Figure 4-17).

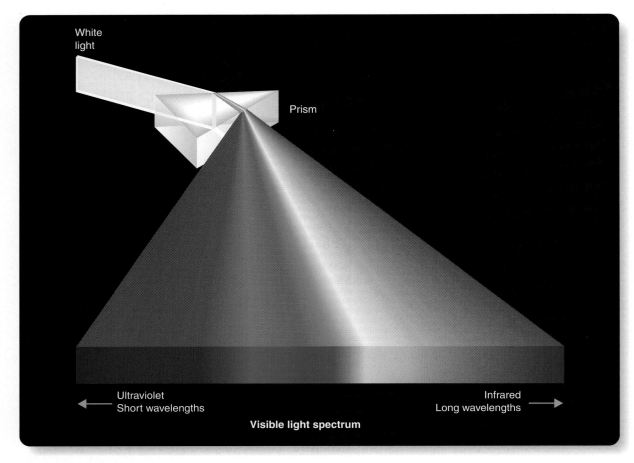

■ Figure 4-15

A prism bends light by an angle that depends on the wavelength of the light. Short wavelengths bend most and long wavelengths least. Thus, white light passing through a prism is spread into a spectrum.

If a telescope observes at far-infrared wavelengths, then it must be cooled. Infrared radiation is emitted by heated objects, and if the telescope is warm it will emit many times more infrared radiation than that coming from a distant object. Imagine trying to look at a dim, moonlit scene through binoculars that are glowing brightly. In a telescope observing near-infrared radiation, only the detector, the element on which the infrared radiation is focused, must be cooled. To observe in the far-infrared, however, the entire telescope must be cooled.

At the short wavelength end of the spectrum, astronomers can observe in the near-ultraviolet. Human eyes do not detect this radiation, but it can be recorded by photographic plates and CCDs. Wavelengths shorter than about 290 nm, the far-ultraviolet, X-ray, and gamma-ray ranges, are completely ab-

■ Figure 4-16

Comet Hale–Bopp hangs in the sky over the 3-m NASA Infrared Telescope atop Mauna Kea. The air at high altitudes is steady and so dry that it is transparent to short-wavelength infrared photons. Infrared astronomers can sometimes observe with the lights on in the telescope dome—their instruments are usually insensitive to visible light. (William Keel)

■ Figure 4-17

SOFIA, the Stratospheric Observatory for Infrared Astronomy, a 2.5-m reflecting telescope installed in a Boeing 747-SP aircraft, is designed to operate for 8 hours at altitudes of 12.5 to 14 km (41,000 to 45,000 ft) above almost all Earth's atmospheric water vapor, able to observe far-infrared radiation from celestial objects that does not reach down to even the highest mountaintops. SOFIA is a joint project between NASA and the German space agency DLR. (SOFIA/USRA/NASA)

sorbed by the ozone layer extending from 20 km to about 40 km above Earth's surface. No mountain is that high, and no balloon or airplane can fly that high, so astronomers cannot observe far-UV, X-ray, and gamma-ray radiation without going into space.

Telescopes in Space

To observe far beyond the ends of the visible spectrum, astronomical telescopes must go above Earth's atmosphere into space. This is very expensive and difficult, but it is the only way to study some important objects. For example, light with visible wavelengths cannot pass through dust clouds in space that are the locations of star and planet formation and that also hide the center of our galaxy. However, radio, infrared, and X-ray telescopes can see into and through those clouds. Black holes are small and difficult to detect, but matter flowing into a black hole emits X rays. Telescopes in space can explore these exciting processes that are invisible from Earth's surface.

One of the most successful space telescopes was the International Ultraviolet Explorer (IUE), launched in 1978. It carried a telescope with a diameter of only 45 cm (18 in.) that was expected to last only a year or two, but it became "the little telescope that could." It made many observations and exciting discoveries until it finally stopped working in 1996. The Infrared Astronomical Satellite (IRAS), a joint project of the United States, the United Kingdom, and the Netherlands, carried a 57-cm telescope cooled almost to absolute zero that made the first map of the entire sky at far-infrared wavelengths during its one year of operation, 1983. Many space telescopes are small satellites like IUE and IRAS that are designed to make specific observations for a short period, but some are large general-purpose telescopes.

Study ■ **The Great Observatories in Space** on pages 72–73 and notice three points:

1 Not only can a telescope in space observe at a wide range of wavelengths, but it is above the atmospheric blurring called seeing. The Hubble Space Telescope observes mostly at visual wavelengths, but it has the advantage of sharp images undistorted by seeing.

2 Telescopes must be specialized for their wavelength range. The Spitzer Space Telescope must have a heat screen to protect it from sunlight, and the Chandra observatory has special optics that focus X rays.

3 Telescopes that collect different types of electromagnetic radiation study different objects and processes in space. For example, warm dust and star formation processes are best studied at infrared wavelengths, whereas X-ray telescopes detect the very-high-temperature gas clouds in some galaxy clusters.

These great observatories in space are controlled from research centers on Earth and are open to proposals from any astronomer with a good idea, but competition is fierce and only the most worthy projects win approval.

Although turbulence in Earth's atmosphere can be partially averaged out in an interferometer, plans are also being made to put interferometers in space. The Space Interferometry Mission (SIM), for example, will work at optical wavelengths and study everything from the cores of erupting galaxies to planets orbiting nearby stars.

Building Scientific Arguments

Why can infrared astronomers observe from high mountaintops, while X-ray astronomers must observe from space?
In this analysis, you find similar consequences springing from different causes. Although both infrared and X-ray telescopes have been put into orbit, they differ dramatically. Infrared radiation is absorbed mainly by water vapor in Earth's atmosphere, which is confined to the lower atmospheric layers. If you built an infrared telescope on top of a high mountain, you would be above most of the water vapor, and you could collect some infrared radiation from celestial objects. The longer-wavelength infrared is absorbed much higher in the atmosphere, so telescopes to observe in the far-infrared must be airborne or go into space. X rays are absorbed in the uppermost layers of the atmosphere, and no mountain is high enough to extend above those layers. To observe the cosmos at X-ray wavelengths, a telescope must go into space.

X-ray and far-infrared telescopes must observe from space, but the Hubble Space Telescope observes in the visual wavelength range. **Then what advantage does Hubble get from being in orbit?**

■   ■   ■

The Great Observatories in Space

1 The Hubble Space Telescope was carried into orbit by the Space Shuttle in 1990. The telescope contains a 2.4-m (96-in.) mirror and can observe from the near-infrared to the near-ultraviolet.

Orbiting above Earth's blurring atmosphere, Hubble is limited only by diffraction in its optics. It can detect details 10 times smaller than Earth-based telescopes.

The telescope, as big as a large bus, has been visited twice by astronauts, who repaired equipment and installed new instruments. Named after Edwin Hubble, the astronomer who discovered the expansion of the universe, the telescope has been tremendously productive observing everything from the weather on Mars to the most distant galaxies visible in the universe.

Hubble image of Mars and its polar cap.

Visual

Hubble image of a nebula around an aging star.

Visual

Hubble image of a dust-filled galaxy.

Visual

2 The Compton Gamma Ray Observatory at left was in orbit from 1991 to 2000. It made observations of very-high-energy photons, helping astronomers understand such violently active objects as neutron stars and black holes.

This far-ultraviolet image of the sun made from space reveals complex structure on the solar surface and clouds of gas being ejected into space. (NASA/SOHO)

Guidepost

Looking Back

Up to this point, you have been thinking about what you can see with your eyes alone or aided by telescopes. In the previous chapter you read what hard work it is to gather light from distant stars using giant telescopes on mountaintops and in space. You also read how several types of instruments are used to analyze the light gathered by telescopes, including spectrographs that spread light into spectra. Now you are ready to see what all the fuss is about.

This Chapter

This chapter explains how light interacts with matter and applies those ideas to let you understand the sun. You will realize how little astronomers would know about the sun were it not for analyses of its spectrum. Just a bit of spectroscopic ingenuity reveals that the brilliance of the sun hides a complex atmosphere of hot gases, churned by powerful storms. Here you will find answers to six essential questions:

How do you know the distance, size, mass, and density of the sun?

How does matter produce light?

What do astronomers see when they observe the sun?

How does matter interact with light to produce spectral lines?

What can you learn from the sun's spectrum?

Why does the sun have a cycle of activity, and how does that affect Earth?

This chapter marks a change in the way you will look at nature. This is the first chapter that applies the methods of science to analyze a celestial body. In this chapter, you become an astrophysicist by using some knowledge of physics to understand the temperature and composition of the sun as well as processes occurring on or near its surface.

Looking Ahead

In the following chapters you will find that the sun is a normal star. You can use the methods of science to understand what is inside the sun and other stars, how they form and evolve, and how they finally end their existences. You can search out secrets of the stars beyond what you can see superficially.

5-1 The Sun—Basic Characteristics

IN ITS GENERAL PROPERTIES, the sun is very simple. It is a great ball of hot gas held together by its own gravity. The tremendously hot gas inside the sun has such a high pressure that it would explode were it not for its own confining gravity. The same gravity would make it collapse into a small, dense body were it not so hot inside. Like a soap bubble, the sun is a simple structure balanced between opposing forces that, if unbalanced, would destroy it. These dramatic statements are also true for other stars, so you can study the sun for insight into the rest of the stars in the universe.

Another reason to study the sun is that life on Earth depends critically on the sun. Very small changes in the sun's luminosity can alter Earth's climate, and a slightly larger change might make Earth uninhabitable. Nearly all of Earth's energy comes from the sun—the energy in oil, gasoline, coal, and even wood is merely stored sunlight. Furthermore, the sun's atmosphere of very thin gas extends past Earth's position, and changes in that atmosphere such as eruptions or magnetic storms can have a direct effect on Earth.

Distance and Size

When you watch the sun set in the west, you see a glowing disk that is 150 million kilometers (93 million miles) from Earth and has a diameter 109 times Earth's (**Celestial Profile 1**). How do you know? Recall from Chapter 3 that, thanks to the work of Johannes Kepler in the 17th century, the relative sizes of the orbits of the planets were already known in Astronomical Units—one AU being equal to the average distance of the sun from Earth. Recall from Chapter 3 (page 78) that the true distance of a nearby object can be calculated from the size of the apparent shift in its position relative to the background as seen from two viewing positions. That shift is called *parallax.*

If you and another astronomer on the opposite side of Earth agree to observe the position of a planet like Venus at the same moment relative to more distant objects, you can measure the planet's parallax shift caused by you and your colleague's different observing locations. The distance of the planet in familiar "earthly" units then can be calculated with some trigonometry. With further analysis, using a map of the solar system and Kepler's third law, you can find the real distance to the sun. From the 17th century onward, astronomers made more and more accurate measurements of this type, especially on rare occasions called **transits of Venus** when that planet can be seen as a tiny dot directly between Earth and the sun (■ Figure 5-1), so the edge of the sun's disk acts as a convenient position marker. The distance from Earth to the sun, combined with the sun's easily measured angular size, gives you the sun's diameter in familiar units.

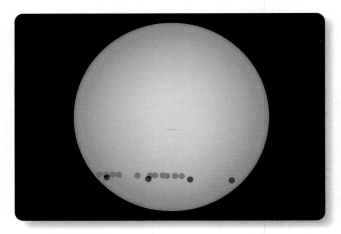

■ Figure 5-1

Comparison of Venus' path across the solar disk as observed from Australia (upper track) and Denmark (lower track) on June 8, 2004. Measurement of the small parallax shift allows the distance to Venus to be calculated based on the known distance between the two observatories. Kepler's third law can then be used to find the distance to the sun. (Astronomy Group—Sønderborg and Learmonth Solar Observatory/ESO/Adapted from the GONG project and Amtsgymnasiet i Sønderborg)

Mass and Density

Once you know the distance to the sun, then Newton's laws (Chapter 3) tell you what the sun's mass needs to be to produce the necessary amount of gravitational force to keep Earth and the other planets in their orbits at their observed speeds. If you know the mass and diameter of the sun, you can make an easy calculation of the sun's **density** (mass per volume). Look at Celestial Profile 1 again and notice the sun's mass, equivalent to 333,000 times the mass of Earth, and its average density—only a little bit denser than water. Although the sun is very large and very massive, the low density and high temperature together tell you that it must be gas from its surface to its center. When you look at the sun you see only the outer layers, the surface and atmosphere, of this vast sphere of gas.

In the following sections of this chapter you will learn how, using the results of laboratory physics experiments on Earth, you can know facts as seemingly unknowable as the temperature and composition of the sun—an object that has never been visited or touched by humans.

5-2 The Origin of Sunlight

THE SUN'S SURFACE GLOWS for the same reason and by the same process that makes the coils in your toaster glow when they are hot. You probably have also noticed that your toaster's coils glow different colors as they heat up. If they are not too hot, the coils are deep red, but as they heat up they grow brighter and yellower. Yellow-hot is hotter than red-hot but not as hot as white-hot. Different parts of the sun's surface also glow slightly different colors depending on their temperatures.

Atoms and Subatomic Particles

Light from hot toaster coils and light from the sun and other stars are all produced by moving electrons. As you know, an **atom** has a massive compact **nucleus** containing positively charged **protons,** usually accompanied by electrically neutral **neutrons.** The nucleus is embedded in a large cloud of relatively low-mass, negatively charged **electrons.** These particles can also exist and move about unattached to an atom.

Charged particles, both protons and electrons, are surrounded by electric fields that they produce. Whenever you change the motion of a charged particle, the change in its electric field spreads outward at the speed of light as electromagnetic radiation. If you run a comb through your hair, you disturb electrons in both hair and comb, producing electric sparks and electromagnetic radiation, which you can sometimes hear as snaps and crackles if you are standing near a radio.

The sun is hot, and there are plenty of electrons zipping around, colliding, changing directions and speeds, and thereby making electromagnetic radiation. Protons can also make electromagnetic radiation, but because electrons are less massive, usually it is electrons that do most of the moving around.

Temperature, Heat, and Blackbody Radiation

The particles inside any object—for example, atoms linked together to form **molecules,** individual atoms, electrons inside atoms or wandering loose—are in constant motion, and in a hot object they are more agitated than in a cool object. You can refer to this agitation as *thermal energy.* When you touch a hot object, you feel **heat** as that thermal energy flows into your fingers. **Temperature** is simply a number related to the average speed of the particles, the intensity of the particle motion.

Astronomers and physicists express temperatures of the sun and other objects on the **Kelvin temperature scale.** Zero degrees Kelvin (written 0 K) is **absolute zero** (−459.7 F), the temperature at which an object contains no thermal energy that can be extracted. Water freezes at 273 K and boils at 373 K. The Kelvin temperature scale is useful in astronomy because it is based on absolute zero and consequently is related directly to the motion of the particles in an object.

Now you can understand why a hot object glows. The hotter an object is, the more motion there is among its particles. The agitated particles, including electrons, collide with each other, and when electrons are accelerated, part of the energy is carried away as electromagnetic radiation. The radiation emitted by an opaque object is called **blackbody radiation,** a name translated from a German term that refers to the way a perfectly opaque emitter and absorber of radiation would behave. At room temperature, such a perfect absorber would look black, but at higher temperatures it would visibly glow. In astronomy you will find the term *blackbody* actually refers to glowing objects.

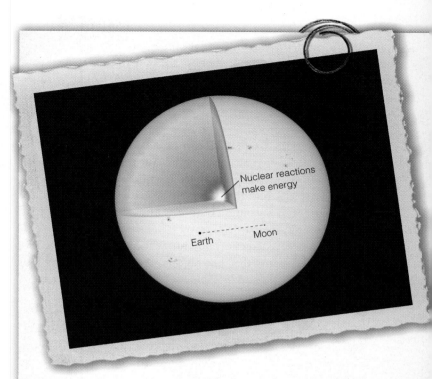

This visible image of the sun shows a few sunspots and is cut away to show the location of energy generation at the sun's center. The Earth–moon system is shown for scale. (Dan Good)

Celestial Profile 1: The Sun
From Earth:

Average distance from Earth	1.00 AU (1.496×10^8 km)
Maximum distance from Earth	1.017 AU (1.521×10^8 km)
Minimum distance from Earth	0.983 AU (1.471×10^8 km)
Average angular diameter	0.53° (32 arc minutes)
Period of rotation	25.38 days at equator
Apparent visual magnitude	−26.74

Characteristics:

Radius	6.96×10^5 km
Mass	1.99×10^{30} kg
Average density	1.41 g/cm^3
Escape velocity at surface	618 km/s
Luminosity	3.83×10^{26} J/s
Surface temperature	5780 K
Central temperature	15×10^6 K
Spectral type	G2 V
Absolute visual magnitude	+4.83

Blackbody radiation is quite common. In fact, it is the type of light emitted by an ordinary incandescent light bulb. Electricity flowing through the filament of the bulb heats it to high temperature, and it glows. You can also recognize the light emitted by a toaster coil as blackbody radiation. Many objects in the sky, including the sun, primarily emit blackbody radiation because they are mostly opaque.

Two simple laws describe how blackbody radiation works. A hot object radiates at all wavelengths, but there is a **wavelength of maximum intensity** at which it radiates the most energy. **Wien's Law** says that the hotter an object is, the shorter is the wavelength of its maximum output. This makes sense because in a hotter object, the particles travel faster, collide more violently, and emit more energetic photons, which have shorter wavelengths. This means that hot objects tend to emit radiation at shorter wavelengths and look bluer than cooler objects. Hot stars look bluer than cool stars.

The **Stefan-Boltzmann Law** says that hotter objects emit more energy than cooler objects of the same size. That makes sense, too, because the hotter an object is, the more rapidly its particle move and the more violent and more frequent are the collisions that produce photons. In later chapters, you will use these two laws to understand stars and other objects.

■ Figure 5-2 shows plots of the intensity of radiation versus wavelength for three objects with different temperatures, illustrating both Wien's law and the Stefan–Boltzmann law. You can see how temperature determines the color of a glowing blackbody. The hotter object has its strongest intensity at shorter wavelengths, so it emits more blue light than red and looks blue. The cooler object emits more red than blue light and consequently looks red. Also, the total area under each curve is proportional to the total energy emitted, so you can see that the hotter object emits more total energy than the cooler objects.

Now you can understand why two famous stars, Betelgeuse and Rigel, have such different colors (see Figure 2-3). According to Wien's law, Betelgeuse is cooler than the sun so it looks red, but Rigel is hotter than the sun and looks blue. A star with the same temperature as the sun would appear yellowish. According to the Stefan–Boltzmann law, Rigel also produces more energy from each square meter of its photosphere than does the sun, which in turn produces more energy from each square meter than does Betelgeuse.

Notice that cool objects may emit little visible radiation but are still producing blackbody radiation. For example, the human body has a temperature of 310 K and emits blackbody radiation mostly in the infrared part of the spectrum. Infrared security cameras can detect burglars by the radiation they emit, and mosquitoes can track you down in total darkness by homing in on your infrared radiation. You have a wavelength of maximum intensity in the infrared part of the spectrum. At your temperature you almost never emit photons with other than infrared wavelengths.

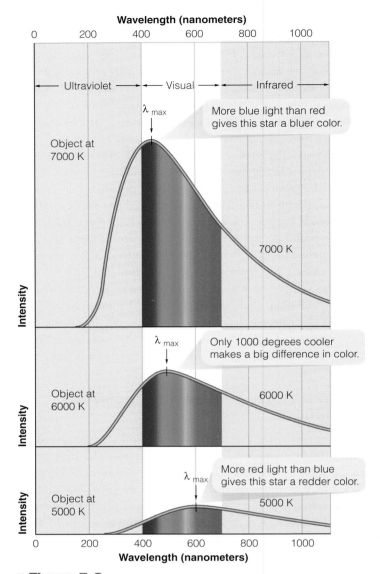

■ **Figure 5-2**

These graphs of blackbody radiation from three objects at different temperatures demonstrate that a hot object radiates more total energy and that the wavelength of maximum intensity is shorter for hotter objects. The hotter object here will look blue to your eyes, while the cooler object will look red.

ThomsonNOW™ Sign in at www.thomsonedu.com and go to ThomsonNOW to see Astronomy Exercises "Blackbody" and "Stefan–Boltzmann Law."

Building Scientific Arguments

The infrared radiation coming out of your ear can tell a doctor your temperature. How does that work?

You know two radiation laws, so your argument must use the right one. Doctors and nurses use a handheld device to measure body temperature by observing the infrared radiation emerging from a patient's ear. You might suspect the device depends on the Stefan–Boltzmann law and measures the intensity of the infrared radiation. A person with a fever will emit more energy than a healthy person. However, a healthy person with a large ear canal would also emit more than a person with a small ear canal, so the measurement of intensity would not necessarily be easy to interpret. The device actually depends on Wien's law in that it measures the "color" of the infrared radiation. A patient with a fever will emit blackbody radiation with a slightly shorter wavelength of maximum intensity, and the infrared radiation emerging from his or her ear will be a tiny bit "bluer" than for a person with a normal temperature.

Astronomers can measure the temperatures of the sun and other stars the same way. Adapt your argument for stars. **Use Figure 5-2 to explain how the colors of stars reveal their temperatures.**

■ ■ ■

(5-3) The Sun's Surface

THE SUN'S DISK LOOKS LIKE a mostly smooth layer of gas. Although the sun seems to have a real surface, it is not solid. In fact, the sun is gaseous from its outer atmosphere right down to its center.

The Photosphere

The apparent surface of the sun is called the **photosphere.** The photosphere is the layer in the sun's atmosphere that is dense enough to emit plenty of light but not so dense that the light can't escape; so the photosphere is the source of most of the sunlight received by Earth. The photosphere is less than 500 km deep. If the sun magically shrank to the size of a bowling ball, the photosphere would be no thicker than a layer of tissue paper wrapped around the ball. When you measure the amount of light with different wavelengths coming from the photosphere and use Wien's law, you find that the average temperature of the photosphere is about 5800 K.

Although the photosphere appears to be substantial, it is really a very-low-density gas, 3000 times less dense than the air you breathe. To find gases as dense as the air at Earth's surface, you would have to descend about 70,000 km below the photosphere, roughly 10 percent of the way to the sun's center. With fantastically efficient insulation, you could fly a spaceship right through the photosphere. The photosphere represents the depth where somebody outside the sun would no longer be able to see the descending spaceship; conversely, onboard the ship, you would be no longer able to see the rest of the universe.

There are regions of the photosphere that appear darker than the rest. These are called **sunspots.** They produce less light than equal-sized pieces of the normal photosphere, and their color is redder than the average, so from both the Wien and Stefan–Boltzmann laws you can conclude that sunspots are cooler than the photosphere; in fact, they are usually about 1000 K cooler. Later in this chapter you will learn about the causes of sunspots and other solar dermatology problems.

Heat Flow in the Sun

At the temperature of 5800 K, every square millimeter of the sun's photosphere must be radiating more energy than a 60-watt light bulb. Simple logic tells you that energy in the form of heat is flowing outward from the sun's interior. With all that energy radiating into space, the sun's surface would cool rapidly if energy did not flow up from inside to keep the surface hot.

As you learn more about the surface and atmosphere of the sun, you will find many phenomena that are driven by this energy flow. Like a pot of boiling soup on a hot stove, the sun is in constant activity as the heat comes up from below. In good photographs, the photosphere has a mottled appearance because it is made up of dark-edged regions called granules, and the visual pattern is called **granulation** (■ Figure 5-3a). Each granule is about the size of Texas and lasts for only 10 to 20 minutes before fading away. Faded granules are continuously replaced by new ones. The color and amount of light from different portions of the granules show, by both Wien's law and the Stefan–Boltzmann law, that granule centers are a few hundred degrees hotter than the edges.

Astronomers recognize granulation as the surface effects of rising and falling currents of gas in and just below the photosphere (the Doppler effect that allows astronomers to measure speeds of these gas currents is discussed later in this chapter). The centers of granules are rising columns of hot gas, and the edges of the granules are cooler, sinking gas. The presence of granulation is clear evidence that energy is flowing upward through the photosphere by a process known as **convection.**

Convection occurs when hot fluid rises and cool fluid sinks, as when, for example, a convection current of hot gas rises above a candle flame. You can watch liquid convection by adding a bit of cool non-dairy creamer to an unstirred cup of hot coffee. The cool creamer sinks, warms, expands, rises, cools, contracts, sinks again, and so on, creating small regions on the surface of the coffee that mark the tops of convection currents. Viewed from above, these regions look much like solar granules.

(5-4) Light, Matter, and Motion

IF LIGHT DID NOT INTERACT WITH MATTER, you would not be able to see these words. In fact, you would not exist, because, among other problems, photosynthesis would be impossible, and there would be no grass, wheat, bread, beef, cheeseburgers, or any other kind of food. The interaction of light and matter

■ Figure 5-3

(a) This ultra-high-resolution image of the photosphere shows granulation. The largest granules here are about the size of Texas. (P. N. Brandt, G. Scharmer, G. W. Simon, Swedish Vacuum Solar Telescope) (b) This model explains granulation as the tops of rising convection currents just below the photosphere. Heat flows upward as rising currents of hot gas and downward as sinking currents of cool gas. The rising currents heat the solar surface in small regions seen from Earth as granules.

makes your life possible, and it also makes it possible for you to understand the universe.

You should begin your study of light and matter by considering electrons that are within atoms. As you have already learned, electrons and other charged particles produce light when they change speed or direction of their motion.

Electron Shells

The electrons are bound to the atom by the attraction between their negative charge and the positive charge on the nucleus. This attraction is known as the **Coulomb force,** after the French physicist Charles-Augustin de Coulomb (1736–1806). A positive **ion** is an atom with missing electrons, meaning, fewer electrons than protons. To **ionize** an atom, you need a certain amount of energy to pull an electron completely away from the nucleus. This energy is the electron's **binding energy,** the energy that holds it to the atom.

The size of an electron's orbit is related to the energy that binds it to the atom. If an electron orbits close to the nucleus, it is tightly bound, and a large amount of energy is needed to pull it away. Consequently, its binding energy is large. An electron orbiting farther from the nucleus is held more loosely, and less energy is needed to pull it away. That means it has less binding energy.

Nature permits atoms only certain amounts (quanta) of binding energy, and the laws that describe how atoms behave are called the laws of **quantum mechanics.** Much of this discussion of atoms is based on the laws of quantum mechanics.

Because atoms can have only certain amounts of binding energy, your model atom can have orbits of only certain sizes, called **permitted orbits.** These are like steps in a staircase: You can stand on the number-one step or the number-two step but not on the number-one-and-one-quarter step. The electron can occupy any permitted orbit but not orbits in between.

The arrangement of permitted orbits depends primarily on the charge of the nucleus, which in turn depends on the number of protons. The number of protons in the nucleus is unique to each element. Consequently, each kind of element has its own pattern of permitted orbits (■ Figure 5-4). Ionized forms of an element have orbital patterns quite different from their un-ionized forms. The arrangement of permitted orbits differs for every kind of atom and ion. **Isotopes** are versions of a given element with different numbers of neutrons. Isotopes of an element have almost but not quite the same pattern of permitted electron orbits as each other because they have the same number of electrons while their nuclei have slightly different masses.

The Excitation of Atoms

Each orbit in an atom represents a specific amount of binding energy, so physicists commonly refer to the orbits as **energy levels.** Using this terminology, you can say that an electron in its smallest and most tightly bound orbit is in its lowest permitted energy level. You could move the electron from one energy level to another by supplying enough energy to make up the difference between the two energy levels. It would be like moving a flowerpot from a low shelf to a high shelf: The greater the distance between the shelves, the more energy you would need to raise the pot. The amount of energy needed to move the electron is the energy difference between the two energy levels.

If you move the electron from a low energy level to a higher energy level, you can call the atom an **excited atom.** That is, you have added energy to the atom by moving its electron. If the electron falls back to the lower energy level, that energy is released. An atom can become excited by collision. If two atoms collide, one or both may have electrons knocked into higher energy lev-

Hydrogen nuclei have one positive charge; the electron orbits are not tightly bound.

Only the innermost orbits are shown.

Boron nuclei have 5 positive charges; the electron orbits are more tightly bound.

Hydrogen

Helium

Boron

■ **Figure 5-4**

The electrons in an atom may occupy only certain permitted orbits. Because different elements have different charges on their nuclei, the elements have different, unique patterns of permitted orbits.

els. This happens very commonly in hot gas, where the atoms move rapidly and collide often.

Another way an atom can get the energy that moves an electron to a higher energy level is to absorb a photon (packet) of electromagnetic radiation. Only a photon with exactly the right amount of energy corresponding to the energy difference between two levels can move the electron from one level to another. If the photon has too much or too little energy, the atom cannot absorb it, and the photon passes right by. Because the energy of a photon depends on its wavelength, only photons of certain wavelengths (certain colors) can be absorbed by a given kind of atom.

■ Figure 5-5 shows the lowest four energy levels of the hydrogen atom along with three photons the atom could absorb. The longest-wavelength (reddest) photon has only enough energy to excite the electron from the first to the second energy level, but the shorter-wavelength (higher-energy, bluer) photons can excite the electron to higher levels. A photon with too much or too little energy cannot be absorbed. Because the hydrogen atom has many more energy levels than shown in Figure 5-5, it can absorb photons of many different wavelengths.

Atoms, like humans, cannot exist in an excited state forever. The excited atom is unstable and must eventually (usually within 10^{-6} to 10^{-9} seconds) give up the energy it has absorbed and return its electron to a lower energy level. The lowest energy level an electron can occupy is called the **ground state.**

When the electron drops from a higher to a lower energy level, it moves from a loosely bound level to one more tightly bound. The atom then has a surplus of energy—the energy difference between the levels—that it can emit as a photon. Study the sequence of events in ■ Figure 5-6 to see how an atom can absorb and emit photons. Jumps of electrons from one orbit to another are sometimes called **quantum leaps.** In casual language that term has come to mean a huge change, but now you see that it represents a very small change indeed. The quantum leap represents a change of electron motion, so electromagnetic radiation is either released or absorbed in the process.

Because each type of atom or ion has its unique set of energy levels, each type absorbs and emits photons with a unique set of wavelengths. As a result, you can identify the elements in a gas by studying the characteristic wavelengths of light absorbed or emitted. It is important to note that the wavelengths (colors) emitted and absorbed by leaping electrons are determined not by the starting or ending energy level of the jump but by the difference between the levels.

The process of excitation and emission is a common sight in urban areas at night. A neon sign glows when atoms of neon gas in the glass tube are excited by electricity flowing through the tube. As the electrons in the electric current flow through the gas, they collide with the neon atoms and excite them. As you have seen, immediately after an atom is ex-

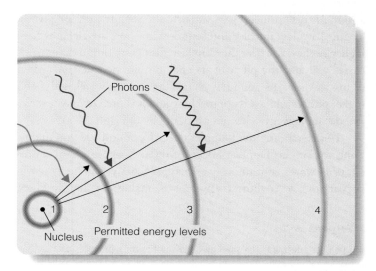

Photons

Nucleus   Permitted energy levels

■ **Figure 5-5**

A hydrogen atom can absorb only those photons that move the atom's electron to one of the higher-energy orbits. Here three different photons are shown along with the change in electron orbit they would produce if they were absorbed.

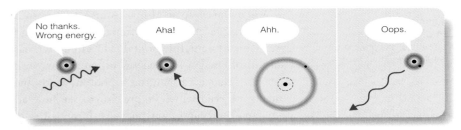

Figure 5-6

An atom can absorb a photon only if the photon has the correct amount of energy. The excited atom is unstable and within a fraction of a second returns to a lower energy level, reradiating the photon in a random direction.

cited, its electron drops back to a lower energy level, emitting the surplus energy as a photon of a certain wavelength. The visible photons emitted by the most common electron jumps within excited neon atoms produce a reddish-orange glow. Street signs of other colors, erroneously called "neon," contain other gases or mixtures of gases instead of pure neon.

The Doppler Effect

The **Doppler effect** is an apparent change in the wavelength of radiation caused by relative motion of a source and observer. Astronomers use it to measure the speed of blobs of gas in the sun's atmosphere toward or away from Earth, as well as speeds of entire stars and galaxies.

When astronomers talk about the Doppler effect, they are talking about small shifts in the wavelength of electromagnetic radiation. The Doppler effect, however, can occur for waves of all types—sound waves, for example. You probably hear the Doppler effect in sound every day without realizing what it is. The pitch of a sound is determined by its wavelength. Sounds with long wavelengths have low pitches, and sounds with short wavelengths have higher pitches. You hear the Doppler effect every time a car or truck passes you and the pitch of its engine noise or emergency siren seems to drop. Its sound is shifted to shorter wavelengths and higher pitches while it is approaching and is shifted to longer wavelengths and lower pitches after it passes by (■ Figure 5-7a).

Understanding the Doppler effect for sound lets you understand the similar Doppler effect for light. Imagine a light source emitting waves continuously as it approaches you. The light will appear to have a shorter wavelength, making it slightly bluer.

Figure 5-7

The Doppler effect. (a) The siren on a moving fire truck produces sounds that move outward (black circles). An observer ahead of the truck hears the siren sound waves closer together, with a higher tone, while an observer behind the truck hears them farther apart, with a lower tone. (b) A moving source of light emits waves that move outward (black circles). An observer in front of the light source observes a shorter wavelength (a blueshift), and an observer behind the light source observes a longer wavelength (a redshift). (c) Earth's orbital motion causes small Doppler shifts in a star's spectrum.

This is called a **blueshift.** A light source moving away from you has a longer wavelength and is slightly redder. This is a **redshift.** Redshifted and blueshifted spectra produced by a moving light source are illustrated in Figure 5-7b.

The terms *redshift* and *blueshift* are used to refer to any range of wavelengths. The light does not actually have to be red or blue, and the terms apply equally to wavelengths in the radio, X-ray, or gamma-ray parts of the spectrum. "Red" and "blue" refer to the relative direction of the shift, not to actual color. Also, note that these shifts are much too small to change the color of a star noticeably, but they are easily detected by changes in the positions of features in a

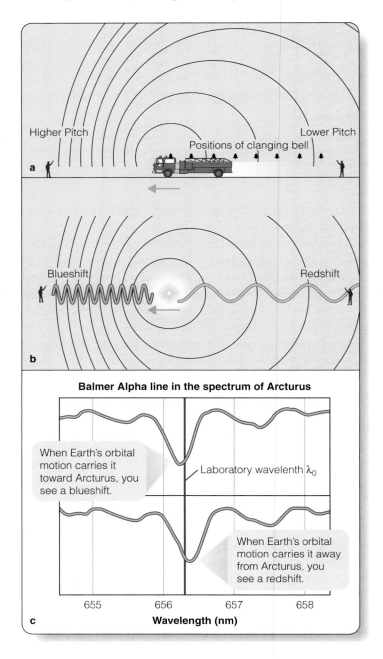

star's spectrum such as spectral lines (discussed in detail in the next section). The Doppler shift, blue or red, reveals the relative motion of wave source and observer—you measure the same Doppler shift if the light source is moving and you are stationary or if the light source is stationary and you are moving.

The amount of change in wavelength depends on the speed of the source. A moving car has a smaller sound Doppler shift than a jet plane, and a slow-moving star has a smaller light Doppler shift than one that is moving at high velocity. You can measure the speed of a star toward or away from you by measuring the size of the Doppler shift of its spectral lines.

Note that the Doppler shift is sensitive only to the part of the velocity directed away from you or toward you, called the **radial velocity (V_r).** A star moving perpendicular to your line of sight, to the left for instance, would have no blueshift or redshift because its distance from Earth would not be decreasing or increasing. Police radar guns use the Doppler effect to measure the speeds of cars. The police park next to the highway and aim their "hair dryers" directly along the road because they can measure only radial velocities, and they want to measure your full velocity along the highway.

ThomsonNOW™ Sign in at www.thomsonedu.com and go to ThomsonNOW to see Astronomy Exercises "Doppler Shift" and "Stellar Rotation."

5-5) The Sun's Atmosphere

SCIENCE IS A WAY OF UNDERSTANDING NATURE, and the spectrum of the sun tells you a great deal about such things as the sun's temperature and the composition and motions of the gases in the solar photosphere and atmosphere. In later chapters you will use spectra to study stars, galaxies, and planets, but you can begin with spectra of the sun.

Formation of Spectra

Spectra of the sun and other stars are formed as light passes from their photospheres outward through their atmospheres. Study **Atomic Spectra** on pages 86–87. Notice three important properties of spectra:

1 There are three kinds of spectra described by three simple rules. When you see one of these types of spectra, you can recognize the arrangement of matter that emitted the light. Dark (absorption) lines in the sun's spectrum are caused by atoms in the sun's (or Earth's) atmosphere between you and the sun's photosphere. The photosphere itself produces a blackbody (continuous) spectrum.

2 The wavelengths of the photons that are absorbed by a given type of atom are the same as the wavelengths of the photons emitted by that type of atom; both are determined by the electron energy levels in the atom. The emitted photons coming from a hot cloud of hydrogen gas have the same wavelengths as the photons absorbed by hydrogen atoms in the sun's atmosphere. Although the hydrogen atom produces many spectral lines from the ultraviolet to the infrared, only three hydrogen lines are visible to human eyes.

3 Most modern astronomy books display spectra as graphs of intensity versus wavelength. Be sure you see the connection between dark absorption lines and dips in the graphed spectrum.

ThomsonNOW™ Sign in at www.thomsonedu.com and go to ThomsonNOW to see Astronomy Exercise "Emission and Absorption Spectra."

The Sun's Chemical Composition

Identifying the elements in the sun's atmosphere by identifying the lines in its spectrum is a relatively straightforward procedure. For example, two dark absorption lines appear in the yellow region of the solar spectrum at the wavelengths 589.0 nm and 589.6 nm. The only atom that can produce this pair of lines is sodium, so the sun must contain sodium. Over 90 elements in the sun have been identified this way. The element helium was known in the sun's spectrum first, before helium (from the Greek word *helios,* meaning "sun") was found on Earth.

However, just because spectral lines that are characteristic of an element are missing, you cannot conclude that the element itself is absent. For example, the hydrogen Balmer lines are weak in the sun's spectrum, yet more than 90 percent of the atoms in the sun are hydrogen. The reason for this apparent paradox is that the sun is too cool to produce strong Balmer lines. At the sun's photosphere temperature, atoms do not usually collide violently enough to knock electrons in hydrogen atoms into the second energy level, which is the necessary starting place for Balmer line absorptions. Spectral lines of other varieties of atoms, for example ionized calcium, are especially easy to observe in the sun's spectrum because the sun is the right temperature to excite those atoms to the energy levels that can produce visible spectral lines, even though those atoms are not very common in the sun.

The effect of temperature on the visibility of spectral lines was first understood by Cecila Payne (later, Payne-Gaposchkin), who was an astronomer doing Ph.D. research work at Harvard Observatory in the 1920s. She used the new techniques of quantum mechanics to derive accurate chemical abundances for the sun and other stars and so was the first person to know that the sun is mostly composed of hydrogen, even though its visible-wavelength hydrogen spectral lines are only moderately strong.

Astronomers must use the physics that describes the interaction of light and matter to analyze a star's spectrum, taking into account the star's temperature, to calculate the amounts of each element present in the star correctly. Such results show that

1 To understand how to analyze a spectrum, begin with a simple incandescent lightbulb. The hot filament emits blackbody radiation, which forms a **continuous spectrum.**

An **absorption spectrum** results when radiation passes through a cool gas. In this case you can imagine that the lightbulb is surrounded by a cool cloud of gas. Atoms in the gas absorb photons of certain wavelengths, which are missing from the spectrum, and you see their positions as dark **absorption lines.** Such spectra are sometimes called **dark-line spectra.**

An **emission spectrum** is produced by photons emitted by an excited gas. You could see **emission lines** by turning your telescope aside so that photons from the bright bulb did not enter the telescope. The photons you would see would be those emitted by the excited atoms near the bulb. Such spectra are also called **bright-line spectra.**

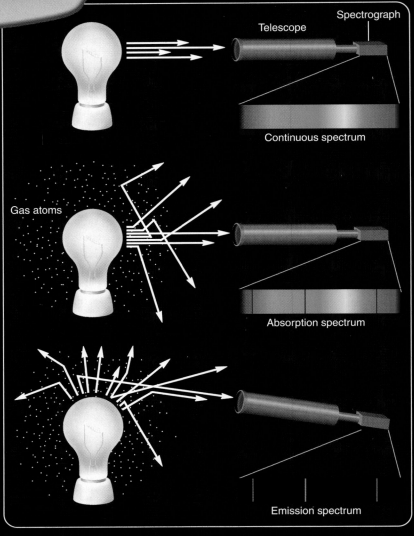

Telescope Spectrograph

Continuous spectrum

Gas atoms

Absorption spectrum

Emission spectrum

1a The spectrum of a star is an absorption spectrum. The denser layers of the photosphere emit black body radiation. Gases in the atmosphere of the star absorb their specific wavelengths and form dark absorption lines in the spectrum.

Absorption spectrum

GUSTAV ROBERT KIRCHHOFF
1824
1887

30
DEUTSCHE BUNDESPOST BERLIN

1b In 1859, long before scientists understood atoms and energy levels, the German scientist Gustav Kirchhoff formulated three rules, now known as **Kirchhoff's laws,** that describe the three types of spectra.

KIRCHHOFF'S LAWS

Law I: **The Continuous Spectrum**

A solid, liquid, or dense gas excited to emit light will radiate at all wavelengths and thus produce a continuous spectrum.

Law II: **The Emission Spectrum**

A low-density gas excited to emit light will do so at specific wavelengths and thus produce an emission spectrum.

Law III: **The Absorption Spectrum**

If light comprising a continuous spectrum passes through a cool, low-density gas, the result will be an absorption spectrum.

954.6 nm
1005.0 nm
1093.8 nm
1281.8 nm
1875.1 nm

Paschen series
(IR)

388.9 nm
397.0 nm
410.2 nm
434.0 nm
486.1 nm
656.3 nm

H_ζ

H_β
H_α

Balmer series
(Visible-UV)

93.8 nm
95.0 nm
97.2 nm
102.6 nm
121.5 nm

Lyman
series
(UV)

Nucleus

2 The electron orbits in the hydrogen atom are shown here as energy levels. When an electron makes a **transition** from one orbit to another, it changes the energy stored in the atom. In this diagram, arrows pointed inward represent transitions that result in the emission of a photon. If the arrows pointed outward, they would represent transitions that result from the absorption of a photon. Long arrows represent large amounts of energy and correspondingly short-wavelength photons.

2a Transitions in the hydrogen atom can be grouped into series—the **Lyman series**, **Balmer series**, **Paschen series**, and the like. Transitions and the resulting spectral lines are identified by Greek letters. Only the first few transitions in the first three series are shown at left.

2b In this drawing (right) of the hydrogen spectrum, emission lines in the infrared and ultraviolet are shown as gray. Only the first three lines of the Balmer series are visible to human eyes.

2c Excited clouds of gas in space emit light at all of the Balmer wavelengths, but you see only the red, blue, and violet photons blending to create the pink color typical of ionized hydrogen.

Visual-wavelength image

AURA/NOAO/NSF

The shorter-wavelength lines in each series blend together.

2000 nm

1500 nm

Paschen lines

Infrared

1000 nm

H_α

Balmer lines

Visible

500 nm

H_β

H_γ

Ultraviolet

100 nm

Lyman lines

3 Modern astronomers rarely work with spectra as bands of light. Spectra are usually recorded digitally, so it is easy to represent them as graphs of intensity versus wavelength. Here the artwork above the graph suggests the appearance of a stellar spectrum. The graph below reveals details not otherwise visible and allows comparison of relative intensities. Notice that dark absorption lines in the spectrum appear as dips in the curve of intensity.

Intensity

H_γ

H_β

H_α

500   600   700
Wavelength (nm)

■ **Table 5-1** | **The Most Abundant Elements in the Sun**

Element	Percentage by Number of Atoms	Percentage by Mass
Hydrogen	91.0	70.9
Helium	8.9	27.4
Carbon	0.03	0.3
Nitrogen	0.008	0.1
Oxygen	0.07	0.8
Neon	0.01	0.2
Magnesium	0.003	0.06
Silicon	0.003	0.07
Sulfur	0.002	0.04
Iron	0.003	0.1

sphere must be an excited, transparent low-density gas viewed with a dark cold background. The density is about 10^8 times less than that of the air you breathe.

Atoms in the lower chromosphere are ionized, and atoms in the higher layers of the chromosphere are even more highly ionized. That is, they have lost more electrons. From this, astronomers can find the temperature in different parts of the chromosphere. Just above the photosphere the temperature falls to a minimum of about 4500 K and then rises to the extremely high temperatures of the corona.

Solar astronomers can take advantage of some of the physics of spectral line formation to study the chromosphere. A photon with a wavelength corresponding to one of the solar atmosphere's strong

nearly all stars have compositions similar to the sun—about 90 percent of the atoms are hydrogen, and about 9 percent are helium, with small traces of heavier elements (■ Table 5-1). It is fair to say that Cecilia Payne, whose thesis has been called the most important doctoral work in the history of astronomy, figured out the true chemical composition of the universe. You will use her results in later chapters when you study the life stories of the stars, the history of our galaxy, and the origin of the universe.

The Chromosphere

Above the photosphere lies the **chromosphere.** Solar astronomers define the lower edge of the chromosphere as lying just above the visible surface of the sun with its upper regions blending gradually with the atmosphere's outermost layer, the **corona,** which is described later in this chapter.

You can think of the chromosphere as being an irregular layer with a depth on average less than Earth's diameter (■ Figure 5-8). Because the chromosphere is roughly 1000 times fainter than the photosphere, you can see it with your unaided eyes only during a total solar eclipse when the moon covers the brilliant photosphere. Then, the chromosphere flashes into view as a thin line of pink just above the photosphere. The word *chromosphere* comes from the Greek word *chroma,* meaning "color." The pink color is produced by combined light of three bright emission lines—the red, blue, and violet Balmer lines of hydrogen.

Astronomers know a great deal about the chromosphere from its spectrum. The chromosphere produces an emission spectrum, and Kirchhoff's second law tells you the chromo-

a

Chromosphere

Photosphere

Corona

b Visual-wavelength image

■ Figure 5-8

(a) A cross section at the edge of the sun shows the relative thickness of the photosphere and chromosphere. Earth is shown for scale. On this scale, the disk of the sun would be more than 1.5 m (5 ft) in diameter. (b) The corona extends from the top of the chromosphere to a great distance above the photosphere. This photograph, made during a total solar eclipse, shows only the inner part of the corona. (Daniel Good)

a Hα image   Spicules   **b** Hα image

■ **Figure 5-9**

H—alpha filtergrams reveal complex structure in the chromosphere, including long, dark filaments and spicules springing from the edges of supergranules larger than twice the diameter of Earth. (NOAA/SEL/USAF; © 1971 NOAO/NSO)

absorption lines is very unlikely to escape from deeper layers and reach Earth. A **filtergram** is a photograph made using light in one of those dark absorption lines. Those photons can only come from high in the solar atmosphere. In this way, filtergrams reveal detail in the upper layers of the chromosphere. In a similar way, an image recorded in the far-ultraviolet or in the X-ray part of the spectrum reveals other structures in the hottest parts of the solar atmosphere.

■ Figure 5-9a shows a filtergram made at the wavelength of the H-alpha Balmer line. This image shows complex structure in the chromosphere, including long, dark **filaments** silhouetted against the brighter surface. **Spicules** are flamelike jets of gas extending upward into the chromosphere and lasting 5 to 15 minutes. Seen at the limb of the sun's disk, these spicules blend together and look like flames covering a burning prairie (Figure 5-9b), but they are not flames at all. Spectra show that spicules are cooler gas from the lower chromosphere extending upward into hotter regions.

Spectroscopic analysis of the chromosphere alerts you that it is a low-density gas in constant motion where the temperature increases rapidly with height. Just above the chromosphere lies even hotter gas.

The Corona

The outermost part of the sun's atmosphere is called the *corona*, after the Greek word for "crown." The corona is so dim that it is not visible in Earth's daytime sky because of the glare of scattered light from the sun's brilliant photosphere. During a total solar eclipse, however, when the moon covers the photosphere, you can see the innermost parts of the corona, as shown in Figure 5-8b.

Observations made with specialized telescopes called **coronagraphs** on Earth or in space can block the light of the photosphere and record the corona out beyond 20 solar radii, almost 10 percent of the way to Earth. Such images reveal that sunspots are linked with features in the chromosphere and corona (■ Figure 5-10).

The spectrum of the corona can tell you a great deal about the coronal gases and simultaneously illustrate how astronomers analyze a spectrum. Some of the light from the corona produces a continuous spectrum that lacks absorption lines. Superimposed on the corona's continuous spectrum are emission lines of highly ionized gases. In the lower corona, the atoms are not as highly ionized as they are at higher altitudes, and this tells you that the temperature of the corona rises with altitude. Just above the chromosphere the coronal temperature is about 500,000 K, but in the outer corona the temperature can be 2,000,000 K or more. Despite that very high temperature, the corona does not produce much light because its density is very low, only 10^6 atoms/cm³ or less. That is about one-trillionth the density of the air you breathe.

Astronomers have wondered for years how the corona and chromosphere can be so hot. Heat flows from hot regions to cool regions, never from cool to hot. So how can the heat from the photosphere, with a temperature of only 5800 K, flow out into the much hotter chromosphere and corona? Observations made by the SOHO satellite show a **magnetic carpet** of looped magnetic fields extending up through the photosphere. Remember that the gas of the chromosphere and corona has a very low density, so it can't resist movement in the magnetic fields. Turbulence below the photosphere seems to flick the magnetic loops back and forth and whip the gas about. That heats the gas. In this instance, energy appears to flow outward in the form of agitation of the magnetic fields.

Not all of the sun's magnetic field loops back toward the sun; some of the field lines lead outward into space. Gas from the solar atmosphere follows along the magnetic fields that point outward and flows away from the sun in a breeze called the **solar wind** that can be considered an extension of the corona. The low-density gases of the solar wind blow past Earth at 300 to 800 km/s with gusts as high as 1000 km/s.

Because of the solar wind, the sun is slowly losing mass, but this is only a minor loss for an object as massive as the sun. The sun loses about 10^7 tons per second, but that is only 10^{-14} of a solar mass per year. Later in life, the sun, like many other stars, will lose mass rapidly. You will see in future chapters how this affects the evolution of stars.

Two nearly simultaneous images show sunspots in the photosphere and excited regions in the chromosphere above the sunspots.

Visual-wavelength image

Ultraviolet

Twisted streamers in the corona suggest magnetic fields.

The corona extends far from the disk.

Background stars

Sun hidden behind mask

Visual image

Sun hidden behind mask

Visual image

■ Figure 5-10

Images of the photosphere, chromosphere, and corona show the relationships among the layers of the sun's atmosphere. The visual-wavelength image shows the sun in white light—that is, as you would see it with your eyes. (SOHO/ESA/NASA)

Do other stars have chromospheres, coronae, and stellar winds like the sun? Ultraviolet and X-ray observations suggest that the answer is yes. The spectra of many stars contain emission lines in the far-ultraviolet that could only have formed in the low-density, high-temperature gases of a chromosphere or corona. Also, many stars are sources of X rays that appear to have been produced by the high temperature gas in coronae. This observational evidence gives astronomers good reason to believe that the sun, for all its complexity, is a typical star.

Building Scientific Arguments

How deeply into the sun can astronomers see?
Scientific arguments usually involve observations, and it is always important to know how observations are made. When you look into the layers of the sun, your sight does not really penetrate into the sun. Rather, your eyes record photons that have escaped from the sun and traveled outward through the layers of the sun's atmosphere. If you observe at a wavelength at the center of a dark absorption line, the photosphere and lower chromosphere are opaque, photons can't escape to your eyes, and the only photons you can see come from the upper chromosphere. What you see are the details of the upper chromosphere. On the other hand, if you observe at a wavelength that is not easily absorbed (a wavelength between spectral lines), the atmosphere is more transparent, and photons from slightly deeper in the photosphere can escape to your eyes.

By choosing the proper wavelength, solar astronomers can observe to different depths from the photosphere up into the chromosphere. But the corona is so thin and the gas below the photosphere so dense that this method doesn't work in these regions. Now it is time to build a new argument. **How can astronomers observe the corona?**

■   ■   ■

5-6 Solar Activity

THE SUN IS NOT QUIET. It is home to slowly changing spots larger than Earth and rapid vast eruptions that dwarf human imagination. All of these seemingly different forms of solar activity have one thing in common—magnetic fields.

Observing the Sun

Solar activity is often visible with even a small telescope, but you should be very careful about observing the sun. It is not safe to look directly at the sun, and it is even more dangerous to look at the sun through any optical instrument such as a telescope, binoculars, or even the viewfinder of a camera that concentrates the sunlight and can cause severe injury. You can safely project an image of the sun onto a screen, or you can use specially designed solar blocking filters.

In the early 17th century, Galileo observed the sun and saw spots on its surface; day by day, he saw the spots moving across the sun's disk. He correctly concluded that the sun is a sphere and is rotating (■ Figure 5-11).

Sunspots

The dark sunspots that you see at visible wavelengths only hint at the complex processes that go on in the sun's atmosphere. To explore those processes, you must turn to the analysis of images and spectra at a wide range of wavelengths.

Study **Sunspots and the Sunspot Cycle** on pages 92–93 and notice five important points:

1 Sunspots are cool spots on the sun's surface caused by strong magnetic fields.

2 Sunspots follow an 11-year cycle not only in the number of spots visible but also in their location on the sun.

3 The Zeeman effect gives astronomers a way to measure the strength of magnetic fields on the sun.

4 Characteristics of the sunspot cycle vary over centuries and appear to affect Earth's climate.

5 Finally, there is clear evidence that sunspots are part of a larger magnetic process that involves other layers of the sun's atmosphere and parts of its interior.

The sunspot groups are merely the visible traces of magnetically active regions. But what causes this magnetic activity? The answer appears to be linked to the cyclical strengthening and weakening of the sun's overall magnetic field.

ThomsonNOW™ Sign in at www.thomsonedu.com and go to ThomsonNOW to see Astronomy Exercises "Zeeman Effect," "Sunspot Cycle I," and "Sunspot Cycle II."

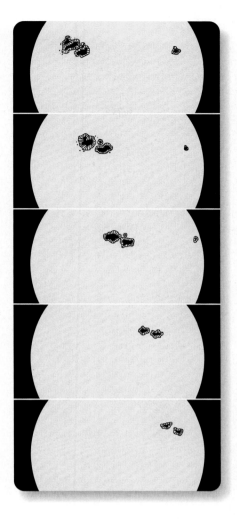

■ **Figure 5-11**

If you sketch the location and structure of sunspots on successive days, you will see the rotation of the sun and gradual changes in the size and structure of sunspots just as Galileo did in 1610.

Insight into the Sun's Interior

Almost no light emerges from below the photosphere, so you can't see into the solar interior. However, solar astronomers can use the vibrations in the sun to explore its depths in a type of analysis called **helioseismology.** Random motions in the sun constantly produce vibrations. Astronomers can detect these vibrations by observing Doppler shifts in the solar surface. These waves make the photosphere move up and down by small amounts—roughly plus or minus 15 km. This covers the surface of the sun with a pattern of rising and falling regions that can be mapped (■ Figure 5-12). In the sun, a vibration with a period of 5 minutes is strongest, but the periods range from 3 to 20 minutes. Just as geologists can study Earth's interior by analyzing vibrations from earthquakes, so solar astronomers can use helioseismology to map the temperature, density, and rate of rotation in the sun's invisible interior.

The Sun's Magnetic Cycle

Sunspots are magnetic phenomena, so the 11-year cycle of sunspots must be caused by cyclical changes in the sun's magnetic field. To explore that idea, you can begin with the sun's rotation.

Sunspots and the Sunspot Cycle

1 The dark spots that appear on the sun are only the visible traces of complex regions of activity. Observations over many years and at a range of wavelengths tell you that sunspots are clearly linked to the sun's magnetic field.

Spectra show that sunspots are cooler than the photosphere with a temperature of about 4240 K. The photosphere has a temperature of about 5800 K. Because the total amount of energy radiated by a surface depends on its temperature raised to the fourth power, sunspots look dark in comparison. Actually, a sunspot emits quite a bit of radiation. If the sun were removed and only an average-size sunspot were left behind, it would be brighter than the full moon.

A typical sunspot is about twice the size of Earth, but there is a wide range of sizes. They appear, last a few weeks to as long as two months, and then shrink away. Usually, sunspots occur in pairs or complex groups.

Size of Earth

NASA

Royal Swedish Academy of Sciences

Umbra

Penumbra

Sunspots are not shadows, but astronomers refer to the dark core of a sunspot as its umbra and the outer, lighter region as the penumbra.

2 The number of spots visible on the sun varies in a cycle with a period of 11 years. At maximum, there are often over 100 spots visible. At minimum, there are very few.

2a Early in the cycle, spots appear at high latitudes north and south of the sun's equator. Later in the cycle, the spots appear closer to the sun's equator. If you plot the latitude of sunspots versus time, the graph looks like butterfly wings, as shown in this **Maunder butterfly diagram,** named after E. Walter Maunder of Greenwich Observatory.

3 Astronomers can measure magnetic fields on the sun using the **Zeeman effect** as shown below. When an atom is in a magnetic field, the electron orbits are altered, and the atom is able to absorb a number of different wavelength photons even though it was originally limited to a single wavelength. In the spectrum, you see single lines split into multiple components, with the separation between the components proportional to the strength of the magnetic field.

Sunspot groups

Magnetic fields around sunspot groups

Ultraviolet filtergram

Magnetic image

J. Harvey/NSO and HAO/NCAR

Simultaneous images

Slit allows light from sunspot to enter spectrograph.

Visual

AURA/NOAO/NSF

Spectral line split by Zeeman effect

3a Images of the sun above show that sunspots contain magnetic fields a few thousand times stronger than Earth's. The strong fields are believed to inhibit gas motion below the photosphere; consequently, convection is reduced below the sunspot, and the surface there is cooler. Heat prevented from emerging through the sunspot is deflected and emerges around the sunspot, which can be detected in infrared images.

Maunder minimum few spots colder winters

Winter severity in London and Paris

Warmer winters

↑ Warm
↓ Cold

4 Historical records show that there were very few sunspots from about 1645 to 1715, a phenomenon known as the **Maunder minimum.** This coincides with a period called the "little ice age," a period of unusually cool weather in Europe and North America from about 1500 to about 1850, as shown in the graph at left. Other such periods of cooler climate are known. The evidence suggests that there is a link between solar activity and the amount of solar energy Earth receives. This link has been confirmed by measurements made by spacecraft above Earth's atmosphere.

Magnetic fields can reveal themselves by their shape. For example, iron filings sprinkled over a bar magnet reveal an arched shape.

M. Seeds

The complexity of an active region becomes visible at short wavelengths.

Far -UV image

SOHO/EIT, ESA and NASA

5 Observations at nonvisible wavelengths reveal that the chromosphere and corona above sunspots are violently disturbed in what astronomers call **active regions.** Spectrographic observations show that active regions contain powerful magnetic fields.
    Arched structures above an active region are evidence of gas trapped in magnetic fields.

Visual-wavelength image

Simultaneous images

Far-UV image

NASA/TRACE

The sun does not rotate as a rigid body. It is a gas from its outermost layers down to its center, and some parts of the sun rotate faster than other parts. The equatorial region of the photosphere rotates faster than do regions at higher latitudes (■ Figure 5-13a). At the equator, the photosphere rotates once every 25 days, but at latitude 45° one rotation takes 28 days. Helioseismology can map the rotation throughout the interior (Figure 5-13b). This phenomenon is called **differential rotation,** and it is clearly linked with the magnetic cycle.

The sun's magnetic field appears to be powered by the energy flowing outward through the moving currents of gas. The gas is highly ionized, so it is a very good conductor of electricity. When an electrical conductor rotates rapidly and is stirred by convection, it can convert some of the energy flowing outward as convection into a magnetic field. This process is called the **dynamo effect,** and it is believed to operate also in the liquid metal of Earth's core to produce Earth's magnetic field.

Helioseismology observations have found evidence that the sun's magnetic field is generated by the dynamo effect at the bottom of the **convective zone** deep under the photosphere. The sun's magnetic cycle is clearly related to how its magnetic field is created.

The magnetic behavior of sunspots provides an insight into how the magnetic cycle works. Sunspots tend to occur in groups or pairs, and the magnetic field around the pair resembles that around a bar magnet with one end magnetic north and the other end magnetic south. At any one time, sunspot pairs south of the sun's equator have reversed polarity compared to those north of the sun's equator. ■ Figure 5-14 illustrates this by showing sunspot pairs south of the sun's equator with magnetic south poles leading and sunspots north of the sun's equator with magnetic north poles leading. At the end of an 11-year sunspot cycle, the new spots appear with reversed magnetic polarity.

The Babcock Model for Solar Activity

The sun's magnetic cycle is not fully understood, but the **Babcock model** (named for its inventor) explains the magnetic cycle as a progressive tangling and then untangling of the solar magnetic field. Because the electrons in an ionized gas are free to move, the gas is a very good conductor of electricity, and any magnetic field in the gas is "frozen" (firmly embedded) in the gas. If the gas moves, the magnetic field must move with it. The sun's magnetic

■ Figure 5-13

(a) In general, the photosphere of the sun rotates faster at the equator than at higher latitudes. If you started five sunspots in a row, they would not stay lined up as the sun rotates. (b) Detailed analyses of the sun's interior rotation from helioseismology reveal regions of slow rotation (blue) and rapid rotation (red). Such studies show that differential rotation occurs inside the sun as well as on its surface. (NASA/SOI)

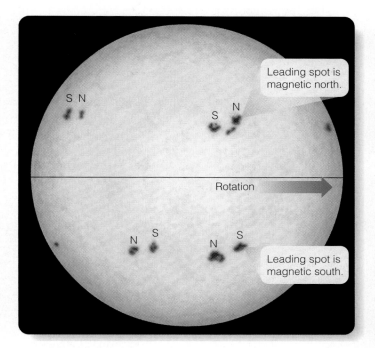

Leading spot is magnetic north.

Rotation

Leading spot is magnetic south.

■ **Figure 5-14**

In sunspot groups, here simplified into pairs of major spots, the leading spot and the trailing spot have opposite magnetic polarity. Spot pairs in the southern hemisphere have reversed polarity relative to those in the northern hemisphere.

field is frozen into its gases, and the differential rotation wraps this field around the sun like a long string caught in a rotating wheel. Rising and sinking gas currents twist the field into ropelike tubes, which tend to float upward. Where these magnetic tubes burst through the sun's surface, sunspot pairs occur (■ Figure 5-15).

The Babcock model explains the reversal of the sun's magnetic field from cycle to cycle. As the magnetic field becomes tangled, adjacent regions of the sun's surface are dominated by magnetic fields that point in different directions. After about 11 years of tangling, the field becomes so complex that adjacent regions of the surface are forced to changing their magnetic field directions to align with neighboring regions. The entire field quickly rearranges itself into a simpler pattern, and differential rotation begins winding it up to start a new cycle. The newly organized field is reversed, and the next sunspot cycle begins with magnetic north replaced by magnetic south. Consequently, the complete magnetic cycle is 22 years long, whereas the sunspot cycle is 11 years long.

This magnetic cycle seems to explain the Maunder butterfly diagram (see page 92). As a sunspot cycle begins, the twisted tubes of magnetic force first begin to float upward and produce sunspot pairs at higher latitude. Later in the cycle, when the field is more tightly wound, the tubes of magnetic force arch up through the surface closer to the equator. As a result, the later sunspot pairs in a cycle appear closer to the equator.

Notice the power of a scientific model. The Babcock model may in fact be incorrect in some details, but it provides a frame-

The Solar Magnetic Cycle

Magnetic field line

Sun

For simplicity, a single line of the solar magnetic field is shown.

Differential rotation drags the equatorial part of the magnetic field ahead.

Differential rotation wraps the sun in many turns of its magnetic field.

Where loops of tangled magnetic field rise through the surface, sunspots occur.

Bipolar sunspot pair

■ **Figure 5-15**

The Babcock model of the solar magnetic cycle explains the sunspot cycle as a consequence of the sun's differential rotation gradually winding up the magnetic field near the base of the sun's outer, convective layer.

work on which to organize your thinking about all the complex solar activity. Even though the models of the sky in Chapter 2 and of atomic energy levels in this chapter are only partially correct, they serve as organizing themes to guide your explorations. Similarly, although the precise details of the solar magnetic cycle are not yet understood, the Babcock model gives you a general picture of the behavior of the sun's magnetic field (**Window on Science 5-1**).

If the sun is truly a representative star, you might expect to find similar magnetic cycles on other stars, but they are too distant for spots to be directly visible. Some stars, however, vary in brightness over a period of days in a way revealing that they are marked

Magnetic Solar Phenomena

1 Magnetic phenomena in the chromosphere and corona, like magnetic weather, result as constantly changing magnetic fields on the sun trap ionized gas to produce beautiful arches and powerful outbursts. Some of this solar activity can affect Earth's magnetic field and atmosphere.

This ultraviolet image of the solar surface was made by the NASA TRACE spacecraft. It shows hot gas trapped in magnetic arches extending above active regions. At visual wavelengths, you would see sunspot groups in these active regions.

1a A **prominence** is composed of ionized gas trapped in a magnetic arch rising up through the photosphere and chromosphere into the lower corona. Seen during total solar eclipses at the edge of the solar disk, prominences look pink because of the three Balmer emission lines. The image below shows the arch shape suggestive of magnetic fields. Seen from above against the sun's bright surface, prominences form dark filaments.

Sacramento Peak Observatory

H-alpha filtergram

1b Quiescent prominences may hang in the lower corona for many days, whereas eruptive prominences burst upward in hours. The eruptive prominences below are many Earth diameters long.

Far-UV image

SOHO, EIT, ESA and NASA

The gas in prominences may be 60,000 to 80,000 K, quite cold compared with the low-density gas in the corona, which may be as hot as a million Kelvin.

Trace/NASA

An ultraviolet
image shows
an active
region
experiencing
a flare.

Far-UV image

Helioseismology
image

2a This sequence of three images of the photosphere taken 5 minutes apart shows sound waves rushing away from a solar flare at 50 km/s. This flare was equivalent to an earthquake 40,000 times stronger than the quake that destroyed San Francisco in 1906. The most powerful flares can be a billion times more powerful than a hydrogen bomb.

Sun

CME

SOHO/LASCO, ESA, and NASA

Visual-wavelength image

2 Solar flares rise to maximum in minutes and decay in an hour. They occur in active regions where oppositely directed magnetic fields meet and cancel each other out in what astronomers call **reconnections.** Energy stored in the magnetic fields is released as short-wavelength photons and as high-energy protons and electrons. X-ray and ultraviolet photons reach Earth in 8 minutes and increase ionization in our atmosphere, which can interfere with radio communications. Particles from flares reach Earth hours or days later as gusts in the solar wind, which can distort Earth's magnetic field and disrupt navigation systems. Solar flares can also cause surges in electrical power lines and damage to Earth satellites.

2b Auroras occur about 130 km above Earth's surface when energy in the solar wind guided by Earth's magnetic field excites gases in the upper atmosphere.

Copyright Jan Curtis

2c Magnetic reconnections can release enough energy to blow large amounts of ionized gas outward from the corona in **coronal mass ejections (CMEs).** These produce violent gusts in the solar wind, and if one blows past Earth it can create electrical currents up to a million megawatts, which flow down into Earth's magnetic poles and excite atoms in Earth's upper atmosphere to emit photons, which are seen as an aurora, as shown above.

X-ray image

Coronal hole

Yohkoh/ISAS/NASA

3 Much of the solar wind comes from **coronal holes,** where the magnetic field does not loop back into the sun. These open magnetic fields allow ionized gas in the corona to flow away as the solar wind. The dark area in this X-ray image at right is a coronal hole.

Confirmation, Confidence, and Consolidation

While many textbooks describe science as the process of testing hypotheses by observation and experiment, you should not think that every astronomer approaches the telescope with the expectation of making an observation that will disprove long-held beliefs and trigger a revolution in science. What is the daily grind of science really about, then?

First, many observations and experiments merely confirm already tested hypotheses. The biologist knows that all worker bees in a hive are sisters, but a careful study of the DNA from different workers further confirms that hypothesis. By repeatedly confirming a hypothesis, scientists build confidence in the hypothesis and may be able to extend it to a wider application. Of course, there is always the chance that a new observation or experiment will disprove the hypothesis, but that is usually very unlikely. Much of the daily grind of science is confirmation.

Another important aspect of routine science is confidence in well-established theories. It is only human nature to hang on to the principles you have come to trust, and that confidence in well-tested scientific principles helps scientists avoid rushing to faulty judgments. For example, claims for perpetual motion machines occasionally crop up in the news, but the world's scientists don't instantly abandon the laws of energy and motion pending an analysis of the latest claim. Of course, if such a claim did prove true, the entire structure of scientific knowledge would come crashing down, but because the known laws of energy and motion have been well tested and no perpetual motion machine has ever been successful, scientists know which way to bet. Like the keel on a ship, confidence in well-tested theories and laws keeps the scientific boat from rocking before every little breeze.

A third aspect of routine science is consolidation, the linking of a hypothesis to other well-studied phenomena. Chemists may understand certain kinds of carbon molecules shaped like rings, but by repeated study they find a carbon molecule shaped like a hollow sphere. To consolidate their findings, they must show that the chemical bonding in the two molecules follows the same rules and that the molecules have certain properties in common. No hypothesis is overthrown, but the chemists consolidate their knowledge and understand carbon molecules better.

The Babcock model of the solar magnetic cycle is an astronomical example of the scientific process. Solar astronomers know that the model explains some solar features but has shortcomings. Although most astronomers don't expect it will ever be necessary to discard the entire model completely, they work through confirmation and consolidation to better understand how the solar magnetic cycle works and how it is related to cycles in other stars.

---

with dark spots believed to resemble sunspots. Other stars have features in their spectra that vary over periods of years, suggesting that they are subject to magnetic cycles much like the sun's cycle. Once again, the evidence tells you that the sun is a normal star.

ThomsonNOW™ Sign in at www.thomsonedu.com and go to ThomsonNOW to see Astronomy Exercise "Convection and Magnetic Fields."

Chromospheric and Coronal Activity

The solar magnetic fields extend high into the chromosphere and corona, where they produce beautiful and powerful phenomena. Study **Magnetic Solar Phenomena** on pages 96–97 and notice three important points:

1 All solar activity is magnetic. You do not experience such events on Earth because Earth's magnetic field is weak and Earth's atmosphere is not ionized, so it is free to move independent of the magnetic field. On the sun, however, the "weather" is a magnetic phenomenon.

2 Tremendous energy can be stored in arches of magnetic field. These are visible near the edge of the solar disk as prominences, and, seen from above, as filaments (Figure 5-9). When that stored energy is released, it can trigger powerful eruptions; and, although these eruptions occur far from Earth, they can affect Earth in dramatic ways, including auroral displays. Auroras, the eerie and pretty northern and southern lights, are produced when gases in Earth's upper atmosphere glow from energy delivered by the solar wind.

3 In some regions of the solar surface, the magnetic field does not loop back. High-energy gas from these regions flows outward and produces much of the solar wind.

You needed the entire range of physical principles presented in this chapter—parallax, Wien's law and the Stefan–Bolzmann law for blackbody radiation, atomic structure, Kirchhoff's laws for the formation of spectra, the Doppler effect, the Zeeman effect—to realize how the sun's surface temperature and composition are known, plus understand solar activity cycles and their effects on Earth. In the next chapter you will learn how the sun fits into the context of other stars.

ThomsonNOW™ Sign in at www.thomsonedu.com and go to ThomsonNOW to see Astronomy Exercise "Auroras."

Building Scientific Arguments

What kind of activity would the sun have if it didn't rotate differentially?

This is a really difficult question because only one star is visible close up. Nevertheless, you can construct a scientific argument by thinking about the Babcock model. If the sun didn't rotate differentially, with its equator traveling faster than do the higher latitudes, the magnetic field might not get wound up, and there might not be a solar cycle. Twisted tubes of magnetic field might not form and rise through the photosphere to produce prominences and flares, although convection might still tangle the magnetic field and produce some activity. Is the magnetic activity that heats the chromosphere and corona driven by differential rotation or by convection?

It is hard to guess; but without differential rotation, the sun might not have a strong magnetic field and high-temperature gas above its photosphere.

This is very speculative, but sometimes in the critical analysis of ideas it helps to imagine a change in a single important factor and try to understand what might happen. For example, redo the argument above. **What do you think the sun would be like if it had no convection inside?**

■   ■   ■

Study and Review

Summary

5-1 | The Sun—Basic Characteristics

How do you know the distance, size, mass, and density of the sun?
■ Observing the parallax shift of Venus in transit against the sun's disk from opposite sides of Earth was the original method for finding the distance to the sun. The diameter of the sun can be calculated from its distance and its angular size.

■ Newton's laws and the motion of Earth and the other planets as they orbit the sun allow the mass of the sun to be determined. Its average density—about the same as the density of water—can be easily calculated from its mass and diameter.

5-2 | The Origin of Sunlight

How does matter produce light?
■ An atom consists of a nucleus surrounded by a cloud of negatively charged electrons. The nucleus is made up of positively charged protons and uncharged neutrons.

■ Motion among charged particles in a solid, a liquid, or a dense gas causes the emission of blackbody radiation. Pure blackbody radiation is a continuous spectrum—a continuous rainbow of color without dark or bright spectral lines.

■ The hotter an object is, the more energy it radiates and the shorter is its wavelength of maximum intensity, λ_{max}. This allows astronomers to estimate the temperature of the sun and other stars from their colors.

5-3 | The Sun's Surface

What do astronomers see when they observe the sun?
■ The photosphere, or visible surface, is the level in the sun from which visible photons most easily escape. Its temperature is about 5800 K.

■ Energy flowing outward from the sun's interior travels as rising currents of hot gas and sinking currents of cool gas in the convective zone just below the photosphere.

■ The granulation of the photosphere is produced by convection currents of gas rising from below.

5-4 | Light, Matter, and Motion

How does matter interact with light to produce spectral lines?
■ Electrons in an atom may occupy various permitted orbits around the nucleus but not orbits in between. The size of an electron's orbit depends on the energy stored in the electron's motion.

■ An electron may be excited to a higher orbit during a collision between atoms, or it may move from one orbit to another by absorbing or emitting a photon of the proper energy.

■ Shifts in the wavelengths of features in spectra of the sun and stars provide clues to the motions of their atoms. For example, when a parcel of gas in the sun's atmosphere is approaching, you observe slightly shorter wavelengths (a blueshift); and when it is receding, you observe slightly longer wavelengths (a redshift). This Doppler effect reveals the radial velocity of the gas, the part of its velocity directed toward or away from Earth.

5-5 | The Sun's Atmosphere

What can you learn from the sun's spectrum?
■ Because orbits of only certain energies are permitted in an atom, photons of only certain wavelengths can be absorbed or emitted. Each kind of atom has its own structure and therefore its own characteristic set of spectral lines.

■ The hydrogen atom has the Lyman series of lines in the ultraviolet, the Balmer series in the visible, and the Paschen series (plus others) in the infrared.

■ If light from a blackbody such as the sun's photosphere passes through a low-density gas such as the sun's atmosphere on its way to your spectrograph, the gas can absorb photons of certain wavelengths. As a result, you detect dark lines in the spectrum at those wavelengths, superimposed on the continuous spectrum of the blackbody source. Such a spectrum is called an absorption spectrum.

- If you look at a low-density gas that is excited to emit photons, you see bright lines in the spectrum. Such a spectrum is called an emission spectrum.

- The spectrum of the sun's atmosphere viewed in front of the photosphere is an absorption spectrum, and the hydrogen lines you see in the visible part of the spectrum are called Balmer lines.

- The sun's spectrum can tell you its chemical composition. The presence of spectral lines of a certain element shows that the element must be present in the sun. However, you must proceed with care because the strengths of lines of each element also depend on the temperature of the gas.

- The solar atmosphere consists of two layers of hot, low-density gas, the chromosphere and the corona.

- The chromosphere is most easily visible during total solar eclipses, when it flashes into view for a few seconds. It is a thin, hot layer of gas just above the photosphere. Its pink color is caused by the Balmer emission lines in its spectrum.

- Filtergrams of the chromosphere reveal spicules and filaments.

- The corona is the sun's outermost atmospheric layer. It is composed of a very-low-density, very hot gas extending far from the visible sun. Astronomers have evidence that its high temperature—2,000,000 K or more—is maintained by effects of a magnetic field extending up through the photosphere and chromosphere—the magnetic carpet.

- Parts of the corona give rise to the solar wind, a breeze of low-density ionized gas streaming away from the sun.

5-6 | Solar Activity

Why does the sun have a cycle of activity, and how does that affect Earth?

- Sunspots seem dark because they are slightly cooler than the rest of the photosphere. The average sunspot is about twice the size of Earth and contains magnetic fields a few thousand times stronger than Earth's.

- Solar astronomers can study the motion, density, and temperature of gases inside the sun by observing the way the solar surface oscillates in a type of analysis known as helioseismology.

- Astronomers can measure magnetic fields on the sun by measuring the splitting of some spectral lines caused by the Zeeman effect.

- The average number of sunspots varies over a period of about 11 years and appears to be related to a magnetic cycle.

- The sunspot cycle does not repeat exactly each cycle, and the decades from 1645 to 1715, known as the Maunder minimum, seem to have been a time when solar activity was very low and Earth's climate was slightly colder.

- The sun rotates differentially with regions far from the equator rotating slower than equatorial regions.

- Alternate sunspot cycles have reversed magnetic polarity, which has been explained by the Babcock model, in which the differential rotation of the sun and internal convection currents wind up the magnetic field. Tangles in the field rise to the surface and cause active regions visible to your eyes as sunspot pairs. When the field becomes strongly tangled, it re-orders itself into a simpler but reversed field, and the cycle starts over.

- Other stars are too far away for starspots to be visible, but spectroscopic observations reveal that many other stars have spots and magnetic fields that follow long-term cycles like the sun's.

- Prominences occur in the chromosphere; their arched shapes show that they are formed of ionized gas trapped in the magnetic field. You see prominences in filtergrams as dark filaments silhouetted against the bright chromosphere.

- Flares are sudden eruptions of X-ray, ultraviolet, and visible radiation plus high-energy atomic particles produced when magnetic fields on the sun interact and reconnect. Flares are important because they can have dramatic effects on Earth, such as communications blackouts and auroras.

- Spacecraft images show long streamers extending from the corona out into space. Coronal mass ejections occur when magnetic fields on the surface of the sun eject bursts of ionized gas that flow outward in the solar wind. Such bursts can produce auroras and other phenomena if they strike Earth.

New Terms

spectral line (p. 76)

transit of Venus (p. 78)

density (p. 78)

atom (p. 79)

nucleus (p. 79)

proton (p. 79)

neutron (p. 79)

electron (p. 79)

molecule (p. 79)

heat (p. 79)

temperature (p. 79)

Kelvin temperature scale (p. 79)

absolute zero (p. 79)

blackbody radiation (p. 79)

wavelength of maximum intensity (p. 80)

Wien's law (p. 80)

Stefan–Boltzmann law (p. 80)

photosphere (p. 81)

sunspot (p. 81)

granulation (p. 81)

convection (p. 81)

Coulomb force (p. 82)

ion, ionize (p. 82)

binding energy (p. 82)

quantum mechanics (p. 82)

permitted orbit (p. 82)

isotope (p. 82)

energy level (p. 82)

excited atom (p. 82)

ground state (p. 83)

quantum leap (p. 83)

Doppler effect (p. 84)

blueshift (p. 84)

redshift (p. 84)

radial velocity V_r (p. 85)

continuous spectrum (p. 86)

absorption spectrum (dark-line spectrum) (p. 86)

absorption line (p. 86)

emission spectrum (bright-line spectrum) (p. 86)

emission line (p. 86)

Kirchhoff's laws (p. 86)

transition (p. 87)

Lyman series (p. 87)

Balmer series (p. 87)

Paschen series (p. 87)

chromosphere (p. 88)

corona (p. 88)

filtergram (p. 89)

filament (p. 89)

spicule (p. 89)

coronagraph (p. 89)

magnetic carpet (p. 89)

solar wind (p. 89)

helioseismology (p. 91)

Maunder butterfly diagram (p. 92)

Zeeman effect (p. 93)

Maunder minimum (p. 93)

active region (p. 93)

differential rotation (p. 94)

dynamo effect (p. 94)

convective zone (p. 94)

Babcock model (p. 94)

prominence (p. 96)

flare (p. 97)

reconnection (p. 97)

aurora (p. 97)

coronal mass ejection (CME) (p. 97)

coronal hole (p. 97)

Review Questions

ThomsonNOW™ Assess your understanding of this chapter's topics with additional quizzing and animations at
www.thomsonedu.com

1. How was the distance to the sun first determined?

2. How is the mass of the sun determined?

3. Define density. How is the density of the sun determined?

4. Why do hot stars look bluer than cool stars?

5. Why does the amount of blackbody radiation emitted depend on the temperature of the object?

6. Why can't you see deeper into the sun than the photosphere?

7. What evidence can you give that granulation is caused by convection?

8. Why is the binding energy of an electron related to the size of its orbit?

9. Why do different atoms have different lines in their spectra?

10. Describe two ways an atom can become excited.

11. What is the difference between an isotope and an ion?

12. Explain why ionized calcium can form absorption lines, but ionized hydrogen cannot.

13. What kind of spectrum does a neon sign produce?

14. How can the Doppler effect explain wavelength shifts in both light and sound?

15. The nebula shown at right contains mostly hydrogen excited to emit photons. What kind of spectrum would you expect this nebula to produce?

T. Rector, University of Alaska, and WIYN/NURO/AURA/NSF

16. Explain why the absence of spectral lines of a given element in the solar spectrum would not necessarily mean that element is absent from the sun.

17. How can astronomers detect structure in the sun's chromosphere?

18. What evidence can you give that the corona has a very high temperature?

19. What heats the chromosphere and corona to high temperature?

20. How are astronomers able to explore the layers of the sun below the photosphere?

21. What evidence can you give that sunspots are magnetic?

22. How does the Babcock model explain the sunspot cycle?

23. What does the shape of a prominence reveal?

24. How can solar flares affect Earth?

25. The upper two images here show two solar phenomena. What are they, and how are they related? How do they differ?

26. The lower image was recorded in the extreme ultraviolet by the SOHO spacecraft. Explain the features you see.

Daniel Good and NOAO

NASA/SOHO

Discussion Questions

1. In what ways is the model of an atom a scientific model? In what ways is it incorrect?

2. What would the spectrum of an auroral display look like? Why?

3. What energy sources on Earth cannot be thought of as stored sunlight?

4. What observations would you make if you were ordered to set up a system that could warn astronauts in orbit of dangerous solar flares? Such a warning system does, in fact, exist.

6 | The Family of Stars

[Love] is the star to every

wandering bark,

Whose worth's unknown,

although his height be taken.

WILLIAM SHAKESPEARE
SONNET 116

S HAKESPEARE COMPARED LOVE
TO A STAR that can be seen easily and even used for
guidance, but whose real nature is utterly unknown. He lived at
about the same time as Galileo and had no idea what stars actually are. To understand the history of the universe, the origin
of Earth, and the nature of our human existence, you need to discover what people in Shakespeare's time did not know—the
real nature of the stars. Unfortunately, it is quite difficult to find out what a star is like. When you look at a star even through
a telescope, you see only a point of light. Real understanding of stars requires careful analysis of starlight. This chapter concentrates on five goals: knowing how far away stars are, how much energy they emit, what their surface temperatures are, how
big they are, and how much mass they contain. By the time you finish this chapter, you will know the family of stars well.

Guidepost

Looking Back

Science is based on measurement, but measurement in astronomy is very difficult. In Chapter 4 you studied the complex telescopes and instruments that astronomers use. In Chapter 5 you learned about the information that can be found in spectra and how astronomers use their understanding of the interaction of light and matter to study a typical star, the sun.

This Chapter

Here you will discover how telescopes, astronomical instruments, and basic physics knowledge can tell us what stars are like. Here you will find answers to six essential questions about stars:

How far away are the stars?

How much energy do stars make?

How can you tell a star's temperature using its spectrum?

How big are stars?

How much matter do stars contain?

What is the typical star like?

With this chapter you leave our sun behind and begin your study of the billions of stars that dot the sky. In a sense, stars are the basic building blocks of the universe. If you hope to understand what the universe is, what our sun is, what our Earth is, and what we are, you must understand stars.

Looking Ahead

Once you know how to find the basic properties of stars, you will be ready to trace the history of the stars from birth to death, a story that begins in the next chapter.

6-1 Star Distances

DISTANCE IS THE MOST IMPORTANT and the most difficult measurement in astronomy, and astronomers have many different ways to find the distances to stars. Each of those ways depends on a simple and direct geometrical method that is much like the method surveyors would use to measure the distance across a river they cannot cross. You can begin by reviewing that method and then apply it to stars.

The Surveyor's Triangulation Method

To measure the distance across a river (■ Figure 6-1), a team of surveyors begins by driving two stakes into the ground. The distance between the stakes is called the *baseline*. The surveyors then choose a landmark on the opposite side of the river, a tree perhaps, thus establishing a large triangle marked by the two stakes and the tree. Using their instruments, the surveyors sight the tree from the two ends of the baseline and measure the two angles on their side of the river. Knowing two angles of this large triangle and the length of the baseline side between them, the surveyors can find the distance across the river by simple trigonometry. For example, if the baseline is 50 meters and the angles are 66° and 71°, they would calculate that the distance from the baseline to the tree is 64 meters.

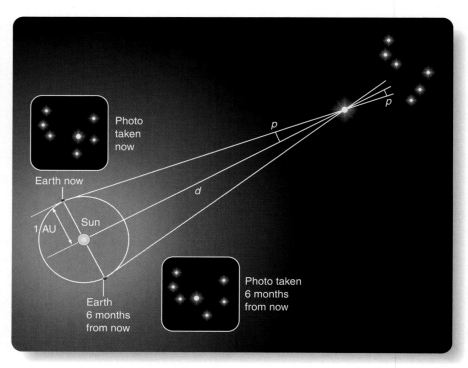

■ Figure 6-2

You can measure the parallax of a nearby star by photographing it from two points along Earth's orbit. For example, you might photograph it now and again in six months. Half of the star's total change in position from one photograph to the other is defined as its stellar parallax, *p*.

The Astronomer's Triangulation Method

To find the distance to a star, astronomers use a very long baseline, the size of Earth's orbit (■ Figure 6-2). If you take a photograph of a nearby star and then wait 6 months, Earth will have moved halfway around its orbit. You can then take another photograph of the star from a slightly different location in space. When you examine the photographs, you will discover that the star is not in exactly the same place in the two photographs.

In Chapter 3 you learned that *parallax* is the term that refers to the common experience of an apparent shift in the position of a foreground object due to a change in the location of the observer's viewpoint. Your thumb, held at arm's length, appears to shift position against a distant background when you look first with one eye and then with the other (see page 36). In that case, the baseline is the distance between your eyes, and the parallax is the angle through which your thumb appears to move when you switch eyes. The farther away you hold your thumb, the smaller the parallax.

Because the stars are so distant, their parallaxes are very small angles, usually expressed in arc seconds. The quantity that astronomers call **stellar parallax (*p*)** is conventionally defined as half the total shift of the star shown in Figure 6-2, in other words, the shift seen across a baseline of 1 AU rather than 2 AU. Astronomers measure the parallax, and surveyors measure the an-

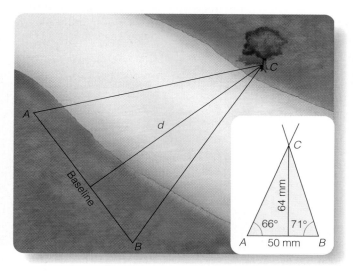

■ Figure 6-1

You can find the distance *d* across a river by measuring the baseline and the angles *A* and *B* and then using trigonometry or constructing a scale drawing of the triangle.

gles at the ends of the baseline, but both measurements tell the same thing—the shape and size of the triangle and thus the distance to the object in question.

Measuring the parallax p is very difficult because it is such a small angle. The nearest star, Alpha Centauri, has a parallax of only 0.76 arc seconds, and more distant stars have even smaller parallaxes. To see how small these angles are, imagine a dime two miles away from you. That dime covers an angle of about 1 arc second. Stellar parallaxes are so small that the first successful measurement of one did not happen until 1838, more than 200 years after the invention of the telescope.

The distances to the stars are so large that it is not convenient to use kilometers or Astronomical Units. When you measure distance via parallax, it is better to use the unit of distance called a **parsec (pc)**. The word *parsec* was created by combining *par*allax and arc *sec*ond. One parsec equals the distance to an imaginary star that has a parallax of 1 arc second. A parsec is 206,265 AU, which equals roughly 3.26 ly (light-years). Parsec units are used more often than light-years by astronomers because parsecs are more directly related to the process of measurement of star distances. However, there are instances in which the light-year is also useful. You will notice that following chapters use either parsecs or light-years as convenience and custom dictate.

The blurring caused by Earth's atmosphere makes star images appear about 1 arc second in diameter, and that makes it difficult to measure parallax. Even if you average together many observations made from Earth's surface, you cannot measure parallax with an uncertainty smaller than about 0.002 arc seconds. Thus, if you measure a parallax of 0.006 arc seconds from an observatory on the ground, your uncertainty will be about 30 percent. If you consider, as astronomers generally do, that 30 percent is the maximum acceptable level of uncertainty, then ground-based astronomers can't measure parallaxes accurately that are smaller than about 0.006 arc seconds. That parallax corresponds to a distance of about 170 pc (550 ly).

In 1989, the European Space Agency launched the satellite Hipparcos to measure stellar parallaxes from above the blurring effects of Earth's atmosphere. That small space telescope observed for four years, and the data were used to produce two parallax catalogs in 1997. One catalog contains 120,000 stars with parallaxes 20 times more accurate than ground-based measurements. The other catalog contains over a million stars with parallaxes as accurate as ground-based parallaxes. Knowing accurate distances from the Hipparcos observations has given astronomers new insights into the nature of stars.

Building Scientific Arguments

Why are parallax measurements made from space better than parallax measurements made from Earth?
When you try to measure parallax, the turbulence in Earth's atmosphere blurs star images and smears them out into blobs

roughly 1 arc second in diameter, even at good observatory sites. It isn't possible to measure shifts in the positions of these fuzzy images smaller than about 0.006 arc seconds, corresponding to a distance of 170 pc, so you can't measure parallax accurately much beyond that distance.

At first you might suppose that a satellite in orbit can measure the parallax of the stars better because the satellite is closer to the stars, but when you recall the immense distances to the stars, you see that being in orbit around Earth doesn't really put the satellite significantly closer to the stars. Rather, what is important is that a satellite is above Earth's atmosphere, so the only blurring in the star images is that produced by diffraction in the telescope optics. In other words, the star images are very sharp, and a satellite in orbit can measure the positions of stars and thus their parallaxes to high accuracy.

If a satellite can accurately measure parallaxes as small as 0.001 second of arc, then how far are the most distant stars it can measure?

■ ■ ■

ThomsonNOW™ Sign in at www.thomsonedu.com and go to **ThomsonNOW** to see Astronomy Exercises "Parallax I" and "Parallax II."

(6-2) Apparent Brightness, Intrinsic Brightness, and Luminosity

YOUR EYES TELL YOU that some stars look brighter than others, and in Chapter 2 you learned about the apparent magnitude scale that refers to stellar apparent brightness. The scale of apparent magnitudes tells you only how bright stars appear to you on Earth, however. To know the true nature of a star you must know its **intrinsic brightness,** a measure of the amount of energy the star produces. An intrinsically very bright star might appear faint if it is far away. Thus, to know the intrinsic brightness of a star, you must take into account its distance.

Brightness and Distance

When you look at a bright light, your eyes respond to the visual-wavelength energy falling on your retinas, which tells you how bright the object appears. Thus, brightness is related to the flux of energy entering your eye. Astronomers and physicists define **flux** as the energy in Joules (J) per second falling on 1 square meter. Recall that a Joule is about the amount of energy released when an apple falls from a table onto the floor. A flux of 1 Joule per second is also known as 1 Watt. The wattage of a light bulb tells you its *intrinsic* brightness. Compare that with the *apparent* brightness of a light bulb, which depends on its distance from you.

If you placed a screen 1 meter square near a light bulb, a certain amount of flux would fall on the screen. If you moved the screen twice as far from the bulb, the light that previously fell on the screen would be spread to cover an area four times larger, and the screen would receive only one-fourth as much light. If you tripled the distance to the screen, it would receive only one-ninth as much light. Thus, the flux you receive from a light source is inversely proportional to the square of the distance to the source. This is known as the inverse square relation (■ Figure 6-3). You first encountered the inverse square relation in Chapter 3, where it was applied to the strength of gravity.

Now you can understand how the apparent brightness of a star depends on its distance. If astronomers know the apparent brightness of a star and its distance from Earth, they can use the inverse square law to correct for distance and find the intrinsic brightness of the star.

Absolute Visual Magnitude

If all the stars were the same distance away, you could compare one with another and decide which was intrinsically brighter or fainter. Of course, the stars are scattered at different distances, and you can't move them around to line them up for comparison. If, however, you know the distance to a star, you can use the

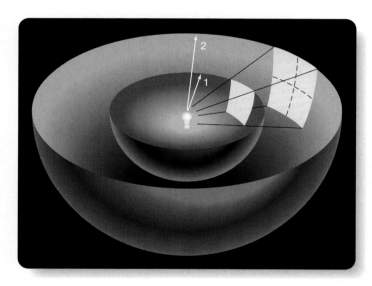

■ **Active Figure 6-3**

The inverse square relation: A light source is surrounded by spheres with radii of 1 unit and 2 units. The light falling on an area of 1 m² on the inner sphere spreads to illuminate an area of 4 m² on the outer sphere. Thus, the brightness of the light source is inversely proportional to the square of the distance.

ThomsonNOW Sign in at www.thomsonedu.com and go to ThomsonNOW to see the Active Figure called "Brightness and Distance."

inverse square relation to calculate the brightness the star would have at some standard distance. Astronomers use 10 pc as the standard distance and refer to the intrinsic brightness of the star as its **absolute visual magnitude (M_V),** which is the apparent visual magnitude that star would have if it were 10 pc away. The subscript V tells you it is a visual magnitude, referring only to the wavelengths of light your eye can see. Other magnitude systems are based on other parts of the electromagnetic spectrum, such as infrared and ultraviolet radiation.

The sun's absolute magnitude is easy to calculate because its distance and apparent magnitude are well known. The absolute visual magnitude of the sun is about 4.8. In other words, if the sun were only 10 pc (33 ly) from Earth (not a great distance in astronomy), it would have an apparent magnitude of 4.8 and look no brighter to your eye than the faintest star in the handle of the Little Dipper.

This path to find the distance to stars has led you to absolute magnitude, a measure of the intrinsic brightness of the stars. You are now ready to reach the next of your five goals in this chapter.

Luminosity

The second goal for this chapter is to find out how much energy the stars emit. With the absolute magnitudes of the stars in hand, you can now compare stars using our sun as a standard. The intrinsically brightest stars have absolute magnitudes of about −8, which means that if such a star were 10 pc away from Earth, you would see it nearly as bright as the moon. Such stars emit over 100,000 times more visible light than the sun.

Absolute visual magnitude refers to visible light, but you want to know the total output including all types of radiation. Hot stars emit a great deal of ultraviolet radiation that you can't see, and cool stars emit plenty of infrared radiation. To add in the energy you can't see, astronomers make a mathematical correction that depends on the temperature of the star. With that correction, astronomers can find the total electromagnetic energy output of a star, which they refer to as its **luminosity (L).**

Astronomers know the luminosity of the sun because they can send satellites above Earth's atmosphere and measure the amount of energy arriving from the sun, adding up radiation of every wavelength, including the types blocked by the atmosphere. Of course, they also know the distance from Earth to the sun very accurately, which is necessary to calculate luminosity. The luminosity of the sun is about 4×10^{26} Watts (Joules per second).

You can express a star's luminosity in two ways. For example, you can say that the star Capella (Alpha Aurigae) is 100 times more luminous than the sun. You can also express this in real energy units by multiplying by the luminosity of the sun. The luminosity of Capella is 4×10^{28} Watts.

When you look at the night sky, the stars look much the same, yet your study of distances and luminosities reveals an astonishing fact. Some stars are almost a million times more luminous than the sun, and some are almost a million times less luminous. Clearly, the family of stars is filled with interesting characters.

6-3 Star Temperatures

YOUR THIRD GOAL IN THIS CHAPTER is to learn about the temperatures of stars. The surprising fact is that stellar spectral lines can be used as a sensitive star thermometer.

Spectral Lines and Temperature

From the discussion of black body radiation in Chapter 5, you know that temperatures of stars can be estimated from their color—red stars are cool, and blue stars are hot. The relative strengths of various spectral lines, however, give much greater accuracy in measuring star temperatures. Recall from Chapter 5 that, for stars, the term *surface* refers to the photosphere, which is the limit of our vision into the star from outside but is not an actual solid surface. Stars typically have surface temperatures of a few thousand or tens of thousands of degrees Kelvin. As you will discover in Chapter 7, the centers of stars are much hotter than their surfaces—many millions of degrees—but the spectra tell only about the outer layers from which the light you see originates.

As you learned in Chapter 5, hydrogen Balmer absorption lines are produced by hydrogen atoms with electrons initially in the second energy level. The strength of these spectral lines can be used to gauge the temperature of a star because scientists know from lab experiments with gasses and radiation, and also from theoretical calculations, that:

- If the surface of a star is as cool as the sun or cooler, there are few violent collisions between atoms to excite the electrons, and most atoms will have their electrons in the ground (lowest) state. These atoms can't absorb photons in the Balmer series. As a result, you should expect to find weak hydrogen Balmer absorption lines in the spectra of very cool stars.

- In the surface layers of stars hotter than about 20,000 K, there are many violent collisions between atoms, exciting electrons to high energy levels or knocking the electrons completely out of most atoms so they become ionized. In this situation, few hydrogen atoms will have electrons in the second energy level to form Balmer absorption lines. As a result, you should also find weak hydrogen Balmer absorption lines in the spectra of very hot stars.

- At an intermediate temperature, roughly 10,000 K, the collisions have the correct amount of energy to excite large numbers of electrons into the second energy level. With many atoms excited to the second level, the gas absorbs Balmer-wavelength photons well and thus produces strong hydrogen Balmer lines.

The strength of the hydrogen Balmer lines thus depends on the temperature of the star's surface layers. Both hot and cool stars have weak Balmer lines, but medium-temperature stars have strong Balmer lines.

■ Figure 6-4 shows the relationship between gas temperature and strength of spectral lines for hydrogen and other substances. Each type of atom or molecule produces spectral lines that are weak at high and low temperatures and strong at some intermediate temperature, although the temperature at which the lines reach maximum strength is different for each type of atom or molecule. Theoretical calculations of the type first made by Cecilia Payne (discussed in Chapter 5) can predict just how strong various spectral lines should be for stars of different temperatures., Astronomers can determine a star's temperature by comparing the strengths of its spectral lines with the predicted strengths.

From stellar spectra, astronomers have found that the hottest stars have surface temperatures above 40,000 K and the coolest about 2000 K. Compare these with the surface temperature of the sun, which is about 5800 K.

Temperature Spectral Classification

You have seen that the strengths of spectral lines depend on the surface (photosphere) temperature of the star. From this you can predict that all stars of a given temperature should have similar spectra. Learning to recognize the pattern of spectral lines produced in the atmospheres of stars of different temperatures means there is no need to do a full analysis every time each type of spectrum is encountered. Time can be saved by classifying stellar spectra rather than analyzing each spectrum individually. Astronomers classify stars by the lines and bands in their spectra, as

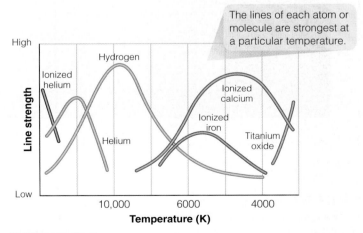

■ **Figure 6-4**

The strength of spectral lines can tell you the temperature of a star. Using the lines of several atoms or molecules gives you the most accurate results.

Spectral Class	Approximate Temperature (K)	Hydrogen Balmer Lines	Other Spectral Features	Naked-Eye Example
O	40,000	Weak	Ionized helium	Meissa (O8)
B	20,000	Medium	Neutral helium	Achernar (B3)
A	10,000	Strong	Ionized calcium weak	Sirius (A1)
F	7,500	Medium	Ionized calcium weak	Canopus (F0)
G	5,500	Weak	Ionized calcium medium	Sun (G2)
K	4,500	Very weak	Ionized calcium strong	Arcturus (K2)
M	3,000	Very weak	TiO strong	Betelgeuse (M2)

shown in ■ Table 6-1. For example, if it has weak Balmer lines and lines of ionized helium, it must be an O star.

The star classification system now used by astronomers was devised at Harvard during the 1890s and 1900s. One of the astronomers there, Annie J. Cannon, personally inspected and classified the spectra of over 250,000 stars. The spectra were first classified in groups labeled A through Q, but some groups were later dropped, merged with others, or reordered. The final classification includes seven main **spectral classes** or types that are still used today: O, B, A, F, G, K, M. This set of star types, called the

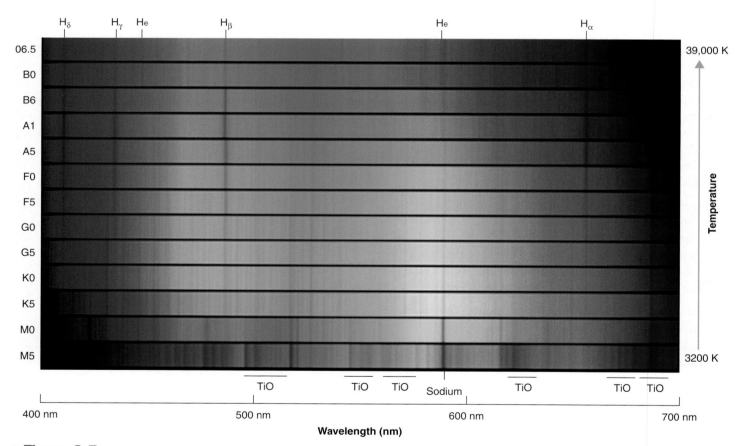

■ Figure 6-5

These spectra show stars from hot O stars at the top to cool M stars at the bottom. The Balmer lines of hydrogen are strongest for about type A0, but the two closely spaced lines of sodium in the yellow are strongest for very cool stars. Helium lines appear only in the spectra of the hottest stars. Notice that the helium line visible in the top spectrum has nearly but not exactly the same wavelength as the sodium lines visible in cooler stars. Bands produced by the molecule titanium oxide are strong in the spectra of the coolest stars. (AURA/NOAO/NSF)

spectral sequence, is important because it is a temperature sequence. The O stars are the hottest, and the temperature continues to decrease down to the M stars, the coolest. For further precision, astronomers divide each spectral class into 10 subclasses. For example, spectral class A consists of the subclasses A0, A1, A2, . . . A8, A9. Next come F0, F1, F2, and so on. These finer divisions define a star's temperature to a precision of about 5 percent. The sun, for example, is not just a G star, but a G2 star, with a temperature of 5800 K. Generations of astronomy students have remembered the spectral sequence by using mnemonics such as "Oh Boy, An F Grade Kills Me," or "Only Bad Astronomers Forget Generally Known Mnemonics." Recently, astronomers have added two more spectral classes, L and T, representing objects cooler than stars, called brown dwarfs, which you will encounter in the next chapter.

■ Figure 6-5 shows color images of 13 stellar spectra ranging from the hottest at the top to the coolest at the bottom. Notice how spectral features change gradually from hot to cool stars. Although these spectra are attractive, astronomers rarely work with spectra as color images. Rather, they display spectra as graphs of intensity versus wavelength with dark absorption lines as dips in the graph (■ Figure 6-6). Such graphs show more detail than photographs. Notice also that the overall curves are similar to blackbody curves. The wavelength of maximum is in the infrared for the coolest stars and in the ultraviolet for the hottest stars. Compare Figures 6-5 and 6-6 and notice how the strength of spectral lines depends on temperature, as indicated in the previous discussion regarding Figure 6-4. It is straightforward to determine a star's temperature from the details of its spectrum.

■ **Figure 6-6**

Modern digital spectra show how stellar spectra depend on spectral class. Here spectra are represented by graphs of intensity versus wavelength, and dark absorption lines appear as sharp dips in intensity. Hydrogen Balmer lines are strongest about A0, while lines caused by ionized calcium (CaII) are strong in K stars. Bands produced by molecules such as TiO (titanium oxide) are strong in the coolest stars. Compare these spectra with Figures 6-4 and 6-5. (Courtesy NOAO, G. Jacoby, D. Hunter, and C. Christian)

6-4 Star Sizes

YOUR FOURTH GOAL in this chapter is to learn about the sizes of stars. Do they all have the same diameter as the sun, or are some larger and some smaller? You certainly can't see their sizes through a telescope; the images of the stars are much too small for you to resolve their disks and measure their diameters. There is a way, however, to find out how big stars really are.

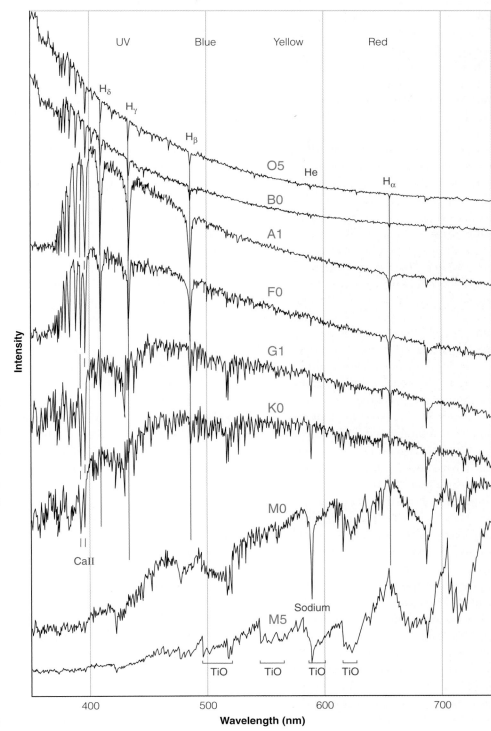

Luminosity, Temperature, and Diameter

The luminosity of a glowing object like a star is determined by its temperature (Stefan-Boltzmann law, Chapter 5) and its surface area. For example, you can eat dinner by candlelight because the candle flame has a small surface area, and, although it is very hot, it cannot radiate much heat because it is small; it has a low luminosity. However, if the candle flame were 12 ft tall, it would have a very large surface area from which to radiate, and, although it might be the same temperatures as a normal candle flame, its luminosity would drive you from the table.

In a similar way, a star's luminosity is proportional to its surface area. A hot star may not be very luminous if it has a small surface area, but it could be highly luminous if it were larger. Even a cool star could be luminous if it had a large surface area. You can use stellar luminosities to determine the diameters of stars if you can separate the effects of temperature and surface area.

The **Hertzsprung-Russell (H–R) diagram,** named after its originators, Ejnar Hertzsprung in the Netherlands and Henry Norris Russell in the United States, is a graph that separates the effects of temperature and surface area on stellar luminosities and enables astronomers to sort the stars according to their diameters. Before discussing the details of the H–R diagram, let's look at a similar diagram you might use to compare automobiles. You can plot a diagram such as ■ Figure 6-7 to show horsepower versus weight for various makes of cars. In so doing, you will find that in general the more a car weighs, the more horsepower it has. Most cars fall somewhere along the normal sequence of cars running from heavy, high-powered cars to light, low-powered models. You might call this the main sequence of cars. Some cars, however, have much more horsepower than normal for their weight—the sport or racing models—and the economy models have less power than normal for cars of the same weight. Just as this diagram helps you understand the different kinds of autos, so the H–R diagram can help you understand different kinds of stars.

An H–R diagram (■ Figure 6-8) has luminosity on the vertical axis and temperature on the horizontal axis. A star is represented by a point on the graph that tells you its luminosity and temperature. Note that in astronomy the symbol $\odot$ refers to the sun. Thus $L_\odot$ refers to the luminosity of the sun, $T_\odot$ refers to the temperature of the sun, and so on. The H–R diagram in Figure 6-8 also contains a scale of spectral types across the top. The spectral type of a star is determined by its temperature, so you can use either spectral type or temperature as the horizontal axis of an H–R diagram.

Astronomers use H–R diagrams so often that they usually skip the words "the point that represents the star." Rather, they will say that a star is located in a certain place in the diagram. Of course, they mean the point that represents the luminosity and temperature of the star and not the star itself. The location of a star in the H–R diagram has nothing to do with the location of the star in space. Furthermore, a star may move in the H–R dia-

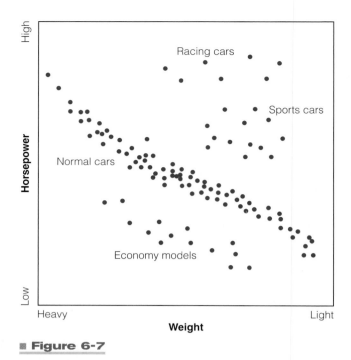

■ **Figure 6-7**

You could analyze automobiles by plotting their horsepower versus their weight and thus reveal relationships between various models. Most would lie somewhere along the main sequence of "normal" cars.

gram as it ages and its luminosity and temperature change, but such motion in the diagram has nothing to do with the star's motion in space.

In an H–R diagram, the location of a point representing a star tells you a great deal about the star. Points near the top of the diagram represent very luminous stars, and points near the bottom represent very-low-luminosity stars. Also, points near the right edge of the diagram represent very cool stars, and points near the left edge of the diagram represent very hot stars. Notice in Figure 6-8 how the artist has used color to represent temperature. Cool stars are red, and hot stars are blue.

The **main sequence** is the region of the H–R diagram running from upper left to lower right. It includes roughly 90 percent of all normal stars, represented by a curved line with dots for stars plotted along it in Figure 6-8. As you might expect, the hot main-sequence stars are more luminous than the cool main-sequence stars. Look again at the classification diagram for cars in Figure 6-7. Vehicles not on the car main sequence have different kinds of engines than main-sequence cars. In Chapter 7 you will find that stars not on the main sequence have different nuclear reactions as their power sources than do main-sequence stars. In addition to temperature, size is important in determining the luminosity of a star. Notice in the H–R diagram that some cool stars lie above the main sequence. Although they are cool, they are luminous, and that must mean they are larger—have more surface area—than main-sequence stars of the same temperature. These are called **giant** stars, and they are roughly 10 to 100 times larger

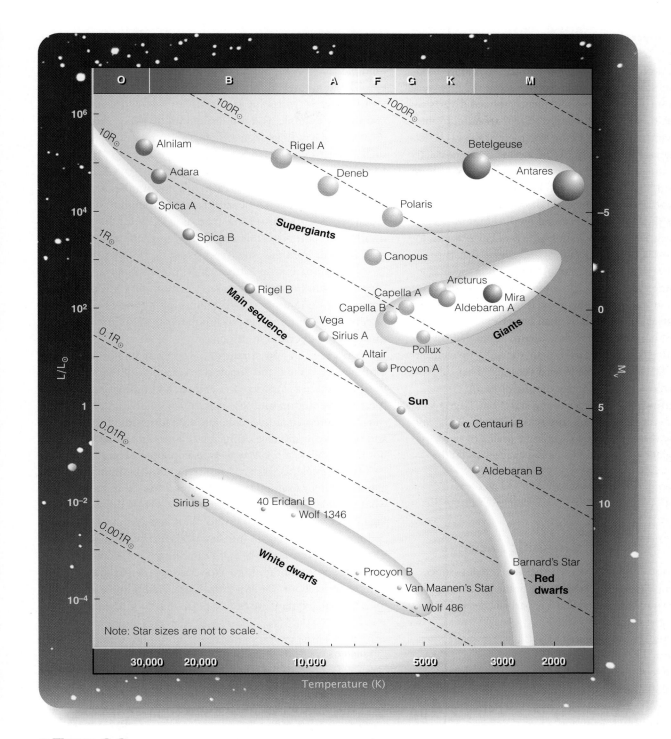

■ Figure 6-8

An H–R diagram showing the luminosity and temperature of many well-known stars. (Individual stars that orbit each other are designated A and B, as in Spica A and Spica B.) The background color in the diagram indicates the temperature of the stars. The sun is a yellow-white G2 star. Most stars fall along a sequence running from hot luminous stars at upper left to cool low luminosity stars at lower right. The exceptions—giants, supergiants, and white dwarfs—are discussed in the text. The dashed lines are lines of constant radius. The star sizes on this diagram are not to scale; try to sketch in the correct sizes for supergiants and white dwarfs using the size of the sun as a guide.

than the sun. The **supergiant** stars at the top of the H–R diagram are as large as 1000 times the sun's diameter.

At the bottom of the H–R diagram are the "economy models," stars that are very low in luminosity because they are very small. At the bottom end of the main sequence, the **red dwarfs** are not only small but also cool, and that gives them low luminosities. In contrast, the **white dwarfs** lie in the lower left of the H–R diagram, and although some white dwarfs are among the hottest stars known, they are so small they have very little surface area from which to radiate and that limits them to low luminosities.

A simple calculation can be made to draw precise lines of constant radius across the H–R diagram based on the luminosities and temperatures at each point as shown in Figure 6-8. For example, locate the line labeled $1R_\odot$ (1 solar radius) and notice that it passes through the point representing the sun. Any star whose point is located along this line has a radius equal to the sun's radius. Notice also that the lines of constant radius slope downward to the right, because cooler stars are always fainter than hotter stars of the same size. These lines of constant radius show dramatically that the supergiants and giants are much larger than the sun. In contrast, white dwarf stars fall near the line labeled $0.01\ R_\odot$. They all have about the same radius—approximately the size of Earth!

Notice the great range of sizes among stars. The largest stars are 100,000 times larger than the tiny white dwarfs. If the sun were a tennis ball, the white dwarfs would be grains of sand, and the largest supergiant stars would be as big as football fields.

ThomsonNOW˙ Sign in at www.thomsonedu.com and go to **ThomsonNOW** to see Astronomy Exercise "Stefan-Boltzmann Law II."

Luminosity Spectral Classification

A star's spectrum also contains clues whether it is a main-sequence star, a giant, or a supergiant. The larger a star is, the less dense its atmosphere is. The widths of spectral lines are partially determined by the density of the gas: If the atoms collide often in a dense gas, their energy levels become distorted, and the spectral lines are broadened. For example, in the spectrum of a main-sequence star, the hydrogen Balmer lines are broad because the star's atmosphere is dense and the hydrogen atoms collide often. In the spectrum of a giant star, the lines are narrower (■ Figure 6-9), because the giant star's atmosphere is less dense and the hydrogen atoms collide less often. In the spectrum of a supergiant star, the Balmer lines are very narrow.

Thus, an astronomer can look closely at a star's spectrum and tell roughly how big it is. Size categories

are called **luminosity classes,** because the size of the star is the dominating factor in determining luminosity. Supergiants, for example, are very luminous because they are very large.

The luminosity classes are represented by the Roman numerals I through V, with supergiants further subdivided into types Ia and Ib. For example, you can distinguish between a bright supergiant (Ia) such as Rigel (Beta Orionis) and a regular supergiant (Ib) such as Polaris, the North Star (Alpha Ursa Minoris). The star Adhara (Epsilon Canis Majoris) is a bright giant (II), Capella (Alpha Aurigae) is a giant (III), and Altair (Alpha Aquilae) is a subgiant (IV). The sun is a main-sequence star (V). The luminosity class notation appears after the spectral type, as in G2 V for the sun. White dwarfs don't enter into this classification because their spectra are very different from the other types of stars.

The approximate positions of the main sequence, giant, and supergiant luminosity classes are shown on the H–R diagram in Figure 6-8. Luminosity classification is subtle and not very accurate, but it is an important technique in modern astronomy because it provides clues to distance

Most stars are too distant to have measurable parallaxes, but astronomers can find the distances to these stars if they can record the stars' spectra and determine their luminosity classes. From spectral type and luminosity class, astronomers can estimate the star's absolute magnitude, compare with its apparent magnitude, and compute its distance. Although this process finds distance and not true parallax, it is called **spectroscopic parallax.**

For example, Betelgeuse (Alpha Orionis) is classified M2 Ia, and its apparent magnitude averages about 0.0 (Betelgeuse is somewhat variable). You can plot this star in an H–R diagram such as Figure 6-8, where you would find that a temperature class of M2 and luminosity class of Ia (bright supergiant) corresponds to a luminosity of about 100,000 $L_\odot$. That information, combined with the star's apparent brightness, allows astronomers to estimate that Betelgeuse is about 350 pc from Earth. The Hipparcos satellite finds the actual distance to be 520 pc, so the distance from the spec-

■ Figure 6-9

These schematic spectra show how the widths of spectral lines reveal a star's luminosity classification. Supergiants have very narrow spectral lines, and main-sequence stars have broad lines. Some spectral lines are more sensitive to this effect than others. Examining the details of a star's spectrum can determine its luminosity classification.

troscopic parallax technique is only approximate. Spectroscopic parallax does give a good first estimate of the distances of stars so far away that their parallax can't easily be measured.

Building Scientific Arguments

What evidence can you cite that giant stars really are bigger than the sun?

You can find stars of the same spectral type as the sun that are clearly more luminous than the sun. Capella, for example, has a G spectral type, so by definition is has about the same surface temperature as the sun. However, Capella is about 100 times more luminous than the sun (100 $L_\odot$). If it has the same surface temperature as the sun but is 100 times more luminous, then it must have a surface area 100 times greater than the sun's. Because the surface area of a sphere is proportional to the square of the diameter, Capella must be 10 times larger in diameter than the sun. That is clear observational evidence that Capella is a giant star.

In Figure 6-11, you can see that Procyon B is a white dwarf only slightly warmer than the sun but about 10,000 times less luminous than the sun (0.0001 $L_\odot$). **Use the same logical procedure followed above to demonstrate that white dwarf stars must be small.**

6-5 Masses of Stars—Binary Stars

YOUR FIFTH GOAL is to find out how much matter stars contain, that is, to know their masses. Do they all contain about the same mass as the sun, or are some more massive and others less?

Gravity is the key to determining mass. Matter produces a gravitational field. Astronomers can figure out how much matter a star contains if they watch an object such as another star move through the star's gravitational field.

Finding the masses of stars involves studying **binary stars,** pairs of stars that orbit each other. Many of the familiar stars in the sky are actually pairs of stars orbiting each other (■ Figure 6-10). Binary systems are common; more than half of all stars are members of binary star systems. Few, however, can be analyzed completely. Many are so far apart that their periods are much too long for practical mapping of their orbits. Others are so close together they are not visible as separate stars.

Binary Stars in General

The key to finding the mass of a binary star is an understanding of orbital motion (see Chapter 3). Each star in a binary system moves in its own orbit around the system's center of mass, the

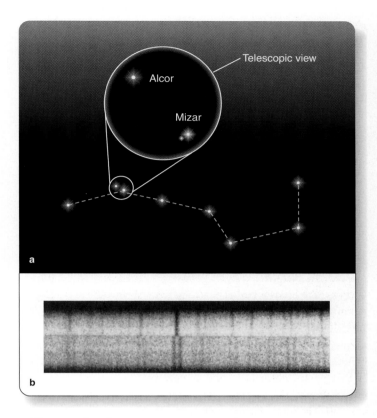

■ **Figure 6-10**

(a) At the bend of the handle of the Big Dipper lies a pair of stars, Mizar and Alcor. Through a telescope you can discover that Mizar has a fainter companion and so is a member of a visual binary system. (b) Spectra of these stars recorded at different times show that Mizar, its faint companion, and the nearby star Alcor are all spectroscopic binaries, a type of binary system discussed in the text. (The Observatories of the Carnegie Institution of Washington)

balance point of the system (see page 51). If one star is more massive than its companion, then the massive star is closer to the center of mass and travels in a smaller orbit, while the lower-mass star whips around in a larger orbit (■ Figure 6-11). The ratio of the masses of the stars in the binary system portrayed in Figure 6-11 is M_A/M_B, which equals r_B/r_A, the inverse of the ratio of the radii of the orbits. For example, if one star in a binary system has an orbit twice as large as the other star's orbit, then it must be half as massive. Getting the ratio of the masses is easy, but that doesn't tell you the individual masses of the stars, which is what you really want to know. That takes one more step.

To find the total mass of a binary star system, you must know the size of the orbits and the orbital period—the length of time the stars take to complete one orbit. The smaller the orbits are and the shorter the orbital period is, the stronger the stars' gravity must be to hold each other in orbit. From the sizes of the orbits and the orbital period, astronomers can use Kepler's third law (see Chapter 3) to figure out how much mass the stars contain in total. Combining that information with the ratio of the masses found from the relative sizes of the orbits reveals the individual masses of the stars.

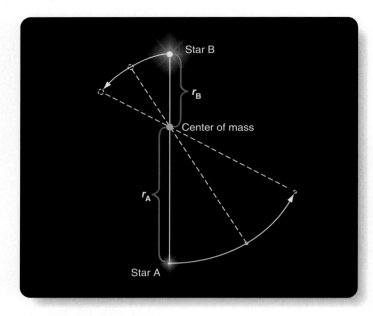

■ Active Figure 6-11

As stars in a binary star system revolve around each other, the line connecting them always passes through the center of mass, and the more massive star is always closer to the center of mass.

ThomsonNOW™ Sign in at www.thomsonedu.com and go to ThomsonNOW to see the Active Figure called "Center of Mass."

Finding the mass of a binary star system is easier said than done. One difficulty is that the true sizes of the star orbits must be measured in units like meters or Astronomical Units in order to find the masses of the stars in units like kilograms or solar masses. Measuring the true sizes of the orbits in turn requires knowing the distance to the binary system. Therefore, the only stars whose masses astronomers know for certain are in binary systems with orbits that have been determined and distances from Earth that have been measured. Other complications are that the orbits of the two stars may be elliptical; also, the plane of their orbits can be tipped at an angle to your line of sight, distorting the apparent shapes of the orbits. Astronomers must find ways to correct for these complications. Notice that finding the masses of binary stars requires a number of steps to get from what can be observed to what astronomers really want to know, the masses. Constructing such sequences of steps is an important part of science (**Window on Science 6-1**).

Three Kinds of Binary Systems

Although there are many different kinds of binary stars, three types are especially important for determining stellar masses. Studying binary stars is also preparation for knowing how to find planets around stars other than our sun, because a star with a planet orbiting around it is like a binary star system with one very small component. Each type of binary star system corre-

sponds to a different technique for finding planets, as you will learn in Chapter 12.

In a **visual binary** system, the two stars are separately visible in the telescope and astronomers can watch the stars orbit each other over periods of years or decades ■ Figure 6-12). From that,

A Visual Binary Star System

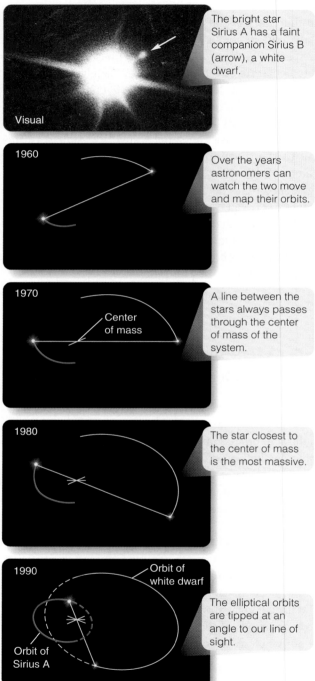

The bright star Sirius A has a faint companion Sirius B (arrow), a white dwarf.

Over the years astronomers can watch the two move and map their orbits.

A line between the stars always passes through the center of mass of the system.

The star closest to the center of mass is the most massive.

The elliptical orbits are tipped at an angle to our line of sight.

■ Figure 6-12

The orbital motion of Sirius A and Sirius B can reveal their individual masses. (Photo: Lick Observatory)

Learning about Nature through Chains of Inference

Scientists can rarely observe the things they really want to know, so they must construct chains of inference. You can't observe the mass of stars directly, so you must find a way to use what you can observe—orbital period and angular separation—and step-by-step figure out the parameters you need to reach your goal. In this case, the goal is the masses of the two stars. You follow the chain of inference from the observable parameters to the unobservable quantities you want to know.

Chains of inference can be mathematical, as when geologists use earthquakes to calculate the temperature and density of Earth's interior. There is no way to drill a hole to Earth's center and lower a thermometer or recover a sample, so the internal temperature and density are not directly observable. Nevertheless, the speed of vibrations from distant earthquakes depends on the temperature and density of the rock they pass through. Geologists can't measure the speed of the vibrations deep inside Earth, but they can measure the delay in the arrival time at different locations on the surface, and that allows them to work their way back to the seismic wave speed and, finally, to the temperature and density deep inside Earth.

Chains of inference can also be nonmathematical. Biologists studying the migration of whales can't follow individual whales for years at a time, but they can observe feeding and mating in different locations; take into consideration food sources, ocean currents, and water temperatures; and construct a chain of inference that leads back to the seasonal migration pattern for whales.

This chapter contains a number of chains of inference because it describes how astronomers know what stars are like. Almost all sciences use chains of inference. If you can link the observable parameters step by step to the final conclusions, you have gained a strong insight into the nature of science.

San Andreas fault: A chain of inference connects earthquakes to conditions inside Earth. (USGS)

---

astronomers can find the orbital period and, if the distance of the system from Earth can be found, the size of the orbits. That is enough to find the masses of the stars.

Many visual binaries have such large orbits their orbital periods are hundreds or thousands of years, and astronomers have not yet seen them complete an entire orbit. Also, many binary stars orbit so close to each other they are not visible as separate stars. Such systems can't be analyzed as a visual binary.

If the stars in a binary system are close together, a telescope view, limited by diffraction and by the atmospheric blurring called seeing, shows a single point of light. Only by looking at a spectrum, which is formed by light from both stars and contains spectral lines from both, can astronomers tell that there are two stars present and not one. Such a system is a **spectroscopic binary** system. Familiar examples of spectroscopic binary systems are the stars Mizar and Alcor in the handle of the Big Dipper (Figure 6-10b).

■ Figure 6-13 shows a pair of stars orbiting each other; the circular orbit appears elliptical because you see it nearly edge-on. If this were a true spectroscopic binary system, you would not see the separate stars. Nevertheless, the Doppler shift would tell you there were two stars orbiting each other. As the two stars move in their orbits, they alternately approach toward and recede from Earth and their spectral lines are Doppler shifted alternately toward blue and then red wavelengths. Noticing pairs of spectral lines moving back and forth across each other would alert you that you were observing a spectroscopic binary. Although spectroscopic binaries are very common, they are not as useful as visual binaries. Astronomers can find the orbital period easily, but they can't find the true size of the orbits because there is no way to find the angle at which the orbits are tipped. That means they can't find the true masses of a spectroscopic binary. All they can find is a lower limit to the masses.

ThomsonNOW Sign in at www.thomsonedu.com and go to **ThomsonNOW** to see Astronomy Exercise "Spectroscopic Binaries."

If the plane of the orbits is nearly edge-on to Earth, then the stars can cross in front of each other as seen from Earth. When one star moves in front of the other, it blocks some of the light, and the star is eclipsed. Such a system is called an **eclipsing binary** system.

Seen from Earth, the two stars are not visible separately. The system looks like a single point of light. But, when one star moves in front of the other star, part of the light is blocked, and the total brightness of the point of light decreases. ■ Figure 6-14 shows a smaller star moving in an orbit around a larger star, first eclipsing the larger star and then being eclipsed as it moves behind. The resulting variation in the brightness of the system is shown as a graph of brightness versus time, a **light curve.**

The light curves of eclipsing binary systems contain plenty of information about the stars, but the curves can be difficult to

A Spectroscopic Binary Star System

Approaching   Receding

A   B

← Blueshift   A   B   Redshift →

Stars orbiting each other produce spectral lines with Doppler shifts.

B

A

← Blueshift   A   B   Redshift →

As the stars circle their orbits, the spectral lines move together.

B

A

← Blueshift   A + B   Redshift →

When the stars move perpendicular to our line of sight, there are no Doppler shifts.

B

A

← Blueshift   B   A   Redshift →

Spectral lines shifting apart and then merging are a sign of a spectroscopic binary.

B   A

← Blueshift   B   A   Redshift →

The size of the Doppler shifts contains clues to the masses of the stars.

An Eclipsing Binary Star System

m

$t \longrightarrow$

A small, hot star orbits a large, cool star, and you see their total light.

m

$t \longrightarrow$

As the hot star crosses in front of the cool star, you see a decrease in brightness.

m

$t \longrightarrow$

As the hot star uncovers the cool star, the brightness returns to normal.

m

$t \longrightarrow$

When the hot star is eclipsed behind the cool star, the brightness drops.

m

$t \longrightarrow$

The depth of the eclipses depends on the surface temperatures of the stars.

■ **Figure 6-13**

From Earth, a spectroscopic binary looks like a single point of light, but the Doppler shifts in its spectrum reveal the orbital motion of the two stars. If the plane of their orbits is oriented very close to edge-on, the stars can pass in front of each other as they orbit and the system will be an eclipsing binary, as illustrated in Figure 6-14.

■ **Figure 6-14**

From Earth, an eclipsing binary looks like a single point of light, but changes in brightness reveal that two stars are eclipsing each other. The light curve, shown here as magnitude versus time, combined with Doppler shift information from spectra, can reveal the size and mass of the individual stars.

analyze. Figure 6-14 shows an idealized example of such a system. Once the light curve of an eclipsing binary system has been accurately observed, astronomers can construct a chain of inference that leads to the masses of the two stars. They can find the orbital period easily and can get spectra showing the Doppler shifts of the two stars. They can find the orbital speed because they don't have to correct for the inclination of the orbits; you know the orbits are nearly edge-on, or there would not be eclipses. From that, astronomers can find the size of the orbits and the masses of the stars.

Earlier in this chapter you learned that luminosity and temperature can be used to determine the radii of stars, but eclipsing binary systems give a way to check those calculations by measuring the sizes of a few stars directly. The light curve shows how long it takes for the stars to cross in front of each other, and multiplying these time intervals by the orbital speeds gives the diameters of the stars. There are complications due to the inclination and eccentricity of orbits, but often these effects can be taken into account, so observations of an eclipsing binary system can directly tell you not only the masses of its stars but also their diameters.

From the study of binary stars, astronomers have found that the masses of stars range from roughly 0.1 solar masses to nearly 100 solar masses. The most massive stars ever found in a binary system have masses of 83 and 82 solar masses. A few other stars are believed to be more massive, 100 solar masses to 150 solar masses, but they do not lie in binary systems, so astronomers can only estimate their masses.

ThomsonNOW™ Sign in at www.thomsonedu.com and go to **ThomsonNOW** to see Astronomy Exercise "Eclipsing Binaries."

Building Scientific Arguments

When you look at the light curve for an eclipsing binary system with total eclipses, how can you tell which star is hotter?

If you assume that the two stars in an eclipsing binary system are not the same size, then you can refer to them as the larger star and the smaller star. When the smaller star moves behind the larger star, you lose the light coming from the total area of the small star. And when the smaller star moves in front of the larger star, it blocks off light from the same amount of area on the larger star. In both cases, the same amount of area, the same number of square meters, is hidden from your sight. Then the amount of light lost during an eclipse depends only on the temperature of the hidden surface, because temperature is what determines how much a single square meter can radiate per second. When the surface of the hotter star is hidden, the brightness will fall dramatically, but when the surface of the cooler star is hidden, the brightness will not fall as much. So you can look at the light curve and point to the

deeper of the two eclipses and say, "That is where the hotter star is behind the cooler star."

But eclipsing binaries also reveal the diameters of the stars. **How would you look at the light curve of an eclipsing binary with total eclipses and find the ratio of the diameters?**

■   ■   ■

6-6 Typical Stars

YOU HAVE ACHIEVED the five goals set at the start of this chapter. You know how to find the distances, luminosities, temperatures, diameters, and masses of stars. Now you can put those data together to paint a family portrait of the stars. As in human family portraits, both similarities and differences are important clues to the history of the family.

Luminosity, Mass, and Density

The H–R diagram is filled with patterns that give you clues as to how stars are born, how they age, and how they die. When you add your data, you see traces of those patterns.

If you label an H–R diagram with the masses of the stars determined by observations of binary star systems, you will discover that the main-sequence stars are ordered by mass (■ Figure 6-15). The most massive main-sequence stars are the hot stars. As you run your eye down the main sequence, you will find lower-mass stars, and the lowest-mass stars are the coolest, faintest main-sequence stars.

Stars that do not lie on the main sequence are not in order according to mass. Some giants and supergiants are massive, while others are no more massive than the sun. All white dwarfs have about the same mass, usually in the narrow range of 0.5 to about 1.0 solar masses.

Because of the systematic ordering of mass along the main sequence, these main-sequence stars obey a **mass-luminosity relation**—the more massive a star is, the more luminous it is (■ Figure 6-16). In fact, the mass-luminosity relation can be expressed as: Luminosity is proportional to mass to the 3.5 power. For example, a star with a mass of 4.0 $M_\odot$ can be expected to have a luminosity of about $4^{3.5}$ or 128 $L_\odot$. Giants and supergiants do not follow the mass-luminosity relation very closely, and white dwarfs not at all. In the next chapters, the main sequence mass-luminosity relation will help you understand how main sequence stars generate their energy.

Though mass alone does not reveal any pattern among giants, supergiants, and white dwarfs, density does. Once you know a star's mass and diameter, you can calculate its average density by dividing its mass by its volume. Stars are not uniform in density but are most dense at their centers and least dense near

ThomsonNOW™ Sign in at www.thomsonedu.com and go to **ThomsonNOW** to see Astronomy Exercises "Mass–Luminosity Relation" and "H–R/Mass-Luminosity 3D Graph."

The white dwarfs have masses about equal to the sun's but are very small, only about the size of Earth. Thus, the matter is compressed to densities of 3,000,000 g/cm³ or more. On Earth, a teaspoonful of this material would weigh about 15 tons.

Density divides stars into three groups. Most stars are main-sequence stars with densities similar to the sun's. Giants and supergiants are very-low-density stars, and white dwarfs are high-density stars. You will see in later chapters that these densities reflect different stages in the evolution of stars.

Surveying the Stars

If you want to know what the average person thinks about a certain subject, you take a survey. If you want to know what the average star is like, you can survey the stars. Such surveys reveal important relationships among the family of stars. Over the years, many astronomers have added their results to the growing collection of data on star distances, luminosities, temperatures, sizes,

■ **Figure 6-15**

The masses of the plotted stars are labeled on this H–R diagram. Notice that the masses of main-sequence stars decrease from top to bottom but that masses of giants and supergiants are not arranged in any ordered pattern.

their surface. The center of the sun, for instance, is about 100 times as dense as water; its density near the photosphere is about 3400 times less dense than Earth's atmosphere at sea level. A star's average density is intermediate between its central and surface densities. The sun's average density is approximately 1 gram per cubic centimeter—about the density of water.

Main-sequence stars have average densities similar to the sun's density. As you learned earlier in the discussion about luminosity classification, giant stars are much larger in diameter than the main-sequence stars but not much larger in mass, so giants have low average densities, ranging from 0.1 to 0.01 g/cm³. The enormous supergiants have still lower densities, ranging from 0.001 to 0.000001 g/cm³. These densities are thinner than the air you breathe, and if you could insulate yourself from the heat, you could fly an airplane through these stars. Only near the center would you be in any danger; scientists can calculate that the material there is very dense.

■ **Figure 6-16**

The mass–luminosity relation shows that the more massive a main-sequence star is, the more luminous it is. The open circles represent white dwarfs, which do not obey the relation.

and masses. They can now analyze those data to search for relationships between these and other parameters.

As the 21st century begins, astronomers are deeply involved in extensive surveys. Remember the earlier mention of the Hipparcos satellite that surveyed the entire sky measuring the parallax of over a million stars? Powerful computers to control instruments and analyze data make such immense surveys possible. For example, the Sloan Digital Sky Survey will eventually map a quarter of the sky, measuring the position and brightness of 100 million stars and galaxies. Also, the Two Micron All Sky Survey (2MASS) has mapped the entire sky at three near-infrared wavelengths. A number of other sky surveys are underway. Astronomers will "mine" these mountains of data for decades to come.

What could you learn about stars from a survey of the stars near the sun? The evidence astronomers have is that the sun is in a typical place in the universe. Therefore, such a survey could reveal general characteristics of the stars. Study **The Family of Stars** on pages 120–121 and notice three important points:

1 Taking a survey is difficult because you must be sure to get an honest sample. If you don't survey enough stars, your results can be biased.

2 Very luminous stars are rare. M dwarfs and white dwarfs are so faint they are difficult to find even near Earth and may be undercounted in surveys.

3 Luminous stars, although they are rare, are easily visible even at great distances. Typical nearby stars have lower luminosity than our sun.

The night sky is a beautiful carpet of stars. Some are giants and supergiants, and some are dwarfs. The family of stars is rich in its diversity. Pretend space is like a very clear ocean in which you are swimming. When you look at the night sky you are seeing mostly "whales," the rare very large and luminous stars, mostly far away. If instead you cast a net near your location, you would mostly catch "sardines," very-low-luminosity M dwarfs that make up most of the stellar population of the universe.

In this chapter you explored the basic properties of stars. Once you found the distance to the stars, you were able to find their luminosities. Knowing their luminosities and temperatures gives you their diameters. Studying binary stars gives you their masses. These are all rather mundane data. But you have now discovered a puzzling situation. The largest and most luminous stars are so rare you might joke that they hardly exist, and the average stars are such small low-mass things they are hard to see even if they are near Earth in space. Why does nature make stars in this peculiar way? To answer that question, you must explore the birth, life, and death of stars. That quest begins in the next chapter.

Building Scientific Arguments

What kind of stars do you see if you look at a few of the brightest stars in the sky?
When you look at the night sky, the brightest stars are mostly giants and supergiants. Most of the bright stars in the constellation Canis Major, for instance, are supergiants. Sirius, the brightest star in the sky, is in Canis Major, but it is a main-sequence star; it looks bright because it is very nearby, not because it is very luminous. In general, the supergiants and giants are so luminous that they stand out and look bright, even though they are not nearby. When you look at a bright star in our sky, you are probably looking at a highly luminous star—a supergiant or a giant.

What kind of star do you see if you examine a few of the nearest stars to the sun?

■   ■   ■

The Family of Stars

1 What is the most common kind of star? Are some rare? Are some common? To answer those questions you must survey the stars. To do so you must know their spectral class, their luminosity class, and their distance. Your census of the family of stars produces some surprising demographic results.

1a You could survey the stars by observing every star within 62 pc of Earth. A sphere 62 pc in radius encloses a million cubic parsecs. Such a survey would tell you how many stars of each type are found within a volume of a million cubic parsecs.

62 pc

Earth

2 Your survey faces two problems.

1. The most luminous stars are so rare you find few in your survey region. There are no O stars at all within 62 pc of Earth.

2. Lower-main-sequence M stars, called red dwarfs, and white dwarfs are so faint they are hard to locate even when they are only a few parsecs from Earth. Finding every one of these stars in your survey sphere is a difficult task.

Spectral Class Color Key

- O and B
- A
- F
- G
- K
- M

The star chart in the background of these two pages shows most of the constellation Canis Major; stars are represented as dots with colors assigned according to spectral class. The brightest stars in the sky tend to be the rare, highly luminous stars, which look bright even though they are far away. Most stars are of very low luminosity, so nearby stars tend to be very faint red dwarfs.

Red dwarf
15 pc

o^2 Canis Majoris
B3Ia   790 pc

Red dwarf
17 pc

δ Canis Majoris
F8Ia   550 pc

σ Canis Majoris
M0Iab   370 pc

η Canis Majoris
B5Ia   980 pc

ϵ Canis Majoris
B2II   130 pc

In this histogram, bars rise from an H–R diagram to represent the frequency of stars in space.

Stars per 10^6 pc^3

Red dwarfs and white dwarfs are the most common kinds of stars.

Red dwarfs

Supergiants

Giants

Main sequence

White dwarfs

2a O and B stars, super-giants, and giants are so rare their bars are not visible in this graph.

Sirius A (α Canis Majoris) is the brightest star in the sky. With a spectral type of A1V, it is not a very luminous star. It looks bright because it is only 2.6 pc away.

Sirius B is a white dwarf that orbits Sirius A. Although Sirius B is not very far away, it is much too faint to see with the unaided eye.

3 Luminous stars are rare but are easy to see. Most stars are very low luminosity. See H–R diagrams at right.

Brightest stars

Supergiants

Main sequence

Giants

Sun

The brightest stars in the sky tend to be highly luminous stars — upper-main-sequence stars, giants, or supergiants. They look bright because they are luminous, not because they are nearby.

Nearest stars

The nearest stars in space tend to be very faint stars — lower-main-sequence red dwarfs or white dwarfs. Nearly all of these stars are faint in the sky even though they are nearby. Only a few are visible to the unaided eye.

Main sequence

Giants

Sun

White dwarfs

Red dwarfs

Temperature (K)

Although white dwarfs and red dwarfs are very common, not a single one is bright enough to be visible to the unaided eye.

Summary

6-1 | Star Distances

How far away are the stars?

■ Distance is critical in astronomy. Finding the luminosities, diameters, and masses of stars requires first finding their distances.

■ Astronomers can measure the distance to nearby stars by observing their parallaxes. Stellar distances are commonly expressed in parsecs. One parsec is 206,265 AU—the distance to an imaginary star whose parallax is 1 arc second.

■ Stars farther away than about 170 pc have parallaxes too small to measure from ground-based observatories.

6-2 | Apparent Brightness, Intrinsic Brightness, and Luminosity

How much energy do stars make?

■ Once you know the distance to a star, you can find its intrinsic brightness, which can be expressed as its absolute magnitude. The absolute magnitude of a star equals the apparent magnitude it would have if it were 10 pc away.

■ The luminosity of a star, which can be found from its absolute magnitude, is a measure of the total energy radiated by the star. Luminosity can be expressed in Watts, but luminosities of stars are often given in units of the luminosity of the sun.

6-3 | Star Temperatures

How can you tell a star's temperature using its spectrum?

■ The hydrogen Balmer lines are weak in cool stars because atoms are not excited out of the ground state. In hot stars, the Balmer lines are weak because atoms are excited to higher orbits or are ionized. Only at intermediate temperatures are the Balmer lines strong.

■ The strengths of many spectral lines in a star's spectrum can be used to tell you its temperature. Stars are classified in the temperature spectral sequence: O, B, A, F, G, K, M.

■ Long after the spectral sequence was created, astronomers discovered L and T objects with temperatures even cooler than the M stars.

6-4 | Star Sizes

How big are stars?

■ The H–R diagram is a plot of luminosity versus surface temperature. It is an important graph in astronomy because it sorts the stars into categories by size.

■ Roughly 90 percent of normal stars, including the sun, fall on the main sequence, with the hotter main-sequence stars being more luminous. The giants and supergiants, however, are much larger and lie above the main sequence in the diagram. Some of the white dwarfs are hot stars, but they fall below the main sequence because they are so small.

■ The large size of the giants and supergiants means their atmospheres have low densities. Giant stars, luminosity class III, have narrow spectral lines, and supergiants, class I, have extremely narrow lines. Class V main-sequence stars have relatively broad spectral lines.

■ Stars so far away that their parallaxes are too small to measure can have their distances estimated by the technique of spectroscopic parallax.

6-5 | Star Masses—Binary Stars

How much matter do stars contain?

■ The only direct way you can find the mass of a star is by studying binary stars. When two stars orbit a common center of mass, astronomers find their masses by observing the period and sizes of their orbits.

■ Few binary star systems are easy to analyze; most are spectroscopic binaries in which the component stars are known only by the alternating Doppler shifts of their spectral lines.

■ Eclipsing binary star systems allow measurement not just of the component stars' masses but also independent checks on their temperatures and diameters.

6-6 | Typical Stars

What is the typical star like?

■ Given the mass and diameter of a star, you can find its average density. On the main sequence, the stars are about as dense as the sun, but the giants and supergiants are very-low-density stars. Some are much thinner than air. The white dwarfs, lying below the main sequence, are tremendously dense.

■ The mass–luminosity relation says that the more massive a main sequence star is, the more luminous it is. Giants and supergiants do not follow this relation in general, and white dwarfs not at all.

■ A survey of stars in the neighborhood of the sun shows that giants and supergiants are rare and that red dwarfs, stars at the low-luminosity end of the main sequence, are the most common type but are also faint and hard to find.

New Terms

stellar parallax *(p)* (p. 104)

parsec (pc) (p. 105)

intrinsic brightness (p. 105)

flux (p. 105)

absolute visual magnitude (M_v) (p. 106)

luminosity *(L)* (p. 106)

spectral classes (p. 108)

spectral sequence (p. 109)

Hertzsprung–Russell (H–R) diagram (p. 110)

main sequence (p. 110)

giant (p. 110)

supergiant (p. 112)

red dwarf (p. 112)

white dwarf (p. 112)

luminosity classes (p. 112)

spectroscopic parallax (p. 112)

binary stars (p. 113)

visual binary system (p. 114)

spectroscopic binary system (p. 115)

eclipsing binary system (p. 115)

light curve (p. 115)

mass–luminosity relation (p. 117)

Review Questions

ThomsonNOW™ Assess your understanding of this chapter's topics with additional quizzing and animations at
www.thomsonedu.com

1. Why are Earth-based parallax measurements limited to the nearest stars?

2. Why was the Hipparcos satellite able to make more accurate parallax measurements than ground-based telescopes?

3. What do the words *absolute* and *visual* mean in the definition of absolute visual magnitude?

4. What does luminosity measure that is different from what absolute visual magnitude measures?

5. Why are hydrogen Balmer lines strong in the spectra of medium-temperature stars and weak in the spectra of hot and cool stars?

6. Explain the connections between Table 6-1 and Figures 6-4, 6-5, and 6-6.

7. Why does the luminosity of a star depend on both its radius and its temperature?

8. How can you be sure that giant stars really are larger than main-sequence stars?

9. Why do astronomers conclude that white dwarfs must be very small?

10. What observations would you make to classify a star according to its luminosity? Why does that method work?

11. Why does the orbital period of a binary star depend on its mass?

12. What observations would you make to study an eclipsing binary star?

13. Why don't astronomers know the inclination of a spectroscopic binary? How do they know the inclination of an eclipsing binary?

14. How do the masses of stars along the main sequence illustrate the mass–luminosity relation?

15. What is the most common kind of star?

16. If you look only at the brightest stars in the night sky, what kind of star are you likely to be observing? Why?

17. If all of the stars in the photo here are members of the same star cluster, then they all have about the same distance. Then why are three of the brightest much redder than the rest? What kind of star are they?

NASA

Discussion Questions

1. If someone asked you to compile a list of the nearest stars to the sun based on your own observations, what measurements would you make, and how would you analyze them to detect nearby stars?

2. Explain how errors in measuring the positions of blurred star images could introduce errors into your calculations of luminosity.

3. The sun is sometimes described as an average star. Is that true? What is the average star really like?

7 | The Structure and Formation of Stars

Jim he allowed [the stars] was made, but I allowed they happened. Jim said the moon could'a laid them; well, that looked kind of reasonable, so I didn't say nothing against it, because I've seen a frog lay most as many, so of course it could be done.

MARK TWAIN
THE ADVENTURES OF HUCKLEBERRY FINN

The stars are not eternal. When you look at the sky, you see hundreds of points of light, and each is an object like the sun held together by its gravity and generating tremendous energy in its core through nuclear reactions. The stars you see tonight are the same stars your parents, grandparents, and great-grandparents saw. Stars change hardly at all in a human lifetime, but they are not eternal. Stars are born, and stars die. This chapter begins that story. ▌ How can you know what the internal processes and life cycles of stars are when you can't see inside them and humans don't live long enough to see them evolve? The answer lies in the methods of science. By constructing theories that describe how nature works and then testing those theories against evidence from observations, you can unravel some of nature's greatest secrets. ▌ In this chapter, you will

▌ Continued on page 126 ▌

Nebula NGC 1333 is filled with newborn stars caught in the act of forming in this false-color infrared image. (NASA/JPL-Caltech/R. A. Gutermuth, Harvard-Smithsonian CfA)

Guidepost

Looking Back

Throughout the preceding chapters you have been building a general understanding of the sky and familiarity with the methods of astronomy. The last chapter provided a detailed family portrait of the stars. You can now describe the basic *external* properties of different kinds of stars.

This Chapter

This is the point where you will really begin to see how the universe works. Here you will be putting together observations and theories to understand how nature makes stars. That will answer six essential questions about stars:

What are the insides of stars like?

How do stars make energy?

What determines the properties of main-sequence stars?

How long can a star survive?

How are stars born?

How do you know that theories of star formation are correct?

Looking Ahead

The most important of the questions above may be the last. Testing theories against evidence is the basic skill required of all scientists, and you will use it over and over in the eight chapters that follow this one. In the next chapter you will learn about what happens at the ends of stars' lives, when they begin to run out of fuel.

see how the flow of energy from inside to the surfaces of the stars balances gravity, making the stars stable, and how nuclear reactions inside stars generate that energy. You will also see how gravity creates new stars from the thin dusty gas between the stars.

(7-1) Stellar Structure

IF THERE IS A SINGLE IDEA IN STELLAR ASTRONOMY that can be called crucial, it is the concept of balance. In this section you will discover that stars are simple, elegant power sources held together by their own gravity balanced by the support of their internal heat and pressure.

Using the basic concept of balance, you can consider the structure of a star. What is meant here by structure is the variation in temperature, density, pressure, and so on between the surface of the star and its center. It will be easier to think about stellar structure if you imagine that the star is divided into concentric shells like those in an onion. You can then discuss the temperature, pressure, and density in each shell. These helpful shells exist only in the imagination; stars do not have separable layers.

The Laws of Conservation of Mass and Energy

The first two laws of stellar structure have something in common—they are both what astronomers and physicists call *conservation* laws. Conservation laws say that certain things cannot be created out of nothing or vanish into nothing. Such conservation laws are powerful aids to help you understand how nature works.

The law of **conservation of mass** is a basic law of nature that can be applied to the structure of stars. It says that the total mass of a star must equal the sum of the masses of its shells. This is like saying the weight of a cake must equal the sum of the weight of its layers.

The law of **conservation of energy** is another basic law of nature. It says that the amount of energy flowing out of the top of a layer in the star must be equal to the amount of energy coming in at the bottom, plus whatever energy is generated within the layer. That means that the energy leaving the surface of the star, its luminosity, must equal the sum of the energies generated in all the layers inside the star. This is like saying that all the new cars driving out of a factory must equal the sum of all the cars made on each of the production lines.

These two laws may seem so familiar or so obvious that they hardly need to be stated, but they are important clues to the structure of stars. The third law of stellar structure is a law about balance.

Hydrostatic Equilibrium

The weight of each layer of a star must be supported by the layer below. (The words *down* or *below* are conventionally used to re-

fer to regions closer to the center of a star.) Picture a pyramid of people in a circus stunt—the people in the top row do not have to hold up anybody else; the people in the next row down are holding up the people in the top row; and so on. In a star that is stable, the deeper layers must support the weight of all of the layers above. The inside of a star is made up of gas, so the weight pressing down on a layer must be balanced by gas pressure in that layer. If the pressure is too low, the weight from above will compress and push down the layer; and if the pressure is too high, the layer will expand and lift the layers above.

This balance between weight and pressure is called **hydrostatic equilibrium.** *Hydro* (from the Greek word for water) tells you the material is a fluid, which by definition includes the gases of a star, and *static* tells you the fluid is stable, neither expanding nor contracting.

■ Figure 7-1 shows this hydrostatic balance in the imaginary layers of a star. The weight pressing down on each layer is shown by lighter red arrows, which grow larger with increasing depth because the weight grows larger. The pressure in each layer is shown by darker red arrows, which must grow larger with increasing depth to support the weight.

The pressure in a gas depends on the temperature and density of the gas. Near the surface, there is little weight pressing down, so the pressure does not need to be high for stability. Deeper in the star, the pressure must be higher, which means that the temperature and density of the gas must also be higher. Hydrostatic equilibrium tells you that stars must have high temperature, pressure, and density inside to support their own weight and be stable.

Although the law of hydrostatic equilibrium can tell you some things about the inner structure of stars, you need one more law to completely describe a star. You need to know how energy flows from the inside to the outside.

ThomsonNOW™ Sign in at **www.thomsonedu.com** and go to ThomsonNOW to see Astronomy Exercise "Hydrostatic Equilibrium."

Energy Transport

The surface of a star radiates light and heat into space and would quickly cool if that energy were not replaced. Because the inside of the star is hotter than the surface, energy must flow outward to the surface, where it radiates away. This flow of energy through each shell determines its temperature, which, as you saw previously, determines how much weight that shell can balance. To understand the structure of a star, you need to understand how energy moves from the center to the surface.

The law of **energy transport** says that energy must flow from hot regions to cooler regions either by conduction, convection, or radiation. Conduction is the most familiar form of heat flow. If you hold the bowl of a spoon in a candle flame, the handle of the spoon grows warmer as heat, in the form of motion among the atoms of the spoon, is conducted from atom to atom up the handle (■ Figure 7-2a). Conduction requires close contact between

Surface

Center

Weight

Pressure

Figure 7-1

The law of hydrostatic equilibrium says the pressure in each layer of a stable star balances the weight pressing down on that layer. Thus, as the weight increases from the surface of a star to its center, the pressure also increases.

Conduction

Convection

Radiation

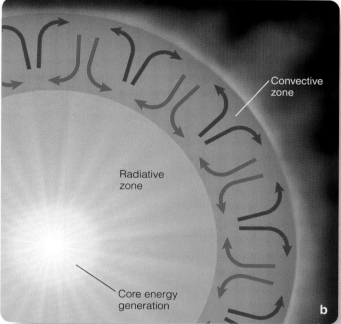

Convective zone

Radiative zone

Core energy generation

b

Active Figure 7-2

(a) The three modes by which energy may be transported from the flame of a candle. (b) Two modes of energy transport within a star like the sun.

ThomsonNOW Sign in at www.thomsonedu.com and go to ThomsonNOW to see Active Figure "Conduction, Convection, and Radiation."

the atoms, so it is not significant in normal stars that do not have high enough densities for conduction even in their centers.

The transport of energy by radiation is another familiar experience. Put your hand near a candle flame, and you can feel the heat. What you actually feel are infrared photons—packets of energy—radiated by the flame and absorbed by your hand (Figure 7-2a). Radiation is the principal means of energy transport in the interiors of most stars. Photons are absorbed and reemitted in random directions over and over as energy works its way from the hot interior toward the cooler surface.

The flow of energy by radiation depends on how difficult it is for photons to move through the gas. The **opacity** of the gas, its resistance to the flow of radiation, depends strongly on its temperature: A hot gas is more transparent than a cool gas. If the opacity is high, radiation cannot flow through the gas easily, and it backs up like water behind a dam. When enough heat builds up, the gas begins to churn as hot gas rises upward and cool gas sinks downward. This heat-driven circulation of a fluid is convection, the third way energy can move in a star. You are familiar with convection; the

Quantitative Thinking with Mathematical Models

One of the most powerful tools in science is the mathematical model, a group of equations carefully designed to describe the behavior of the object that scientists want to study. Astronomers make mathematical models of stars using only four equations, but other systems are much more complicated and may require many more equations.

Many sciences use mathematical models. Medical scientists have created mathematical models of the nerves that control the heart, and physicists have made mathematical models of the inside of an atomic nucleus. Economists have mathematical models of certain aspects of systems such as the municipal bond market, and Earth scientists have mathematical models of Earth's atmosphere. In each case, the mathematical model allows the scientists to study something that is difficult to study in reality. The model can reveal regions scientists cannot observe, speed up a slow process, slow down a fast process, or allow scientists to perform experiments that would be impossible in reality. Astronomers, for example, can change the abundance of different chemical elements in a model star to see how its structure depends on its composition. Mathematical modeling is one of the most important research methods in astronomy, even though it often requires exceptionally large and fast computers.

As is true for any scientific model, a mathematical model is only as reliable as the assumptions that go into its creation. In Window on Science 2-1, you saw that the celestial sphere is an adequate model of the sky for some purposes but doesn't work if you try to extend it too far. The same is true of mathematical models. You can think of a mathematical model as a numerical expression of one or more theories. Although such models can be very helpful, they are always based on theories and assumptions and so must be compared with the real world at every opportunity. As always in science, there is no substitute for a careful comparison with reality.

---

rising wisp of smoke above a candle flame is carried by convection. Energy is carried upward in these convection currents as rising hot gas (red in Figure 7-2) and also as sinking cool gas (blue in Figure 7-2). Figure 7-2b shows a cross-section of the sun in which zones of radiative and convective energy transport are indicated.

The four laws of stellar structure are summarized in ■ Table 7-1. These laws, properly understood, can tell you how stars are born, how they live, and how they die.

Stellar Models

The laws of stellar structure, described in general terms in the previous sections, can be written as mathematical equations. By solving those equations, astronomers can make a mathematical model of the inside of a star. To start one of these models requires knowing the type of information you learned in Chapter 6 about the external properties of a star: its diameter, luminosity, mass, surface temperature, atmospheric density, and composition.

If you want to build a model of a star, you could divide the star into, say, 100 concentric shells and then write down the four equations of stellar structure for each shell. You would then have 400 equations that would have 400 unknowns, namely, the temperature, density, mass, and energy flow in each shell. Solving 400 equations simultaneously is not easy, and the first such solutions, done by hand before the invention of electronic computers, took months of work. Now a properly programmed computer can solve the equations in a few seconds and print a table of numbers that represent the conditions in each shell of the star. Such a table is called a **stellar model** (Window on Science 7-1).

The table shown in ■ Figure 7-3 is a model of the sun. The bottom line, for radius equal to 0.00, represents the center of the sun, and the top line, for radius equal to 1.00, represents the surface. The other lines tell you the temperature and density in each shell, the mass inside each shell, and the fraction of the sun's luminosity flowing outward through the shell. The bottom line tells you that the temperature at the center of the sun must be over 15 million Kelvin in order for the sun to be stable. At such a high temperature, the gas is very transparent, and energy flows as radiation. Nearer the surface, the temperature is lower, the gas is more opaque, and energy moves by convection. This is confirmed by evidence of convection visible on the sun's surface (see Chapter 5).

Stellar models also let you look into a star's past and future. In fact, you could use models as time machines to follow the evolution of stars over billions of years. To look into a star's future, for instance, you could use a stellar model to determine how fast the star uses its fuel in each shell. As the fuel is consumed, the chemi-

---

> ### ■ Table 7-1 | The Four Laws of Stellar Structure

I. Conservation of mass	Total mass equals the sum of shell masses.
II. Conservation of energy	Total luminosity equals the sum of energy generated in each shell.
III. Hydrostatic equilibrium	The weight on each layer is balanced by the pressure in that layer.
IV. Energy transport	Energy moves from hot to cool regions by conduction, radiation, or convection.

$R/R_\odot$	T (10^6 K)	Density (g/cm3)	$M/M_\odot$	$L/L_\odot$
1.00	0.006	0.00	1.00	1.00
0.90	0.60	0.009	0.999	1.00
0.80	1.2	0.035	0.996	1.00
0.70	2.3	0.12	0.990	1.00
0.60	3.1	0.40	0.97	1.00
0.50	4.9	1.3	0.92	1.00
0.40	5.1	4.1	0.82	1.00
0.30	6.9	13.	0.63	0.99
0.20	9.3	36.	0.34	0.91
0.10	13.1	89.	0.073	0.40
0.00	15.7	150.	0.000	0.00

$$\frac{dM}{dr} = 4\pi r^2 \rho$$

$$\frac{dL}{dr} = 4\pi r^2 \rho e$$

$$\frac{dP}{dr} = -\frac{GM}{r^2}\rho$$

$$\frac{dT}{dr} = \frac{-3}{16\pi ac}\frac{\bar{\kappa}\rho}{T^3}\frac{L}{r^2}$$

Surface

Convective zone

Radiative zone

Center

■ Figure 7-3

A stellar model is a table of numbers that represent conditions inside a star. Such tables can be computed using the four laws of stellar structure, shown here in mathematical form. The table in this figure describes the sun. (Illustration design by M. A. Seeds.)

cal composition of the gas changes, the opacity changes, and the amount of energy generated declines. By calculating the rate of these changes, you could predict what the star will look like at some time in the future. You could repeat that process over and over and follow the evolution of the star step-by-step as it ages.

Stellar astronomy has made great advances since the 1960s, when high-speed computers began to be available that could calculate models of the structure and evolution of stars. The summary of star formation later in this chapter is based on thousands of stellar models. You will continue to rely on theoretical models as you study main-sequence stars later in this chapter and the deaths of stars in the next chapter.

7-2 Nuclear Fusion in the Sun and Stars

ASTRONOMERS OFTEN USE THE WRONG WORDS to describe energy generation in the sun and stars. Astronomers will say, "The star ignites hydrogen burning." You normally use the word *ignite* to mean "catch on fire" and *burn* to mean "consume by fire." What goes on inside stars isn't really burning in the usual sense.

The sun is a normal star, and it is powered by nuclear reactions that occur near its center. Later in this section you will learn some of the evidence that this is so. The energy produced keeps a star's in-

terior hot, and the gas is totally ionized. That is, the electrons are not attached to the atomic nuclei, and the gas is a soup of rapidly moving particles colliding with each other at high velocity.

When astronomers discuss nuclear reactions inside stars, they refer to atomic nuclei and not to atoms. How exactly can the nucleus of an atom yield energy? The answer lies in the forces that hold the nuclei together.

Nuclear Binding Energy

The sun generates its energy by breaking and reconnecting the bonds between the particles *inside* atomic nuclei. That is quite different from the way we generate energy by burning wood in a fireplace. The process of burning extracts energy by breaking and reconnecting chemical bonds between atoms in the fuel. Chemical bonds are formed by electrons on the outsides of atoms, and you saw in Chapter 5 that the electrons are bound to the atoms by the electromagnetic force. Thus chemical energy originates in the electromagnetic force.

There are only four known forces in nature: the force of gravity, the electromagnetic force, and the strong and weak **nuclear forces.** The weak nuclear force is involved in the radioactive decay of certain kinds of nuclear particles, and the strong force binds together atomic nuclei. Thus nuclear energy originates in the strong nuclear force.

Nuclear power plants on Earth generate energy through **nuclear fission** reactions that split uranium nuclei into less massive fragments. A uranium nucleus contains a total of 235 protons and neutrons, and it splits into a range of fragments containing roughly half as many particles. Because the fragments produced are more tightly bound than the uranium nuclei, energy is released during uranium fission (■ Figure 7-4).

Stars make energy in **nuclear fusion** reactions that combine light nuclei into heavier nuclei. The most common reaction, including that in the sun, fuses hydrogen nuclei (single protons) into helium nuclei (two protons and two neutrons). Because the nuclei produced are more tightly bound than the original nuclei, energy is released. Notice in Figure 7-4 that both fusion and fission reactions move downward in the diagram, meaning they release the binding energy of nuclei and create more tightly bound nuclei.

Hydrogen Fusion

The sun fuses four hydrogen nuclei to make one helium nucleus. Because one helium nucleus contains 0.7 percent less mass than four hydrogen nuclei, some mass vanishes in the process. In fact, that mass is converted to energy, and you could figure out how much by using Einstein's famous equation, $E = mc^2$. For example, converting 1 kg (2.2 lb) of matter completely into energy would produce an enormous amount of energy, 9×10^{16} Joules, comparable to the amount of energy a big city like Philadelphia uses in a year.

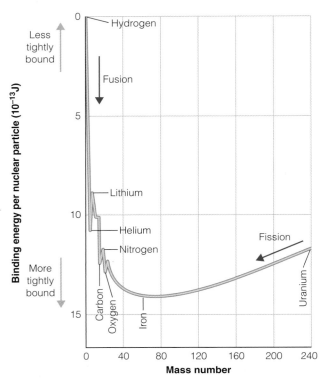

■ Figure 7-4

The red line in this graph shows the binding energy (the energy that holds an atomic nucleus together) for different types of atoms as it depends on atomic mass (the number of protons and neutrons in an atom's nucleus). Both fission and fusion nuclear reactions convert fuel nuclei to product nuclei that are lower in the diagram (arrows), with more tightly bound nuclei. Iron has the most tightly bound nucleus, so no nuclear reactions can begin with iron and release energy.

Making one helium nucleus releases only a small amount of energy, hardly enough to raise a housefly 1/40 of a millimeter into the air. Only by concentrating many reactions in a small region can nature produce significant amounts of energy. The sun has a voracious appetite and needs 10^{38} reactions per second, transforming 5 million tons of mass into energy every second, just to replace the energy pouring into space from its surface. That might sound as if the sun is losing mass at a furious rate, but in its entire 10-billion-year lifetime, the sun will convert less than 0.07 percent of its mass into energy.

Fusion reactions can occur only when the nuclei of two atoms get very close to each other. Because atomic nuclei carry positive charges, they repel each other with an electrostatic force called the Coulomb force. Physicists commonly refer to this repulsion between nuclei as the **Coulomb barrier.** To overcome this barrier, atomic nuclei must have violent collisions that are rare unless the gas is very hot, in which case the nuclei move at high speeds. (You recall that an object's temperature is just a measure of the average speed with which its particles move.)

Thus nuclear reactions in the sun take place only near the center, where the gas is hot and dense. A high temperature ensures that collisions between nuclei are violent enough to overcome the Coulomb barrier, and a high density ensures that there are enough collisions, and thus enough reactions, to produce enough energy to keep the sun stable.

You can symbolize this process with a simple nuclear reaction:

$$4\ ^1\mathrm{H} \longrightarrow {}^4\mathrm{He} + \text{energy}$$

In this equation, $^1\mathrm{H}$ represents a proton, the nucleus of a hydrogen atom, and $^4\mathrm{He}$ represents the nucleus of a helium atom. The superscripts indicate the total numbers of protons and neutrons in each nucleus.

The actual steps in the process are more complicated than this simple summary suggests. Instead of waiting for four hydrogen nuclei to collide simultaneously, which would be a highly unlikely event, the process can proceed by steps in a series of reactions called the proton–proton chain.

The **proton–proton chain** is a series of three nuclear reactions that builds a helium nucleus by adding protons one at a time. This process is efficient at temperatures above 10,000,000 K. The sun, for example, manufactures its energy in this way. Recall from the previous section that models of the interior of the sun based on its overall stability indicate the central temperature is about 15,000,000 K.

The three reactions in the proton–proton chain are:

$$^1\mathrm{H} + {}^1\mathrm{H} \longrightarrow {}^2\mathrm{H} + e^+ + \nu$$
$$^2\mathrm{H} + {}^1\mathrm{H} \longrightarrow {}^3\mathrm{He} + \gamma$$
$$^3\mathrm{He} + {}^3\mathrm{He} \longrightarrow {}^4\mathrm{He} + {}^1\mathrm{H} + {}^1\mathrm{H}$$

In the first reaction, two hydrogen nuclei (two protons) combine, and one changes into a neutron, to result in a heavy hydrogen nucleus called **deuterium,** while emitting a particle called a *positron* (a positively charged electron, symbolized by *e+*) and another called a **neutrino** (*ν*). In the second reaction, the heavy hydrogen nucleus absorbs another proton and, with the emission of a gamma ray (*γ*), becomes a lightweight helium nucleus (^{3}He). Finally, two light helium nuclei combine to form a normal helium nucleus and two hydrogen nuclei. Because the last reaction needs two ^{3}He nuclei, the first and second reactions must each occur twice for each ^{4}He produced (■ Figure 7-5). The net result of this chain reaction is the transformation of four hydrogen nuclei into one helium nucleus plus energy.

All main-sequence stars fuse hydrogen into helium to generate energy. The sun and smaller stars fuse hydrogen by the proton–proton chain. Upper-main-sequence stars, more massive than the sun, fuse hydrogen by a more efficient process called the **CNO (carbon-nitrogen-oxygen) cycle.** The CNO cycle begins with a carbon nucleus and transforms it first into a nitrogen nucleus, then into an oxygen nucleus, and then back to a carbon nucleus. The carbon is unchanged in the end, but along the way four hydrogen nuclei are fused to make a helium nucleus plus energy, just as in the proton–proton chain. A carbon nucleus has six times more positive electric charge than hydrogen, so the Coulomb barrier is higher than for combining two protons. Temperatures higher than 16,000,000 K are required to make the CNO cycle work: The center of the sun is not quite hot enough for this reaction. In stars more massive than about 1.1 solar masses the cores are hot enough, and the CNO cycle dominates over the slower proton–proton chain.

In both the proton–proton chain and the CNO cycle, energy appears in the form of gamma rays, positrons, and neutrinos. The gamma rays are photons that are absorbed by the surrounding gas before they can travel more than a few centimeters. This heats the gas. The positrons produced in the first reaction combine with free electrons, and both particles vanish, converting their mass into more gamma rays. Thus, the positrons also help keep the center of the star hot. The neutrinos, however, are particles that travel at nearly the speed of light and almost never interact with other particles. The average neutrino could pass unhindered through a lead wall more than 1 light-year thick. Consequently, the neutrinos do not help heat the gas but race out of the star, carrying away roughly 2 percent of the energy produced.

It is time to ask the critical question that lies at the heart of science: What is the evidence to support this theoretical explanation of how the sun and other stars make energy? The search for that evidence will introduce you to one of the great triumphs of modern astronomy.

ThomsonNOW™ Sign in at **www.thomsonedu.com** and go to Thomson-NOW to see Astronomy Exercise "Nuclear Fusion."

Neutrinos from the Sun's Core

Nuclear reactions in the sun's core produce floods of neutrinos that rush out of the sun and off into space. If you could detect these neutrinos, you could probe the sun's interior.

Because neutrinos almost never interact with atoms, neutrinos are extremely hard to detect. You never feel the flood of over 10^{12} solar neutrinos that flows through your body every second. Even at night, neutrinos from the sun rush through Earth as if it weren't there, up through your bed, through you, and onward into space. Certain particle reactions, however, can be triggered by a neutrino of the right energy.

Beginning in the 1960s astronomers found several ways to detect solar neutrinos. Initial results, however, detected too few neutrinos—about one-third as many as predicted by models of the sun's interior. Later experiments were finally able to solve the puzzle and confirm that the original models were correct. Physicists know of three kinds of neutrinos, which they call "flavors." The first neutrino experiments could detect (or taste) only one flavor, electron-neutrinos. Theory hinted that the electron-neutrinos produced in the core of the sun might change back and forth among the three flavors as they rushed out through the sun and across space to Earth. Observations begun in 2000 confirm this explana-

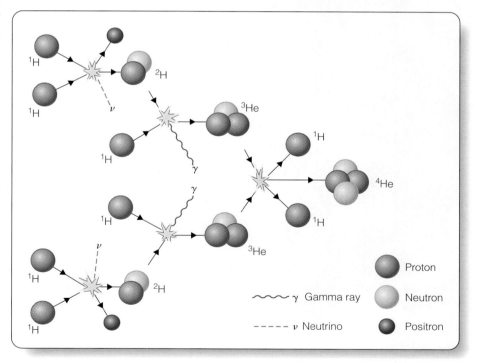

■ **Figure 7-5**

The proton–proton chain combines four protons (at left) to produce one helium nucleus (at right). Energy appears as gamma rays, positrons that combine with electrons to convert their mass into energy, and neutrinos. Neutrinos escape, carrying away about 2 percent of the energy.

tion (■ Figure 7-6). Two-thirds of the electron neutrinos produced in the sun transform into muon- and tau-neutrinos, which most detectors cannot count, before they reach Earth. You can be confident that models of the sun's interior are basically correct because they predict the true rate of proton–proton fusion in the sun's core.

The Pressure–Temperature Thermostat

Nuclear reactions in stars manufacture energy and heavy atoms under the supervision of a natural thermostat that keeps the reactions from erupting out of control. That thermostat is the relation between gas pressure and temperature.

In a star, the nuclear reactions generate just enough energy to balance the inward pull of gravity. Consider what would happen if the reactions began to produce too much energy. The star balances gravity by generating energy, so the extra energy flowing out of the star would force it to expand. The expansion would lower the central temperature and density and slow the nuclear reactions until the star regained stability. Thus the star has a built-in regulator that keeps the nuclear reactions from occurring too rapidly.

The same thermostat keeps the reactions from dying down. Suppose the nuclear reactions began making too little energy.

■ Figure 7-6

The Sudbury Neutrino Observatory is a globe 12 meters in diameter containing water rich in deuterium in place of ordinary hydrogen. Buried 2100 m (6800 ft) deep in an Ontario mine, it can detect all three flavors of neutrinos and confirms that neutrinos change back and forth between the flavors. (Photo courtesy of SNO.)

Then the star would contract slightly, increasing the central temperature and density and increasing the nuclear energy generation.

The overall stability of a star depends on the relation between gas pressure and temperature. If the material of the star has the property of normal gases—for which an increase or decrease in temperature produces a corresponding change in pressure—then the nuclear reaction pressure–temperature thermostat can function properly and contribute to the stability of the star. In the next section of this chapter you will see how this thermostat accounts for the relation between mass and luminosity for main-sequence stars.

Building Scientific Arguments

Why does nuclear fusion require that the gas be very hot? Only under certain conditions do the nuclei of atoms fuse (combine) together to form a new nucleus. Inside a star, the gas is ionized, which means the electrons have been stripped off the atoms and the nuclei are bare and have a positive charge. For hydrogen fusion, the nuclei are single protons.

These atomic nuclei repel each other because of their positive charges, so they must collide with each other violently to overcome that repulsion and get close enough together to fuse. Nuclear fusion requires that the gas have a very high temperature so that the atoms are moving rapidly. If the gas is cooler than about 5 million K, hydrogen can't fuse because the protons don't collide violently enough to overcome the repulsion of their positive charges.

The sun requires high temperature to make energy by nuclear fusion. **Why does it also need high density?**

■ ■ ■

7-3 Main-Sequence Stars

YOU CAN UNDERSTAND THE DEEPEST SECRETS OF THE STARS by looking at the most obvious feature of the H–R diagram—the main sequence. Roughly 90 percent of all the stars that make energy by nuclear fusion are main-sequence stars, and they obey a simple relationship that is the key to understanding the life and death of stars.

The Mass–Luminosity Relation

You learned in Chapter 6 that observations of the temperature and luminosity of stars show that main-sequence stars obey a simple rule—the more massive a star is, the more luminous it is. That rule is the key to understanding the stability of main-sequence stars. In fact, the mass–luminosity relation is predicted by the theories of stellar structure, giving astronomers direct observational confirmation of those theories.

To understand the mass–luminosity relation, you must consider both the law of hydrostatic equilibrium, which says that pressure balances weight, and the pressure–temperature thermostat, which regulates energy production. A star that is more massive than the sun has more weight pressing down on its interior, so the interior must have a high pressure to balance that weight. That means the massive star's automatic pressure–temperature thermostat must keep the gas in its interior hot and the pressure high. A star less massive than the sun has less weight on its interior and thus needs less internal pressure; therefore, its pressure–temperature thermostat is set lower. In other words, massive stars are more luminous because they must make more energy to support their larger weight. If they were not so luminous, they would not be stable.

Brown Dwarfs

The mass–luminosity relation also tells you why the main sequence has a lower end, a minimum mass. Objects with masses less than about 0.08 of the sun's mass cannot raise their central temperature high enough to begin hydrogen fusion. Called **brown dwarfs,** such objects are only about ten times the diameter of Earth—about the

same size as Jupiter—and may still be warm from the processes of formation, but they do not generate energy by hydrogen fusion. They have contracted as much as they can and are slowly cooling off.

Brown dwarfs fall in the gap between low-mass M stars and massive planets like Jupiter. They would look dull orange or red to your eyes—which is why they are labeled "brown"—because they emit most of their energy in the infrared. The warmer brown dwarfs fall in spectral class L and the cooler in spectral class T (see Chapter 6). Because they are so small and cool, brown dwarfs are very low-luminosity objects and difficult to find. Nevertheless, hundreds are known, and they may be as common as M stars.

The Life of a Main-Sequence Star

While a star is on the main sequence, it is stable, so you might think its life would be totally uneventful. But a main-sequence star balances its gravity by fusing hydrogen, and as the star gradually uses up its fuel, that balance must change. Thus, even the stable main-sequence stars are changing as they consume their hydrogen fuel.

Recall that hydrogen fusion combines four nuclei into one. As a main-sequence star fuses its hydrogen, the total number of particles in its interior decreases. Each newly made helium nucleus can exert only the same pressure as a hydrogen nucleus. Because the gas has fewer nuclei, its total pressure is less. This unbalances the gravity–pressure stability, and gravity squeezes the core of the star more tightly. As the core contracts, its temperature increases and the nuclear reactions run faster, releasing more energy. This additional energy flowing outward forces the outer layers to expand. Therefore, as a main-sequence star slowly turns hydrogen into helium in its core, the core contracts and heats up while the outer parts of the star become larger, the star becomes more luminous, and its surface cools down.

These gradual changes during the lifetimes of main-sequence stars mean that the main sequence is not a sharp line across the H–R diagram but rather a band. Stars begin their stable lives fusing hydrogen on the lower edge of this band, which is known as the **zero-age main sequence (ZAMS),** but gradual changes in luminosity and surface temperature move the stars upward and slightly to the right, as shown in ■ Figure 7-7. By the time they reach the upper edge of the main sequence, they have exhausted nearly all the hydrogen in their centers. If you precisely measure and plot the luminosity and temperature of main-sequence stars on the H–R diagram, you will find them at various positions within the main-sequence band, indicating how much hydrogen has been converted to helium in their cores. Therefore, you can use the position of a star in the band, combined with stellar evolution models, as one way to estimate the star's age.

The sun is a typical main-sequence star; and, as it undergoes these gradual changes, Earth will suffer. When the sun began its main-sequence life about 5 billion years ago, it was only about 70 percent as luminous as it is now, and by the time it leaves the main sequence in another 5 billion years, the sun will have twice its

■ Figure 7-7

Newborn stars reach stability at the lower edge of the main sequence, the zero-age main sequence (ZAMS). As a star converts hydrogen in its core into helium, it moves slowly across the main sequence, becoming slightly more luminous and slightly cooler. Once a star consumes all of the hydrogen in its core, it can no longer remain a stable main-sequence star. More massive stars age rapidly, but less massive stars use up the hydrogen in their cores more slowly and have longer main-sequence lives.

present luminosity. Long before that, the rising luminosity of the sun will raise Earth's average temperature, melt the polar caps, modify Earth's climate, and ultimately boil away and destroy Earth's oceans. Life on Earth will probably not survive these changes in the sun, but humans have a billion years or more to prepare.

The average star spends 90 percent of its life on the main sequence. This explains why 90 percent of all true stars are main-sequence stars—you are most likely to see a star during that long, stable period while it is on the main sequence. The number of years a star spends on the main sequence depends on its mass (■ Table 7-2). Massive stars consume fuel rapidly and live short lives, but low-mass stars conserve their fuel and shine for billions of years. For example, a 25-solar-mass star will exhaust its hydrogen and die in only about 7 million years. The sun has enough fuel to last about 10 billion years. The red dwarfs, although they have little fuel, use it up very slowly and may be able to survive for 100 billion years or more.

ThomsonNOW™ Sign in at www.thomsonedu.com and go to ThomsonNOW to see Astronomy Exercise "Mass–Lifetime Relation."

Building Scientific Arguments

Why is there a mass–luminosity relation?
Sometimes building a scientific argument to answer a simple question can be highly revealing. Observational evidence shows that the luminosity of a main-sequence star depends on its mass: The more massive a star is, the more luminous it is. You can understand this if you remember that a star supports itself by being hot inside. The high temperature produces a high pressure, which balances the weight of the layers pressing inward. This is called hydrostatic equilibrium. A massive star has a stronger gravitational field, and it has more mass to support, so the pressure needed to keep it from collapsing is very high. That means the internal temperature must be high, and that means the star will generate energy rapidly. All that energy, leaking outward through to the sur-

face of the star and into space, makes the star more luminous.

Now extend your argument. **Why do massive stars live shorter lifetimes than low-mass stars?**

■   ■   ■

7-4 The Birth of Stars

THE KEY TO UNDERSTANDING STAR FORMATION is the correlation between young stars, short-lived stars, and interstellar clouds of gas and dust. Where you find the youngest groups of stars, as well as stars with very short lifetimes, you also find large clouds of gas and dust illuminated by the hottest and brightest of the new stars (■ Figure 7-8). This should lead you to suspect that stars form from such clouds, just as raindrops condense from the water vapor in a thundercloud.

The Interstellar Medium

A common misperception is to imagine that space is empty—a vacuum. In fact, if you glance at Orion's sword on a winter evening, you can see the Great Nebula in Orion, a glowing cloud of gas and dust. That cloud and others show that the space between the stars is filled with low-density gas and dust called the **interstellar medium (ISM)**.

About 75 percent of the mass of interstellar gas is hydrogen, and 25 percent is helium; there are also traces of carbon, nitrogen, oxygen, calcium, sodium, and heavier atoms. Notice that this is very similar to the composition of the sun (see Chapter 5) and other stars. Roughly 1 percent of the ISM's mass is made up of microscopic particles called **interstellar dust.** The dust particles are about the size of the particles in cigarette smoke. The dust seems to be made mostly of carbon and silicates (rocklike minerals) mixed with or coated with frozen water plus some organic compounds.

This interstellar material is not uniformly distributed through space; it consists of a complex tangle of cool, dense clouds pushed and twisted by currents of hot, low-density gas. Although the cool clouds contain only 10 to 1000 atoms/cm³ (fewer than in any laboratory vacuum on Earth), astronomers refer to them as dense clouds in contrast with the hot, low-density gas that fills the spaces between clouds. That thin gas contains only about 0.1 atom/cm³.

How do you know there is an interstellar medium, and how do you know its properties? In some cases, the interstellar medium is easily visible as clouds of gas and dust; for example, the Great Nebula in Orion. Astronomers call such a cloud a **nebula** from the Latin word for "cloud." Such nebulae (plural) give clear evidence of an interstellar medium. Study **Three Kinds of Nebulae** on pages 136–137 and notice three important points:

❶ First, very hot stars can excite clouds of gas and dust to emit light, and this reveals that the clouds contain mostly hydrogen gas at very low densities.

■ Table 7-2 I Main-Sequence Stars

Spectral Type	Mass (Sun = 1)	Luminosity (Sun = 1)	Approximate Years on Main Sequence
O5	40	405,000	1×10^6
B0	15	13,000	11×10^6
A0	3.5	80	440×10^6
F0	1.7	6.4	3×10^9
G0	1.1	1.4	8×10^9
K0	0.8	0.46	17×10^9
M0	0.5	0.08	56×10^9

■ Figure 7-8

Young stars are found in clouds of gas and dust from which they have been born. The nebula N44 is 170,000 ly from Earth in a nearby galaxy, and the Horsehead Nebula is only about 1500 ly distant in our own galaxy. Gas in both nebulae is excited to glow by hot, young stars. Interstellar dust is visible as dark, twisting clouds seen against the bright background gas. (N44: ESO; Horsehead Nebula: NOAO/Nigel Sharp)

2 Second, where dusty clouds reflect the light of stars you see evidence that the dust in the clouds is made up of very small particles.

3 The third thing to notice is that some dense clouds of gas and dust are dark and are detectable only where they are silhouetted against background regions filled with stars or bright nebulae.

If a cloud is less dense, the starlight may be able to penetrate it, and stars can be seen through the cloud; but the stars look dimmer because the dust in the cloud scatters some of the light. Because shorter wavelengths are scattered more easily than longer wavelengths, the redder photons are more likely to make it through the cloud, and the stars look redder than they should for their respective spectral types. This effect is called **interstellar reddening** (■ Figure 7-9). The same physical process makes the setting sun look red. The fact that distant stars are dimmed and reddened by intervening gas and dust is clear evidence of an interstellar medium.

Although you have been imagining individual nebulae, the thin gas and dust of the interstellar medium also fills the spaces between the nebulae. You can see evidence of that because the gas forms interstellar absorption lines in the spectra of distant stars

(■ Figure 7-10). As starlight travels through the interstellar medium, gas atoms of elements such as calcium and sodium absorb photons of certain wavelengths, producing absorption lines. You can be sure those lines originate in the interstellar medium because they appear in the spectra of O and B stars—stars too hot to have visible calcium and sodium absorption lines in their own atmospheres. Also, the narrowness of the interstellar lines indicates they could not have been formed in the hot atmospheres of stars—the line widths indicate a very small range of Doppler shifts, meaning the atoms are moving at speeds corresponding to temperatures of only 10 to 50 K.

Observations at nonvisible wavelengths provide valuable evidence about the interstellar medium. Infrared observations can directly detect dust in the interstellar medium. Although the dust grains are very small and very cold, there are huge numbers of grains in a cloud, and each grain emits infrared radiation at long wavelengths (■ Figure 7-11a). Furthermore, some molecules in the cold gas emit in the infrared, so infrared observations can detect very cold clouds of gas. Also, infrared light penetrates ISM dust better than shorter-wavelength radiation, so astronomers can use infrared cameras to look into and through interstellar clouds that are opaque to visible light. X-ray observations can detect regions of very hot gas apparently produced by exploding stars (Figure 7-11b). Ra-

Three Kinds of Nebulae

1 **Emission nebulae** are produced when a hot star excites the gas near it to produce an emission spectrum. The star must be hotter than about B1 (25,000 K). Cooler stars do not emit enough ultraviolet radiation to ionize the gas. Emission nebulae have a distinctive pink color produced by the blending of the red, blue, and violet Balmer lines. Emission nebulae are also called **HII regions,** following the custom of naming gas with a roman numeral to show its state of ionization. HI is neutral hydrogen, and HII is ionized.

In an HII region, the ionized nuclei and free electrons are mixed. When a nucleus captures an electron, the electron falls down through the atomic energy levels, emitting photons at specific wavelengths. Spectra indicate that the nebulae have compositions much like that of the sun – mostly hydrogen. Emission nebulae have densities of 100 to 1000 atoms per cubic centimeter, better than the best vacuums produced in laboratories on Earth.

Visual-wavelength image

European Southern Observatory

2 A **reflection nebula** is produced when starlight scatters from a dusty nebula. Consequently, the spectrum of a reflection nebula is just the reflected absorption spectrum of starlight. Gas is surely present in a reflection nebula, but it is not excited to emit photons. See image below.

Reflection nebulae NGC 1973, 1975, and 1977 lie just north of the Orion nebula. The pink tints are produced by ionized gases deep in the nebulae.

2a Reflection nebulae look blue for the same reason the sky looks blue. Short wavelengths scatter more easily than long wavelengths. See image below.

Sunlight enters Earth's atmosphere

Blue photons are scattered more easily than longer wavelengths and blue photons enter your eyes from all directions, making the sky look blue.

Visual wavelength image

AATB

2b The blue color of reflection nebulae at left shows that the dust particles must be very small in order to preferentially scatter the blue photons. Interstellar dust grains must have diameters ranging from 0.001 mm down to 1 nm or so.

The hottest star in the Pleiades star cluster is Merope, a B3 star. It is not hot enough to ionize the gas so you see a reflection nebula rather than an emission nebula.

Merope

Visual-wavelength image

2c A dusty reflection nebula is located very close to the star Merope above. The glare from the star is caused by internal reflections in the telescope, but the wispy nature of the nebula is real. The intense light from the star is pushing the dust particles away and may destroy the little nebula over the next few thousand years.

Caltech

Merope

Visual   NASA

Reflection

Emission

Trifid Nebula

The Milky Way in Sagittarius contains two nebulae that dramatically demonstrate the difference between emission and reflection nebulae.

Emission

Visual

Lagoon Nebula

Daniel Good

3 **Dark nebulae** are dense clouds of gas and dust that obstruct the view of more distant stars. Some are generally round, but others are twisted and distorted, as shown at the left, suggesting that even when there are no nearby stars to ionize the gas or produce a reflection nebula, there are breezes and currents pushing through the interstellar medium.

Dark Nebula Barnard 86

Star Cluster NGC6520

Hubble Heritage Team, NASA

Visual

Visual   AATB

Northern Coalsack

Cygnus

Milky Way

Great Rift

The Horsehead Nebula in Orion is a dark nebula silhouetted against a more distant emission nebula.

Large dark nebulae obstruct the view of more distant stars and form holes and rifts along the Milky Way. The Great Rift extends from Cygnus to Sagittarius.

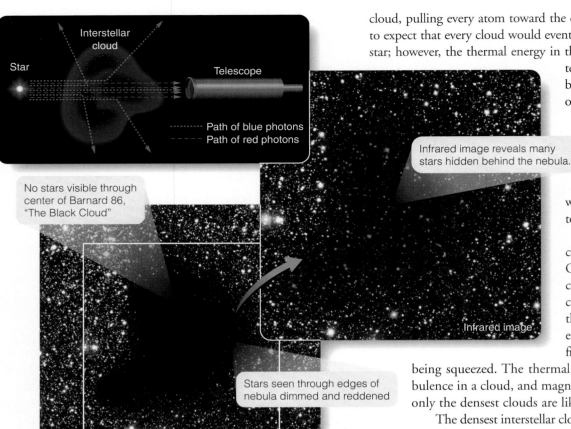

No stars visible through center of Barnard 86, "The Black Cloud"

Interstellar cloud

Star

Telescope

············· Path of blue photons
- - - - - Path of red photons

Infrared image reveals many stars hidden behind the nebula.

Infrared image

Stars seen through edges of nebula dimmed and reddened

Visual-wavelength image

■ **Figure 7-9**

Interstellar reddening makes stars seen through a cloud of gas and dust look redder than they should because shorter wavelengths are more easily scattered. If the gas and dust is especially dense, no stars are visible through the cloud at visual wavelengths except near the edges. At longer near-infrared wavelengths, many stars can be detected behind the cloud. (ESO)

dio observations reveal the emissions of specific molecules in the interstellar medium—the equivalent of emission lines in visible light. Such studies show that some of the atoms in space have linked together to form molecules.

There is no shortage of evidence concerning the interstellar medium. Clearly, space is not empty, and from the correlation between young stars and clouds of gas and dust you can suspect that stars form from these clouds.

ThomsonNOW™ Sign in at www.thomsonedu.com and go to ThomsonNOW to see Astronomy Exercise "Why Is the Sky Blue?"

The Formation of Stars from the Interstellar Medium

To study the formation of stars, you can continue comparing theory with evidence. The theory of gravity predicts that the combined gravitational attraction of the atoms in a cloud of gas will shrink the

cloud, pulling every atom toward the center. That might lead you to expect that every cloud would eventually collapse and become a star; however, the thermal energy in the cloud resists collapse. Interstellar clouds are very cold; but, even at a temperature of only 10 K, the average hydrogen atom moves about 0.5 km/s (1100 mph). That much thermal motion would make the cloud drift apart if gravity were too weak to hold it together.

Other factors can help a cloud resist its own gravity. Observations show that clouds are turbulent with currents of gas pushing through and colliding with each other. Also, magnetic fields in clouds may resist being squeezed. The thermal motion of the atoms, turbulence in a cloud, and magnetic fields resist gravity, and only the densest clouds are likely to contract.

The densest interstellar clouds contain from 10^3 to 10^5 atoms/cm^3 and have temperatures as low as 10 K. They include a few hundred thousand to a few million solar masses. In such dense clouds, hydrogen can exist as molecules (H_2) rather than as atoms. These very densest parts of the ISM are called **molecular clouds,** and the largest are called giant molecular clouds. Although hydrogen molecules cannot be detected by radio telescopes, the clouds can be mapped by the emission lines of carbon monoxide molecules (CO) present in small amounts in the gas. By now you may have realized the main point: Stars can form inside molecular clouds

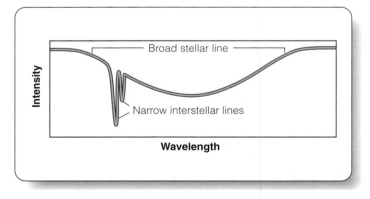

Broad stellar line

Intensity

Narrow interstellar lines

Wavelength

■ **Figure 7-10**

Interstellar absorption lines can be recognized easily because they are much narrower than the spectral lines produced in stellar atmospheres. The multiple interstellar lines in this spectrum are produced by separate interstellar clouds with slightly different radial velocities and are seen superimposed on a broad stellar line. (Adapted from a diagram by Binnendijk.)

■ Figure 7-11

Infrared and X-ray observations reveal different parts of the interstellar medium. (a) This far-infrared image shows the infrared cirrus, wispy clouds of very cold dust gas spread all across the sky. (NASA/IPAC/Deborah Levine) (b) The Cygnus Superbubble, detectable in X rays, triggers star formation where it pushes into the surrounding gas. (Steve Snowden, Max-Planck Institute for Extraterrestrial Physics, Germany)

when the densest parts of the clouds become unstable and contract under the influence of their own gravity.

Most clouds do not appear to be gravitationally unstable and will not contract to form stars on their own. However, a stable cloud colliding with a **shock wave** (the astronomical equivalent of a sonic boom) can be compressed and disrupted into fragments. Theoretical calculations show that some of these fragments can become dense enough to collapse under the influence of their own gravity and form stars (■ Figure 7-12).

Supernova explosions (exploding stars described in the next chapter) produce shock waves that compress the interstellar medium, and recent observations show young stars forming at the edges of such shock waves. Another source of shock waves may be the birth of very hot stars. A massive star is so luminous and hot that it emits vast amounts of ultraviolet photons. When such a star is born, the sudden blast of light, especially ultraviolet radiation, can ionize and drive away nearby gas, forming a shock wave that could compress nearby clouds and trigger further star formation. Even the collision of two interstellar clouds can produce a shock wave and trigger star formation. Some of these processes are shown in ■ Figure 7-13.

Although these are important sources of shock waves, the dominant trigger of star formation in our galaxy may be the spiral pattern itself. In Chapter 1, you learned that our galaxy contains spiral arms. As interstellar clouds encounter these spiral arms, the clouds are compressed, and star formation can be triggered (see Chapter 9).

Once begun, star formation can spread like a grass fire. Both high-mass and low-mass stars form together, but when the massive stars form, their intense radiation or eventual supernova explosions push back the surrounding gas and compress it. This compression in turn can trigger the formation of more stars, some of which will be massive. Thus, a few massive stars can drive a continuing cycle of star formation in a giant molecular cloud.

A collapsing cloud of gas does not form a single star; because of instabilities, the cloud breaks into fragments, producing 10 to 1000 stars. Stars held in a stable group by their combined gravity are called a **star cluster.** In contrast, a **stellar association** is a group of stars that are not gravitationally bound to one another. The stars in an association drift away from each other in a few million years. The youngest associations are rich in young stars, including O and B stars. O and B stars have lifetimes so short in astronomical terms (Table 7-2) that they must be located close to their birthplaces, so these stars serve as brilliant signposts pointing to regions of star formation.

The Formation of Protostars

You might be wondering how the unimaginably cold gas of an interstellar cloud can heat up to form a star. The answer is gravity.

Once part of a cloud is triggered to collapse, gravity draws each atom toward the center. At first the atoms fall unopposed; they hardly ever collide with each other. In this free-fall contraction, the atoms pick up speed as they fall until, by the time the gas becomes dense enough for the atoms to collide often, they are traveling very fast. Collisions convert the inward velocities of the atoms into random motions, and you remember that temperature is a measure of the random velocities of the atoms in a gas. Thus, the temperature of the gas goes up.

Shock Wave Triggers Star Formation

A shock wave (red) approaches an interstellar gas cloud.

The shock wave passes through and compresses the cloud.

The densest parts of the cloud become gravitationally unstable.

Contracting regions of gas give birth to stars.

■ **Figure 7-12**

In this summary of a computer model, an interstellar gas cloud is triggered into star formation by a passing shock wave. The events summarized here might span about 6 million years.

The initial collapse of the gas forms a dense core of gas, and as more gas falls in, a warm **protostar** develops, buried deep in the dusty gas cloud that continues to contract, although much more slowly than free-fall. A protostar is an object that will eventually become a star. Theory predicts that protostars are luminous red objects larger than main-sequence stars, with temperatures ranging from a few hundred to a few thousand degrees K.

Throughout its contraction, the protostar converts its gravitational energy into thermal energy. Half of this thermal energy radiates into space, but the remaining half raises the internal temperature. As the internal temperature climbs, the gas becomes ionized, changing into a mixture of positively charged atomic nu-

clei and free electrons. When the center gets hot enough, nuclear reactions begin generating enough energy to replace the radiation leaving the surface of the star. The protostar then halts its contraction and becomes a stable, main-sequence star.

The time a protostar takes to contract from a cool interstellar gas cloud to a main-sequence star depends on its mass. The more massive the star, the stronger its gravity and the faster it contracts. The sun took about 30 million years to reach the main sequence, but a 15-solar-mass star can contract in only 100,000 years. Conversely, a star of 0.2 solar mass takes 1 billion years to reach the main sequence.

Observations of Star Formation

By understanding what the interstellar medium is like and by understanding how the laws of physics work, astronomers have been able to work out the story of how stars must form. But you can't accept a scientific theory without testing it, and that means you must compare the theory with the evidence.

The protostar stage is predicted to be less than 0.1 percent of a star's total lifetime, so, although that is a long time in human terms, you cannot expect to find many stars in the protostar stage. Furthermore, protostars form deep inside clouds of dusty gas that absorb any light the protostar might emit. Only when the protostar is hot enough to drive away its enveloping cloud of gas and dust does it become easy to observe at wavelengths your eye can see.

Although astronomers cannot easily observe protostars at visible wavelengths, protostars can be detected in the infrared. Protostars are cooler than stars and radiate as blackbodies predominantly at infrared wavelengths. The dust of the surrounding interstellar cloud absorbs light from the protostar and grows warm, and that warm dust radiates additional large amounts of infrared radiation. Infrared radiation also can penetrate the dust to reach us. Thus, infrared observations reveal many bright sources of radiation that are protostars buried in interstellar clouds.

You can be sure that star formation is going on right now because astronomers find regions containing massive stars with lifetimes so short they must still be in their birthplaces. Small very dense zones of gas and dust called **Bok globules** within larger nebulae probably represent the very first stages of star formation. Observations of jets coming from hidden protostars show that protostars are often surrounded by disks of gas and dust (■ Figure 7-14). These jets appear to produce small flickering nebulae called **Herbig–Haro objects.** The same regions contain many stars with temperatures and luminosities that correspond to model predictions of the properties of very young stars. These include **T Tauri stars,** protostars with about the mass of the sun that show signs of active chromospheres as you might expect from young rapidly rotating stars with strong magnetic dynamos (see Chapter 5). T Tauri stars are also often surrounded by thick disks of gas and dust that are very bright at infrared wavelengths.

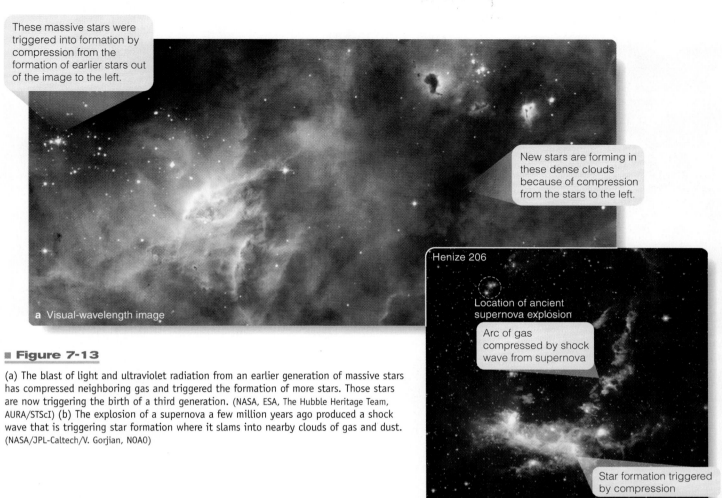

These massive stars were triggered into formation by compression from the formation of earlier stars out of the image to the left.

a Visual-wavelength image

New stars are forming in these dense clouds because of compression from the stars to the left.

Henize 206

Location of ancient supernova explosion

Arc of gas compressed by shock wave from supernova

Star formation triggered by compression

b Infrared image

■ **Figure 7-13**

(a) The blast of light and ultraviolet radiation from an earlier generation of massive stars has compressed neighboring gas and triggered the formation of more stars. Those stars are now triggering the birth of a third generation. (NASA, ESA, The Hubble Heritage Team, AURA/STScI) (b) The explosion of a supernova a few million years ago produced a shock wave that is triggering star formation where it slams into nearby clouds of gas and dust. (NASA/JPL-Caltech/V. Gorjian, NOAO)

These observations all confirm the theoretical models of contracting protostars. Although star formation still holds many puzzles, the general process seems clear. In at least some cases, interstellar gas clouds are compressed by passing shock waves, and the clouds' gravity, acting unopposed, draws the matter inward to form protostars. Now you know how to make a star: Just find some way to compress more than 0.08 solar masses of ISM material into a small enough region of space, and gravity will do the rest of the job.

The Orion Nebula

On a clear winter night, you can see with your unaided eye the Great Nebula of Orion as a fuzzy wisp in Orion's sword. With binoculars or a small telescope it is striking, and through a large telescope it is breathtaking. At the center lie four brilliant blue-white stars known as the Trapezium, the brightest of a cluster of a few hundred stars. Surrounding the stars are the glowing filaments of a nebula more than 8 pc across. Like a great thundercloud illuminated from within, the churning currents of gas and dust suggest immense power. The significance of the Orion Nebula lies hidden, figuratively and literally, beyond the visible nebula. The region is ripe with star formation.

Evidence of Young Stars

You should not be surprised to find star formation in Orion. The constellation is a brilliant landmark in the winter sky because it is marked by hot blue stars. These stars are bright not because

HH34S   NASA   Visual

Gas jet

Young star

Visual   HH34N

■ **Figure 7-14**

At the center of this image, a newborn star is emitting powerful jets to left and right. Where the jets strike the interstellar medium, they produce Herbig–Haro objects. The inset shows how irregular the jet is. Such jets can be over a light-year long and contain gas traveling at 100 km/s or more. (Richard Mundt, Calar Alto 3.5m telescope)

tar Formation
the Orion Nebula

1 The visible Orion Nebula shown below is a pocket of ionized gas on the near side of a vast, [du]sty molecular cloud that fills much of the southern [pa]rt of the constellation Orion. The molecular cloud [ca]n be mapped by radio telescopes. To scale, the [clo]ud would be many times larger than this page. As [th]e stars of the Trapezium were born in the cloud, their [ra]diation has ionized the gas and pushed it away. [W]here the expanding nebula pushes into the larger [mo]lecular cloud, it is compressing the gas (see [di]agram at right) and may be triggering the formation [of] the protostars that can be detected at infrared [wa]velengths within the molecular cloud.

[Hu]ndreds of stars lie within the nebula, [bu]t only the four brightest, those in the [Tra]pezium, are easy to see with a [sm]all telescope. A fifth star, at the [na]rrow end of the Trapezium, [ma]y be visible on nights [of] good seeing.

[Th]e cluster of stars in [th]e nebula is less than [1]million years old. [Th]is must mean [th]e nebula is [sim]ilarly young.

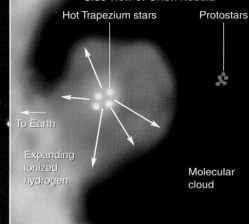

Side view of Orion Nebula

Hot Trapezium stars

Protostars

To Earth

Expanding ionized hydrogen

Molecular cloud

Infrared

The near-infrared image above reveals about 50 low-mass, very cool stars that must have formed recently.

Trapezium

X-ray

Roughly 1000 young stars with hot chromospheres appear in this X-ray image of the Orion Nebula.

Visual-wavelength image
Daniel Good

Photons with enough energy to ionize H

Energy radiated by O6 star

Energy radiated by B1 star

2 Of all the stars in the Orion Nebula, only one is hot enough to ionize the gas. Only photons with wavelengths shorter than 91.2 nm can ionize hydrogen. The second-hottest stars in the nebula are B1 stars, and they emit little of this ionizing radiation. The hottest star, however, is an O6 star 30 times the mass of the sun. At a temperature of 40,000 K, it emits plenty of photons with wavelengths short enough t[o] ionize hydrogen. Remove that one star, and the nebula would turn off its emission.

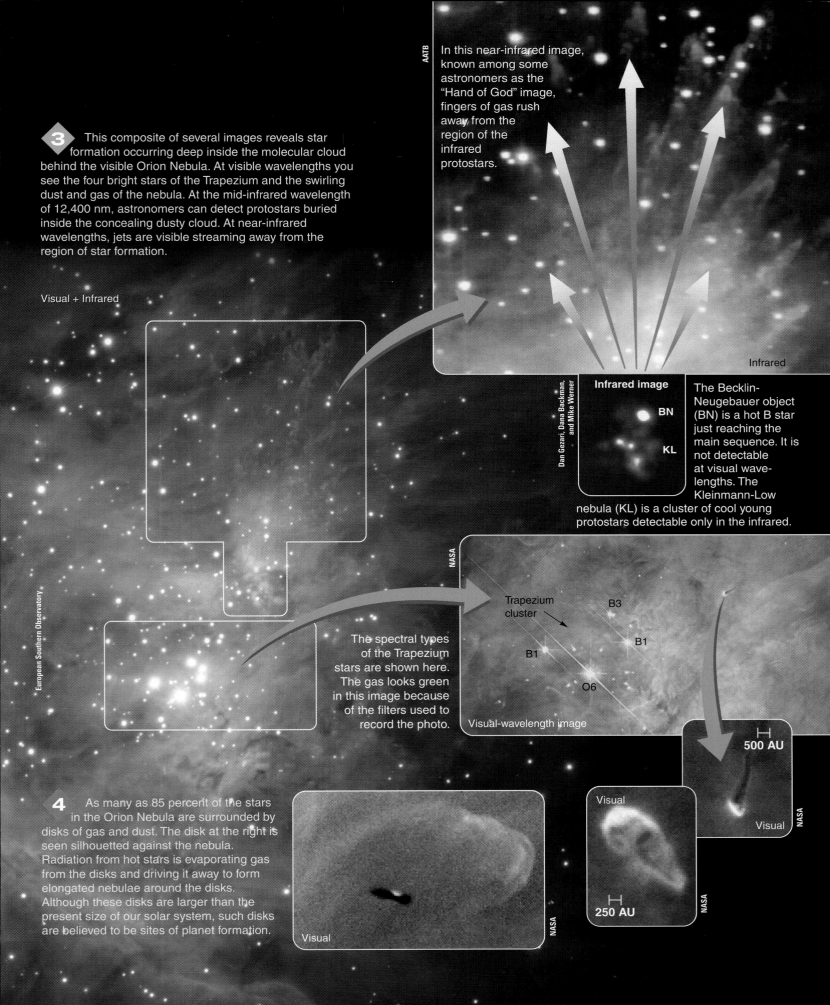

3 This composite of several images reveals star formation occurring deep inside the molecular cloud behind the visible Orion Nebula. At visible wavelengths you see the four bright stars of the Trapezium and the swirling dust and gas of the nebula. At the mid-infrared wavelength of 12,400 nm, astronomers can detect protostars buried inside the concealing dusty cloud. At near-infrared wavelengths, jets are visible streaming away from the region of star formation.

Visual + Infrared

European Southern Observatory

In this near-infrared image, known among some astronomers as the "Hand of God" image, fingers of gas rush away from the region of the infrared protostars.

AATB

Infrared

Infrared image

Dan Gezari, Dana Backman, and Mike Werner

BN

KL

The Becklin-Neugebauer object (BN) is a hot B star just reaching the main sequence. It is not detectable at visual wavelengths. The Kleinmann-Low nebula (KL) is a cluster of cool young protostars detectable only in the infrared.

NASA

Trapezium cluster

B3

B1

B1

O6

The spectral types of the Trapezium stars are shown here. The gas looks green in this image because of the filters used to record the photo.

Visual-wavelength image

500 AU

Visual

Visual

NASA

NASA

4 As many as 85 percent of the stars in the Orion Nebula are surrounded by disks of gas and dust. The disk at the right is seen silhouetted against the nebula. Radiation from hot stars is evaporating gas from the disks and driving it away to form elongated nebulae around the disks. Although these disks are larger than the present size of our solar system, such disks are believed to be sites of planet formation.

Visual

250 AU

NASA

they are nearby but because they are tremendously luminous. These O and B stars cannot live more than a few million years, so you can conclude that they must have been born astronomically recently, near where they are seen now. Furthermore, the constellation contains large numbers of T Tauri stars, which are known to be young. Orion is rich with young stars.

The history of star formation in the constellation of Orion is written in its stars. The stars at Orion's west shoulder are about 12 million years old, whereas the stars of Orion's belt are about 8 million years old. The stars of the Trapezium at the center of the Great Nebula are no older than 2 million years. Apparently, star formation began near the west shoulder, and the massive stars that formed there triggered the formation of the stars you see in Orion's belt. That star formation may have triggered the formation of the stars you see in the Great Nebula. Like a grass fire, star formation has swept across Orion from northwest to southeast.

Study **Star Formation in the Orion Nebula** on pages 142–143 and notice four points:

1 The nebula you see is only a small part of a vast, dusty molecular cloud. You see the nebula because the stars born within it have ionized the nearby gas and driven it outward, breaking out of the much larger molecular cloud.

2 Also notice that a single very hot star is almost entirely responsible for ionizing the gas.

3 Notice how infrared observations reveal clear evidence of active star formation deeper in the molecular cloud just to the northwest of the Trapezium.

4 Finally, notice that many stars visible in the Orion Nebula are surrounded by disks of gas and dust. Such disks do not last long and are clear evidence that the stars are very young.

Observations of stars in the Orion nebula show that some of the gas and dust in the cloud from which a star forms settles into a disk. Make special note of these disks because planets may form in them. In Chapter 12, you will study how our solar system formed in the disk of gas and dust that encircled the young sun.

Building Scientific Arguments

What evidence can you cite that stars are forming right now?

This is a very good example of building a scientific argument because you must cite different forms of evidence to confirm a theory. First, you should note that some extremely luminous stars can't live very long, so when you see such stars, such as the hot, blue stars in Orion, you know they must have formed in the last few million years. Star formation must be a continuous process, or you would not see any of these short-lived stars. Second, you have more direct evidence when you look at T Tauri stars which have characteristics expected of young solar-type stars and are often embedded in disks of gas and dust. In fact, you see entire associations of T Tauri stars as well as other associations of short-lived O stars. Many regions of gas and dust contain bright, hot stars that are blowing the nebulae apart. Those stars must have formed recently.

There seems to be no doubt that star formation is an ongoing process, and you can revise your argument to discuss how it happens. **What evidence can you cite that protostars are often surrounded by disks of gas and dust?**

Study and Review

Summary

7-1 | Stellar Structure

What are the insides of stars like?

■ In a star, the weight of each layer pressing inward must be balanced by the pressure inside pushing outward. This is called hydrostatic equilibrium.

■ Deep layers of a star must be hotter and denser and have higher pressure than the less-deep layers in order for the star to be stable.

■ Energy in a star must flow from the hot interior to the cool exterior by conduction, convection, or radiation. Conduction is not usually important inside stars.

■ Much of what astronomers know about stars is based on detailed mathematical models of the interior of stars.

7-2 | Nuclear Fusion in the Sun and Stars

How do the stars make energy?

■ Stars generate energy near their centers, where the temperature and density are high enough for nuclear fusion reactions to combine hydrogen nuclei to make helium nuclei.

■ Energy generated in a star's core flows outward either as photons moving through a radiative zone or as rising currents of hot gas and sinking currents of cool gas in a convective zone.

■ Stars of the sun's mass or less make energy via the proton–proton chain; more massive main-sequence stars fuse hydrogen into helium in the CNO cycle.

- Observations of too few neutrinos coming from the sun's core are now explained by the changing of neutrinos back and forth among three different types called "flavors." The neutrinos confirm that the sun makes its energy by hydrogen fusion.

- The relationship between pressure and temperature, called the pressure–temperature thermostat, ensures that stars generate just enough energy to be stable.

7-3 | Main-Sequence Stars

What determines the properties of main-sequence stars?
- The luminosity of a star depends on its mass. More massive stars have more weight to support, and their pressure–temperature thermostats must make more energy. That makes them more luminous.

- The main sequence has a lower end (a minimum mass) because objects less massive than 0.08 solar masses cannot get hot enough to begin hydrogen fusion. Such objects become brown dwarfs.

How long can a star survive?
- How long a star can remain on the main sequence depends on its mass. The more massive a star is, the faster it uses up its hydrogen fuel. A 25-solar-mass star will exhaust its hydrogen and die in only about 7 million years, but the sun is expected to last for 10 billion years.

7-4 | The Birth of Stars

How are stars born?
- Stars form from the gas and dust of the interstellar medium.

- Astronomers know there is an interstellar medium because they can see emission nebulae, reflection nebulae, and dark nebulae. They can also detect interstellar absorption lines in the spectra of distant stars.

- The dust in the interstellar medium makes distant stars look fainter and redder than expected.

- In large, dense, cool clouds of gas, hydrogen can exist as molecules rather than as atoms. These molecular clouds are sites of star formation. Such clouds can be triggered to collapse by collision with a shock wave that compresses and fragments the gas cloud, producing a cluster of protostars.

How do you know that theories of star formation are correct?
- A contracting protostar is large and cooler than a main-sequence star. The dust in the cocoon absorbs the protostar's light and re-radiates it as infrared radiation. Many infrared sources are known that are probably protostars.

- As a protostar grows hot enough to begin hydrogen fusion at its core, it settles onto the main sequence.

- T Tauri stars may be in later stages of formation as the surrounding dust and gas are absorbed by the protostar or driven into space. Very young star clusters contain large numbers of T Tauri stars.

- Some protostars have been found emitting jets of gas. Where these flows strike existing clouds of gas, astronomers can see nebulae called Herbig–Haro objects. The jets are evidently focused by disks of gas and dust around the protostars.

- The Great Nebula in Orion is an active region of star formation. The bright stars in the center of the nebula formed within the last few million years, and infrared telescopes detect protostars buried inside the molecular cloud that lies behind the visible nebula.

New Terms

conservation of mass (p. 126)

conservation of energy (p. 126)

hydrostatic equilibrium (p. 126)

energy transport (p. 126)

opacity (p. 127)

stellar model (p. 128)

nuclear force (p. 129)

nuclear fission (p.129)

nuclear fusion (p. 130)

Coulomb barrier (p. 130)

proton–proton chain (p. 130)

deuterium (p. 130)

neutrino (p. 131)

CNO (carbon–nitrogen–oxygen) cycle (p. 131)

brown dwarf (p. 133)

zero-age main sequence (ZAMS) (p. 133)

interstellar medium (ISM) (p. 134)

interstellar dust (p. 134)

nebula (p. 134)

interstellar reddening (p. 135)

emission nebula (p. 136)

HII region (p. 136)

reflection nebula (p. 136)

dark nebula (p. 137)

molecular cloud (p. 138)

shock wave (p. 139)

star cluster (p. 139)

stellar association (p. 139)

protostar (p. 140)

Bok globule (p. 140)

Herbig–Haro object (p. 140)

T Tauri star (p. 140)

Review Questions

ThomsonNOW™ Assess your understanding of this chapter's topics with additional quizzing and animations at **www.thomsonedu.com**

1. What opposing forces are balanced in a stable star?

2. Why does the CNO cycle require a higher temperature than the proton–proton chain?

3. How does the pressure–temperature thermostat control the nuclear reactions inside stars?

4. Step by step, explain how energy flows from the sun's core to Earth.

5. Why is there a mass–luminosity relation for stars on the main sequence?

6. Why is there a lower limit to the mass of a main-sequence star?

7. Why does a star's life expectancy depend on its mass?

8. What evidence can you cite that the space between the stars is not empty?

9. Why would an emission nebula near a hot star look red, while a reflection nebula near its star looks blue?

10. Why do astronomers rely heavily on infrared observations to study star formation?

11. What observational evidence can you cite that star formation spreads from place to place?

12. How are Herbig–Haro objects related to star formation? T Tauri stars?

13. What evidence can you cite that star formation is happening right now in the Orion Nebula?

14. The image above to the right shows two nebulae, one pink in the background and one black in the foreground. What kind of nebulae are these, and how are they related to star formation?

15. The star at right appears to be ejecting a jet of gas. What is happening to this star?

NASA and the Hubble Heritage Team, STScI/AURA

STScI and NASA

Discussion Questions

1. When you see distant streetlights through smog, they look dimmer and redder than they do normally. But when you see the same streetlights through fog or falling snow, they look dimmer but not redder. Use your knowledge of the interstellar medium to discuss the relative sizes of the particles in smog, fog, and snow compared with the wavelength of light.

2. If planets form as a natural by-product of star formation, which do you think are more common—stars or planets?

8 | The Deaths of Stars

Natural laws have no pity.

Robert Heinlein
THE NOTEBOOKS OF LAZARUS LONG

GRAVITY IS PATIENT—so patient it can kill stars. Astronomers occasionally see what seems to be a new star in the sky that grows brighter, then fades away after a few weeks or a year. You will discover that a **nova,** an apparently new star in the sky, is in fact produced by an eruption around a stellar remnant, and that a **supernova,** a particularly luminous and long-lasting nova, is caused by the violent explosive death of a massive star. Modern astronomers find a few novae (plural of *nova*) each year, but supernovae (plural) are so rare that only one or two happen each century in our galaxy. ▌ In the previous chapter, you saw that stars resist their own gravity by generating energy through nuclear fusion. The energy keeps their interiors hot, and the resulting high pressure balances gravity and prevents the star from collapsing. Stars, however, have

▌ Continued on page 148 ▌

Guidepost

Looking Back

You are beginning to understand stars. The preceding chapter described how stars are born from clouds of dust gas in the interstellar medium and how nuclear fusion maintains high temperatures and pressures inside stars, which keeps them from collapsing under the influence of their own gravity. Stars are the lighting system of the universe, but they can last only as long as their fuel supplies.

This Chapter

Now you are ready to consider the fate of the stars. Here you will see how stars die and what kinds of remains are left behind. The details of their life and death stories, and the types of remnants they produce, all depend on the mass of the individual star. Stars with masses like that of the sun end as white dwarfs, but more massive stars leave behind the strangest beasts in the cosmic zoo, neutron stars and black holes. Some matter from dying stars escapes back into the interstellar medium and is incorporated into new stars and the planets that form with them. The deaths of stars are part of a great cycle of stellar birth and death that includes the sun, Earth, and you. In this chapter you will find the answers to six essential questions:

What happens to a star when it uses up the last of the hydrogen fuel in its core?

How will the sun and stars smaller than the sun die?

What happens if an evolving star is in a binary system?

How do massive stars die?

What are neutron stars, and how do you know they really exist?

What are black holes, and how do you know they really exist?

The life and death stories of stars are important because Earth depends on one star, the sun. Perhaps even more important, the lives and deaths of previous generations of stars created the atomic elements of which Earth and you are made. If those stars hadn't lived and died, you would not exist. This chapter will also show you more examples of how astronomers combine observations and theory to understand nature.

Looking Ahead

This chapter ends the story of individual stars, but it does not end the story of stars. In the next chapter, you will begin exploring the giant communities in which stars live—the galaxies.

limited fuel. When they exhaust their fuel, gravity wins, and the stars die.

The mass of a star is critical in determining its fate. Massive stars use up their nuclear fuel at a furious rate and die after only a few million years. In contrast, the lowest-mass stars use their fuel sparingly and may be able to live hundreds of billions of years. Stars with different masses lead dramatically different lives and die in different ways.

8-1 Giant Stars

A MAIN-SEQUENCE STAR GENERATES ITS ENERGY BY NUCLEAR FUSION reactions that combine hydrogen to make helium. A star remains on the main sequence for a time span equal to 90 percent of its total existence as an energy-generating star. When the hydrogen is exhausted, however, the star begins to evolve rapidly. It swells into a giant star and then begins to fuse helium into heavier elements. A star can remain in this giant stage for only about 10 percent of its total lifetime; then it must die. The giant-star stage is the first step in the death of a star.

Expansion into a Giant

The nuclear reactions in a main-sequence star's core produce helium. Helium can fuse into heavier elements only at temperatures higher than 100,000,000 K, and no main-sequence star has a core that hot, so helium accumulates at the star's center like ashes in a fireplace. Initially, this helium ash has little effect on the star, but as hydrogen is exhausted and the stellar core becomes almost pure helium, the star's ability to generate nuclear energy is reduced. Because the energy generated at the center is what opposes gravity and supports the star, the core begins to contract as soon as the energy generation starts to die down.

Although the core of helium ash cannot generate nuclear energy, it can grow hotter because it contracts and converts gravitational energy into thermal energy (see Chapter 7). The rising temperature heats the unprocessed hydrogen just outside the core—hydrogen that was never previously hot enough to fuse. When the temperature of the surrounding hydrogen becomes high enough, hydrogen fusion begins in a spherical layer, called a shell, surrounding the exhausted core of the star. Like a grass fire burning outward from an exhausted campfire, the hydrogen-fusion shell creeps outward, leaving helium ash behind and increasing the mass of the helium core.

The flood of energy produced by the hydrogen-fusion shell pushes toward the surface, heating the outer layers of the star and forcing them to expand dramatically (■ Figure 8-1). Stars like the sun become **giant stars** 10 to 100 times the present diameter of the sun, and the most massive stars become **supergiant stars** as much as 1000 times larger than the sun.

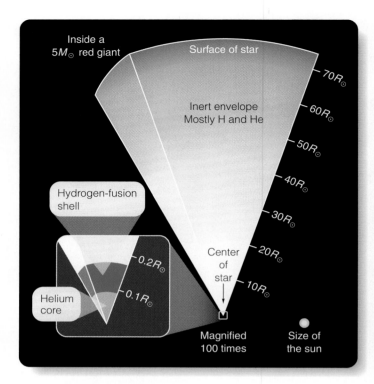

■ Figure 8-1

When a star runs out of hydrogen at its center, the core contracts to a small size, becomes very hot, and begins nuclear fusion in a shell (blue). The outer layers of the star expand and cool. The red giant star shown here has an average density much lower than the air at Earth's surface. Here $M_\odot$ stands for the mass of the sun, and $R_\odot$ stands for the radius of the sun. (Illustration design by M. Seeds)

The expansion of a star to giant or supergiant size cools the star's outer layers, and so the stars move toward the upper right in the H-R diagram (■ Figure 8-2). Look at Figure 6-8 and notice that some of the most familiar stars, such as Aldebaran, Betelgeuse and Polaris, are giants or supergiants.

Although the energy output of the hydrogen-fusion shell can force the envelope of the star to expand, it cannot stop the contraction of the helium core. Because the core is not hot enough to fuse helium, gravity squeezes it to a relatively tiny size. If you represented a typical giant star as being the size of a baseball stadium, its helium core would be only about the size of a baseball, yet would contain about 10 percent of the star's mass.

Helium Fusion

As a star becomes a giant, fusing hydrogen in a shell, the inert core of helium ash contracts and grows hotter. When the core finally reaches a temperature of 100,000,000 K, helium nuclei can begin fusing to make carbon nuclei.

The ignition of helium in the core changes the structure of the star. The star now makes energy in two locations by two dif-

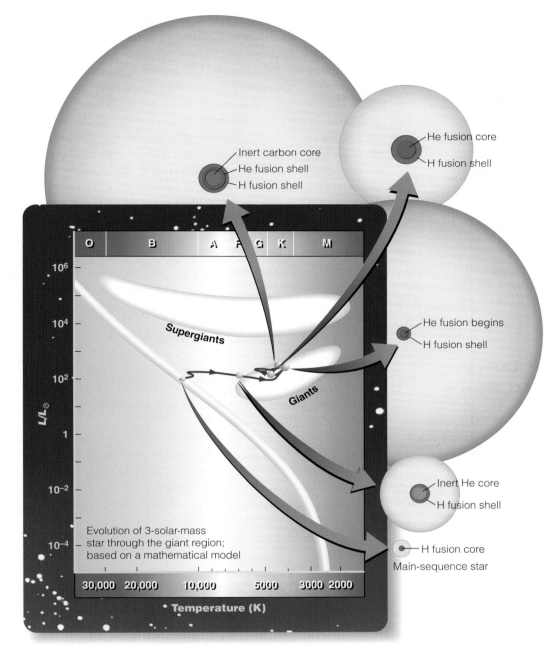

When a main-sequence star exhausts the hydrogen in its core, it evolves rapidly to the right in the H–R diagram as it expands to become a cool giant. These changes are caused at first by the ignition of a hydrogen-fusion shell. When helium ignites in the core, the star contracts and heats up slightly, but when helium begins to fuse in a shell, the star again expands and cools.

Inert carbon core
He fusion shell
H fusion shell

He fusion core
H fusion shell

O B A F G K M

10^6

10^4

Supergiants

10^2

He fusion begins
H fusion shell

$L/L_\odot$

1

Giants

10^{-2}

Inert He core
H fusion shell

10^{-4}

Evolution of 3-solar-mass
star through the giant region;
based on a mathematical model

H fusion core
Main-sequence star

30,000 20,000   10,000   5000   3000 2000

Temperature (K)

now makes energy in two fusion shells, so it quickly expands and its surface cools once again. The point that represents the star in the H–R diagram moves back to the right, completing a loop to become a red giant again (see Figure 8-2). The approximate rule is that, if the core of a post-main sequence star is "dead" (has no nuclear reactions), the star is a red giant, and if the core is "alive" (has fusion reactions), the star is a yellow giant.

Now you can understand why giant stars are relatively rare (see page 121). A star spends about 90 percent of its lifetime on the main sequence and only 10 percent as a giant star. At any moment you look, only a fraction of the visible stars will be passing through the red and yellow giant stages.

What happens to a star after helium fusion depends on its mass, but in any event it cannot survive long: It must eventually collapse and end its career as a star. The remainder of this chapter will trace the details of this process of stellar death, but before you begin that story, you must ask the

ferent processes, helium fusion in the core and hydrogen fusion in the surrounding shell. The energy flowing outward from the core halts the contraction of the core, and at the same time the star's envelope contracts and grows hotter. Consequently, the point that represents the star in the H–R diagram moves downward, corresponding to lower luminosities, and to the left, corresponding to higher surface temperatures, to a region above the main sequence called the **horizontal branch** (see Figure 8-2). Astronomers sometimes refer to stars with those characteristics as "yellow giants."

Helium fusion produces carbon and oxygen that accumulate in an inert core. Once again, the core contracts and heats up, and a helium-fusion shell ignites below the hydrogen-fusion shell. The star

most important question in science: What is the evidence? What evidence shows that stars actually evolve as theories predict? You can find the answers in observations of star clusters.

Star Clusters: Evidence of Evolution

Astronomers look at star clusters and say, "Aha! Evidence to solve a mystery." The stars in a star cluster all formed at about the same time and from the same cloud of gas, so they must be about the same age and composition. Each star cluster freeze-frames and makes visible a moment in stellar evolution. The differences you see among stars in one cluster must arise from differences in their masses.

Study **Star Cluster H–R Diagrams** on pages 152–153 and notice three important points:

1 There are two kinds of star clusters, but they are similar in the way their stars evolve. You will learn more about these clusters in the next chapter.

2 You can estimate the age of a star cluster by observing the distribution of the points that represent its stars in the H–R diagram.

3 Finally, a star cluster's H–R diagram shape is governed by the evolutionary path the stars take. By comparing clusters of different ages, you can visualize how stars evolve almost as if you were watching a film of a star cluster evolving over billions of years.

Were it not for star clusters, astronomers would have little confidence in the theories of stellar evolution. Star clusters make that evolution visible and assure astronomers that they really do understand how stars are born, live, and die.

8-2 The Deaths of Lower-Main-Sequence Stars

CONTRACTING STARS HEAT UP BY CONVERTING GRAVITATIONAL ENERGY into thermal energy. Low-mass stars have little gravitational energy, so when they contract, they can't get very hot. This limits the nuclear fuels they can ignite. In Chapter 7, you saw that objects less massive than 0.08 solar mass cannot sustain hydrogen fusion. Consequently, this section will concentrate on stars more massive than 0.08 solar mass but no more than about 4 times the mass of the sun.

Structural differences divide the lower-main-sequence stars into two subgroups, very-low-mass red dwarfs and medium-mass stars such as the sun. The critical difference between the two groups is the extent of interior convection.

Red Dwarfs

Stars less massive than about 0.4 solar mass—the red dwarfs—have two important differences from more massive stars. First, because they have very small masses, they have very little weight to support. Their pressure–temperature thermostats are set low, and they consume their hydrogen fuel very slowly. The discussion of the life expectancies of stars in Chapter 7 predicted that red dwarfs have very long lives.

The red dwarfs have a second advantage because they are completely convective. That means they are stirred by circulating currents of hot gas rising from the interior and cool gas sinking inward, extending all the way from their cores to their surfaces. The interiors of these stars are mixed like a pot of soup that is constantly stirred as it cooks. Hydrogen is consumed and helium

accumulates uniformly throughout the star, which means the star is not limited only to the fuel in its core. It can use all of its hydrogen to prolong its life on the main sequence.

Because a red dwarf is mixed by convection, it cannot develop an inert helium core surrounded by unprocessed hydrogen; therefore it never ignites a hydrogen shell and cannot become a giant star. Rather, nuclear fusion converts hydrogen into helium, but the helium does not fuse into heavier elements because the star can never get hot enough. What astronomers know about stellar evolution indicates that these red dwarfs should use up nearly all of their hydrogen and live very long lives on the lower main sequence. They could survive for a hundred billion years or more. Of course, astronomers can't test this part of their theories because the universe is only 13.7 billion years old (see Chapter 11), so not a single red dwarf has died of old age anywhere in the universe. Every red dwarf that has ever been born is still glowing today.

Medium-Mass (Sunlike) Stars

Stars like the sun can ignite hydrogen and helium and become giants; but, if they contain less than 4 solar masses, they cannot get hot enough to ignite carbon, the next fuel after helium. Note that this mass limit is uncertain, as are many of the masses quoted here. The evolution of stars is highly complex, and such parameters are difficult to specify.

There are two keys to the evolution of these sunlike stars, the lack of complete mixing and mass loss. The interiors of medium-mass stars are not completely mixed because, unlike the red dwarf stars, they are not completely convective. The helium ash accumulates in an inert helium core surrounded by unprocessed hydrogen. As described earlier, when this core contracts, the unprocessed hydrogen ignites in a shell and swells the star into a red giant. During red giant stages stars tend to lose mass due to strong stellar winds. When the helium core finally ignites, the star becomes a yellow giant. When the core fills up with carbon–oxygen ash, that is the end of fusion for stars in the medium-mass range because they have masses too low to make their cores hot enough to ignite carbon or oxygen fusion. The carbon–oxygen core is the dead end for these stars.

All of this discussion is based on theoretical models of stars and a general understanding of how stars evolve. Does it really happen? Astronomers need observational evidence to confirm this theoretical discussion, and the gas that is expelled from these giant stars gives visible evidence that sunlike stars have gone through these stages of nucleosynthesis and do indeed die in this way.

Planetary Nebulae

When a medium-mass star like the sun becomes a distended giant, its atmosphere becomes cool and consequently more opaque. Light has to push against it to escape. At the same time, the fusion shells become thin and unstable, and they begin to flare, which pushes the atmosphere outward. An aging giant can expel its outer atmosphere in repeated surges to form one of the most beautiful objects in astronomy, a **planetary nebula,** so called because the first ones discovered looked like the greenish-blue disks of planets such as Uranus or Neptune. In fact, a planetary nebula has nothing to do with a planet. It is composed of ionized gases expelled by a dying star (■ Figure 8-3).

You can understand what planetary nebulae are like by using simple observations and theoretical methods explained in Chapter 5 such as Kirchhoff's laws and the Doppler effect. Although real nebulae are quite complex, a simple model of a slow stellar wind followed by a fast wind (■ Figure 8-4) explains their structures fairly well and provides a way to organize the observed phenomena. The complexities and asymmetries seen in planetary nebulae may be due to repeated expulsions of expanding shells and oppositely directed jets, much like bipolar flows observed coming from protostars. Observations and stellar evolution models indicate the central star of a planetary nebula star finally must contract and become a white dwarf.

White Dwarfs

As you have just learned, medium-mass stars die by ejecting gas into space and contracting into white dwarfs. In Chapter 6, you surveyed the stars and learned that white dwarfs are the second most common kind of star (see page 121). Only red dwarfs are more abundant. The billions of white dwarfs in our galaxy must be the remains of medium-mass stars.

The first white dwarf discovered was the faint companion to the well-known star Sirius, the brightest star in the sky. Sirius is a visual binary star, the most luminous member of which is Sirius A. The white dwarf, Sirius B, is 10,000 times fainter than Sirius A. The orbital motions of the stars (see Figure 6-12) reveal that the white dwarf's mass is 0.98 solar mass, and its blue-white color tells you that its surface is hot, about 25,000 K. Although it is very hot, it has a very low luminosity, so it must have a small surface area. In fact, it is about the size of Earth. Dividing its mass by its volume reveals that it is very dense—about 2×10^6 g/cm³. On Earth, a teaspoonful of Sirius B material would weigh more than 11 tons. These basic observations and simple physics lead to the conclusion that white dwarfs are astonishingly dense.

A normal star is supported by energy flowing outward from its core, but a white dwarf cannot generate energy by nuclear fusion. It has exhausted its hydrogen and helium fuel and produced carbon and oxygen. As the star collapses into a white dwarf, it converts gravitational energy into thermal energy, and its interior becomes very hot, but it cannot get hot enough to fuse carbon into heavier elements.

The contraction of a white dwarf compresses the gases in its interior to such high densities that quantum mechanical laws become important and the electrons in the gas cannot get closer together. Such a gas is termed **degenerate matter,** and it takes on two properties that are important in understanding the structure and evolution of dying stars. A degenerate gas is millions of times harder to compress than solid steel, and the pressure in the gas no longer depends on the temperature. Unlike a normal star,

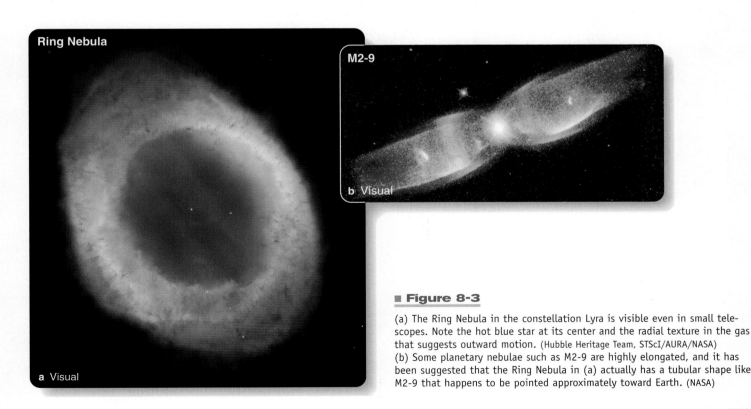

Ring Nebula

a Visual

M2-9

b Visual

■ Figure 8-3

(a) The Ring Nebula in the constellation Lyra is visible even in small telescopes. Note the hot blue star at its center and the radial texture in the gas that suggests outward motion. (Hubble Heritage Team, STScI/AURA/NASA)
(b) Some planetary nebulae such as M2-9 are highly elongated, and it has been suggested that the Ring Nebula in (a) actually has a tubular shape like M2-9 that happens to be pointed approximately toward Earth. (NASA)

An **open cluster** is a collection of 10
1000 stars in a region about 25 pc in
er. Some open clusters are quite small
me are large, but they all have an open,
rent appearance because the stars are
wded together.

In a star cluster each star
follows its orbit around the
center of mass of the cluster.

AURA/NOAO/NSF

Visual-wavelength image

**Open Cluster
The Jewel Box**

1a A **globular cluster** can contain 10^5 to 10^6
stars in a region only 10 to 30 pc in diameter.
The term "globular cluster" comes from the word
"globe," although globular cluster is pronounced
like "glob of butter." These clusters are nearly
spherical, and the stars are much closer together
than the stars in an open cluster.

**Globular Cluster
47 Tucanae**

Visual-wavelength image

ATTB

Astronomers can construct an H–R diagram for
a star cluster by plotting a point to represent the
luminosity and temperature of each star.

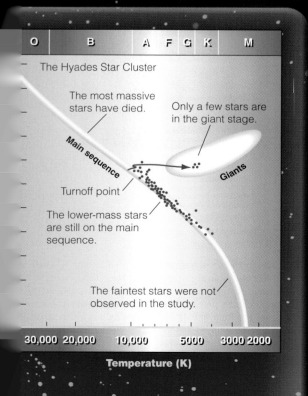

| O | B | A | F | G | K | M |

The Hyades Star Cluster

The most massive
stars have died.

Only a few stars are
in the giant stage.

Main sequence

Giants

Turnoff point

The lower-mass stars
are still on the main
sequence.

The faintest stars were not
observed in the study.

30,000 20,000 10,000 5000 3000 2000

Temperature (K)

2 The H–R diagram of a star cluster can make the evolution of stars
visible. The key is to remember that all of the stars in the star cluster
have the same age but differ in mass. The H–R diagram of a star cluster
provides a snapshot of the evolutionary state of the stars at the time you
happen to be alive. The diagram here shows the 650-million-year-old star
cluster called the Hyades. The upper main sequence is missing because
the more massive stars have died, and our snapshot catches a few
medium-mass stars leaving the main sequence to become giants.

As a star cluster ages, its main sequence grows shorter like a candle
burning down. You can judge the age of a star cluster by looking at the
turnoff point, the point on the main sequence where stars evolve to the
right to become giants. Stars at the **turnoff point** have lived out their lives
and are about to die. Consequently, the life expectancy of the stars at the
turnoff point equals the age of the cluster.

ThomsonNOW

Sign in at www.thomsonedu.com and go to ThomsonNOW
to see Active Figure "Cluster Turnoff" and notice how the
shape of a cluster's H–R diagram changes with time.

3 From theoretical models of stars, you could construct a film to show how the H–R diagram of a star cluster changes as it ages. You can then compare theory (left) with observation (right) to understand how stars evolve. Note that the time step for each frame in this film increases by a factor of 10.

Highest-mass stars evolving. Low-mass stars still contracting.

10^6 y

10^7 y

Upper main sequence stars have died.

10^8 y

10^9 y

Only the lower mass stars remain on the main sequence.

10^{10} y

NGC2264
Age 10^6 yr

Faintest stars not observed.

Pleiades
Age 10^8 yr

Turnoff point

Faintest stars not observed.

M67
Age 4×10^9 yr

Turnoff point

Faintest stars not observed.

Temperature (K)

NGC 2264 is a very young cluster still embedded in the nebula from which it formed. Its lower-mass stars are still contracting, and it is rich in T Tauri stars.

Visual   AATB

The nebula around the Pleiades is produced by gas and dust through which the cluster is passing. Its original nebula dissipated long ago.

Visual   Caltech

M67 is an old open cluster. In photographs, such clusters have a uniform appearance because they lack hot, bright stars. Compare with the Jewel Box on the opposite page.

Visual   Nigel Sharp, Mark Hanna/AURA/NOAO/NSF

Globular cluster H–R diagrams resemble the last frame in the film, which tells you that globular clusters are very old.

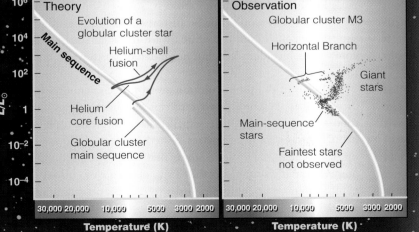

Theory

Evolution of a globular cluster star

Main sequence

Helium-shell fusion

Helium core fusion

Globular cluster main sequence

Temperature (K)

Observation

Globular cluster M3

Horizontal Branch

Giant stars

Main-sequence stars

Faintest stars not observed

Temperature (K)

3a The H–R diagrams of globular clusters have very faint turnoff points showing that they are very old clusters. The best analysis suggests these clusters are about 11 billion years old.

The horizontal branch stars are giants fusing helium in their cores and then in shells. The shape of the horizontal branch outlines the evolution of these stars.

The main-sequence stars in globular clusters are fainter and bluer than the zero-age main sequence. Spectra reveal that globular cluster stars are poor in elements heavier than helium, and that means their gases are less opaque. Therefore energy can flow outward more easily, which makes the stars slightly smaller and hotter. Again the shape of star cluster H–R diagrams illustrates principles of stellar evolution.

Slow stellar wind
from a red giant

Fast wind from
exposed interior

The gases of the slow wind
are not easily detectable.

You see a planetary nebula
where the fast wind
compresses the slow wind.

a

b

■ **Figure 8-4**

The process that produces planetary nebulae involves two stellar winds.
(a) First, as an aging giant, the star gradually blows away its outer layers
in a slow wind of cool gas that is not easily visible. (b) Once the hot interior of the star is exposed, it ejects a high-speed wind that can overtake
and compress the gas of the earlier slow wind like a snowplow, while UV
radiation from the hot remains of the central star excites the gases to
glow like a giant spectral line emission lamp.

which is supported by ordinary gas pressure, a white dwarf is
supported against its own gravity by the resistance to compression of a degenerate gas.

Clearly, a white dwarf is not a true star. It generates no nuclear energy, is almost completely degenerate matter, and, except
for a thin layer at its surface, contains no gas. Instead of calling a
white dwarf a "star," you could call it a **compact object.** Later
sections of this chapter discuss two other types of compact object, neutron stars and black holes.

A white dwarf 's future is bleak. As it radiates energy into
space, its temperature gradually falls, but it cannot shrink any
smaller because its degenerate electrons cannot get closer together. This degenerate matter is a very good thermal conductor,
so heat flows to the surface and escapes into space, and the white
dwarf gets fainter and cooler, moving downward and to the right
in the H–R diagram. Because the white dwarf contains a tremendous amount of heat, it needs billions of years to radiate that heat
through its small surface area. The coolest white dwarfs in our
galaxy are about the temperature of the sun.

Perhaps the most interesting thing astronomers have learned
about white dwarfs has come from mathematical models. The
equations predict that degenerate electron pressure cannot support
an object with more than about 1.4 solar masses. A white dwarf
with that mass would have such strong gravity that its radius would
shrink to zero. This is called the **Chandrasekhar limit** after
Subrahmanyan Chandrasekhar, the astronomer who calculated it.

This seems to imply that a star more massive
than 1.4 solar masses could not become a
white dwarf unless it got rid of the extra mass
in some way.

Can stars lose substantial amounts of
mass? Observations provide clear evidence
that young stars have strong stellar winds,
and aging giants and supergiants also lose
mass. This suggests that stars more massive
than the Chandrasekhar limit can eventually end up as white dwarfs if they reduce
their mass under the limit. Theoretical
models show that stars that begin life with
as much as 8 solar masses could lose mass
fast enough to reduce their mass so low
they can collapse to form white dwarfs
with masses below 1.4 solar masses. With
mass loss, a wide range of medium-mass
stars can eventually die as white dwarfs.

The Fate of the Sun and the End of Earth

Astronomy is about you. Although this chapter has been discussing the deaths of medium-mass stars, it has also been discussing the future of this planet Earth. The sun is a
medium-mass star and must eventually die by becoming a giant,
possibly producing a planetary nebula, and collapsing into a
white dwarf. That will spell the end of Earth.

Mathematical models of the sun suggest that it may survive
for another 5 billion years or so, but it is already growing more
luminous as it fuses hydrogen into helium. In a few billion years,
it will exhaust hydrogen in its core and swell into a giant star
about 100 times its present radius. That giant sun will be about
as large as the orbit of Earth, so that will mark the end of this
world. Whether the expanding sun becomes large enough to totally engulf Earth, its growing luminosity will certainly evaporate
Earth's oceans, drive away the atmosphere, and even vaporize
much of Earth's crust and drive the vapor away into space.

While it is a giant star, the sun will lose mass into space. This
mass loss is a relatively gentle process, so any cinder that might
remain of Earth would not be disturbed. Of course, when the
sun finally collapses into a white dwarf, the solar system will become a much colder place, but not much of Earth will remain by
then.

Models of the sun's evolution are not precise enough to predict whether its white dwarf remnant will become hot enough
soon enough to light up its expelled gas and create a planetary
nebula before that gas disperses. Whether the expelled gas lights
up or not, it would include atoms that were once part of Earth.
Some research also suggests that a star needs a close binary companion to speed up its spin in order to create a planetary nebula.

The sun, of course, has no close stellar companion. This is an area of active research, and there are as yet no firm conclusions.

There is no danger that the sun will explode as a nova; it has no binary companion (see the next section of this chapter). And, as you will see, the sun is definitely not massive enough to die the violent supernova death of the most massive stars.

The most important lesson of astronomy is that you are part of the universe and not just an observer. The atoms you are made of are destined to return to the interstellar medium in just a few billion years. You, and the planet Earth, began as stardust and will end as stardust.

Building Scientific Arguments

What produced the energy that makes a planetary nebula glow?
To build this scientific argument, you must think about the earlier history of the central star. The low-density gas in a planetary nebula emits light at certain wavelengths to produce an emission spectrum. The gas of the nebula is excited by ultraviolet radiation from the hot central object in the planetary nebula. That object is hot because it was once the core of the star that ejected its surface to create the planetary nebula. Furthermore, the inside of the star was hot because nuclear fusion reactions in and around the core fused hydrogen into helium and then helium into carbon. The light you see coming from a planetary nebula is energy that was originally locked inside the nuclei of hydrogen and helium atoms that became nuclear fuel.

The goal of science is to tell the story of natural objects. You have traced the life stories of medium-mass stars to their ultimate deaths as white dwarfs. Now revise your argument. **What has become of the hydrogen that was once contained in the core of a star that is now a white dwarf?**

■   ■   ■

(8-3) The Evolution of Binary Systems

STARS IN BINARY SYSTEMS CAN EVOLVE independently of each other if they orbit at a large distance from each other. In this situation, one of the stars can swell into a giant and collapse without disturbing its companion. Other binary stars are as close to each other as 0.1 AU, and when one of those stars begins to swell into a giant, its companion can suffer in peculiar ways.

Mass Transfer and Accretion Disks

Binary stars can sometimes interact by transferring mass from one star to the other. Of course, the gravitational field of each star holds its mass together, but the gravitational fields of the two stars, combined with the rotation of the binary system, define a dumbbell-shaped volume called the **Roche lobes** around the pair of stars. Matter inside a star's Roche lobe is gravitationally bound to the star, but matter outside the lobe can be transferred to the other star or lost completely from the system. If the stars are close together, the Roche lobes are relatively small and can interfere with the evolution of the stars. When an evolving star in a close binary system expands so far that it fills its Roche lobe, matter can flow from that star's lobe into the other lobe and onto the other star (■ Figure 8-5).

Matter flowing from one star to another cannot fall directly into the star. Rather, because of conservation of angular momentum, it must flow into a whirling disk around the star. **Angular**

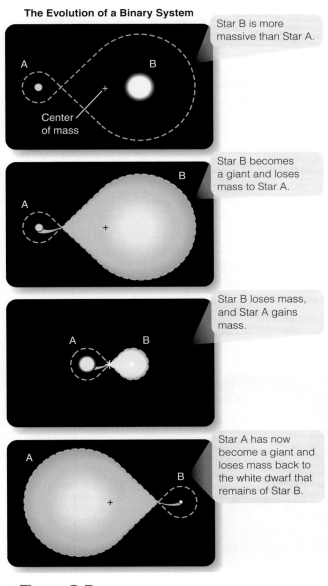

The Evolution of a Binary System

Star B is more massive than Star A.

A   B

Center of mass

Star B becomes a giant and loses mass to Star A.

A   B

Star B loses mass, and Star A gains mass.

A   B

Star A has now become a giant and loses mass back to the white dwarf that remains of Star B.

A   B

■ **Figure 8-5**

A pair of stars orbiting close to each other can exchange mass and modify each other's evolution.

momentum refers to the tendency of a rotating object to continue rotating. All rotating objects possess some angular momentum, and in the absence of external forces, an object maintains (conserves) its total angular momentum. An ice dancer takes advantage of conservation of angular momentum by starting a spin slowly with arms extended and then drawing them in. As her mass becomes concentrated closer to her axis of rotation, she spins faster. The same effect causes the slowly circulating water in a bathtub to spin faster in a whirlpool as it approaches the drain.

Mass transferred from one star to another in a binary must conserve its angular momentum. Thus it must flow into a rapidly rotating whirlpool called an **accretion disk** around the second star (■ Figure 8-6). The gas in the disk grows hot due to friction and tidal forces (see Chapter 2) and eventually falls onto the second star. If that second star, the one receiving the matter lost from its companion, is a compact object like a white dwarf, the gas in the accretion disk can become very compressed. The gas temperature can exceed a million K, producing X rays. In addition, the matter accumulating on the white dwarf can eventually cause a violent explosion called a nova.

Novae

At the beginning of this chapter you saw that the word *nova* refers to an apparently new star that appears in the sky for a while and then fades away. Modern astronomers know that a nova is not a new star but an old star flaring up. After a nova fades, astronomers can photograph the spectrum of the remaining faint point of light. Invariably, they find a normal star and a white dwarf in a close binary system. A nova is evidently an explosion involving a white dwarf.

Observational evidence can tell you how nova explosions occur. As the explosion begins, spectra show blueshifted absorption lines that tell you the gas is dense and coming toward you at a

few thousand kilometers per second. After a few days, the spectral lines change to emission lines, which tells you the gas has thinned, but the blueshifts remain, showing that a cloud of debris has been ejected into space.

Nova explosions occur when mass transfers from a normal star into an accretion disk around a white dwarf. As the matter loses its angular momentum in the accretion disk, it settles inward onto the surface of the white dwarf and forms a layer of unused nuclear fuel—mostly hydrogen. As the layer deepens, it becomes denser and hotter until the hydrogen fuses in a sudden explosion that blows the surface off the white dwarf. Although the expanding cloud of debris contains less than 0.0001 solar mass, it is hot, and its expanding surface area makes it very luminous. Nova explosions can become 100,000 times more luminous than the sun. As the debris cloud expands, cools, and thins over a period of weeks and months, the nova fades from view.

The explosion of its surface hardly disturbs the white dwarf and its companion star. Mass transfer quickly resumes, and a new layer of fuel begins to accumulate. How fast the fuel builds up depends on the rate of mass transfer. Accordingly, you can expect novae to repeat each time an explosive layer accumulates. Many novae take thousands of years to build an explosive layer, but some take only decades.

8-4 The Deaths of Massive Stars

YOU HAVE SEEN THAT LOW- AND MEDIUM-MASS STARS DIE relatively quietly as they exhaust their hydrogen and helium, and then some push away their surface layers to form planetary nebulae. In contrast, massive stars live spectacular lives (■ Figure 8-7) and destroy themselves in violent explosions.

Nuclear Fusion in Massive Stars

Stars on the upper main sequence have too much mass to die as white dwarfs, but their evolution begins much like that of their lower-mass cousins. They consume the hydrogen in their cores, ignite hydrogen shells, and become giants or, for the most massive stars, supergiants. Their cores contract and fuse helium first in the core and then in a shell, producing a carbon–oxygen core.

Unlike medium-mass stars, the massive stars finally can get hot enough to ignite carbon fusion at a temperature of about 1 billion Kelvin. Carbon fusion produces more oxygen plus neon. As soon as the carbon is exhausted in the core, the core contracts, and carbon ignites in a shell. This pattern of core ignition and shell ignition continues with a series of heavier nuclei as fusion fuel, and the star develops a layered structure as shown in Figures 8-2 and 8-7, with a hydrogen-fusion shell surrounding a helium-fusion shell surrounding a carbon-fusion shell surrounding . . . and so on. At higher temperatures than carbon fusion, nuclei of

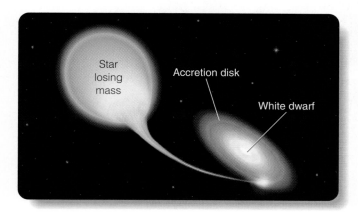

■ **Figure 8-6**

Matter falling into a compact object forms a whirling accretion disk. Friction and tidal forces can make the disk very hot.

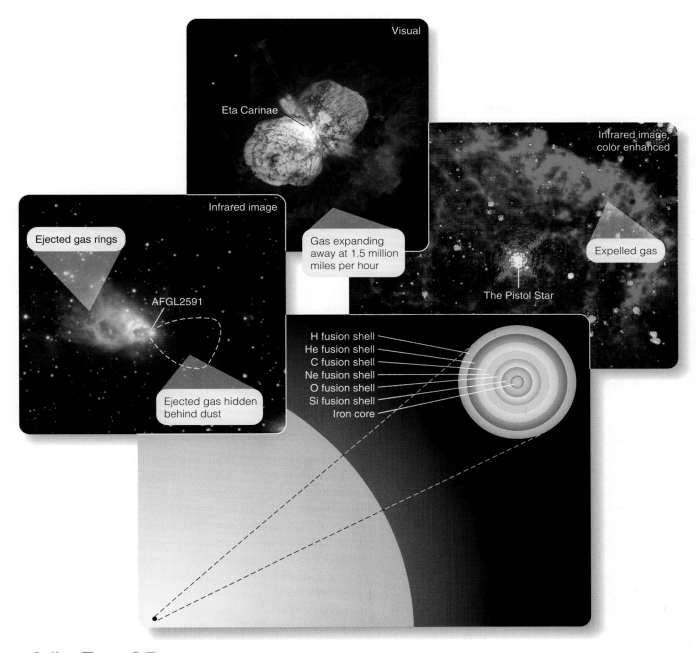

Visual

Eta Carinae

Infrared image

Ejected gas rings

AFGL2591

Ejected gas hidden behind dust

Gas expanding away at 1.5 million miles per hour

Infrared image, color enhanced

Expelled gas

The Pistol Star

H fusion shell
He fusion shell
C fusion shell
Ne fusion shell
O fusion shell
Si fusion shell
Iron core

■ **Active Figure 8-7**

Massive stars live fast and die young. The three shown here are among the most massive stars known, containing 100 solar masses or more. They are rapidly ejecting gas into space. The centers of these massive stars develop Earth-size cores (magnified 100,000 times in this figure) composed of concentric layers of gases undergoing nuclear fusion. The iron core at the center leads eventually to a star-destroying supernova explosion. (AFGL2591: Gemini Observatory/NSF/C. Aspin; Eta Carinae and the Pistol star: NASA)

ThomsonNOW™ **Sign in at www.thomsonedu.com and go to ThomsonNOW to see the Active Figure "Inside Stars."**

oxygen, neon, and magnesium fuse to make silicon and sulfur, and at even higher temperatures silicon can fuse to make iron.

The fusion of the nuclear fuels in this series goes faster and faster as the massive star evolves rapidly. The amount of energy released per fusion reaction decreases as the mass of the types of atoms involved increases. To support its weight and remain stable, a star must fuse oxygen much faster than it fused hydrogen. Also, there are fewer nuclei in the core of the star by the time heavy nuclei begin to fuse. Four hydrogen nuclei made one helium nucleus, and three helium nuclei make one carbon, so there

■ Table 8-1 | Heavy-Element Fusion in a 25-$M_\odot$ Star

Fuel	Time	Percentage of Lifetime
H	7,000,000 years	93.3
He	500,000 years	6.7
C	600 years	0.008
O	0.5 years	0.000007
Si	1 day	0.00000004

are 12 times fewer nuclei of carbon available for fusion than there were hydrogen nuclei. This means the heavy elements are used up, and fusion goes very quickly in massive stars (■ Table 8-1). Hydrogen fusion can last 7 million years in a 25-solar-mass star, but that same star will fuse its oxygen in 6 months and its silicon in just one day.

Supernova Explosions of Massive Stars

Theoretical models of evolving stars combined with nuclear physics allow astronomers to describe what happens inside a massive star when the last nuclear fuels are exhausted. It begins with iron nuclei and ends in cosmic violence.

Silicon fusion produces iron, the most tightly bound of all atomic nuclei (see Figure 7-4). Nuclear fusion releases energy only when less tightly bound nuclei combine into a more tightly bound nucleus. Once the gas in the core of the star has been converted to iron, there are no further nuclear reactions that can release energy. The iron core is a dead end in the evolution of a massive star.

As a star develops an iron core, energy production begins to decline, and the core contracts. Nuclear reactions involving iron begin, but they remove energy from the core, causing it to contract even further. Once this process starts, the core of the star collapses inward in less than a tenth of a second.

The collapse of a giant star's core after iron fusion starts is calculated to happen so rapidly that the most powerful computers are unable to predict the details. Thus, models of supernova explosions contain many approximations. Nevertheless, the models predict exotic nuclear reactions in the collapsing core that should produce a flood of neutrinos (see Chapter 7). In fact, for a short time the core produces more energy per second than all of the stars in all of the visible galaxies in the universe, and 99 percent of that energy is in the form of neutrinos. This flood of neutrinos carries large amounts of energy out of the core, allowing the core to collapse further. The models also predict that the collapsing core of the star must quickly become a neutron star or

a black hole, the subjects of the next sections of this chapter, while the envelope of the star is blasted outward.

To understand how the inward collapse of the core can produce an outward explosion, you can think about a traffic jam. The collapse of the innermost part of the degenerate core allows the rest of the core to fall inward, and this creates a tremendous traffic jam as all of the nuclei fall toward the center. The position of the traffic jam, called a shock wave, begins to move outward as more in-falling material encounters the jam. The torrent of neutrinos, as well as energy flowing out of the core in sudden violent convective turbulence, help drive the shock wave outward. Within a few hours, the shock wave bursts outward through the surface of the star and blasts it apart.

The supernova seen from Earth is the brightening of the star as its envelope is blasted outward by the shock wave. As months pass, the cloud of gas expands, thins, and fades, but the rate at which it fades tells astronomers more about the death throes of the star. The rate at which the supernova fades matches the rate at which radioactive nickel and cobalt decay, so the explosion must produce great abundances of those atoms. The radioactive cobalt decays into iron, so destruction of iron in the core of the star is followed by the production of iron through nuclear reactions in the expanding outer layers.

Types of Supernovae

In studying supernovae in other galaxies, astronomers have noticed that there are a number of different types. **Type I supernovae** have no hydrogen lines in their spectra, and astronomers have thought of at least two ways a supernova could occur without involving much hydrogen (see below). **Type II supernovae**, in contrast, have spectra containing hydrogen lines and appear to be produced by the collapse and explosion of a massive star, the process discussed in the previous section.

A type Ia supernova is believed to occur when a white dwarf in a binary system receives enough mass to exceed the Chandrasekhar limit and collapse. The collapse of a white dwarf is different from the collapse of a massive star because the core of the white dwarf contains usable fuel. As the collapse begins, the temperature and density shoot up, and the carbon–oxygen core begins to fuse in violent nuclear reactions. In a few seconds, the carbon–oxygen interior is entirely consumed, and the outermost layers are blasted away in a violent explosion that, at its brightest, is about six times more luminous than a type II supernova. The white dwarf is entirely destroyed; no neutron star or black hole is left behind. This explains why no hydrogen lines are seen in the spectrum of a type Ia supernova explosion—white dwarfs contain very little hydrogen.

The less common type Ib supernova is believed to occur when a massive star in a binary system loses its hydrogen-rich outer layers to its companion star. The remains of the massive star could develop an iron core and collapse, as described in the previous section, producing a supernova explosion that lacked hydrogen lines in its spectrum.

Astronomers working with the largest and fastest computers are using modern theory to try to understand supernova explosions. The companion to theory is observation, so you should ask what observational evidence supports this story of supernova explosions.

Observations of Supernovae

In AD 1054, Chinese astronomers saw a "guest star" appear in the constellation known in the Western tradition as Taurus the Bull. The star quickly became so bright it was visible in the daytime, and then, after a month, it slowly faded, taking almost two years to vanish from sight. When modern astronomers turned their telescopes to the location of the guest star, they found a peculiar nebula now known as the Crab Nebula for its many-legged shape. In fact, the legs of the Crab Nebula are filaments of gas that are moving away from the site of the explosion at about 1400 km/s. Comparing the radius of the nebula, 1.35 pc, with its velocity of expansion reveals that the nebula began expanding nine or ten centuries ago, just when the guest star made its appearance. The Crab Nebula is clearly the remains of the supernova seen in AD 1054 (■ Figure 8-8). In the next section, you will meet the neutron star found at the center of the Crab Nebula.

The blue glow of the Crab Nebula is produced by **synchrotron radiation.** This form of electromagnetic radiation, unlike black body radiation, is produced by rapidly moving electrons spiraling through magnetic fields and is common in the nebulae produced by supernovae. In the case of the Crab Nebula, the electrons travel so fast they emit visual wavelengths. In most such nebulae, the electrons move slower, and the synchrotron radiation is at radio wavelengths.

Supernovae are rare. Only a few have been seen with the naked eye in recorded history. Arab astronomers saw one in AD 1006, and the Chinese saw one in AD 1054. European astronomers observed two—one in AD 1572 (Tycho's supernova) and one in AD 1604 (Kepler's supernova). In addition, the guest stars of AD 185, 386, 393, and 1181 may have been supernovae.

In the centuries following the invention of the astronomical telescope in 1609, no supernova was bright enough to be visible to the naked eye. Then, in the early morning hours of February 24, 1987, astronomers around the world were startled by the discovery of a naked-eye supernova still growing brighter in the southern sky (■ Figure 8-9). The supernova, known officially as SN1987A, is 53,000 pc away in the Large Magellanic Cloud, a small satellite galaxy to our own Milky Way Galaxy. This first

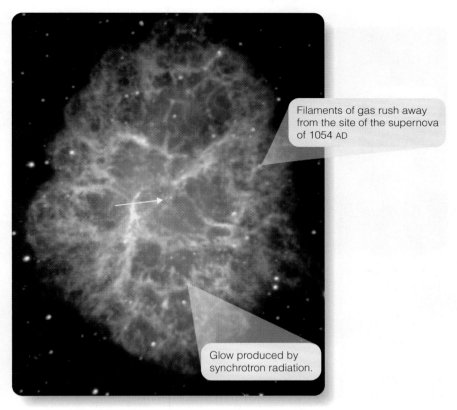

Filaments of gas rush away from the site of the supernova of 1054 AD

Glow produced by synchrotron radiation.

■ Figure 8-8

The Crab Nebula is located in the constellation Taurus the Bull, just where Chinese astronomers saw a brilliant guest star in AD 1054. High-speed electrons produced by the central neutron star spiral through magnetic fields and produce the foggy glow of synchrotron radiation that fills the nebula. The pulsar itself (arrow) is detectable at visual wavelengths. The star just to the right of the pulsar lies much closer to Earth and is not in the Crab Nebula. (Caltech)

naked-eye supernova in 383 years has given astronomers a ringside seat for the most spectacular event in stellar evolution.

One observation of SN1987A is critical in that it confirms the theory of core collapse. At 2:35 AM EST on February 23, 1987, nearly 4 hours before the supernova was first seen, a blast of neutrinos swept through Earth. Instruments buried in a salt mine under Lake Erie and in a lead mine in Japan, though designed for another purpose, recorded 19 neutrinos in less than 15 seconds. Neutrinos are so difficult to detect that the 19 neutrinos actually detected mean that some 10^{17} neutrinos must have passed through the detectors in those 15 seconds. Furthermore, the neutrinos were arriving from the direction of the supernova. Thus, astronomers conclude that the burst of neutrinos was released when the iron core collapsed, and the supernova was first seen at visual wavelengths hours later when the shock wave blasted the star's surface into space.

Most supernovae are seen in distant galaxies, and careful observations allow astronomers to compare types (■ Figure 8-10). Type Ia supernovae, caused by the collapse of white dwarfs, are more luminous at maximum brightness and decline rapidly at first and then more slowly. Type II supernovae, produced by the

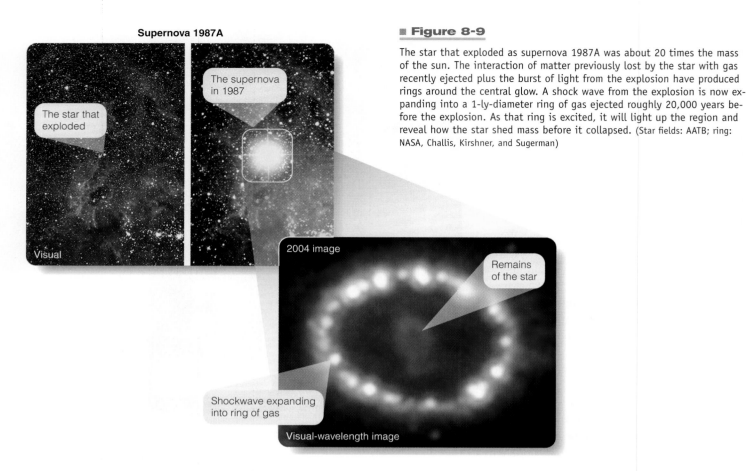

Supernova 1987A

The star that exploded

The supernova in 1987

Visual

2004 image

Remains of the star

Shockwave expanding into ring of gas

Visual-wavelength image

■ **Figure 8-9**

The star that exploded as supernova 1987A was about 20 times the mass of the sun. The interaction of matter previously lost by the star with gas recently ejected plus the burst of light from the explosion have produced rings around the central glow. A shock wave from the explosion is now expanding into a 1-ly-diameter ring of gas ejected roughly 20,000 years before the explosion. As that ring is excited, it will light up the region and reveal how the star shed mass before it collapsed. (Star fields: AATB; ring: NASA, Challis, Kirshner, and Sugerman)

collapse of massive stars, are not as bright at maximum, and they decline in a more irregular way.

SN1987A was a type II supernova, although its light curve shown in Figure 8-10 is not typical. Models indicate that most type II supernovae are caused by the collapse of red supergiants, but SN1987A was produced by the explosion of a hot, blue supergiant. Astronomers hypothesize that this star was once a red supergiant but later contracted and heated up slightly, becoming bluer before it exploded.

Although the supernova explosion fades to obscurity in a year or two, an expanding shell of gas marks the site of the explosion. The gas, originally expelled at 10,000 to 20,000 km/s, may carry away a fifth of the mass of the star. The collision of that expanding gas with the surrounding interstellar medium can sweep up even more gas and excite it to produce a **supernova remnant,** the nebulous remains of a supernova explosion. ■ Figure 8-11 show images of 4 supernova remnants; images made at other than visible wavelengths are displayed in false color (see Chapter 4).

Supernova remnants look quite delicate and do not survive very long—a few tens of thousands of years—before they gradually mix with the interstellar medium and vanish. The Crab Nebula is a young remnant, only about 950 years old and about 8.8 ly in diameter. Older remnants can be larger. Some supernova remnants are visible only at radio and X-ray wavelengths. They have become too tenuous to emit detectable light, but the collision of the expanding hot gas with the interstellar medium can generate radio and X-ray radiation. You learned in Chapter 7 that the compression of the interstellar medium by expanding supernova remnants can also trigger star formation.

Gravity always wins. However a star lives, theory predicts it must eventually die and leave behind one of three types of final

■ **Figure 8-10**

Type I supernovae decline in brightness rapidly at first and then more slowly, but type II supernovae pause for about 100 days before beginning a steep decline. Supernova 1987A was odd in that it did not rise directly to maximum brightness. These light curves have been adjusted to the same maximum brightness.

The supernova remnant called the Cygnus Loop is 5000 to 10,000 years old and 80 ly in diameter.

Visible light produced by gas expanding into surrounding interstellar medium

Supernova remnant N132D is 3000 years old and 80 ly in diameter. It is 180,000 ly from Earth.

Shock waves in expanding gas heat it to millions of degrees, and it emits X rays.

Cassiopeia A (Cas A) is a supernova remnant about 300 years old and 10 ly in diameter.

X-ray image

Cas A is also bright at radio wavelengths.

Visual-wavelength image

Infrared radiation from dust condensing out of the gas

Infrared image

Radio image

■ **Figure 8-11**

A supernova remnant is an expanding bubble of hot gas created by a supernova explosion. As the remnant expands and pushes into neighboring gas, it can emit radiation at many wavelengths. (Cygnus Loop: Caltech; SN1006: NASA/CXC/J. Hughes et al.; Cas A: NASA/CXC/GSFC/U. Hwang et al.)

remnant—a white dwarf, neutron star, or black hole. These compact objects are small monuments to the power of gravity. Almost all of the energy available has been squeezed out of compact objects, and you find them in their final, high-density states.

8-5 Neutron Stars

A **NEUTRON STAR** CONTAINS A LITTLE OVER 1 SOLAR MASS compressed to a radius of about 10 km. Its density is so high that the matter is stable only as a fluid of pure neutrons. Two questions should occur to you immediately. First, how could any theory

predict such a strange object? And second, how do you know they exist?

Theoretical Prediction of Neutron Stars

The subatomic particles called neutrons were discovered in a laboratory in 1932, and physicists quickly realized that because neutrons spin much like electrons, a gas of neutrons could become degenerate and therefore nearly incompressible. Just two years later, in 1934, two astronomers, Walter Baade and Fritz Zwicky, suggested that some of the most luminous novae in the historical record were not regular novae but were caused by the collapse plus explosion of a massive star in a cataclysm they named a "super-

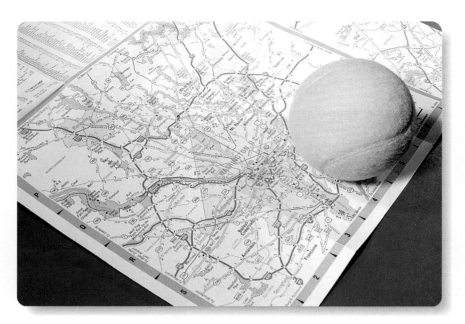

■ Figure 8-12

A tennis ball and a road map illustrate the relative size of a neutron star. Such an object, containing slightly more than the mass of the sun, would fit with room to spare inside the beltway around Washington, D.C. (Photo by M. Seeds)

nova." If the collapsing core is more massive than the Chandrasekhar limit of 1.4 solar masses, then the weight is too great to be supported by degenerate electron pressure, and the core cannot become a stable white dwarf. The collapse would force protons to combine with electrons and become neutrons. The envelope of the star would be blasted away in a supernova explosion, and the core of the star would be left behind as a small tremendously dense sphere of neutrons that Zwicky called a "neutron star."

Mathematical models predict that a neutron star will be only 10 or so kilometers in radius (■ Figure 8-12) and will have a density of about 10^{14} g/cm^3. That is roughly the density of atomic nuclei, and you can think of a neutron star as matter with all of the empty space squeezed out of it. On Earth, a sugar-cube-sized lump of this material would weigh 100 million tons, the mass of a small mountain.

Simple physics, the physics you have used in previous chapters to understand normal stars, predicts that neutron stars should (1) spin rapidly, perhaps 100 to 1000 rotations per second; (2) be hot, with surface temperatures of millions of degrees K; and (3) have strong magnetic fields, up to a trillion times stronger than the sun's or Earth's magnetic fields. For example, the collapse of a massive star's core would greatly increase its spin rate by conservation of angular momentum. Other processes during core collapse should create high temperature and magnetic field strength. Despite their high temperature, neutron stars should be difficult to detect because of their tiny size.

What is the maximum mass for a stable neutron star? In other words, is there an upper limit to the mass of neutron stars

like the Chandrasekhar limit that defines the maximum mass of a white dwarf star? That is difficult to answer because physicists don't know enough about the properties of pure neutron material. It can't be made in a laboratory, and theoretical calculations in this case are very difficult. The most widely accepted results suggest that a neutron star can't be more massive than 2 to 3 solar masses. An object more massive than that can't be supported by degenerate neutron pressure, so it would collapse, presumably becoming a black hole.

What size stars will end their lives with supernova explosions that leave behind neutron star corpses? Theoretical calculations suggest that stars that begin life on the main sequence with 8 to about 15 solar masses will end up as neutron stars. Stars more massive than about 15 solar masses are believed to form black holes when they die.

The Discovery of Pulsars

In November 1967, Jocelyn Bell, a graduate student at Cambridge University in England, found a peculiar pattern in the data from a radio telescope. Unlike other radio signals from celestial bodies, this was a series of regular pulses (■ Figure 8-13). At first she and the leader of the project, Anthony Hewish, thought the signal was interference, but they found it day after day at the same celestial latitude and longitude. Clearly, it was cosmic in origin.

Another possibility, that it came from a distant civilization, led them to consider naming it LGM for Little Green Men. Within a few weeks, the team found three more objects in other parts of the sky pulsing with different periods. The objects were clearly natural, and the team dropped the name LGM in favor of **pulsar**—a contraction of "pulsing star." The pulsing radio source Bell had observed with her radio telescope was the first known pulsar. Hewish received the Nobel Prize in physics for this work, and Bell (now Bell Burnell) has been remarkably gracious about that.

■ Figure 8-13

The 1967 detection of regularly spaced pulses in the output of a radio telescope led to the discovery of pulsars. This record of the radio signal from the first pulsar, CP1919, contains regularly spaced pulses (marked by ticks). The period is 1.33730119 seconds.

As more pulsars were found, astronomers argued over their nature. The pulses, which typically last only about 0.001 second, gave astronomers an important clue. The pulse length places an upper limit on the size of the object producing the pulse. This is a very important principle in astronomy—an object cannot change its brightness significantly in an interval shorter than the time light takes to cross its diameter. If pulses from pulsars are no longer than 0.001 seconds, then the objects cannot be larger than 0.001 light seconds, or 300 km (190 mi) in diameter, smaller than white dwarfs, which makes neutron stars the only reasonable explanation.

The missing link between pulsars and neutron stars was found in late 1968, when astronomers discovered a pulsar at the heart of the Crab Nebula (see Figure 8-8). The Crab Nebula is a supernova remnant, which agrees nicely with Zwicky and Baade's prediction that some supernovae should produce a neutron star. The short pulses and the discovery of the pulsar in the Crab Nebula are strong evidence that pulsars are neutron stars.

The modern model of a pulsar has been called the **lighthouse model** (■ Figure 8-14). The name *pulsar* is, in a sense, inaccurate: A pulsar does not pulse (vibrate) but rather emits beams of radiation that sweep around the sky as the neutron star rotates, like a rotating lighthouse light. A name of "spinar" would have been more accurate. The mechanism that produces the beams involves extremely high energies and strong electric and magnetic fields and is not fully understood. Over 1000 pulsars are now known. There may be many more which are undetected because their beams never point toward Earth.

The Evolution of Pulsars

Neutron stars are not simple objects, and modern astronomers need knowledge of frontier physics to understand them. Nevertheless, the life story of pulsars can be worked out to some extent. When a pulsar first forms, it may be spinning as many as 100 times a second. The energy it radiates into space ultimately comes from its energy of rotation, so as it blasts beams of radiation outward, its rotation slows. Judging from their pulse periods and rates at which they slow down, the average pulsar is apparently only a few million years old and the oldest has an age of about 10 million years. Presumably, neutron stars older than that rotate too slowly to generate detectable radio beams. Supernova remnants last only about 50,000 years before they mix into the interstellar medium and disappear, so most pulsars have long outlived the remnants in which they were originally embedded. You can expect that a young neutron star should emit especially strong beams of radiation powered by its rapid rotation. The Crab Nebula provides an example of such a system. Only about 950 years old and spinning 30 times per second, the Crab pulsar is so powerful that astronomers can detect photons of radio, infrared, visible, X-ray, and gamma-ray wavelengths from it.

The explosion of Supernova 1987A in February 1987 probably formed a neutron star. You can draw this conclusion because

Neutron Star Rotation with Beams

The pulsing of a lighthouse is actually caused by the rotation of beams of light.

As in the case of Earth, the magnetic axis of a neutron star could be inclined to its rotational axis.

The rotation of the neutron star will sweep its beams around like beams from a lighthouse.

While a beam points roughly toward Earth, observers detect a pulse.

While neither beam is pointed toward Earth, observers detect no energy.

■ **Figure 8-14**

A neutron star contains a powerful magnetic field and spins very rapidly. The spinning magnetic field generates a tremendously powerful electric field, and the field causes the creation of electron-positron pairs which accelerate through the magnetic field, emit photons in the directions of their motion, and thereby produce powerful beams of electromagnetic radiation emerging from the neutron star's magnetic poles.

a burst of neutrinos was detected passing through Earth, and theory predicts that the collapse of a massive star's core into a ball of neutrons would produce such a burst of neutrinos. The neutron star initially should be hidden at the center of the expanding shells of gas ejected into space by the supernova explosion, but as the gas continues to expand and become thinner, you can expect that astronomers might eventually be able to detect it. As of this writing, almost 20 years after the explosion, no neutron star has been detected in the SN1987A remnant, but astronomers continue to watch the site hoping to find the youngest pulsar known.

Binary Pulsars

One reason pulsars are so fascinating is the extreme conditions found in spinning neutron stars. To see even more extreme natural processes, you have only to look at the pulsars that are members of binary systems. These pulsars are of special interest because astronomers can learn more about the neutron star and about the behavior of matter in unusual circumstances by studying them.

Binary pulsars can be sites of tremendous violence because of the strength of gravity at the surface of a neutron star. Matter falling onto a neutron star can release titanic amounts of energy. If you dropped a single marshmallow onto the surface of a neutron star from a distance of 1 AU, it would hit with an impact equivalent to a 3-megaton nuclear warhead. Even a small amount of matter flowing from a companion star to a neutron star can generate high temperatures and release X rays and gamma rays.

As an example of such an active system, examine Hercules X-1 (■ Figure 8-15). It emits pulses of X rays with a period of about 1.2 seconds, but every 1.7 days the pulses vanish for a few hours. Hercules X-1 seems to contain a 2-solar-mass star and a neutron star that orbit each other with a period of 1.7 days. Matter flowing from the normal star into an accretion disk around the neutron star reaches temperatures of millions of degrees and emits a powerful X-ray glow, some of which is in beams that sweep around with the rotating neutron star. Earth receives a pulse of X rays every time a beam points this way. The X rays shut off completely every 1.7 days when the neutron star is eclipsed behind the normal star. Hercules X-1 is a complex system with many different high-energy processes going on simultaneously, and this quick analysis only serves to illustrate how complex and powerful such binary systems are during mass transfer.

A binary system in which both objects are neutron stars was discovered in 1974 when radio astronomers Joseph Taylor and Russell Hulse noticed that the pulse period of the pulsar PSR1913+16 grew longer and then shorter in a cycle that takes 7.75 hours. Taylor and Hulse realized that must be the binary orbital period of the pulsar. They analyzed the system with the same techniques used to study spectroscopic binary stars (see Chapter 6) to find that PSR1913+16 consists of two neutron stars separated by a distance roughly equal to the radius of our sun. The masses of the two neutron stars are each about 1.4 solar masses, in good agreement with models of neutron stars and how they are created.

A nice surprise was hidden in the motion of PSR1913+16. In 1916 Einstein published his **general theory of relativity** that

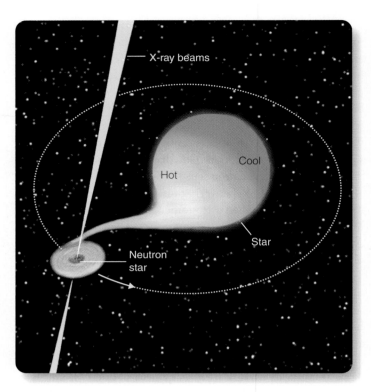

■ Figure 8-15

In Hercules X-1, matter flows from a star into an accretion disk around a neutron star producing X rays, which heat the near side of the star to 20,000 K compared with only 7000 K on the far side. X rays turn off when the neutron star is eclipsed behind the star.

described gravity as a curvature of space-time. Einstein realized that any rapid change in a gravitational field should spread outward at the speed of light as **gravitational radiation.** Gravity waves themselves have not been detected yet, but Taylor and Hulse were able to show that the orbital period of the binary pulsar is slowly growing shorter because the stars are gradually spiraling toward each other exactly at the rate expected if they radiate orbital energy away as gravitational radiation. Taylor and Hulse won the Nobel Prize in 1993 for their work with binary pulsars.

The Fastest Pulsars

This discussion of pulsars suggests that newborn pulsars should blink rapidly, and old pulsars should blink slowly, but a few that blink the fastest may be quite old. A number of **millisecond pulsars** have been found, so called because their pulse periods are almost as short as a millisecond (0.001 s). The energy stored in a neutron star rotating at this rate is equal to the total energy of a supernova explosion, and these pulsars generally have weak magnetic fields consistent with advanced age, so it seemed difficult at first to understand their rapid rotation. Astronomers hypothesized that an old neutron star could gain rotational energy from a companion star in a binary system. Some of the millisecond

The Impossibility of Proof in Science

No scientific theory or hypothesis can be proved correct. You can test a hypothesis over and over by performing experiments or making observations, but you can never prove that the hypothesis is absolutely true. It is always possible that you have misunderstood the hypothesis or the evidence, and the next observation you make might disprove the hypothesis.

For example, you might propose the hypothesis that the sun is mostly calcium and iron vapor because those elements are known to be abundant on Earth. You might test the hypothesis by looking at the calcium and iron lines in the solar spectrum, and the strength of the lines would suggest your hypothesis is right. Although your observation has confirmed your hypothesis, it has not proven the hypothesis is right. You might confirm the hypothesis many times in many ways before you realized that at the temperature of the sun, iron and calcium atoms absorb photons much more efficiently than hydrogen. Although the hydrogen lines are weak in the sun's spectrum, a careful analysis shows that most of the atoms in the sun are hydrogen and not iron or calcium.

The nature of scientific thinking can lead to two common misconceptions. Sometimes nonscientists will say, "You scientists just want to tear everything down—you don't believe in anything." Scientists test a hypothesis over and over to test its worth. If a hypothesis survives many tests, scientists begin to have confidence it is true, and the hypothesis "graduates" to being considered a theory.

The second common misconception arises when nonscientists say, "You scientists are never sure of anything." Again, the scientist knows that no hypothesis can be proved correct. That the sun will rise tomorrow is very likely, and scientists have great confidence in that theory. But it is still only a theory. People will say of an idea they dislike, "That is only a theory," as if a theory were simply a random guess. In fact, a theory is a hypothesis that has been well tested many times and so has become

It is only a theory, but astronomers have tremendous confidence that the sun is made almost entirely of hydrogen and helium. (SOHO/MDI)

a truth in which all scientists have great confidence. Yet you can never prove that any hypothesis or theory is absolutely true.

---

pulsars are caught in the act of receiving matter from companions in a fashion that should speed the pulsar's rotation to the observed high rates, so the hypothesis seems to be confirmed.

"Show me," say scientists; and, in the case of neutron stars, the evidence seems very strong. Of course, you can never prove a scientific hypothesis or theory is absolutely true (**Window on Science 8-1**), but the evidence for neutron stars is so strong that astronomers have great confidence that they really do exist. Other theories that describe how they emit beams of radiation and how they form and evolve are less certain, but continuing observations at many wavelengths are expanding the understanding of these last embers of massive stars. In fact, precise observations have turned up objects no one expected.

Pulsar Planets

Because a pulsar's period is so precise, astronomers can detect tiny variations by comparison with atomic clocks. When astronomers checked pulsar PSR1257+12, they found variations in the period of pulsation (■ Figure 8-16a) analogous to the variations caused by the orbital motion of the binary pulsar but much smaller. When these variations were interpreted as Doppler shifts, it became evident that the pulsar was being orbited by at least two objects with planetlike masses of about 4

and 3 Earth masses. The gravitational tugs of the planets make the pulsar wobble about the center of mass of the system by about 800 km, and that produces the tiny changes in period that are observed (Figure 8-16b).

Astronomers greeted this discovery with both enthusiasm and skepticism. As usual, they looked for ways to test the hypothesis. Simple gravitational theory predicts that the planets should interact and slightly modify each other's orbit. When the data were analyzed, that interaction was found, further confirming the hypothesis that the variations in the period of the pulsar are caused by planets. In fact, more observations and analyses revealed the presence of a third planet of about twice the mass of Earth's moon, and a fourth planet of only 3 percent the mass of Earth's moon. This illustrates the astonishing precision of studies based on pulsar timing.

Astronomers wonder how a neutron star can have planets. The inner three planets that orbit PSR1257+12 are closer to the pulsar than Venus is to the sun. Any planets that orbit a star would be lost or vaporized when the star exploded. Furthermore, a star about to explode as a supernova would be a large giant or a supergiant, and planets only a few AU distant would be inside such a large star and could not survive. It seems more likely that these planets are the remains of a stellar companion that was devoured by the neutron star. In fact, PSR1257+12 is very fast

Figure 8-16

(a) The dots in this graph are observations showing that the period of pulsar PSR1257+12 varies from its average value by a fraction of a billionth of a second. The blue line shows the variation that would be produced by planets orbiting the pulsar. (b) As the planets orbit the pulsar, they cause it to wobble by less than 800 km, a distance that is invisibly small in this diagram. (Adapted from data by Alexander Wolszczan)

(162 pulses per second), suggesting that it was spun up in a binary system.

Building Scientific Arguments

Why are neutron stars detectable at X-ray wavelengths?
This argument draws together a number of ideas you know from previous chapters. First, you should remember that a neutron star is very hot because of the heat released when the core of the original star contracted to a radius of only 10 km. It could easily have a surface temperature of 1,000,000 K, and Wien's law (Chapter 5) tells you that such an object will radiate most intensely at very short wavelengths: X rays. However, you know that the total luminosity of a star depends on both its surface temperature and its surface area, and a neutron star is so small it can't radiate much energy. X-ray telescopes have found such neutron stars, but they are not easy to locate.

There is, however, a second way a neutron star can radiate X rays. If a normal star in a binary system loses mass to a neutron star companion, the inflowing matter will form a very hot accretion disk that can radiate intense X rays easily detectable by X-ray telescopes orbiting above Earth's atmosphere. Further, any matter that hits the surface of the neutron star will arrive with so much energy that it will be heated to very high temperatures and radiate X rays.

Now build a new argument as if you were seeking funds for a research project. **What observations would you make to determine whether a newly discovered pulsar was young or old, single or a member of a binary system, alone or accompanied by planets?**

■   ■   ■

8-6 Black Holes

YOU HAVE STUDIED WHITE DWARFS AND NEUTRON STARS, two of the three end states of dying stars. Now it's time to think about the third end state—black holes.

Although the physics of black holes is difficult to discuss without sophisticated mathematics, simple logic is sufficient to predict that they should exist. The problem is to use their predicted properties and attempt to confirm that they exist. What objects observed in the heavens could be real black holes? More difficult than the search for neutron stars, the quest for black holes has nevertheless met with success.

To begin, you must consider a simple question. How fast must an object travel to escape from the surface of a celestial body? The answer will lead to black holes.

Escape Velocity

In Chapter 3 you learned that the escape velocity is the initial velocity an object needs to escape from a celestial body. Whether you are discussing a baseball leaving Earth or a photon leaving a collapsing star, the escape velocity depends on two things, the mass of the celestial body and the distance from the center of mass to the escapee's starting point. If the celestial body has a large mass, its gravity is strong, and you need a high velocity to escape; but if you begin your journey farther from the center of mass, the velocity needed is less. For example, to escape from Earth, a spaceship would have to leave Earth's surface at 11 km/s (25,000 mph), but if you could launch spaceships from the top of a tower 1000 miles high, the escape velocity would be only 10 km/s (22,000 mph).

An object massive enough and/or small enough could have an escape velocity greater than the speed of light. Relativity says that nothing can travel faster than the speed of light, so even photons would be unable to escape. Such a small, massive object could never be seen because light could not leave it. This was first noted by British astronomer Reverend John Mitchell in 1783, long before Einstein and relativity.

ThomsonNOW™ Sign in at www.thomsonedu.com and go to ThomsonNOW to see Astronomy Exercise "Black Hole."

Schwarzschild Black Holes

If the core of a star collapses and contains more than about 3 solar masses, no force can stop the collapse. When the object reaches the size of a white dwarf, the collapse continues because degenerate electrons cannot support that much weight. It also cannot stop when it reaches the even smaller size of a neutron star because degenerate neutrons also cannot support that weight. No force remains to stop the object from collapsing to zero radius.

As an object collapses, its density and the strength of its surface gravity increase; and if an object collapses to zero radius, its density and gravity become infinite. Mathematicians call such a point a **singularity,** but in physical terms it is difficult to imagine an object of zero radius. Some theorists believe that a singularity is impossible and that the laws of quantum physics must somehow halt the collapse at some subatomic radius roughly 10^{20} times smaller than a proton. Astronomically, it seems to make little difference.

If the contracting core of a star becomes small enough, the escape velocity in a region around it is so large that no light can escape. You can receive no information about the object or about the region of space near it. Such a region is called a **black hole.** Note that the term *black hole* refers to a volume of space, not just the singularity at the region's center. If the core of an exploding star collapsed to create a black hole, the expanding outer layers of the star could produce a supernova remnant, but the core would vanish without a trace.

Albert Einstein's general theory of relativity treats space and time as a single entity—space-time. His equations showed that gravity could be described as a curvature of space-time, and almost immediately astronomer Karl Schwarzschild found a way to solve the equations to describe the gravitational field around a single, nonrotating, electrically neutral lump of matter. That solution contained the first general relativistic description of a black hole, and nonrotating, electrically neutral black holes are now known as Schwarzschild black holes. In recent decades, theorists such as Roy Kerr and Stephen Hawking have found ways to apply the sophisticated mathematical equations of the general theory of relativity and quantum mechanics to black holes that are rotating and have electrical charges. For this discussion the differences are minor, and you may proceed as if all black holes were Schwarzschild black holes.

Schwarzschild's solution shows that if matter is packed into a small enough volume, then space-time curves back on itself. Objects can still follow paths that lead into the black hole, but no path leads out, so nothing can escape, not even light. Consequently, the inside of the black hole is totally beyond the view of an outside observer. The **event horizon** is the boundary between the isolated volume of space-time and the rest of the universe, and the radius of the event horizon is called the **Schwarzschild radius,** R_S—the radius within which an object must shrink to

become a black hole (■ Figure 8-17) and the point of no return for any object falling in later.

Although Schwarzschild's work was highly mathematical, his conclusions were quite simple. The size of a black hole, its Schwarzschild radius, is simply proportional to its mass. A 3-solar-mass black hole will have a Schwarzschild radius of about 9 km, a 10-solar-mass black hole will have a Schwarzschild radius of 30 km, and so on. Note that even a very massive black hole would not be very large—just a few miles across.

It is a common misconception to think of black holes as giant vacuum cleaners that will suck up everything in the universe. A black hole is just a gravitational field, and at a reasonably large distance its gravity is no greater than that of a normal object of similar mass. If the sun were replaced by a 1-solar-mass black hole, the orbits of the planets would not change at all. The gravity of a black hole becomes extreme only when you approach close to it. ■ Figure 8-18 illustrates this by representing gravitational fields as curvature of the fabric of space-time. Physicists like to graph the strength of gravity around a black hole as curvature in a flat sheet. The graphs look like funnels in which the depth of the funnel indicates the strength of the gravitational field, but black holes themselves are not shaped like funnels; they

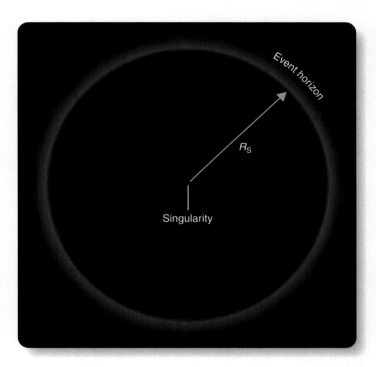

Event horizon

R_S

Singularity

■ Active Figure 8-17

A black hole forms when an object collapses to a small size (perhaps to a singularity) and the escape velocity becomes so great light cannot escape. The boundary of the black hole is called the event horizon because any event that occurs inside is invisible to outside observers. The radius of the black hole R_S is the Schwarzschild radius.

ThomsonNOW™ Sign in at www.thomsonedu.com and go to ThomsonNOW to see the Active Figure called "Schwarzschild Radius."

Gravitational field around
a 5-solar-mass star

— Surface of star

■ **Figure 8-18**

If you fell into the gravitational field of a star, you would hit the star's surface before you fell very far. Because a black hole is so small, you could fall much deeper into its gravitational field and eventually cross the event horizon. At a distance, the two gravitational fields are the same.

Gravitational field around
a 5-solar-mass black hole

To the event horizon

are spheres or spheroids. In Figure 8-18, you should note that the strength of the gravitational field around the black hole becomes extreme only if you venture too close.

This chapter discusses black holes that might originate from the deaths of massive stars. In later chapters, you will encounter black holes whose masses might exceed a million solar masses, located in the centers of galaxies.

Leaping into a Black Hole

Before you can search for real black holes, you must understand what theory predicts about the behavior of a black hole. To explore that idea, you can imagine that you leap, feet-first, into a Schwarzschild black hole.

If you were to leap into a black hole of a few solar masses from a distance of an astronomical unit, the gravitational pull would not be very large, and you would fall slowly at first. Of course, the longer you fell and the closer you came to the center, the faster you would travel. Your wristwatch would tell you that you fell for about two months before you reached the event horizon.

Your friends who stayed behind would see something different. They would see you falling more slowly as you came closer to the event horizon because, as explained by general relativity, time slows down in curved space-time. This is known as **time dilation.** In fact, your friends would never actually see you cross the event horizon. To them you would fall more and more slowly until you seemed hardly to move. Generations later, your descendants could focus their telescopes on you and see you still inching closer to the event horizon. You, however, would have sensed no slowdown and would conclude that you had crossed the event horizon after about two months.

Another relativistic effect would make it difficult to see you with normal telescopes. As light travels out of a gravitational field,

it loses energy, and its wavelength grows longer. This is known as the **gravitational redshift.** Although you would notice no effect as you fell toward the black hole, your friends would need to observe at longer and longer wavelengths in order to detect you.

While these relativistic effects seem merely peculiar, other effects would be quite unpleasant. Imagine again that you are falling feet-first toward the event horizon of a black hole. You would feel your feet, which would be closer to the black hole, being pulled in more strongly than your head. This is a tidal force, and at first it would be minor. As you fell closer, however, the tidal force would become very large. Another tidal force would compress you as your left side and your right side both fell toward the center of the black hole. For any black hole with a mass like that of a star, the tidal forces would crush you laterally and stretch you longitudinally long before you reached the event horizon (■ Figure 8-19). Needless to say, this would render you inoperative as a thoughtful observer.

Science fiction books, movies, and TV shows sometimes suggest that you can travel through the universe by jumping into a black hole in one place and popping out of another somewhere far across space. That might make for good science fiction, but tidal forces would make it an unpopular form of transportation even if it worked. You would certainly lose your luggage.

Your imaginary leap into a black hole is not frivolous. You now know how to find a black hole: Look for a strong source of X rays that may be from matter compressed and stressed just before disappearing across the event horizon.

The Search for Black Holes

An isolated black hole is totally invisible because nothing can escape from the event horizon. But a black hole into which matter is flowing would be a source of X rays. Of course, X rays can't es-

■ **Figure 8-19**

Leaping feet-first into a black hole. A person of normal proportions (left) would be distorted by tidal forces (right) long before reaching the event horizon around a typical black hole of stellar mass. Tidal forces would stretch the body lengthwise while compressing it laterally.

cape from *inside* the event horizon, but X rays emitted by the heated matter flowing into the black hole can escape if the X rays are emitted *before* the matter crosses the event horizon. An isolated black hole in space will not have much matter flowing into it, but a black hole in a binary system might receive a steady flow of matter transferred from its companion star. This suggests you can search for black holes by searching among X-ray binaries.

You have learned about X-ray binaries such as Hercules X-1 that contain a neutron star, and they emit X rays much as a binary containing a black hole should. You can tell the difference in two ways. If the compact object emits pulses, you know it is a neutron star. Otherwise, you might check the mass of the object. If the mass of the compact object is greater than about 3 solar masses, the object cannot be a neutron star, and you can conclude that it must be a black hole.

The first X-ray binary suspected of harboring a black hole was Cygnus X-1. It contains a supergiant B0 star and a compact

object orbiting each other with a period of 5.6 days. Matter flows from the B0 star as a strong stellar wind, and some of that matter enters a hot accretion disk around the compact object (■ Figure 8-20). The accretion disk is about 5 times larger in diameter than the orbit of Earth's moon, and the inner few hundred kilometers of the disk have a temperature of about 2 million Kelvin—hot enough to radiate X rays. The compact object is invisible, but Doppler shifts in the spectrum reveal the motion of the B0 star around the center of mass of the binary. From the geometry of the orbit, astronomers were able to calculate the mass of the compact object—at least 3.8 solar masses, well above the maximum for a neutron star.

As X-ray telescopes have found more X-ray objects, the list of black hole candidates has grown to a few dozen. Some of these objects are shown in ■ Table 8-2. Each candidate is a compact object surrounded by a hot accretion disk in a close X-ray binary system without regular pulsations. A few of the binary systems are easier to analyze than others, but, in the end, it has become clear that some of these objects are too massive to be neutron stars. As you saw in Window on Science 8-1, absolute proof is not possible in science, but the evidence is now overwhelming: Black holes really do exist.

Another way to confirm that black holes are real is to search for evidence of their distinguishing characteristic—event horizons—and that search also has been successful. In one study, astronomers selected twelve X-ray binary systems, six of which

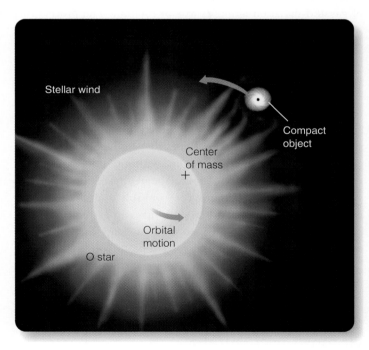

■ **Figure 8-20**

The X-ray source Cygnus X-1 consists of a supergiant B star and a compact object orbiting each other. Gas from the B star's stellar wind flows into the hot accretion disk around the compact object, and astronomers detect X rays from the disk.

seemed to contain neutron stars and six of which were believed to contain black holes. Using X-ray telescopes, the astronomers could see flares of energy as blobs of matter fell into the accretion disks and spiraled inward. In the six systems believed to contain neutron stars, the astronomers could also detect bursts of energy when the blobs of matter finally fell onto the surfaces of the neutron stars. In the six systems believed to contain black holes, however, the blobs of matter spiraled inward through the accretion disks and vanished without final bursts of energy. Evidently, those blobs of matter had approached the event horizons and become undetectable due to time dilation and gravitational redshift. This is dramatic evidence that event horizons are real.

Energy from Compact Objects—Jets

As you can see, it is a common misconception to think that it is impossible to get any energy out of a black hole. In a later chapter, you will meet galaxies radiating vast amounts of energy from massive black holes, so you should pause here to see how a compact object and an accretion disk can eject powerful jets and beams of energy.

Whether a compact object is a black hole or a neutron star, it has a strong gravitational field. Any matter flowing into that field is accelerated inward; and, because it must conserve angular momentum, it flows into an accretion disk made so hot by friction that the inner regions can emit X rays and gamma rays. Somehow the spinning accretion disk can emit powerful beams of gas and radiation along its axis of rotation. The process isn't well understood, but it seems to involve magnetic fields that get caught in the accretion disk and are twisted into tightly wound tubes that squirt gas and radiation out of the disk and confine it in narrow beams.

This process is similar to the bipolar outflows ejected by protostars (see Chapter 7), but it is much more powerful. One example of this process is an X-ray binary called SS433. Its optical spectrum shows sets of spectral lines that are Doppler-shifted by about one-fourth the speed of light, with one set shifted to the red and one set shifted to the blue. Apparently, SS433 is a binary system in which a compact object (probably a black hole) pulls matter from its companion star and forms an extremely hot accretion disk. Jets of high-temperature gas blast away in beams aimed in opposite directions. SS433 is a prototype that illustrates how the gravitational field around a compact object can produce powerful beams of radiation and matter.

Energy from Compact Objects— Gamma-Ray Bursts

During the 1960s the United States put a series of satellites in orbit to watch for bursts of gamma rays coming from Earth indicating nuclear weapons tests that would be violations of an

A Hypernova Explosion

The collapsing core of a massive star drives its energy along the axis of rotation because. . .

the rotation of the star slows the collapse of the equatorial regions.

Within seconds, the remaining parts of the star fall in.

Beams of gas and radiation strike surrounding gas and generate beams of gamma rays.

The gamma-ray burst fades in seconds, and a hot accretion disk is left around the black hole.

■ **Active Figure 8-21**

The collapse of the cores of extremely massive stars can produce hypernova explosions, which are believed to be the source of gamma-ray bursts longer than 2 seconds. (NASA/Skyworks Digital)

Thomson**NOW**˙ Sign in at www.thomosnedu.com and click on Movie Gallery to see a movie of a "Hypernova."

Table 8-2 | Nine Black Hole Candidates

Object	Location	Companion Star	Orbital Period	Mass of Compact Object
Cygnus X-1	Cygnus	B0 supergiant	5.6 days	$>3.8\ M_\odot$
LMC X-3	Dorado	B3 main-sequence	1.7 days	$\sim 10\ M_\odot$
A0620-00	Monocerotis	K main-sequence	7.75 hours	$10 \pm 5\ M_\odot$
V404 Cygni	Cygnus	K main-sequence	6.47 days	$12 \pm 2\ M_\odot$
J1655-40	Scorpius	F–G main-sequence	2.61 days	$6.9 \pm 1\ M_\odot$
QZ Vul	Vulpecula	K main-sequence	8 hours	$10 \pm 4\ M_\odot$
4U 1543-47	Lupus	A main-sequence	1.123 days	$2.7–7.5\ M_\odot$
V4641 Sgr	Sagittarius	B supergiant	2.81678 days	$8.7–11.7\ M_\odot$
XTEJ1118+480	Ursa Major	K main-sequence	0.170113 days	$>6\ M_\odot$

international treaty. The experts were startled when the satellites detected about one **gamma-ray burst** coming from space per day. The Compton Gamma Ray Observatory launched in 1991 discovered that gamma-ray bursts were occurring all over the sky and not from any particular region. Starting in 1997, new satellites in orbit were able to detect gamma-ray bursts, determine their location in the sky, and immediately alert astronomers on the ground. When telescopes swiveled to image the locations of the bursts, they detected fading glows that resembled supernovae, and that has led to the conclusion that long gamma-ray bursts are produced by a certain kind of supernova explosion.

Theoretical calculations indicate that a star more massive than some threshold around 15 or 20 solar masses will collapse and become a black hole when its nuclear fuel is exhausted. Models show that the collapsing star would conserve angular momentum and spin very rapidly, and that would slow the collapse of the equatorial parts of the star. The poles of the star would fall in quickly, and that would focus beams of intense radiation and ejected gas blasting out along the axis of rotation. Such an eruption has been called a **hypernova** (■ Figure 8-21). If either of those beams happens to point in the right direction, Earth would receive a powerful gamma-ray burst. The evidence seems clear that at least some of the gamma-ray bursts are produced by hypernovae. Massive stars explode as hypernovae only rarely in any one galaxy, but the gamma-ray bursts they produce are so powerful that astronomers can detect these explosions among a vast number of galaxies. Other gamma-ray bursts may be produced by the merger of two neutron stars or a neutron star and a black hole, and yet others by sudden shifts in the crusts of highly magnetized neutron stars.

Incidentally, if a gamma-ray burst occurred only 1000 ly from Earth, the gamma rays would shower Earth with radiation equivalent to a 10,000-megaton nuclear blast. (The largest bombs ever made were a few tens of megatons in size.) The gamma rays could create enough nitric oxide in the atmosphere to produce intense acid rain and would destroy the ozone layer and expose life on Earth to deadly levels of solar ultraviolet radiation. Gamma-ray bursts can occur relatively near the Earth as often as every few 100 million years and could be one of the causes of the mass extinctions that show up in the fossil record.

Summary

8-1 | Giant Stars

What happens to a star when it uses up the last of the hydrogen fuel in its core?

■ As a star uses up the last of its hydrogen, the nuclear reactions die down, and the core contracts and heats up. Hydrogen fusion begins in a spherical layer around the core—a hydrogen-fusion shell.

■ Energy flowing from the hydrogen-fusion shell swells the star into a cool giant 10 to 100 times larger in diameter than the sun. Massive stars swell to become supergiants up to 1000 times larger in diameter than the sun.

■ Eventually contraction of the star's core ignites helium, first in the core and later in a shell. If the star is massive enough, it can eventually fuse carbon and other elements.

■ You can see evidence of stellar evolution in the H–R diagrams of star clusters. Stars in a cluster begin their evolution at the same time but evolve at different rates, depending on their masses. The most massive leave the main sequence first and are followed later by progressively less-massive stars.

■ You can estimate the age of a star cluster from the position of the turnoff point in its H–R diagram—identifying the hottest, bluest stars that are still on the main sequence.

8-2 | The Deaths of Lower-Main-Sequence Stars

How will the sun and stars smaller than the sun die?

■ Red dwarfs less massive than about 0.4 solar mass are completely mixed by convection and will have very little hydrogen left when they die. They cannot ignite a hydrogen-fusion shell, so they cannot become giant stars. They will remain on the main sequence for many times longer than the present age of the universe.

■ Medium-mass stars between about 0.4 and 4 solar masses, including the sun, become cool giants and fuse helium but will never be hot enough in their cores to fuse carbon.

■ Some medium-mass stars, up to about 4 solar masses, produce planetary nebulae as they become white dwarfs.

8-3 | The Evolution of Binary Systems

What happens if an evolving star is in a binary star system?

■ Mass can be transferred in a binary system from an evolving star through an accretion disk around the receiving star. Accretion disks can become hot enough to emit light and even X rays.

■ Mass transferred onto the surface of a white dwarf can build up a layer of fuel that erupts in a nova explosion. A white dwarf can erupt as a nova repeatedly so long as mass transfer continues to form new layers of fuel.

8-4 | The Deaths of Massive Stars

How do massive stars die?

■ A star more massive than about 8 solar masses cannot lose mass fast enough to reduce to a mass low enough to end as a white dwarf. Such massive stars must die more violent deaths.

■ The massive stars on the upper main sequence fuse nuclear fuels up to iron but cannot generate further nuclear energy because iron is the most tightly bound of all atomic nuclei. When an aging massive star forms an iron core, the core collapses and triggers an explosion known as a type II supernova.

■ A type Ia supernova can occur when mass transferred onto a white dwarf increases its mass above the Chandrasekhar limit and it collapses suddenly, fusing all of its remaining nuclear fuel at once.

■ A type Ib supernova occurs when a massive star in a binary system loses its outer layers of hydrogen before it explodes.

■ One way to distinguish between the types of supernovae is that spectra of type II supernovae contain hydrogen lines, but the spectra of type Ia and type Ib supernovae do not.

■ A supernova expels an expanding shell of gas called a supernova remnant. The supernova of AD 1054 produced a supernova remnant known as the Crab Nebula.

8-5 | Neutron Stars

What are neutron stars, and how do you know they actually exist?

■ When a supernova explodes, the core collapses to very small size. Theory predicts that the collapsing core cannot support itself as a white dwarf if its mass is greater than 1.4 solar masses, the Chandrasekhar limit. If its mass lies between 1.4 solar masses and about 3 solar masses, it can halt its contraction and form a neutron star.

■ A neutron star is supported by degenerate neutron pressure. Theory predicts that a neutron star should be about 10 km in radius, spin very fast because it conserves angular momentum as it contracts, have a surface temperature of millions of degrees K, and have a powerful magnetic field.

■ Pulsars, rapidly pulsing radio sources, were discovered in 1967 and were eventually understood to be spinning neutron stars. The discovery of a pulsar in the supernova remnant called the Crab Nebula was a key link in the story.

■ Pulsars are evidently spinning neutron stars that emit beams of radiation from their magnetic poles. As they spin, they sweep the beams around the sky like lighthouses. If the beams sweep across Earth, astronomers detect pulses.

■ A neutron star slows its spin rate as it radiates its energy into space.

■ Dozens of pulsars have been found in binary systems, and those objects allow astronomers to estimate the masses of the neutron stars. Such masses are consistent with the predicted masses of neutron stars.

■ In some binary systems, mass flows into a hot accretion disk around the neutron star and causes the emission of X rays.

■ One binary pulsar system is famous for allowing measurements that confirm in detail Einstein's theory of general relativity.

■ The fastest pulsars, called millisecond pulsars, appear to be old pulsars that have been spun up to high speed by mass flowing from binary companions.

■ Planets have been found orbiting at least one neutron star. They may be the remains of a companion star that was mostly devoured by the neutron star.

8-6 | Black Holes

What are black holes, and how do you know they really exist?

■ If the collapsing core of a supernova has a mass greater than 3 solar masses, then it must contract to a very small size—perhaps to a singularity, an object of zero radius. Near such an object, gravity is so strong that not even light can escape, and the region is called a black hole.

- The outer boundary of a black hole is the event horizon. No object or event inside the horizon is detectable from outside. The radius of the event horizon is the Schwarzschild radius, amounting to only a few kilometers for a black hole of stellar mass.

- If you were to leap into a black hole, your friends who stayed behind would see two relativistic effects. They would see your clock slow relative to their own clock because of time dilation. Also, they would see the light from you redshifted to longer wavelengths. You would not notice these effects, but you would feel powerful tidal forces that would deform and heat you.

- To search for black holes, astronomers must look for binary star systems in which mass flows into a compact object and emits X rays. If the mass of the compact object is greater than about 3 solar masses, then the object is presumably a black hole. A number of such objects have been located.

- Black holes and neutron stars at the center of accretion disks can eject powerful beams of radiation and gas. Such beams have been detected.

- Gamma-ray bursts appear to be related to violent events involving neutron stars and black holes. Long bursts appear to arise during hypernovae, the collapse of massive stars to form black holes. Short gamma-ray bursts may be produced by the merger of binary compact objects such as neutron-star pairs or neutron-star/black-hole pairs, or by shifts in neutron star crusts.

New Terms

nova (p. 146)

supernova (p. 146)

giant star (p. 148)

supergiant star (p. 148)

horizontal branch (p. 149)

planetary nebula (p. 151)

degenerate matter (p. 151)

open cluster (p. 152)

globular cluster (p. 152)

turnoff point (p. 152)

compact object (p. 154)

Chandrasekhar limit (p. 154)

Roche lobe (p. 155)

angular momentum (p. 155)

accretion disk (p. 156)

type I supernova (p. 158)

type II supernova (p. 158)

synchrotron radiation (p. 159)

supernova remnant (p. 160)

neutron star (p. 161)

pulsar (p. 162)

lighthouse model (p. 163)

general theory of relativity (p. 164)

gravitational radiation (p. 164)

millisecond pulsar (p. 164)

singularity (p. 167)

black hole (p. 167)

event horizon (p. 167)

Schwarzschild radius (R_s) (p. 167)

time dilation (p. 168)

gravitational redshift (p. 168)

gamma-ray burst (p. 171)

hypernova (p. 171)

Review Questions

ThomsonNOW™ Assess your understanding of this chapter's topics with additional quizzing and animations at
www.thomsonedu.com

1. Why does helium fusion require a higher temperature than hydrogen fusion?

2. How does contraction of an inert helium core trigger the ignition of a hydrogen-fusion shell?

3. Why does the expansion of a star's envelope make it cooler but more luminous?

4. How can star clusters confirm astronomers' theories of stellar evolution?

5. Why don't red dwarfs become giant stars?

6. Why can't a white dwarf contract as it cools?

7. Why can't a white dwarf have a mass greater than 1.4 solar masses?

8. How can a star of as much as 8 solar masses form a white dwarf when it dies?

9. How does the inward collapse of the core of a massive star produce an outward explosion?

10. What is the difference between a supernova explosion and a nova explosion?

11. What is the difference between type I and type II supernovae?

12. The star cluster in the photo at the right contains many hot, blue, luminous stars. Sketch its H–R diagram and discuss its probable age.

NASA/Walborn, Maiz-Apellániz, and Barbá

13. What processes caused a medium-mass star to produce the nebula at the right? The nebula is now about 0.1 ly in diameter and still expanding. What will happen to it?

NASA/Hubble Heritage Team/STScI/AURA

14. The image at right combines X-ray (blue), visible (green), and radio (red) images. Observations show the sphere is expanding at a high speed and is filled with very hot gas. What kind of object produced this nebula? Roughly how old do you think it must be?

NASA/CAC/SAO/ CSIRO/ATNF/ATCA

15. How are neutron stars and white dwarfs similar? How do they differ?

16. Why is there an upper limit to the mass of neutron stars?

17. Why do you expect neutron stars to spin rapidly?

18. If neutron stars are hot, why aren't they very luminous?

19. What evidence can you cite that pulsars are neutron stars?

20. How does the lighthouse model explain pulsars?

21. Why would astronomers at first assume that the first millisecond pulsar discovered was young?

22. How can a neutron star in a binary system generate X rays?

23. How can a black hole emit X rays?

24. What evidence can you cite that black holes really exist?

25. How can mass transfer into a compact object produce jets of high-speed gas?

26. Discuss the possible causes of gamma-ray bursts.

27. The X-ray image at above right shows the supernova remnant G11.2–0.3 and its central pulsar. How old do you think this system is? Discuss the appearance of this system a million years from now.

NASA/McGill University, V. Kaspi et al.

28. What is happening in the artist's impression at the right? How would you distinguish between a neutron star and a black hole in such a system?

CXC/M. Weiss

Discussion Questions

1. False-color radio images and time-exposure photographs of astronomical images show aspects of nature you can never see with unaided eyes. Can you think of common images in newspapers or on television that reveal phenomena you cannot see?

2. In your opinion, has the existence of neutron stars been tested sufficiently to be called a theory, or should it be called a hypothesis? What about the existence of black holes?

3. Why wouldn't an accretion disk orbiting a giant star get as hot as an accretion disk orbiting a compact object?

9 The Milky Way Galaxy

The Stars Are Yours.

JAMES S. PICKERING

THE STARS ARE YOURS was the title of a popular astronomy book written by James S. Pickering in 1948. The point of the title is that the stars belong to everyone equally, and you can enjoy the stars as if you owned them. ▌ You live inside one of the largest of the star systems that fill the universe. Our Milky Way Galaxy is over 80,000 ly in diameter and contains over 100 billion stars. As you read this chapter, you will be learning about your home galaxy, but you will also be learning how the stars of the galaxy have cooked up the atoms heavier than helium. You know already how the cores of massive stars make the heavier atoms up to atomic number 26, iron, and how supernovae blast those atoms back into space and add even heavier atoms made only during the supernova explosion. In this

▌ Continued on page 176 ▌

Guidepost

Looking Back

In previous chapters, you learned how stars are born from clouds of gas and dust, how they evolve as they consume their nuclear fuels, and how they die. You discovered that most stars live long, quiet lives but that the most massive stars die in violent supernova explosions, leaving behind neutron stars or black holes as their corpses.

This Chapter

Now you are ready to see stars in their vast communities called galaxies. This chapter discusses our home galaxy, the Milky Way Galaxy, and attempts to answer four essential questions:

How do you know you live in a galaxy?

How do you know ours is a spiral galaxy, and what are the spiral arms?

How did our galaxy form and evolve?

What lies at the very center of our galaxy?

In this chapter you will see more examples of how scientists use evidence and theory to understand nature. In some cases the evidence seems contradictory and the theories incomplete because the adventure of discovery is not yet finished.

Looking Ahead

In the chapter that follows, you will meet some of the billions of galaxies that fill the depths of the universe. Understanding the Milky Way Galaxy, big as it is, is only a step in understanding the universe as a whole.

chapter you will see how the stars in the Milky Way Galaxy have, generation after generation, made the atoms now in your body.

9-1 The Discovery of the Galaxy

IT SEEMS ODD TO SAY ASTRONOMERS DISCOVERED SOMETHING that is all around you; but, until the early 20th century, no one knew what the Milky Way was.

It isn't obvious that you live in a galaxy. You are inside, and you see nearby stars scattered in all directions over the sky, whereas the more distant clouds of stars in the galaxy make a faint band of light circling the sky (■ Figure 9-1). The ancient Greeks named that band *galaxies kuklos,* the "milky circle." The Romans changed the name to *via lactea,* "milky road" or "milky way." It was not until early in the 20th century that astronomers understood that humans live inside a great wheel of stars and that the universe is filled with other such star systems. Drawing on the Greek word for milk, astronomers called these star systems *galaxies.*

The Great Star System

Galileo's telescope revealed that the glowing Milky Way was made up of stars, and later astronomers realized that the great cloud of stars in which the sun is located, which they called the star system, must be wheel shaped. If the star system were spherical, for example, you would see stars scattered more or less uniformly in all directions over the sky. Only a wheel or disk shape would look, from the inside, like the Milky Way band encircling the sky.

The English astronomers Sir William Herschel (1738–1822) and his sister Caroline Herschel (1750–1848) attempted to gauge the true shape of the star system by counting stars in 683 different directions in the sky. In directions where they saw more stars, they assumed, the star system extended further into space. They concluded that the star system has a disk shape with some noticeable "holes" lacking stars around its edges (■ Figure 9-2). Modern astronomers know that those apparent holes are caused by dense clouds of gas and dust blocking the view of more distant stars. The Herschels counted similar numbers of stars in most directions around the Milky Way and concluded that the sun and Earth are near the center of the star system.

As the 20th century began, astronomers still believed that the sun was located near the center of a wheel-shaped star system that they estimated was about 15,000 ly in diameter. How humanity realized the truth about the size of the galaxy, Earth's location in it, and the galaxy's location in a much larger universe of other galaxies, is an adventure that begins with a woman studying stars that pulsate and leads to a man studying star clusters. The next section will follow that story because it is an important historical moment in human history and because it illustrates how scientists build on the work of their predecessors and step-by-step refine their ideas about the natural world.

The Size of the Milky Way

It is a common misconception of poets that the stars are eternal and unchanging, but astronomers have known for centuries

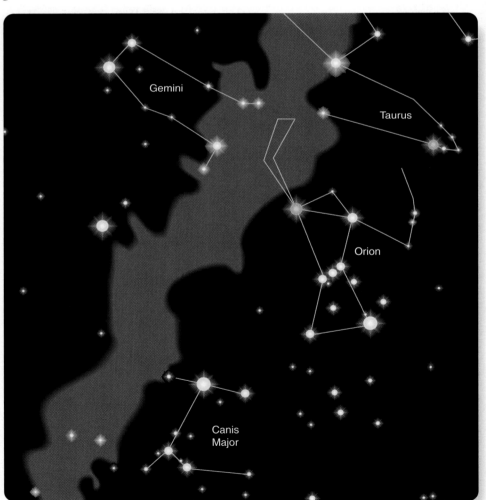

Gemini

Taurus

Orion

Canis Major

■ Figure 9-1

Nearby stars look bright, and peoples around the world group them into constellations. Nevertheless, the vast majority of the stars in our galaxy merge into a faintly luminous path that circles the sky, the Milky Way. This artwork shows the location of a portion of the Milky Way near a few bright winter constellations. (See Figure 9-5a and the star charts at the front and back of this book to further locate the Milky Way in your sky.)

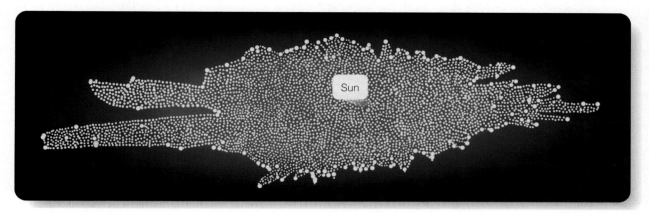

■ Figure 9-2

In 1785, William Herschel published this diagram showing the star system as a thick disk seen edge-on, with the sun located near the center of the star system.

that some stars change in brightness. Of course, novae and supernovae burst into view, grow brighter, and then fade, but many other variable stars actually pulsate like beating hearts. The period of pulsation is the time it takes a star to complete a cycle from bright to faint to bright again. You will learn next how the properties of some types of variable stars allowed astronomers to measure the size of the galaxy.

In 1912, Henrietta Leavitt (1868–1921) was studying a star cloud in the southern sky known as the Small Magellanic Cloud. On her photographic plates she found many variable stars, and she noticed that the brightest had the longest periods. Because all of the variables were in the same cloud at nearly the same distance she concluded that there was a relationship between the pulsation periods and luminosities (true total power output) of those variable stars.

The stars Leavitt saw, **Cepheid variable stars,** are named after the first such star discovered, Delta Cephei. They are giant and supergiant stars with pulsation periods of 1 to 60 days that lie near the top (the high-luminosity end) of a region of the H–R diagram known as the **instability strip** (■ Figure 9-3a). As stars evolve, and the points that represent their temperatures and luminosities move in the H-R diagram, they can cross into the instability strip and start pulsating; they stop pulsating when they evolve out of the strip. Massive stars are larger and pulsate slower, just as large bells vibrate slower and have deeper tones. Lower-mass stars are less luminous and, being smaller, pulsate faster. This explains why, as first noticed by Leavitt, the long-period Cepheids are more luminous than the short-period Cepheids. That is now known as the **period–luminosity relation,** shown graphically in ■ Figure 9-4. You may be interested to learn that the North Star, Polaris, is a Cepheid variable with a pulsation period of 4 days.

ThomsonNOW™ Sign in at www.thomsonedu.com and go to ThomsonNOW to see Astronomy Exercise "Cepheid Variable."

Star Clusters and the Center of the Galaxy

A young astronomer named Harlow Shapley (1885–1972) began the discovery of the true nature of the Milky Way when he noticed that different kinds of star clusters have different distributions in the sky. In Chapter 8 you met two types of star clusters, open clusters and globular clusters (see page 152). Open clusters are concentrated along the Milky Way. Globular clusters are widely scattered, but Shapley noticed that the globular clusters were more common toward the constellations Sagittarius and Scorpius (■ Figure 9-5a).

Shapley assumed that the concentration of globular clusters is controlled by the combined gravitational field of all the stars in the galaxy. In that case, he realized, he could study the size and extent of the galaxy by studying the globular clusters. To do that, he needed to measure the distances to as many globular clusters as possible.

Globular clusters are much too far away to have measurable parallaxes, but they do contain variable stars. Shapley knew of Leavitt's work on these stars, which depended on their relative, rather than true, luminosities. Cepheids are rare, and there are none near enough to have measurable parallaxes, so their true luminosities were not then known. Shapley realized that he could calculate the distance to the globular clusters if he found the true luminosities of the Cepheid variable stars in the clusters.

Through a statistical process involving measurements of position shifts due to their motions through space, called **proper motions,** Shapley was able to find the average distances of a few of the nearest Cepheids, and from that their average luminosities. That meant he could replace Leavitt's apparent magnitudes with absolute magnitudes on the period–luminosity diagram (as shown in Figure 9-4). Astronomers say that Shapley **calibrated** the variable stars for distance determination (**Window on Science 9-1**).

The Calibration of One Parameter to Find Another

Astronomers often say that Shapley "calibrated" the Cepheids for the determination of distance, meaning that he did all the detailed background work so that the Cepheids could be used to find distances. Other astronomers could then use the Cepheids without repeating the calibration.

Calibration is actually very common in science. Chemists, for instance, have carefully calibrated the colors of certain compounds against acidity. They can quickly measure the acidity of a solution by dipping into it a slip of paper containing the indicator compound and looking at the color. They don't have to repeat the careful calibration of the paper slips every time they measure acidity.

Engineers in steel mills have calibrated the color of molten steel against its temperature. They can use a handheld device that measures the color of the blackbody radiation from a ladle of molten steel and then looks up the temperature from a calibrated table. They don't have to repeat the calibration every time. Astronomers have made the same kind of color–temperature calibration for stars.

As you read about any science, notice how calibrations are used to simplify common measurements. But notice, too, how important it is to get the calibration right. An error in calibration can throw off every measurement made with that calibration. Some of the biggest errors in science have been errors of calibration.

■ **Figure 9-3**

The more massive a star is, the more luminous it is. Massive stars become larger when they leave the main sequence, and those larger stars pulsate with longer periods when they pass through the instability strip. Because both luminosity and period depend on mass, there is a relationship between period of pulsation and luminosity.

Finally, having calibrated the Cepheids, Shapley could find the distance to the globular star clusters. He could identify the variable stars he found in the clusters and find their apparent magnitudes from his photographs. Comparison of apparent and absolute magnitudes gave him the distance to the star cluster based on the inverse square law for light (see Chapter 6).

Shapley finally plotted the direction and distance to the globular clusters and found that they form a great swarm that is not centered on the sun. Instead, the center of the swarm of clusters lies many thousands of light-years in the direction of Sagittarius. By analogy, if you see a bunch of tall buildings, and they are all together in one direction away from you, you might conclude that downtown is over there, and you are in the suburbs. Evidently the center of the star system is in Sagittarius and far away. The star system is much bigger than anyone had suspected (Figure 9-5b). You live not near the center

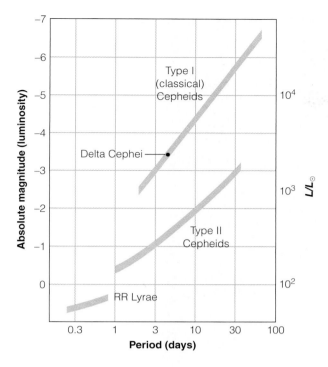

The period–luminosity diagram is a graph of the luminosity of variable stars versus their periods of pulsation. Because the diagram is used for comparison with apparent magnitudes for distance calculations, it is most convenient with absolute magnitudes on the vertical axis to express luminosity. Modern astronomers know that there are two types of Cepheids plus other similar types of variable stars, something that astronomers in the early 20th century could not recognize in their limited data, causing some errors in the first determinations of the size of the Milky Way.

luminated by recently formed brilliant, blue, massive stars and has an overall relatively blue color.

The diameter of the disk and the position of the sun are difficult to determine accurately. Gas and dust block the view in the plane of the galaxy, so astronomers cannot see to the center or to the edge easily, and the outer edge of the disk is not well defined. Most recent studies suggest the sun is about 8.5 kpc from the center, where 1 kpc is a **kiloparsec,** or 1000 pc. The sun and Earth seem to be about two-thirds of the way from the center to the edge, so the diameter of our galaxy appears to be about 25 kpc or about 80,000 ly, but that isn't known to better than 10 percent precision. This is the diameter of the luminous part of our galaxy, the part you would see from a distance. You will learn later that strong evidence shows that our galaxy is much larger than that, but the outer parts are not luminous.

Observations made at other than visual wavelengths can help astronomers peer through the dust and gas. Infrared and radio photons have wavelengths long enough to be unaffected by the dust. Thus, a map of the sky at long infrared wavelengths reveals the disk of our galaxy (■ Figure 9-7).

The most striking features of the disk component are the **spiral arms**—long curves of bright stars, star clusters, gas, and dust. Such spiral arms are easily visible in other galaxies, and you will see later how astronomers found that our own galaxy has a spiral pattern.

The second component of our galaxy is the **spherical component,** which includes all matter in our galaxy scattered in a roughly spherical distribution around the center. This includes a large halo and the nuclear bulge.

The **halo** is a spherical cloud of thinly scattered stars and globular star clusters. It contains only about 2 percent as many stars as the disk of the galaxy and has very little gas and dust. Thus, with no raw material available, no new stars are forming in the halo. In fact, the halo stars are mostly old, cool giants or dim lower-main-sequence stars plus, as revealed by recent careful studies, old white dwarfs that are difficult to detect. Astronomers can map the halo of our galaxy by studying the more easily detected giant stars.

The **nuclear bulge** is the dense cloud of stars that surrounds the center of our galaxy. It has a radius of about 2 kpc and is slightly flattened. It is hard to observe because thick dust in the disk scatters and absorbs radiation of visible wavelengths, but ob-

of a small star system but in the suburbs of a very big wheel of stars, a galaxy.

Why did astronomers before Shapley think humanity lives near the center of a small star system? The answer: Space is filled with gas and dust that dims the view of distant stars. When you look toward the band of the Milky Way, you can see only the neighborhood near the sun. Most of the star system is invisible and, like a traveler in a fog, you seem to be at the center of a small region. Shapley was able to see the globular clusters at greater distances because they lie outside the Milky Way plane and are not dimmed very much by the gas and dust.

Building on Shapley's work, other astronomers began to suspect that some of the faint patches of light visible through telescopes were other galaxies like our own. In 1923, Edwin Hubble photographed individual stars in the Andromeda Galaxy, and in 1924 he identified Cepheids there, allowing its distance to be estimated. As a result, it became clear that our galaxy is just one in a universe filled with galaxies.

An Analysis of the Galaxy

Our galaxy, like many others, contains two primary components—a disk and a sphere. ■ Figure 9-6 shows these components and other features discussed in this section.

The **disk component** consists of all matter confined to the plane of rotation—that is, everything in the disk itself. This includes stars, open star clusters, and nearly all of the galaxy's gas and dust. As you will learn in more detail in the next section, because the disk contains lots of gas and dust it is the site of most of the star formation in the galaxy. Consequently, the disk is il-

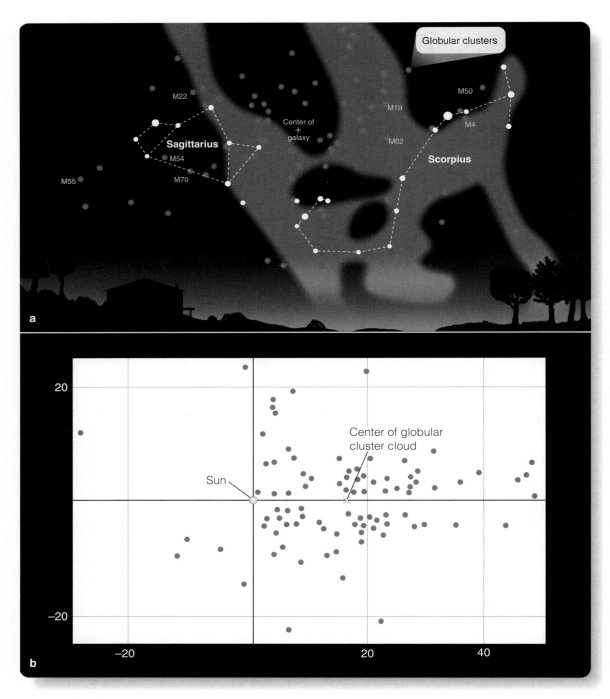

■ Figure 9-5

(a) Nearly half of cataloged globular clusters (red dots) are located in or near Sagittarius and Scorpius. A few of the brighter globular clusters, labeled with their catalog designations, are visible with binoculars or small telescopes. Constellations are shown as they appear above the southern horizon on a summer night as seen from latitude 40° N, typical for most of the United States. (b) A "side view" of the Milky Way Galaxy's globular clusters. Shapley's study of the clusters and their distances showed that they were not centered on the sun, which is located at the origin of this graph, but rather formed a great cloud centered far away in the direction of Sagittarius. Distances are given in thousands of parsecs and correspond to Shapley's original calibration that produced values more than 2× larger than the modern calibration.

80,000 ly

Sun   Nuclear bulge   Disk

Halo

Globular cluster

■ **Figure 9-6**

An artist's conception of our Milky Way Galaxy, seen face-on and edge-on. Note the position of the sun and the distribution of globular clusters in the halo. Hot blue stars light up the spiral arms. Only the inner halo is shown here. At this scale, the entire halo would be larger than a dinner plate.

servations at longer wavelengths can penetrate the dust. The bulge seems to contain little gas and dust, and there is not much star formation there. Most of the stars in the nuclear bulge are old, cool stars like those in the halo.

The Mass of the Galaxy

The vast numbers of stars in the disk, halo, and nuclear bulge lead to a basic question: How massive is the galaxy?

When you needed to find the masses of stars, you studied the orbital motions of the pairs of stars in binary systems. To find the mass of the galaxy, you must look at the orbital motions of the stars within the galaxy. Every star in the galaxy follows an orbit around the center of mass of the galaxy. In the disk, the stars follow parallel, circular orbits, and astronomers say the disk of the galaxy rotates. That rotation can allow an estimate of the mass of the galaxy. Consequently, any discussion of the mass of the galaxy is also a discussion of the rotation of the galaxy and the orbits of the stars within the galaxy.

Astronomers can find the orbits of stars by finding how they move. Of course, the Doppler effect reveals a star's radial velocity (see Chapter 5). In addition, if astronomers can measure the distance to a star and its proper motion, they can find the velocity of the star perpendicular to the radial direction. Combining all of this information, astronomers can find the shape of the star's orbit.

The orbital motions of the stars in the halo are strikingly different from those in the disk (■ Figure 9-8). In the halo, each star and globular cluster follows its own randomly tipped elliptical orbit. These orbits carry the stars and clusters far out into the spherical halo, where they move slowly, but when they fall back into the inner part of the galaxy, their velocities increase. Motions in the halo do not resemble a general rotation but are more like the random motions of a swarm of bees. In contrast, the stars in the disk of the galaxy move in the same direction in nearly circular orbits that lie in the plane of the galaxy. The sun is a disk star and follows a nearly circular orbit around the galaxy that always remains within the disk.

You can use the orbital motion of the sun to find the mass of the galaxy inside the sun's orbit. By observing the radial velocity of distant galaxies in various directions around the sky, astronomers can tell that the sun moves about 220 km/s in the direction of Cygnus, carrying Earth and the other planets of our solar system along with it. Because its orbit is a circle with a radius of 8.5 kpc, you can divide the circumference of the orbit by the velocity and find that the sun completes a single orbit in about 240 million years.

If you think of the sun and the center of mass of our galaxy as two objects orbiting each other, you now have enough information to determine the mass of the galaxy. The Milky Way Galaxy must have a mass of about 100 billion solar masses. This estimate is uncertain for a number of reasons. First, you don't know the radius of the sun's orbit with great certainty. Astronomers estimate the radius as 8.5 kpc, but they could be wrong by 10 percent or more, and the radius gets cubed in the calculation, so it makes a big difference. Second, this estimate of the mass includes only the mass inside the sun's orbit. Mass spread uniformly outside the sun's orbit will not affect its orbital motion. Thus 100 billion solar masses is a lower limit for the mass of the galaxy, but no one knows exactly how much to increase the estimate to include the rest of the mass in the galaxy that lies outside the sun's orbit.

The measured motion of the stars shows that the disk does not rotate as a solid body. Each star follows its own orbit, and stars in some regions have shorter or longer orbital periods than the sun. This is called differential rotation. (Recall that you met the term *differential rotation* in Chapter 5 where it was applied to

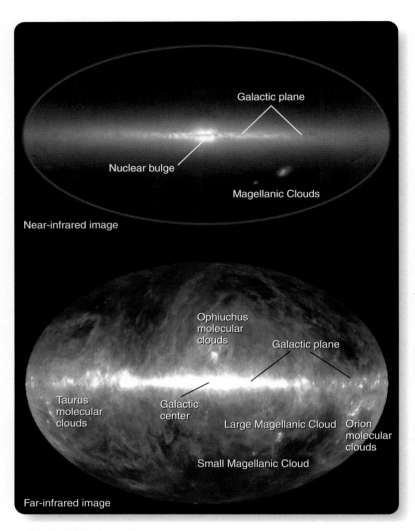

■ Active Figure 9-7

In these infrared images, the entire sky has been projected onto ovals with the center of the galaxy at the center of each oval. The Milky Way extends from left to right. In the near-infrared image, the nuclear bulge is prominent, and dust clouds block the view along the Milky Way. At far-infrared wavelengths, the dust emits significant black body radiation and glows brightly. (Near-IR: 2MASS; ; Far-IR: DIRBE image courtesy Henry Freudenreich)

ThomsonNOW™ Sign in at www.thomsonedu.com and go to ThomsonNOW to see the Active Figure called "Explorable Milky Way."

the sun.) A graph of the orbital velocity of stars at various orbital radii in the galaxy is called a **rotation curve** (■ Figure 9-9). If all of the mass in the galaxy were concentrated at its center, then orbital velocity would be high near the center and would decline away from the center. This kind of motion has been called *Keplerian motion* because it follows Kepler's third law, as in the case of our solar system, where nearly all of the mass is in the sun.

Of course, you already know the galaxy's mass is not all concentrated at its center, but if most of the mass were inside the orbit of the sun, then orbital velocities should decline at greater distances. Many observations confirm, however, that velocities

do not decline and may actually increase at greater distance; this observation shows that larger and larger orbits enclose more and more mass. Although it is difficult to determine a precise edge to the visible galaxy, it seems clear that large amounts of matter are located beyond what seems to be the limit of the galaxy's luminous matter.

The evidence is clear that extra mass lies in an extended halo sometimes called a **dark halo.** It may extend up to 10 times farther than the edge of the visible disk and could contain up to two trillion solar masses. Some small fraction of this mass is made up of low-luminosity stars and white dwarfs, but most of the matter is not producing any light. Astronomers call it **dark matter** and conclude that it must be some as yet unknown form of matter. The following two chapters will return to the question of dark matter. It is one of the fundamental problems of modern astronomy.

Building Scientific Arguments

If gas and dust block the view, how do astronomers know how big our galaxy is?

Because scientific arguments depend ultimately on evidence, they must explain how scientists know what they know. The gas and dust in our galaxy block the view only in the plane of the galaxy. When telescopes look away from the plane of the galaxy, they look out of the gas and dust and can see to great distances. The globular clusters are scattered through the halo with a strong concentration in the direction of Sagittarius.

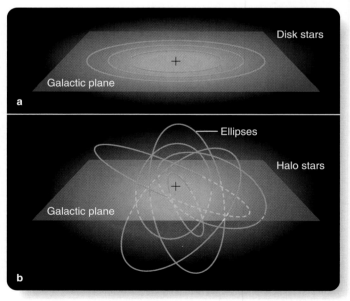

■ Figure 9-8

(a) Stars in the galactic disk have nearly circular orbits that lie in the plane of the galaxy. (b) Stars in the halo have randomly oriented, elliptical orbits.

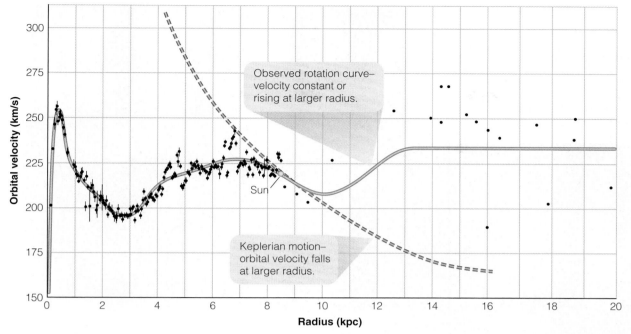

■ **Figure 9-9**

The rotation curve of our galaxy is plotted here as orbital speed versus radius. Data points show measurements made by radio telescopes. Observations outside the orbit of the sun are much more uncertain, and the data points scatter widely. Orbital speeds do not decline outside the orbit of the sun, as you would expect if most of the mass of the galaxy were concentrated toward the center. Rather, the curve is approximately flat at great distances, suggesting that the galaxy contains significant mass outside the orbit of the sun. (Adapted from a diagram by Françoise Combes)

When astronomers look above or below the plane of the galaxy, they can see those clusters, and careful observations reveal variable stars in the clusters. By using a modern calibration of the period–luminosity diagram, they can find the distance to those clusters. If they assume that the distribution of the clusters is controlled by the gravitation of the galaxy as a whole, then they can find the distance to the center of the galaxy by finding the center of the distribution of globular clusters.

In any logical argument, it is important to ask, "How do we know?" and that is especially true in science. Create a new argument: **What measurements and assumptions reveal the total mass of our galaxy?**

■ ■ ■

9-2 Spiral Arms and Star Formation

THE MOST STRIKING FEATURE OF GALAXIES like the Milky Way is the system of spiral arms that wind outward through the disk and contain swarms of hot, blue stars; clouds of dust and gas; and young star clusters. These young objects hint that the spiral arms involve star formation. As you try to understand the spiral arms, you face two questions. First, how can you be sure our galaxy has spiral arms when Earth is embedded inside the galaxy and your view is obscured by gas and dust? Second, why doesn't the differential rotation of the galaxy destroy the arms? The answers to both questions involve star formation.

Tracing the Spiral Arms

Studies of other galaxies show that spiral arms contain hot, blue stars. Thus, one way to study the spiral arms of our own galaxy is to locate these stars. Fortunately, this is not difficult, since O and B stars are often found in associations and, being very luminous, are easy to detect across great distances. Unfortunately, at these great distances their parallax is too small to measure, so their distances must be found by other means, usually by spectroscopic parallax (see Chapter 6).

O and B associations in the sky are not located randomly but reveal parts of three spiral arms near the sun, which have been named for the prominent constellations through which they pass (■ Figure 9-10). Objects used to map spiral arms are called **spiral tracers.** O and B associations are good spiral tracers because they are bright and easy to see at great distances. Other tracers include young open clusters, clouds of hydrogen ionized

(a) Many of the galaxies in the sky are disk shaped, and most of those galaxies have spiral arms. Images of other galaxies show that spiral arms are marked by hot, luminous stars that must be very young; and this should make you suspect that spiral arms are related to star formation. (NASA, Hubble Heritage Team) (b) Gas and dust block your view of most of the disk of our galaxy, but near the sun, young O and B stars fall along bands that appear to be segments of spiral arms.

by hot stars (emission nebulae), and certain higher-mass variable stars.

Notice that all spiral tracers are young objects. O stars, for example, live for only a few million years. If their orbital velocity is about 250 km/s, they cannot have moved more than about 500 pc since they formed. This is less than the width of a spiral arm. Because they don't live long enough to move away from the spiral arms, they must have formed there. The youth of spiral tracers provides an important clue about spiral arms. Somehow they are associated with star formation.

Radio telescopes can detect emission from clouds of cool neutral hydrogen gas. Radio maps of the galaxy disk, combined with optical and infrared data, allow astronomers to deduce the spiral pattern of our galaxy. The segments near the sun are part of a spiral pattern that continues throughout the disk. However, the maps show that the spiral arms are rather irregular and interrupted by branches, spurs, and gaps. The stars in Orion, for example, appear to be a detached segment of a spiral arm, a spur. There are significant sources of error in the mapping methods, but many of the irregularities along the arms seem real, and images of nearby spiral galaxies show similar features (Figure 9-10a). Studies comparing all of the available data on our galaxy's spiral pattern with patterns seen in other galaxies do not necessarily agree, although the newest models suggest that the nuclear bulge in our galaxy is elongated into a bar (■ Figure 9-11). You will see in the next chapter that such bars are common in spiral galaxies.

Spiral tracers show that the arms contain young objects, and that suggests active star formation. The radio maps confirm this suspicion by showing that the material needed to make stars, hydrogen gas, is abundant in spiral arms.

ThomsonNOW™ Sign in at www.thomsonedu.com and go to ThomsonNOW to see Astronomy Exercise "Milky Way Galaxy."

Star Formation in Spiral Arms

Having mapped the spiral pattern, you can ask, "What are spiral arms?" You can be sure that they are not physically connected structures. Like a kite string caught on a spinning hubcap, such arms would be wound up and pulled apart by differential rotation within a few tens of millions of years. Yet spiral arms are common in disk-shaped galaxies and must be reasonably permanent features.

The most prominent theory about spiral arms is called the **density wave theory**, which proposes that the arms are waves of compression that move around the galaxy triggering star formation. The density wave is a bit like a traffic jam behind a truck moving slowly along a highway. Seen from an airplane overhead, the jam seems a permanent, though slow-moving, feature. But individual cars overtake the jam from behind, slow down, move up through the jam, wait their turn, pass the truck, and resume speed along the highway. Similarly, clouds of gas overtake the

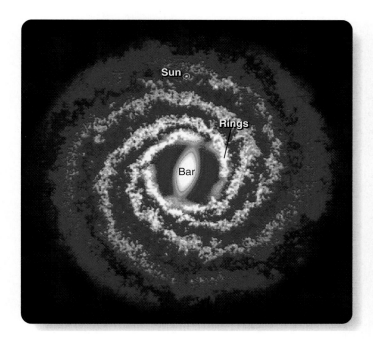

■ Figure 9-11

This model of our galaxy is based on far-infrared observations of dust and includes four spiral arms with no branches plus a central bar and ring. (Courtesy Henry Freudenreich)

The Spiral Density Wave

Orbiting gas clouds overtake the spiral arm from behind.

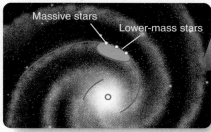

The compression of a gas cloud triggers star formation.

Massive stars are highly luminous and light up the spiral arm.

The most massive stars die quickly.

Low-mass stars live long lives but are not highly luminous.

■ Figure 9-12

According to the density wave theory, star formation occurs as gas clouds pass through spiral arms.

spiral density wave, become compressed in the "traffic jam," and eventually move out in front of the arm, leaving the slower-moving density wave behind.

Of course, star formation will occur where the gas clouds are compressed. Stars pass through the spiral arms unaffected, like bullets passing through a wisp of fog, but large clouds of gas slam into the spiral density wave from behind and are suddenly compressed (■ Figure 9-12). You saw in Chapter 7 that sudden compression could trigger the formation of stars in a gas cloud. Thus, new stars should form along the spiral arms. The spiral arms are not wound up by differential rotation because they are patterns, not physically connected structures.

The brightest stars, the O and B stars, live such short lives that they never travel far from their birthplace and are found only along the arms. Their presence is what makes the spiral arms glow so brightly, due to both their own light and the emission from clouds of gas excited by the stars' UV radiation (■ Figure 9-13a). Lower-mass stars, like the sun, live longer and have time to move out of the arms and continue their journey around the galaxy. The sun may have formed in a star cluster about 5 billion years ago when a gas cloud smashed into a spiral arm. Since that time, the sun has escaped from its birth cluster and made about 20 trips around the galaxy, passing through spiral arms many times.

The density wave theory is very successful in explaining general properties of spiral galaxies, but it has two problems. First, what stimulates the formation of the spiral pattern in the first place? Theorists calculate that minor fluctuations in the galaxy's disk shape or gravitational interactions with passing galaxies may be able to start a density wave. Second, the density wave theory does not account for the branches and spurs in the spiral arms of our own and other galaxies. The solution to this second problem may lie in a process that sustains star formation once it begins.

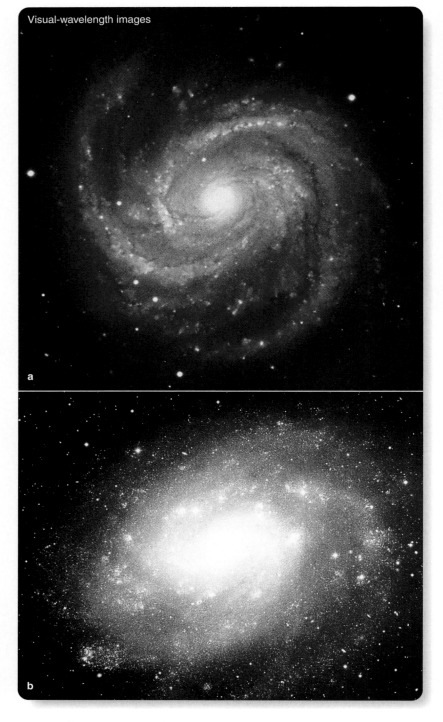

Visual-wavelength images

a

b

■ **Figure 9-13**

(a) Some galaxies are dominated by two spiral arms; but, even in these galaxies, minor spurs and branches are common. The spiral density wave can generate the two-armed, grand-design pattern, but another process called self-sustained star formation may be responsible for the irregularities. (b) Many spiral galaxies do not appear to have two dominant spiral arms. Multiple spurs and branches suggest that star formation is proceeding rapidly in such galaxies. (Anglo-Australian Telescope Board)

protostars buried deep in the dark interstellar clouds behind the nebula (see pages 142–143), seems to be a region of self-sustaining star formation.

Astronomers have calculated models indicating that the differential rotation of the galaxy can drag the inner edge of a star-forming region ahead and let the outer edge lag behind to produce a cloud of star formation shaped like a segment of a spiral arm. Self-sustaining star formation plus differential rotation may produce the branches and spurs so prominent in some galaxies (Figure 9-13b), including our own, but only the spiral density wave can generate the beautiful two-armed grand spiral patterns.

Building Scientific Arguments

Why can't astronomers use solar-type stars as spiral tracers?
Sometimes the timing of events is the critical factor in a scientific argument. In this case, you need to think about the evolution of stars and their orbital periods around the galaxy. Stars like the sun live about 10 billion years, but the sun's orbital period around the galaxy is 240 million years. The sun almost certainly formed when a gas cloud passed through a spiral arm, but since then the sun has circled the galaxy many times and has passed through spiral arms often. That means the sun's present location has nothing to do with the location of any spiral arms. An O star, however, lives only a few million years. It is born in a spiral arm and lives out its entire lifetime before it can leave the spiral arm. Short-lived stars such as O stars are found only in spiral arms and prominently mark the locations of the arms, but G stars are found all around the galaxy.

The spiral arms of our galaxy must make it beautiful in photographs taken from a distance, but Earthlings are trapped inside it. Create an argument based on evidence: **How do astronomers know that the spiral arms mapped out near Earth by spiral tracers actually extend across the disk of our galaxy?**

■   ■   ■

Star formation can control the shape of spiral patterns if the birth of stars in a cloud of gas can cause the birth of more new stars. Massive stars evolve so quickly that their lifetimes are only an instant in the history of a galaxy, and then they explode as supernovae. You learned in Chapter 7 that bursts of luminosity and jets from newborn stars and the expanding gases of supernova explosions can compress neighboring clouds of gas and trigger more star formation. This process is known as **self-sustaining star formation.** The Orion complex, consisting of the Great Nebula in Orion plus the

9-3 The Origin and History of the Milky Way

JUST AS PALEONTOLOGISTS RECONSTRUCT THE HISTORY OF LIFE on Earth from the fossil record, astronomers try to reconstruct our galaxy's past from the fossil it left behind as it formed and evolved. That fossil is the spherical component of the galaxy. The stars in the halo formed when the galaxy was young. The chemical composition and the distribution of these stars can provide clues to how our galaxy formed.

The Age of the Milky Way

To begin, you should ask yourself how old our galaxy is. That question is easy to answer because you already know how to find the age of star clusters. But there are uncertainties that make that easy answer hard to interpret.

The oldest open clusters (see page 152) have ages of about 9 billion years. These ages are determined by analyzing the turnoff point in the cluster H–R diagram (see Chapter 8) and are somewhat uncertain. Nevertheless, from open clusters you can get a rough age for the disk of our galaxy of at least 9 billion years.

Globular clusters have faint turnoff points in their H–R diagrams and are clearly old, but finding their ages accurately is difficult. Clusters differ slightly in chemical composition, which must be accounted for in calculating the stellar models from which ages are determined. Also, to find the age of a cluster, astronomers must know the distance to the cluster. Precise parallaxes from the Hipparcos satellite have allowed astronomers to increase the precision of the Cepheid variable stars' calibration, and careful studies with the newest large telescopes have refined the cluster H–R diagrams. Globular cluster ages seem to average about 11 billion years, with the oldest being a bit over 13 billion years old. The halo of our galaxy therefore seems to be at least 13 billion years old, older than the disk.

Stellar Populations

In the 1940s, astronomers realized that there were two types of stars in the galaxy. The stars they were most accustomed to studying were located in the disk, such as the stars near the sun. These they called **population I** stars. The second type, called **population II** stars, are usually found in the halo, in globular clusters, or in the nuclear bulge. In other words, the two stellar populations are associated respectively with the disk and spherical components of the galaxy.

The stars of the two populations fuse nuclear fuels and evolve in nearly identical ways. They differ only in their abundance of atoms heavier than helium, atoms that astronomers refer to collectively as **metals.** (Note that this is definitely not the way the word *metal* is commonly used by nonastronomers.) Population I stars are relatively metal rich, containing 2 to 3 percent metals, while population II stars are metal poor, containing only about 0.1 percent metals or less. The metal content of a star defines its population.

Population I stars, sometimes called *disk population stars,* have circular orbits in the plane of the galaxy and are relatively young stars that formed within the last few billion years. The sun is a population I star. Population II stars belong to the spherical component of the galaxy and are sometimes called the *halo population stars.* These stars have randomly tipped orbits with a wide range of shapes. A few follow circular orbits, but most follow elliptical orbits. The population II stars are all lower-mass main-sequence stars or giants. They are old stars. The metal-poor globular clusters are part of the halo population. Since the discovery of stellar populations, astronomers have realized that there is a gradation between populations, illustrated in ■ Table 9-1.

Why do the disk and halo stars have different metal abundances? The two types of star must have formed at different stages in the life of the galaxy, at times when the chemical composition of the galaxy differed. This is a clue to the history of our galaxy, but to use the clue, you must understand the cycle of element building.

The Element-Building Cycle

The atoms of which you are made were created in a process that spanned a number of generations of stars. The process that built the chemical elements over the history of our galaxy led to the possibility of Earth and life on Earth.

You saw in Chapter 8 how elements heavier than helium but lighter than iron are built up by nuclear reactions inside evolving stars. Atoms more massive than iron are made only by short-lived nuclear reactions that occur during a supernova explosion. This explains why lower-mass atoms like carbon, nitrogen, and oxygen are more common and why atoms more massive than iron—such as gold, silver, platinum, and uranium—are so rare and, often, valuable. ■ Figure 9-14a shows the abundance of the chemical elements, but notice that the graph has an exponential scale. To get a feeling for the true abundance of the elements, you should draw this graph using a linear scale as in Figure 9-14b. With that scale you see how rare the elements heavier than helium really are.

Most of the matter in stars is hydrogen and helium. Other elements, including carbon, nitrogen, oxygen, and the rest of what astronomers call metals, were cooked up inside stars. When the galaxy first formed, there should have been no metals because stars had not yet manufactured any. Judging from the composition of stars in the oldest clusters, the gas from which the galaxy originally condensed must have contained about 90 percent hydrogen atoms and 10 percent helium atoms. (The hydrogen and helium came from the big bang that began the universe, which you will study in Chapter 11.)

The first stars to form from this gas were metal poor, and now, 13 billion years later, their spectra still show few metal lines. Of course, they may have manufactured many atoms heavier than helium in their cores, but because the stars' interiors are not

■ Table 9-1 | Stellar Populations

	Population I		Population II	
	Extreme	**Intermediate**	**Intermediate**	**Extreme**
Location	Spiral arms	Disk	Nuclear bulge	Halo
Metals (%)	3	1.6	0.8	Less than 0.8
Shape of orbit	Circular	Slightly elliptical	Moderately elliptical	Highly elliptical
Average age (yr)	100 million and younger	0.2–10 billion	2–10 billion	10–13 billion

a

b

■ **Figure 9-14**

The abundance of the elements in the universe. (a) When the elements are plotted on an exponential scale, you see that elements heavier than iron are about a million times less common than iron and that all elements heavier than helium (referred to by astronomers as "metals") are quite rare. (b) The same data plotted on a linear scale provide a more realistic impression of how rare the metals are. Carbon, nitrogen, and oxygen make small peaks near atomic mass 15, and iron is just visible in the graph.

mixed, those heavy atoms stay trapped at the centers of the stars where they were produced and do not affect the spectra (■ Figure 9-15). The population II stars in the halo are the survivors of an earlier generation of stars that formed in the galaxy.

Most of the first stars evolved and died, and various types of star death throes, including supernovae, enriched the interstellar gas with metals. Succeeding generations of stars formed from gas clouds that were more enriched, and each generation added to the enrichment with its death. By the time the sun formed, roughly 5 billion years ago, the element-building process had added about 1.6 percent metals. Since then, the metal abundance has increased further, and stars forming at the present time incorporate 2 to 3 percent metals and become extreme population I stars. Thus metal abundance varies between populations because of the production of heavy atoms in successive generations of stars as the galaxy aged.

The oxygen atoms you are breathing, the carbon atoms in your flesh, the calcium atoms in your bones were all created in the interiors of red and yellow giant stars hot enough to sustain fusion reactions beyond helium. The gold and silver atoms in your jew-

elry and dental fillings, the iodine in medicine, the mercury (quicksilver) atoms in old-fashioned thermometers were all created in a few moments during supernova explosions, because those are the only places in the universe hot enough to make these types of atoms. This one idea is perhaps the most significant idea to take away from an introductory astronomy course and textbook—you, and everything on Earth, are made of stardust.

The History of the Milky Way Galaxy

The lack of metals in the spherical component of the galaxy tells you it is very old, a fossil left behind by the galaxy when it was young and drastically different from its present disk shape. The study of element building and stellar populations leads to the fundamental question, "How did our galaxy form?"

In the 1950s, astronomers began to develop a hypothesis, sometimes called the **monolithic collapse model,** to explain the formation of our galaxy. Recent observations are forcing a major reevaluation and revision of that traditional hypothesis.

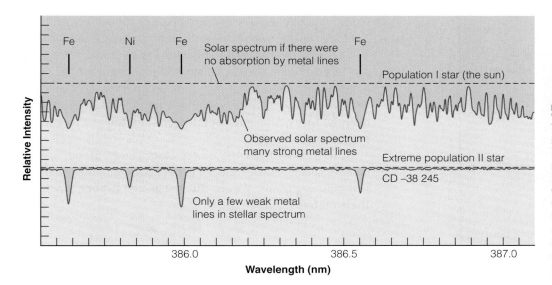

Figure 9-15

The difference between population I stars and population II stars is dramatic. (a) A graph of a population I star's spectrum reveals overlapping absorption lines of metals completely blanketing the continuum. (b) The lower spectrum is that of an extremely metal-poor star with only a few, weak metal lines of iron (Fe) and nickel (Ni). This population II star has about 10,000 times less metal content than the sun. (Adapted from an ESO illustration)

The traditional hypothesis says that the galaxy formed from a single large cloud of turbulent gas over 13 billion years ago. Stars and star clusters that formed from this material went into randomly shaped and randomly tipped orbits. These first stars were metal poor because no stars had existed earlier to enrich the gas with metals. In this way, the initial contraction of a large, turbulent gas cloud produced the spherical component of the galaxy.

The second stage of this hypothesis accounts for the disk component. The turbulent motions would eventually have canceled out, leaving the cloud with uniform rotation. A rotating, low-density cloud of gas cannot remain spherical. A star is spherical because its high internal pressure balances its gravity, but in a low-density cloud, the pressure cannot support the weight. Like a blob of pizza dough spun in the air, such a cloud must flatten into a disk (Figure 9-16).

This contraction into a disk took billions of years, with the metal abundance gradually increasing as generations of stars were born from the gradually flattening gas cloud. The stars and globular clusters that formed first in the halo would not have been affected by the motions of the gas and would have been left behind by the cloud as it collapsed and flattened. Later generations of stars formed in flatter distributions. The gas distribution in the galaxy now is so flat that the youngest stars are confined in a thin disk only about 100 parsecs thick. These stars are metal rich and have nearly circular orbits.

This monolithic collapse hypothesis accounts for many of the Milky Way's properties. Advances in technology, however, have improved astronomical observation; and, beginning in the 1980s, contradictions arose. For example, not all globular clusters have the same age, but, surprisingly, the younger clusters seem to be in the outer halo. In contrast, the monolithic collapse hypothesis says that the halo formed first and predicts that the clusters within it should either have a uniform age or the most distant ones should be slightly older. Another problem is that the oldest stars are observed to be metal poor but not completely metal free. There must have

been at least a few massive stars to create these metals from a generation before the formation of the oldest stars now seen in the halo.

Can the monolithic collapse model be modified to explain these observations? Perhaps the galaxy began with the contraction of a gas cloud to form the central bulge, and the halo later accumulated from gas clouds that had been slightly enriched in metals by an early generation of massive stars. That first generation of stars would have formed from almost pure hydrogen and helium gas, which astronomers calculate would have formed only stars with very high masses. Those stars would have lived very short lives, made metals, and died in supernovae explosions; and none would be left today. This would explain the metals in the oldest stars surviving today.

There is also evidence that entire small galaxies were captured by the growing Milky Way, and that fresh gas was added to the disk over time scales much longer than can be explained by one overall gravitational collapse. If our galaxy absorbed a few small but partially evolved galaxies, then some of the globular clusters in the halo may be hitchhikers that originally belonged to the captured galaxies. This could explain the range of globular cluster ages and compositions. (You will see in the next chapter that such galaxy mergers do occur.)

The problem of the formation of our galaxy is frustrating because the theories are incomplete, but astronomers are still gathering observations and testing hypotheses. You see an older theory that has proven to be inadequate to explain the observations, and you see astronomers attempting to refine the observations and devise new theories. The metal abundances and ages of the stars in our galaxy seem to be important clues, but metal abundance and age do not tell the whole story.

Building Scientific Arguments

Why do metal-poor stars have the most elongated orbits?
A good argument makes connections between ideas, and

A spherical cloud of turbulent gas gives birth to the first stars and star clusters.

The rotating cloud of gas begins to contract toward its equatorial plane.

Stars and clusters are left behind in the halo as the gas cloud flattens.

New generations of stars have flatter distributions.

The disk of the galaxy is now very thin.

■ **Figure 9-16**

The traditional "monolithic collapse" model for the origin of our galaxy begins with a spherical gas cloud that flattens into a disk.

sometimes those connections are not obvious. Certainly, the metal abundance of a star cannot affect its orbit, so an analysis must not confuse cause and effect with the relationship between these two factors. Both chemical composition and orbital shape depend on a third factor—age. The oldest stars are metal poor because they formed before there had been many supernova explosions to create and scatter metals into the interstellar medium. Those stars formed long ago when the galaxy was young and motions were not organized into a disk, and they tended to take up randomly shaped orbits, many of which are quite elongated. Consequently, today, the most metal-poor stars tend to follow the most elongated orbits.

Nevertheless, even the oldest stars known in our galaxy contain some metals. They are metal poor, not metal free. Adjust your argument. **Where did these metal-poor stars get their metals?**

■ ■ ■

9-4 The Nucleus

THE MOST MYSTERIOUS REGION OF OUR GALAXY is its very center, the nucleus. At visual wavelengths, this region is totally hidden by gas and dust that dim the light by 30 magnitudes (■ Figure 9-17). If a trillion (10^{12}) photons of light left the center of the galaxy on a journey to Earth, only one would make it through the gas and dust. The longer-wavelength infrared photons are scattered much less often: One in every ten makes it to Earth. Consequently, visual-wavelength observations reveal nothing about the nucleus, but it can be observed at longer wavelengths such as in the infrared and radio parts of the spectrum.

The Center of the Galaxy

Harlow Shapley's study of globular clusters placed the center of our galaxy in Sagittarius, and the first radio maps of the sky showed a powerful radio source located in Sagittarius. The first infrared map of the nuclear bulge made by Eric Becklin in 1968 showed the location of intense radiation where the stars are most crowded together, identifying the gravitational center of the galaxy. Later high-resolution radio maps of that stellar center revealed a complex collection of radio sources, with one, **Sagittarius A*** (abbreviated Sgr A* and pronounced "sadge A-star"), lying at the expected location of the galactic core.

Observations show that Sgr A* is only a few astronomical units in diameter but is a powerful source of radio energy. The tremendous amount of infrared radiation coming from the central area appears to be produced by crowded stars and by dust warmed by those stars. But what could be as relatively tiny as Sgr A* and produce so much radio energy?

Study **"Sagittarius A*"** on pages 192–193 and notice three important points:

❶ First, observations at radio wavelengths reveal complex structures near Sgr A* caused by magnetic fields and by

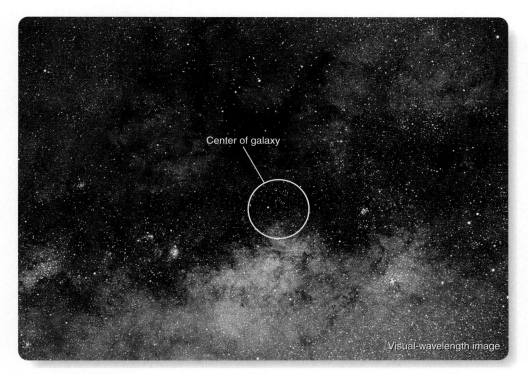

Center of galaxy

Visual-wavelength image

■ **Figure 9-17**

Looking toward Sagittarius, you see nothing to suggest this is the center of the galaxy. Gas and dust block your view. Only the distribution of globular clusters told Shapley the sun lies far from the center of the star system. (Daniel Good)

rapid star formation. Supernova remnants show that massive stars have formed there recently and died explosively.

2 Also, the center is very crowded. Tremendous numbers of stars heat the dust and produce strong infrared radiation.

3 Finally, there is evidence that Sgr A* is a supermassive black hole into which gas is flowing.

Astronomers continue to test the hypothesis that the center of our galaxy contains a supermassive black hole. Such an object is sufficient to explain the observations, but is it necessary? Is there some other way to explain what is observed? For example, astronomers have suggested that gas flowing toward the center of the galaxy could trigger tremendous bursts of star formation. Such theories have been considered and tested against the evidence, but none appears to be adequate to explain all the observations. So far, the only theory that seems adequate is that our galaxy's nucleus is home to a supermassive black hole.

Meanwhile, observations are allowing astronomers to improve their models. For instance, Sgr A* is not as bright in X rays as it should be if it had a hot accretion disk with matter constantly flowing into the black hole. Observations of X-ray and infrared flares lasting only a few hours suggest that mountain-size blobs of matter may occasionally fall into the black hole and be heated and ripped apart by tidal forces (see Chapter 8). The black hole may be mostly dormant and lack a fully developed hot accretion disk because the rate of matter flow into it is relatively low at the present time.

Such a supermassive black hole could not be the remains of a single dead star. It contains much too much mass. It probably formed when the galaxy first formed over 13 billion years ago. In the next chapter you will learn that supermassive black holes are commonly found in the centers of galaxies.

Building Scientific Arguments

Why do astronomers think the center of our galaxy contains a large mass?
Because scientific arguments hinge on evidence, they often involve discussions of measurement—one of the keys to science. The best way to measure the mass of an astronomical object is to watch something orbit around it. You can then use Kepler's third law to find the mass inside the orbit. Because of gas and dust, astronomers can't see to the center of our galaxy at visual wavelengths, but infrared observations can detect individual stars orbiting Sgr A*. In the case of the star S0-2, for example, astronomers can see the shape of its orbit and determine its orbital period. The stars that orbit Sgr A* have small orbits and very short periods, and a detailed calculation shows that Sgr A* contains roughly 2.6 million solar masses.

Now build a new argument to analyze a different measurement. **Why does the strong radiation at wavelengths longer than 4 microns (4000 nm) imply that vast numbers of stars are crowded into the center?**

■   ■   ■

Sagittarius A*

1 The constellation of Sagittarius is so filled with stars and with gas and dust you can see nothing at visual wavelengths of the center of our galaxy.

The image below is a wide-field radio image of the center of our galaxy. Many of the features are supernova remnants (SNR), and a few are clouds of star formation. Peculiar features such as threads, the Arc, and the Snake may be gas trapped in magnetic fields. At the center lies Sagittarius A, the center of our galaxy.

Arc

Radio image

NRAO/AUI/NSF

The radio map above shows Sgr A and the Arc filaments, 50 parsecs long. The image was made with the VLA radio telescope. The contents of the white box are shown on the opposite page.

Sgr D HII

Sgr D SNR

SNR 0.9 + 0.1

Apparent angular size of the moon for comparison

Sgr B2

Sgr B1

New SNR 0.3 + 0.0

Threads

The Cane

Arc

Background galaxy

2 Infrared photons with wavelengths longer than 4 microns (4000 nm) come almost entirely from warm interstellar dust. The radiation at these wavelengths coming from Sagittarius is intense, and that indicates that the region contains lots of dust and is crowded with stars that warm the dust.

Sgr A

Threads

Radio image

NRL

Sgr C

The Pelican

Coherent structure?

Snake

Sgr E

2MASS

SNR 359.1 − 00.5

Infrared image

1a This high-resolution radio image of Sgr A (the white boxed area on the opposite page) reveals a spiral swirl of gas around an intense radio source known as Sgr A*, the presumed central object in our galaxy. About 3 pc across, this spiral lies in a low-density cavity inside a larger disk of neutral gas. The arms of the spiral are thought to be streams of matter flowing into Sgr A* from the inner edge of the larger disk (drawing at right).

Sgr A*

Radio image

N. Killeen and Kwok-Yung Lo

Evidence of a Black Hole at the Center of Our Galaxy

3 Since the middle 1990s, astronomers have been able to use large infrared telescopes and active optics to follow the motions of stars orbiting around Sgr A*. A few of those orbits are shown here. The size and period of the orbit allows astronomers to calculate the mass of Sgr A* using Kepler's third law. The orbital period of the star SO-2, for example, is 15.2 years and the semimajor axis of its orbit is 950 AU. The combined motions of the observed stars suggest that Sgr A* has a mass of 2.6 million solar masses.

The Chandra X-ray Observatory has imaged Sgr A* and detected over 2000 other X-ray sources in the area.

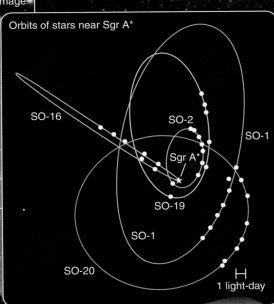

NASA/CXC/MIT/F.K. Baganoff et al.

Infrared Image

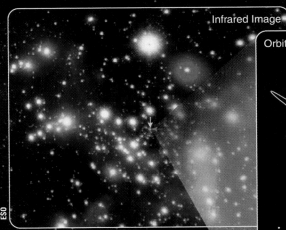

ESO

Orbits of stars near Sgr A*

SO-16

SO-2

SO-1

Sgr A*

SO-19

SO-1

SO-20

1 light-day

At its closest, SO-2 comes within 17 light-hours of Sgr A*. Alternative theories that Sgr A* is a cluster of stars, of neutron stars, or of stellar black holes are eliminated. Only a single black hole could contain so much mass in so small a region.

3a A black hole with a mass of 2.6 million solar masses would have an event horizon smaller than the smallest dot in this diagram. A slow dribble of only 0.0002 solar masses of gas per year flowing into the black hole could produce the observed energy. A sudden increase as when a star falls in could produce a violent eruption.

The evidence of a massive black hole at the center of our galaxy seems conclusive. It is much too massive to be the remains of a dead star, however, and astronomers conclude that it probably formed as the galaxy first took shape.

Our solar system is half a light-day in diameter.

Summary

9-1 | The Discovery of the Galaxy

How do you know you live in a galaxy?

- The hazy band of the Milky Way is our wheel-shaped galaxy seen from within, but its size and shape are not obvious. William and Caroline Herschel counted stars over the sky to show that it seemed to be shaped like a disk with the sun near the center, but they could not see very far into space because of interstellar gas and dust.

- In the early 20th century, Harlow Shapley calibrated Cepheid variable stars to find the distance to globular clusters and demonstrated that our galaxy is much larger than the part you can see easily and that the sun is not at the center.

- Modern observations suggest that our galaxy contains a disk component about 80,000 ly in diameter and that the sun is two-thirds of the way from the center to the visible edge. The nuclear bulge around the center and an extensive halo containing old stars and little gas and dust make up the spherical component.

- The mass of the galaxy can be found from its rotation curve. Kepler's third law tells astronomers the galaxy contains over 100 billion solar masses, and the rising rotation curve at great distance from the center shows that the halo must contain much more mass. Because that material is not emitting detectable electromagnetic radiation, astronomers call it dark matter.

9-2 | Spiral Arms and Star Formation

How do you know ours is a spiral galaxy, and what are the spiral arms?

- The most massive stars live such short lives that they don't have time to move from their place of birth. Because they are concentrated along the spiral arms, astronomers conclude that the spiral arms are sites of star formation.

- Spiral arms can be traced through the sun's neighborhood by using spiral tracers such as O and B stars; but to extend the map over the entire galaxy, astronomers must use infrared and radio observations to see through the gas and dust.

- The spiral density wave theory suggests that the spiral arms are regions of compression that move around the disk. When an orbiting gas cloud overtakes the compression wave, the gas cloud forms stars. Another process, self-sustaining star formation, may act to modify the arms with branches and spurs as the birth of massive stars triggers the formation of more stars by compressing neighboring gas clouds.

9-3 | The Origin and History of the Milky Way

How did our galaxy form and evolve?

- The oldest open star clusters indicate that the disk of our galaxy is only about 9 billion years old. The oldest globular clusters appear to be at least 13 billion years old, so our galaxy must have formed at least 13 billion years ago.

- Stellar populations are an important clue to the formation of our galaxy. The first stars to form in our galaxy, termed population II stars, were poor in elements heavier than helium—elements that astronomers call metals. As generations of stars manufactured metals and spread them back into the interstellar medium, the metal abundance of more recent generations increased. Population I stars, including the sun, are richer in metals than population II stars.

- Because the halo is made up of population II stars and the disk is made up of population I stars, astronomers conclude that the halo formed first and the disk later. The monolithic collapse model, that the galaxy formed from a single, roughly spherical cloud of turbulent gas and gradually flatted into a disk, has been amended to include mergers with other galaxies and later infalling gas contributing to the disk.

9-4 | The Nucleus

What lies at the very center of our galaxy?

- The nucleus of the galaxy is not visible at visual wavelengths, but radio, infrared, and X-ray radiation can penetrate the gas and dust in space. These wavelengths reveal crowded central stars and warmed dust.

- The very center of the Milky Way Galaxy is marked by a radio source, Sagittarius A*. The source must be less than a few astronomical units in diameter, but the motion of stars around the center shows that it must contain roughly 2.6 million solar masses. A supermassive black hole is the only object that could contain so much mass in such a small space.

New Terms

Cepheid variable star (p. 177)

instability strip (p. 177)

period–luminosity relation (p. 177)

proper motion (p. 177)

calibration (p. 177)

disk component (p. 179)

kiloparsec (kpc) (p. 179)

spiral arm (p. 179)

spherical component (p. 179)

halo (p. 179)

nuclear bulge (p. 179)

rotation curve (p. 182)

dark halo (p. 182)

dark matter (p. 182)

spiral tracer (p. 183)

density wave theory (p. 184)

self-sustaining star formation (p. 186)

population I (p. 187)

population II (p. 187)

metals (p. 187)

monolithic collapse model (p. 188)

Sagittarius A* (Sgr A*) (p. 190)

Review Questions

ThomsonNOW™ Assess your understanding of this chapter's topics with additional quizzing and animations at
www.thomsonedu.com

1. Why isn't it possible to locate the center of our galaxy from the appearance of the Milky Way?

2. Why is there a period–luminosity relation for Cepheid variable stars?

3. How can astronomers use Cepheid variable stars to find distances?

4. Why didn't astronomers before Shapley realize how large the galaxy is?

5. How can astronomers find the mass of the galaxy?

6. What evidence is there that our galaxy has an extended halo of dark matter?

7. Why do spiral tracers have to be short lived?

8. What evidence is there that the density wave theory is not fully adequate to explain spiral arms in our galaxy?

9. How do astronomers know how old our galaxy is?

10. Why do astronomers conclude that stars with metal-poor spectra are older than stars with metal-rich spectra?

11. How do the orbits of stars around the Milky Way Galaxy help explain its origin?

12. What evidence contradicts the monolithic collapse model for the origin of our galaxy?

13. What is the evidence that the center of our galaxy contains a massive black hole?

14. Why must astronomers use infrared telescopes to observe the motions of stars around Sgr A*?

NASA/Hubble Heritage Team/STScI/ AURA

15. Why does the galaxy shown at the right have so much dust in its disk? How big do you suppose the halo of this galaxy really is?

NASA/Hubble Heritage Team and A. Riess, STScI

16. Why are the spiral arms in the galaxy at the right blue? What color would the halo be if it were bright enough to see in this photo?

Discussion Questions

1. How would this chapter be different if interstellar matter didn't absorb starlight?

2. Are there any observations you could make with the Hubble Space Telescope that would allow you to better understand Sgr A*?

10 | Galaxies

LESS THAN A CENTURY AGO, astronomers did not even understand that galaxies exist. Nineteenth-century telescopes revealed faint nebulae scattered among the stars, some with spiral shapes. Astronomers argued about the nature of these nebulae, but it was not until the 1920s that astronomers understood that many of those nebulae are other galaxies much like our own, and it was not until recent decades that astronomical telescopes could reveal the variety, intricacy, and beauty of the galaxies (■ Figure 10-1). ▌ In this chapter, you will try to understand how galaxies form and evolve, and you will discover that the amount of gas and dust in a galaxy is a critical clue. You will also discover that interactions between galaxies can influence their structure and evolution.

▌ Continued on page 198 ▌

Guidepost

Looking Back

In the preceding chapter, you learned about our Milky Way Galaxy, an important object to Earthlings but only one out of billions of galaxies visible in the sky. The Milky Way Galaxy is a normal spiral galaxy, somewhat larger than average, with a history that is probably typical of other large galaxies. However, your understanding of galaxies from the study of a single example, the Milky Way, is no more complete than your understanding of humanity would be from knowing a single person.

This Chapter

This chapter will expand your horizon to discuss the different types of galaxies, their range of properties, and their complex histories. Many galaxies, including the Milky Way, have nuclei that are active, powered by supermassive black holes. Here you can expect answers to these important questions:

What types of galaxies exist?

How do astronomers measure the distances to galaxies, and how does that allow the sizes, luminosities, and masses of galaxies to be determined?

Why are there different kinds of galaxies, and how do galaxies evolve?

What is the energy source for active galaxies, what can trigger the activity, and what does that reveal about the history of galaxies?

The answers to these questions will lead you to insights about how the Milky Way galaxy in which you live fits into the universe as a whole.

Looking Ahead

In the next chapter, you will study the universe as a whole. Galaxies, active galaxies, and dark matter are the final puzzle pieces you need before you try to understand the birth and evolution of the entire universe.

M83 12 million ly

Dust clouds glow red in this infrared image.

Infrared image

ESO510-G13
150 million ly

Dusty disk of galaxy warped by interaction with another galaxy

New stars forming in dust clouds.

Visual-wavelength image

■ **Figure 10-1**

A century ago, photos of galaxies looked like spiral clouds of haze. Modern images of these relatively nearby galaxies reveal newborn stars and clouds of gas and dust. (M83: ESO; ESO510-G13: Hubble Heritage Team, AURA/STScI/NASA)

❸ Finally, some galaxies are highly irregular in shape, and those galaxies tend to be rich in gas and dust.

You might also wonder what proportion of the galaxies are elliptical, spiral, and irregular, but that is a difficult question to answer. In some catalogs of galaxies, about 70 percent are spiral, but that is the result of what scientists call a selection effect (**Window on Science 10-1**). Spiral galaxies contain hot, bright stars and are consequently very luminous and easy to see. From careful studies astronomers conclude that ellipticals are actually more common than spirals and that irregulars make up about 25 percent of all galaxies. Among spiral galaxies, about two-thirds are barred spirals.

Different kinds of galaxies have different colors, depending mostly on how much star formation is happening in them. Spirals and irregulars usually contain plenty of young stars, including massive, hot, luminous O and B stars. They produce most of the light and give spirals and irregulars a distinct blue tint. In contrast, elliptical galaxies usually have few young stars. The most luminous stars in ellipticals are red giants, which give those galaxies a red tint.

How many galaxies are there? A research effort called GOODS (Great Observatories Origins Deep Survey) has used the Hubble Space Telescope, the Chandra X-Ray Observatory, the Spitzer Space Telescope, and the XXM-Newton X-Ray Telescope, plus the largest ground-based telescopes, to study two selected areas in the northern and southern skies. The GOODS images reveal tremendous numbers of galaxies (■ Figure 10-2). There are good reasons to believe that the two regions of the sky chosen for study are typical, so it seems likely that the entire sky is carpeted with galaxies. The result of the GOODS research is an estimate that at least 100 billion galaxies would be visible if today's telescopes were used for an all-sky census.

Before you can build theories, however, you need to gather some basic data concerning galaxies. You can classify the different kinds of galaxies and discover their basic properties—diameter, luminosity, and mass. Once you know the typical properties of galaxies, you will be ready to theorize about their origin and evolution. You also will learn about some of the most energetic events in the universe: The energy pouring out of the nuclei of certain galaxies is enormously greater than mere supernova explosions.

10-1 The Family of Galaxies

Astronomers classify galaxies according to their shape using a system developed in the 1920s by Edwin Hubble (after whom the Hubble Space Telescope is named). Creating a system of classification is a fundamental technique in science.

Study **Galaxy Classification** on pages 200–201 and notice three important points:

❶ Many galaxies have no disk, no spiral arms, and almost no gas and dust. These elliptical galaxies range from huge giants to small dwarfs.

❷ Disk-shaped galaxies usually have spiral arms and contain gas and dust, although some have very little. Many spiral galaxies have a barred structure.

10-2 Measuring the Properties of Galaxies

LOOKING BEYOND THE EDGE of our Milky Way Galaxy, astronomers find many billions of galaxies. What are the properties of these star systems? What are the diameters, luminosities, and masses of galaxies? Just as in your study of stellar characteristics (see Chapter 6), the first step in your study of galaxies is to find out how far away they are. Once you know a galaxy's distance, its size and luminosity are relatively easy to find. Later in this section, you will see that finding the masses of galaxies is more difficult, but the results are quite interesting.

Selection Effects in Science

Many different kinds of science depend on selecting objects to study. Scientists studying insects in the rain forest, for example, must choose which ones to catch. They can't catch every insect they see, so they might decide on some way to select the ones they do catch. If they are not careful, a selection effect could bias their data and lead them to incorrect conclusions without their ever knowing it. The reason selection effects are dangerous in science is that they can have powerful influences without always being evident. Scientists can avoid selection effects only by carefully designing a research project.

For example, suppose you decide to measure the speed of cars on a highway. There are too many cars to measure every one, so you reduce the workload and measure the speed of only red cars. It is quite possible that this selection criterion will mislead you because people who buy red cars may be more likely to be younger and drive faster. Should you measure only brown cars? No, because you might suspect that only older, more sedate people would buy a brown car. Should you measure the speed of any car in front of a truck? Perhaps you should pick any

car following a truck? Again, you may be selecting cars that are traveling a bit faster or a bit slower than normal. Only by very carefully designing your experiment can you be certain that the cars you measure are traveling at speeds truly representative of the entire population of cars.

Astronomers face the danger of selection effects quite often. Very luminous stars are easier to see at great distances than faint stars. Spiral galaxies are brighter, bluer, and more noticeable than elliptical galaxies. What astronomers see through a telescope depends on what they notice, and that is powerfully influenced by selection effects.

Scientists engaged in observation must spend considerable time designing their experiments. They must be careful to observe an unbiased sample if they expect to make logical deductions from their results. The scientists in the rain forest, for example, should not catch and study only the red insects. Often, the most brightly colored insects are poisonous (or at least taste bad) to predators. Catching only brightly colored insects could produce a highly biased sample of the insect

Visual-wavelength image

Things that are bright and beautiful, such as spiral galaxies, may attract a disproportionate amount of attention. Scientists must be aware of such selection effects. (Hubble Heritage Team/STScI/AURA/NASA)

population. Careful scientists must plan their work with great care and avoid any possible selection effects.

Visual-wavelength image

■ **Figure 10-2**

Very long time exposures called "deep images" reveal faint galaxies billions of light-years from Earth. Presumably, the entire sky is carpeted with galaxies like this. (NASA/ESA/GOODS Team and M. Giavalisco, STScI)

1 **Elliptical galaxies** are round or elliptical, contain no visible gas and dust, and have few or no hot, bright stars. They are classified with a numerical index ranging from 1 to 7; E0s are round, and E7s are highly elliptical. The index is calculated from the largest and smallest diameter of the galaxy used in the following formula and rounded to the nearest integer.

$$\frac{10(a - b)}{a}$$

Outline of an
E6 galaxy

AURA/NOAO/NSF

Visual-wavelength image

Anglo-Australian Telescope Board

The Leo 1 dwarf elliptical galaxy is not many times bigger than a globular cluster.

.Visual

M87 is a giant elliptical galaxy classified E1. It is a number of times larger in diameter than our own galaxy and is surrounded by a swarm of over 500 globular clusters.

Anglo-Australian Telescope Board

Sa

2 **Spiral galaxies** contain a disk and spiral arms. Their halo stars are not visible, but presumably all spiral galaxies have halos. Spirals contain gas and dust and hot, bright O and B stars, as shown at right. The presence of short-lived O and B stars alerts us that star formation is occurring in these galaxies. Sa galaxies have larger nuclei, less gas and dust, and fewer hot, bright stars. Sc galaxies have small nuclei, lots of gas and dust, and many hot, bright stars. Sb galaxies are intermediate.

Visual   NGC 3623

Sb

BAR

ThomsonNOW™

Sign in at www.thomsonedu.com and go to ThomsonNOW to see Active Figure "Galaxy Types" and review the classification of galaxies.

NGC 3627   Visual

Sc

2a Roughly 2/3 of all spiral galaxies are **barred spiral galaxies** classified SBa, SBb, and SBc. They have an elongated nucleus with spiral arms springing from the ends of the bar, as shown at left. Our own galaxy is a barred spiral.

NGC 1365   Visual

NGC 2997   Visual

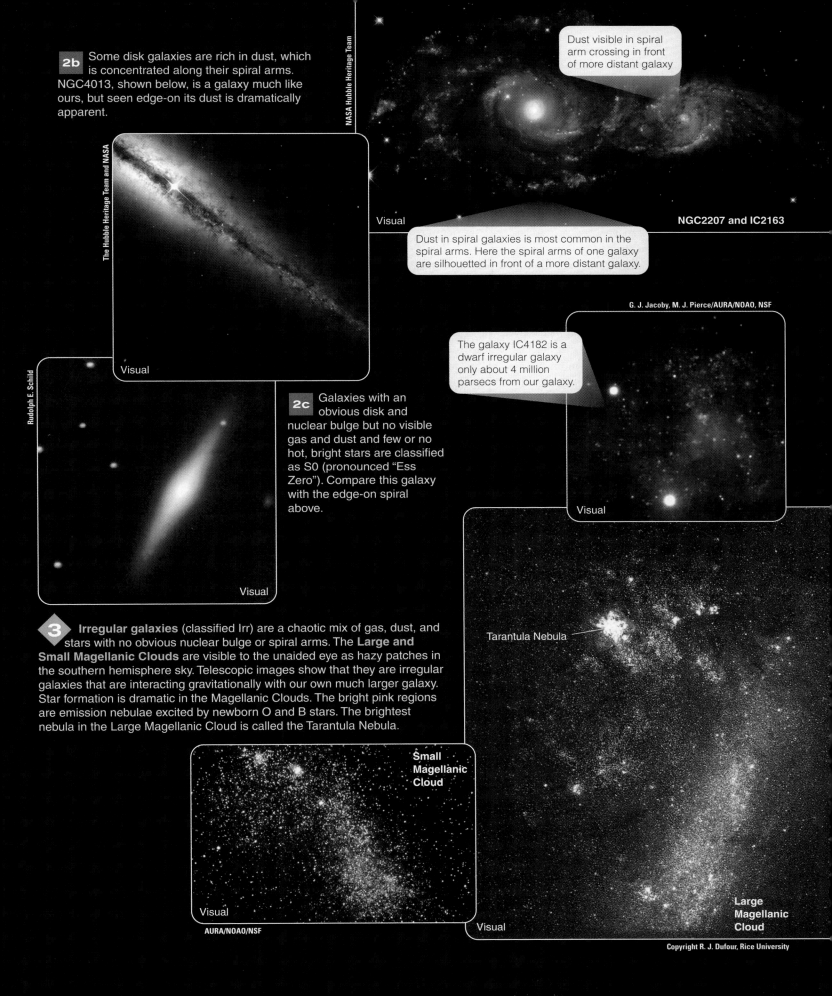

2b Some disk galaxies are rich in dust, which is concentrated along their spiral arms. NGC4013, shown below, is a galaxy much like ours, but seen edge-on its dust is dramatically apparent.

Dust visible in spiral arm crossing in front of more distant galaxy

Visual

Visual

NGC2207 and IC2163

Dust in spiral galaxies is most common in the spiral arms. Here the spiral arms of one galaxy are silhouetted in front of a more distant galaxy.

The galaxy IC4182 is a dwarf irregular galaxy only about 4 million parsecs from our galaxy.

2c Galaxies with an obvious disk and nuclear bulge but no visible gas and dust and few or no hot, bright stars are classified as S0 (pronounced "Ess Zero"). Compare this galaxy with the edge-on spiral above.

Visual

Visual

3 **Irregular galaxies** (classified Irr) are a chaotic mix of gas, dust, and stars with no obvious nuclear bulge or spiral arms. The **Large and Small Magellanic Clouds** are visible to the unaided eye as hazy patches in the southern hemisphere sky. Telescopic images show that they are irregular galaxies that are interacting gravitationally with our own much larger galaxy. Star formation is dramatic in the Magellanic Clouds. The bright pink regions are emission nebulae excited by newborn O and B stars. The brightest nebula in the Large Magellanic Cloud is called the Tarantula Nebula.

Tarantula Nebula

Small Magellanic Cloud

Large Magellanic Cloud

Visual

Visual

Distance

The distances to galaxies are so large that it is not convenient to express them in light-years, parsecs, or even kiloparsecs. Instead, astronomers use the unit **megaparsec (Mpc),** or 1 million pc. One Mpc equals 3.26 million ly, or approximately 3×10^{19} km (2×10^{19} miles).

To find the distance to a galaxy, astronomers must search among its stars, nebulae, and star clusters for familiar objects whose luminosity they know. Such objects are called **standard candles.** If you can find a standard candle in a galaxy, you can judge its distance.

Cepheid variable stars are reliable standard candles because their period is related to their luminosity (see Figure 9-4). If you know the period of the star's variation, you can use the period–luminosity diagram to learn its absolute magnitude. Then, by comparing its absolute and apparent magnitudes, you can find its distance. ■ Figure 10-3 shows a galaxy in which the Hubble Space Telescope detected Cepheids. Even with the Hubble Space Telescope, Cepheids are not detectable much beyond 100 million ly (30 Mpc), so astronomers must search for less common but brighter distance indicators and calibrate them using nearby galaxies containing detectable Cepheids.

When a supernova explodes in a distant galaxy, astronomers rush to observe it. Studies show that type Ia supernovae, those caused by the collapse of a white dwarf, all reach about the same luminosity at maximum (which make them more like "standard bombs" than standard candles). By searching for Cepheids and other distance indicators in nearby galaxies where type Ia supernovae have occurred, astronomers have been able to calibrate these supernovae. As a result, when type Ia supernovae are seen in more distant galaxies, astronomers can measure the apparent brightness at maximum and compare that with the known luminosity of these supernovae to find their distances. Because type Ia supernovae are much brighter than Cepheids, they can be seen in galaxies at great distances. The drawback is that supernovae are rare, and none may occur during your lifetime in a galaxy you might be studying. You will see in Chapter 11 how these distance indicators have reshaped our understanding of the history of the universe.

Notice how astronomers use calibration (see Window on Science 9-1) to build a distance scale reaching from the nearest galaxies to the most distant visible galaxies. Often astronomers refer to this as the **distance ladder** because each step depends on the steps below it. The foundation of the distance scale rests on understanding the luminosities of stars, which ultimately rests on measurements of stellar parallax (see Chapter 6).

Telescopes as Time Machines

The most distant visible galaxies are a little over 10 billion ly (3000 Mpc) away, and at such distances you see an effect like time travel. When you look at a galaxy millions of light-years away, you do not see it as it is now but as it was millions of years ago when its light began the journey toward Earth. When you look at a distant galaxy, you look into the past by an amount called the **look-back time,** a time in years equal to the distance to the galaxy in light-years.

The look-back time to nearby objects is usually not significant. The look-back time to the moon is only 1.3 seconds, to the sun 8 minutes, and to the nearest star about 4 years. The Andromeda Galaxy has a look-back time of about 2 million years, but that is a mere eye blink in the lifetime of a galaxy. When astronomers look at more distant galaxies, the look-back time becomes an appreciable part of

Visual-wavelength image

April 23
May 4
May 9
May 16
May 20
May 31

■ **Figure 10-3**

The vast majority of spiral galaxies are too distant for Earth-based telescopes to detect Cepheid variable stars. The Hubble Space Telescope, however, can locate Cepheids in some of these galaxies, for example in the bright spiral galaxy M100. From a series of images taken on different dates, astronomers can locate Cepheids (inset), determine the period of pulsation, and measure the average apparent brightness. They can then deduce the distance to the galaxy—51 million ly for M100. (J. Trauger, JPL; Wendy Freedman, Observatories of the Carnegie Institution of Washington; and NASA) the age of the universe. You will see evidence in Chapter 11 that the universe began about 13.7 billion years ago. When astronomers observe the most distant visible galaxies, they are looking back over 10 billion years to a time when the universe may have been significantly different.

The Hubble Law

Although astronomers find it difficult to measure the distance to a galaxy precisely, they often estimate such distances using a simple relationship. Early in the 20th century, astronomers noticed that the lines in galaxy spectra are generally shifted toward longer wavelengths—redshifted (see Chapter 5). These redshifts imply that the galaxies are receding from Earth.

In 1929, the American astronomer Edwin Hubble published a graph that plotted the apparent velocity of recession versus distance for a number of galaxies. The points in the graph fell along a straight line (■ Figure 10-4). This relation between apparent velocity of recession and distance is known as the **Hubble law,** and the slope of the line is known as the **Hubble constant,** symbolized by capital *H.*

The Hubble law is important in astronomy for two reasons. It is taken as evidence that the universe is expanding, a subject to be discussed in detail in Chapter 11. The present chapter discusses the Hubble law because astronomers use it to estimate the distance to galaxies. The distance to a galaxy can be found by dividing its apparent velocity of recession by the Hubble constant. This is a very useful calculation, because it is usually possible to obtain a spectrum of a galaxy and measure its redshift even if it is too far away to have observable standard candles. Obviously, knowing the value of the Hubble constant precisely is important.

Edwin Hubble's original measurement of *H* was too large because of errors in his measurements of the distances to galaxies. Later astronomers have struggled to measure this important

constant. The most precise measurements of the Hubble constant, made using the Hubble Space Telescope, yield a value for *H* of about 70 km/s/Mpc with an uncertainty of about 5 percent. This means that a galaxy at a distance of 1 Mpc from the Milky Way is receding from us at a rate of 70 km/s, a galaxy 2 Mpc away is receding at 140 km/s, and so on.

Note that the redshifts of galaxies are not really Doppler shifts, even though astronomers often express the red shifts in kilometers per second as if they were true velocities. In Chapter 11 you will discover a much more sophisticated way of understanding these redshifts. For the moment, you can just keep in mind that modern astronomers interpret redshifts of galaxies as caused by the expansion of the universe, and the Hubble law allows them to estimate the distance to a galaxy from its redshift.

ThomsonNOW™ Sign in at www.thomsonedu.com and go to ThomsonNOW to see Astronomy Exercise "Hubble Relations."

Galaxy Diameters and Luminosities

The distance to a galaxy is the key to finding its diameter and its luminosity. With even a modest telescope and a CCD camera, you could photograph a galaxy and measure its angular diameter. If you know the distance to the galaxy, you can then find its linear diameter. Also, if you measure the apparent brightness of the galaxy, you can use the distance to find its luminosity, as you learned regarding stars in Chapter 6.

The results of such observations show that galaxies differ dramatically in size and luminosity. Irregular galaxies tend to be small, 1 to 25 percent the diameter of our galaxy, and of low luminosity. Although they are common, they are easy to overlook. Our Milky Way Galaxy is large and luminous compared with most spiral galaxies, though astronomers know of a few spiral galaxies that are even larger and more luminous. Elliptical galaxies cover a wide range of diameters and luminosities. The largest, called giant ellipticals, are five or more times the diameter of our Milky Way Galaxy, but many so-called dwarf elliptical galaxies are only 1 per cent the diameter of our galaxy.

Clearly, the diameter and luminosity of a galaxy do not determine its type. Some small galaxies are irregular, and some are elliptical. Some large galaxies are spiral, and some are elliptical. Other factors must influence the evolution of galaxies.

Galaxy Masses

Although the mass of a galaxy is difficult to determine, it is an important quantity. It tells you how much matter the galaxy contains, which provides clues to the galaxy's origin and evolution.

The most precise method for measuring the mass of a galaxy is called the **rotation curve method** (■ Figure 10-5). It requires knowing: (1) the true sizes of the orbits of stars or gas

■ Figure 10-4

Edwin Hubble's first diagram of the apparent velocities of recession and distances of galaxies did not probe very deeply into space, and the vertical axis were later recalibrated. It did show, however, that the galaxies are receding at speeds related to their distances.

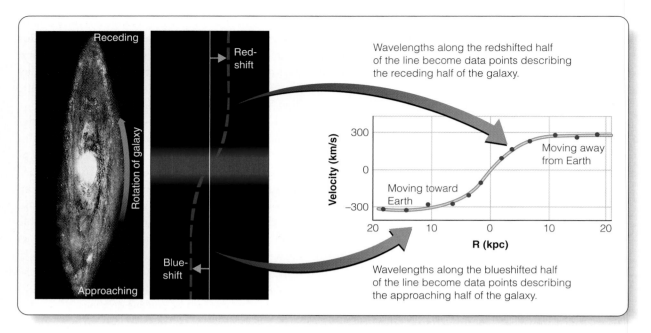

Wavelengths along the redshifted half of the line become data points describing the receding half of the galaxy.

Moving away from Earth

Moving toward Earth

Wavelengths along the blueshifted half of the line become data points describing the approaching half of the galaxy.

■ Figure 10-5

In this artist's conception, the astronomer has placed the image of the galaxy over a narrow slit so that light from the galaxy can enter the spectrograph and produce a spectrum. A very short segment of the spectrum shows an emission line redshifted on the receding side of the rotating galaxy and blueshifted on the approaching side. Converting these Doppler shifts into velocities, the astronomer can plot the galaxy's rotation curve (right).

clouds within a galaxy, which in turn requires knowing the distance of that galaxy; and (2) the orbital speeds of the stars or gas clouds, measured from the Doppler shifts of their spectral lines. That is enough information to use Kepler's third law (see Chapter 9) and find the mass of the part of the galaxy contained within the star orbits with measured sizes and speeds. The rotation curve method works only for galaxies near enough to be well resolved. More distant galaxies appear so small that astronomers cannot measure the radial velocity at different points across the galaxy and must use other, less precise, methods to estimate masses.

The masses of galaxies cover a wide range. The smallest contain about 10^{-6} as much mass as the Milky Way, and the largest contain as much as 50 times more than the Milky Way.

Dark Matter in Galaxies

Given the size and luminosity of a galaxy, astronomers can make a rough guess as to the amount of matter it should contain. Astronomers know how much light stars produce, and they know about how much matter there is between the stars, so it should be possible to estimate very roughly the mass of a galaxy from its luminosity. When astronomers measure the masses of galaxies, however, they often find that the measured masses are much larger than expected from the luminosities of the galaxies. You

discovered this effect in the previous chapter when you studied the rotation curve of our own galaxy and concluded that it must contain large amounts of dark matter, especially in its outer regions (see Figure 9-9). Astronomer Vera Rubin found in the 1960s that this also seems to be true of most nearby galaxies. Measured masses of galaxies amount to 10 to 100 times more mass than you can see.

Dark matter is difficult to detect, and it is even harder to explain. Some astronomers have suggested that dark matter consists of low-luminosity white dwarfs and brown dwarfs scattered through the halos of galaxies. Searches for white dwarfs and brown dwarfs in the halo of our galaxy have found a few but not enough to make up most of the dark matter. The dark matter can't be hidden in vast numbers of black holes and neutron stars, because astronomers don't see the X rays these objects would emit. The evidence indicates there is 10 to 100 times more dark matter than visible matter in galaxies, and if there were that many black holes they would produce X rays that would be easy to detect. Furthermore, recent images from the Chandra X-ray observatory indicate that a collision between two galaxy clusters caused their gas and dark matter components to separate.

Because observations imply that the dark matter can't be composed of familiar objects or material, astronomers are forced to conclude that the dark matter is made up of unexpected forms of matter. Until recently neutrinos were thought to be massless, but studies now suggest they have a very small mass. Thus they can be part of the dark matter, but their masses are too low to make up all of the dark matter. There must be some other undiscovered form of matter in the universe that is detectable only by its gravitational field.

Dark matter remains one of the fundamental unresolved problems of modern astronomy. Observations of galaxies and clusters of galaxies reveal that 90 to 95 percent of the matter in the universe is dark matter. The universe you see—the kind of matter that you and the stars are made of—has been compared to the foam on an invisible ocean. You will return to this problem in Chapter 11 when you try to understand how dark matter affects the nature of the universe, its past, and its future.

Supermassive Black Holes in Galaxies

Doppler shift measurements show that the stars near the centers of many galaxies are orbiting very rapidly. To hold stars in such small, short-period orbits, the centers of those galaxies must contain masses of a million to a few billion solar masses, yet no object is visible. The evidence seems to require that the nuclei of many galaxies contain supermassive black holes. You saw in Chapter 9 that the Milky Way contains a supermassive black hole at its center. Evidently that is typical of galaxies. It is a common misconception that the orbits of stars throughout a galaxy are controlled by the central black hole. The masses of those black holes, large as they may seem, are negligible compared with a galaxy's mass. The 2.6-million-solar-mass black hole at the center of the Milky Way Galaxy contains only a thousandth of one per cent of the total mass of the galaxy.

Building Scientific Arguments

Why do you have to know the distance to a galaxy to find its mass?
A scientific argument must proceed step-by-step with no gaps. This is a good example. To find the mass of a galaxy, you need to know the size of the orbits and the orbital periods of stars at the galaxy's outer edge and then use Kepler's third law to find the mass inside the orbits. Measuring the orbital velocity of the stars as the galaxy rotates is easy if you can obtain a rotation curve from a spectrum. But you must also know the radii of the stars' orbits in linear units, for example, meters or kilometers or astronomical units. That is where the distance to the galaxy comes in. To find the radii of the orbits, you must know the distance to the galaxy; that allows you to convert the angular size of the orbits into linear units. If your measurement of the distance to the galaxy isn't accurate, you will get inaccurate radii for the orbits and will compute an inaccurate mass for the galaxy.

Many different measurements in astronomy depend on the calibration of the distance scale. Build a step-by-step argu-

ment to analyze the following problem. **What would happen to your measurements of the diameters and luminosities of the galaxies if astronomers discovered that the Cepheid variable stars were slightly more luminous than had been believed?**

■ ■ ■

10-3 The Evolution of Galaxies

YOUR GOAL IN THIS CHAPTER HAS BEEN TO BUILD a theory to explain the evolution of galaxies. In Chapter 9, you learned about one model that describes the origin of our own Milky Way Galaxy, plus recent modifications to that model. Presumably, other galaxies formed similarly. But why did some galaxies become spiral, some elliptical, and some irregular? Clues to that mystery lie in the clustering of galaxies.

Clusters of Galaxies

The distribution of galaxies is not entirely random. Galaxies tend to occur in clusters ranging from a few to thousands. Deep photographs made with the largest telescopes reveal clusters of galaxies scattered out to the limits of visibility. This clustering of the galaxies can help you understand their evolution.

For this discussion, you can sort clusters of galaxies into two groups, rich and poor. **Rich galaxy clusters** contain over a thousand galaxies, mostly elliptical, crowded into a spherical volume about 3 Mpc (10^7 ly) in diameter. The Coma cluster (located 100 Mpc from Earth in the direction of the constellation Coma Berenices) is an example of a rich cluster (■ Figure 10-6). It contains at least 1000 galaxies, mostly E and S0 types. Close to its center are a giant elliptical galaxy and a large S0 galaxy. Rich clusters often contain one or more giant elliptical galaxies at their centers.

Poor galaxy clusters contain fewer than 1000 galaxies and are irregularly shaped and less crowded toward the center. Our own **Local Group,** which contains the Milky Way, is a good example of a poor cluster (■ Figure 10-7). It contains a few dozen members scattered irregularly through a volume slightly over 1 Mpc in diameter. Of the brighter galaxies, 14 are elliptical, 3 are spiral, and 4 are irregular.

Classifying galaxy clusters into rich and poor clusters reveals a fascinating and suggestive clue to the evolution of galaxies. In general, rich clusters tend to contain 80 to 90 percent E and S0 galaxies and few spirals. Poor clusters contain a larger percentage of spirals; and, among isolated galaxies, those that are not in clusters, 80 to 90 percent are spirals. This suggests that a galaxy's environment is important in determining its structure and has led

■ Figure 10-6

(a) The Coma cluster of galaxies contains at least 1000 galaxies and is especially rich in E and S0 galaxies. Two giant galaxies lie near its center. Only the central area of the cluster is shown in this image. If the cluster were visible in the sky, it would span eight times the diameter of the full moon. (L. Thompson/NOAO) (b) In false colors, this X-ray image of the Coma cluster shows it filled and surrounded by hot gas. The box outlines the area of the cluster shown in part (a). (Simon D. M. White, Ulrich G. Briel, and J. Patrick Henry)

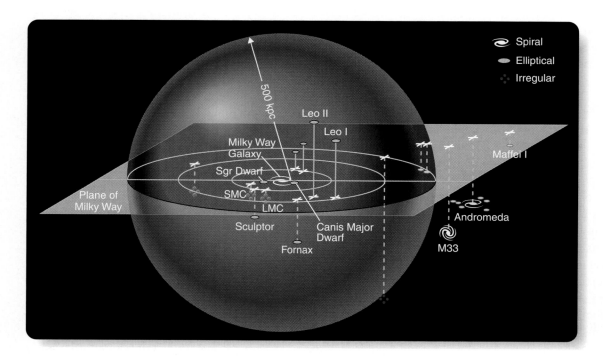

■ Figure 10-7

The Local Group. Our Milky Way Galaxy is located at the center of this diagram. The vertical lines giving distances from the plane of the Milky Way are solid above the plane and dashed below.

astronomers to suspect that the secrets to galaxy evolution lie in collisions between galaxies.

Colliding Galaxies

Astronomers are finding more and more evidence to show that galaxies collide, interact, and merge. In fact, collisions among galaxies may dominate their evolution.

You should not be surprised that galaxies collide with each other. The average separation between galaxies is only about 20 times their diameter, so galaxies should bump into each other fairly often, astronomically speaking. In comparison, stars almost never collide because the typical separation between stars is about 10^7 times their diameter. A collision between two stars is about as likely as a collision between two gnats flitting about in a baseball stadium.

Study **Interacting Galaxies** on pages 208–209 and notice four important points:

1 Interacting galaxies can distort each other with tides, producing tidal tails and shells of stars. Interactions may trigger the formation of spiral arms. Large galaxies can even absorb smaller galaxies.

2 Galaxy interactions can trigger star formation.

3 Evidence within galaxies in the form of motions and multiple nuclei reveals that many have suffered past interactions and mergers.

4 There are beautiful ring galaxies—bull's-eyes left behind by high-speed collisions.

Evidence of galaxy mergers is all around you. Our Milky Way Galaxy is a cannibal galaxy, snacking on the two Magellanic Clouds as they orbit around it. Its tides are also pulling apart two other small satellite galaxies, the Sagittarius and Canis Major Dwarf galaxies, producing great streamers of stars wrapped around the Milky Way. Almost certainly, our galaxy has dined on other small galaxies in the past.

The Origin and Evolution of Galaxies

The test of any scientific understanding is whether you can put all the evidence and theory together to tell the history of the objects studied. Can you describe the origin and evolution of the galaxies? Just a few decades ago, it would have been impossible, but the evidence from space telescopes and new-generation telescopes on Earth combined with advances in computer modeling and theory allow astronomers to outline the story of the galaxies.

Elliptical galaxies appear to be the product of galaxy mergers, which triggered star formation that used up the gas and dust. In fact, astronomers see star formation being stimulated to high levels in many galaxies (■ Figure 10-8). These **starburst galaxies** are very luminous in the infrared because a collision has triggered a burst of star formation that is heating the dust. The warm dust

■ Figure 10-8

Rapid star formation: (a) NGC1569 is a starburst galaxy filled with clouds of young stars and supernovae. At least some starbursts are triggered by interactions between galaxies. (ESA/NASA/P. Anders) (b) The inner parts of M64, known as the "Black Eye Galaxy," are filled with dust produced by rapid star formation. Radio Doppler shift observations show that the inner part of the galaxy rotates backward compared to the outer part of the galaxy, a product of a merger. Where the counterrotating parts of the galaxy collide, star formation is stimulated. (NASA/Hubble Heritage Team/AURA/STScI)

NGC1569   UV + visual image

M64 The Black Eye Galaxy

a

b   Visual + infrared image

Interacting Galaxies

Tidal Distortion

Small galaxy passing near a massive galaxy.

1 When two galaxies collide, they can pass through each other without stars colliding because the stars are so far apart relative to their sizes. Gas clouds and magnetic fields do collide, but the biggest effects may be tidal. Even when two galaxies just pass near each other, tides can cause dramatic effects, such as long streamers called **tidal tails.** In some cases, two galaxies can merge and form a single galaxy.

Gravity of a second galaxy represented as a single massive object

1a When a galaxy swings past a massive object such as another galaxy, tides are severe. Stars near the massive object try to move in smaller, faster orbits while stars further from the massive object follow larger, slower orbits. Such tides can distort a galaxy or even rip it apart.

Galaxy interactions can stimulate the formation of spiral arms

In this computer model, two uniform disk galaxies pass near each other.

The small galaxy passes behind the larger galaxy so they do not actually collide.

Tidal forces deform the galaxies and trigger the formation of spiral arms.

The upper arm of the large galaxy passes in front of the small galaxy.

Allen Beechel

A photo of the well-known Whirlpool Galaxy resembles the computer model.

Visual

NOAO

Visual false-color image

The Mice are a pair of galaxies whipping around each other and being distorted.

NOAO

Computer model of the Mice

Allen Beechel

1b The merger of galaxies is called **galactic cannibalism.** Models show that merging galaxies spiral around their common center of mass while tides rip stars away and form shells.

Computer model

François Schweizer and Alar Toomre

Shells of stars

NOAO

Visual enhanced image

Such shells have been found around elliptical galaxies such as NGC5128. It is peculiar in many ways and even has a belt of dusty gas. The shells revealed in this enhanced image are evidence that the giant galaxy has cannibalized at least one smaller galaxy. The giant galaxy itself may be the result of the merger of two large galaxies.

2   The collision of two galaxies can trigger firestorms of star formation as gas clouds are compressed. Galaxies NGC4038 and 4039 have been known for years as the Antennae because the long tails visible in Earth-based photos resemble the antennae of an insect. Hubble Space Telescope images reveal that the two galaxies are blazing with star formation. Roughly a thousand massive star clusters have been born.

Spectra show that the galaxy is 10 to 20 times richer in elements like magnesium and silicon than the Milky Way galaxy. Such metals are produced by massive stars and spread by supernova explosions.

The Antennae

Giant star cluster

Brad Whitmore, STScI/NASA

Ground-based visual image

Hubble Space Telescope visual image

3   Evidence of past galaxy mergers shows up in the motions inside some galaxies. NCG7251 is a highly distorted galaxy with tidal tails in this ground-based image.

Visual

This Hubble Space Telescope image of the core of the galaxy reveals a small spiral spinning backward in the heart of the larger galaxy.

This counter rotation suggests that NCG7251 is the remains of two oppositely rotating galaxies that merged about a billion years ago.

An X-ray image of the Antennae shows clouds of very hot gas heated by supernovae exploding 30 times more often than in our own galaxy.

X-ray image

NASA/CXC/SAO/ G. Fabbiano et al.

3a   Radio evidence of past mergers: Doppler shifts reveal the rotation of the spiral galaxy M64. The upper part of the galaxy has a redshift and is moving away from Earth, and the bottom part of the galaxy has a blueshift and is approaching. A radio map of the core of the galaxy reveals that it is rotating backward. This suggests a merger long ago between two galaxies that rotated in opposite directions.

Rotation of galaxy M64

Redshift

Blueshift

Robin Braun, NRAO/AUI/NSF

Evidence of galactic cannibalism: Giant elliptical galaxies in rich clusters sometimes have multiple nuclei, thought to be the densest parts of smaller galaxies that have been absorbed and only partly digested.

Multiple nuclei

Visual false-color image

4   Galaxy AM 0644-741 below was once a normal spiral galaxy, but now it is a **ring galaxy.** Mathematical models show that such galaxies are produced by high-speed collisions in which a smaller galaxy passes through another galaxy almost perpendicular to the disk. Ring galaxies tend to have nearby companions, suggestive evidence of past encounters.

Visual-wavelength image

Hubble Heritage Team, NASA/ESA/AURA/STScI

reradiates the energy in the infrared. The Antennae Galaxies (page 209) contain over 15 billion solar masses of hydrogen gas and will become a starburst galaxy as their ongoing merger continues to trigger rapid star formation.

A few collisions and mergers could leave a galaxy with no gas and dust from which to make new stars. Astronomers now suspect that most large ellipticals are formed by the merger of two or more galaxies. The dwarf ellipticals, too small to be formed by mergers, and also irregulars, may be fragments left over by the merger of larger galaxies.

In contrast, spirals seem never to have suffered major collisions. Their thin disks are delicate and would be destroyed by tidal forces in a collision with a massive galaxy. Also, they retain plenty of gas and dust and continue making stars. Our Milky Way Galaxy has evidently never merged with another large galaxy, although the Milky Way, as well as its large spiral neighbor the Andromeda Galaxy, both seem to have cannibalized smaller galaxies.

Barred spiral galaxies also may be the product of tidal interactions. Mathematical models show that bars are not stable and should eventually dissipate. It may take tidal interactions with other galaxies to regenerate the bars. Because well over half of all spiral galaxies have bars, you can suspect that these tidal interactions are common. Our Milky Way Galaxy is probably a barred spiral (see Figure 9-11), and interaction with its two Magellanic Cloud companions, or the more distant but very massive Andromeda Galaxy, could be the cause.

Other processes can alter galaxies. The S0 galaxies, which have disks and bulges like a spiral galaxy but no spiral arms, may have lost much of their gas and dust moving through the gas trapped in the dense clusters to which they belong. For example, X-ray observations show that the Coma cluster contains thin, hot gas between the galaxies (Figure 10-6). A galaxy orbiting through that gas would encounter a tremendous wind that could strip its gas and dust away.

Observations with the largest and most sophisticated telescopes take astronomers back to the age of galaxy formation. At great distances the look-back time is so large that they see the universe as it was soon after the galaxies began to form. There were more spirals then and fewer ellipticals. The observations show that galaxies were closer together then; about 33 percent of all distant galaxies are in close pairs, but only 7 percent of nearby (in other words, present-day) galaxies are in pairs. The observational evidence clearly supports the hypothesis that galaxies have evolved by merger.

The evolution of galaxies is not a simple process. A good theory helps you understand how nature works, and astronomers are just beginning to understand the exciting and complex story of the galaxies. Nevertheless, it is already clear that galaxy evolution has some resemblance to a pie-throwing contest and is just about as neat.

10-4) Active Galaxies and Quasars

MANY GALAXIES HAVE POWERFUL ENERGY SOURCES in their nuclei that in some cases produce powerful jets and other outbursts. These are called **active galaxies.** By looking far away and back in time, astronomers have discovered that the origin of active galaxy energy sources and outbursts is closely related to the formation and history of galaxies.

The first type of active galaxy was discovered in the 1950s and named **radio galaxies** because these galaxies are sources of unusually strong radio waves. By the 1970s, astronomers had put space telescopes in orbit and discovered that radio galaxies are generally bright at many other wavelengths. The flood of energy pouring out of active galaxies originates almost entirely in their nuclei, which are referred to as **active galactic nuclei (AGN).**

Seyfert Galaxies

In 1943, astronomer Carl K. Seyfert conducted a study of spiral galaxies in which he noted that about 2 percent of spirals have small, highly luminous nuclei in their bulges (■ Figure 10-9). Today, these **Seyfert galaxies** are recognized by the peculiar spectra of these luminous nuclei that contain broad emission lines of highly ionized atoms. Emission lines come from low-density gas; the presence of ionized atoms is evidence that the gas is very hot; broad spectral lines indicate large Doppler shifts produced by high gas velocities. The velocities of gas clouds at the centers of Seyfert galaxies are roughly 10,000 km/s, about 30 times greater than velocities at the center of normal galaxies. Something violent is happening in the cores of Seyfert galaxies.

Astronomers later discovered that the brilliant nuclei of Seyfert galaxies change brightness rapidly, in only a few hours or minutes, especially at X-ray wavelengths. As you saw in Chapter 8, an astronomical body cannot change its brightness significantly in a time shorter than the time it takes light to cross its diameter. If a Seyfert nucleus can change in a few minutes, then it cannot be larger in diameter than a few light-minutes. For comparison, the distance from Earth to the sun is 8 light minutes. Yet, despite their small size, the brightest Seyfert nuclei emit a hundred times more energy than the entire Milky Way Galaxy. Something in the centers of these galaxies not much bigger than Earth's orbit produces a galaxy's worth of energy.

Seyfert nuclei are three times more common in interacting pairs of galaxies than in isolated galaxies. Also, about 25 percent of Seyfert galaxies have peculiar shapes suggesting tidal interactions with other galaxies (■ Figure 10-10). This statistical evidence (**Window on Science 10-2**) hints that Seyfert galaxies may have been triggered into activity by collisions or interactions with companions. Some Seyferts are observed to be expelling matter in oppositely directed jets, a geometry you have

It Wouldn't Stand Up in Court: Statistical Evidence

Notice that some scientific evidence is statistical. For example, you might note that Seyfert galaxies are three times more likely to have a nearby companion than a normal galaxy is. This is statistical evidence because you can't be certain that any single Seyfert galaxy will prove to have a companion. Yet the probability is higher than if it were a normal galaxy, and that leads you to suspect that interactions between galaxies are involved. Such statistical evidence can tell you something in general about the cause of Seyfert eruptions.

Statistical evidence is common in science, but it is inadmissible as evidence of guilt in a criminal trial. No judge would allow you to argue that the defendant must be guilty of robbing a store in the middle of the night because the defendant is an astronomer and astronomers are statistically more likely to be awake at night. The American legal system is based on the concept of reasonable doubt, so statistics can't be applied to prove guilt. But you could use statistics to learn something about the general sleep habits of astronomers.

You can use statistical evidence in science if you do not demand that the statistics demonstrate anything conclusively about any single example. To continue the example of Seyfert galaxies, you can't demand that the statistics predict that any specific galaxy has a companion. Rather, you can use the statistical evidence to gain a general insight into the cause of active galactic nuclei. That is, you are trying to understand galaxies in general, not to convict any single galaxy.

Of course, if you surveyed only a few galaxies, your statistics might not be very good, and a critic might be justified in making the common complaint, "Oh, that's only statistics." For example, if you surveyed only four galaxies and found that three had companions, your statistics would not be very reliable. But if you surveyed 1000 galaxies and found that 750 had companions, your statistics could be very good indeed, and your conclusions could be highly significant.

Scientists can use statistical evidence if it passes two tests. It cannot be used to draw conclusions about specific cases, and it must

Visual

Guilt by association is statistical evidence and can be applied to galaxies. (NASA)

be based on large enough samples so the statistics are significant. With these restrictions, statistical evidence can provide deep insights into how nature works.

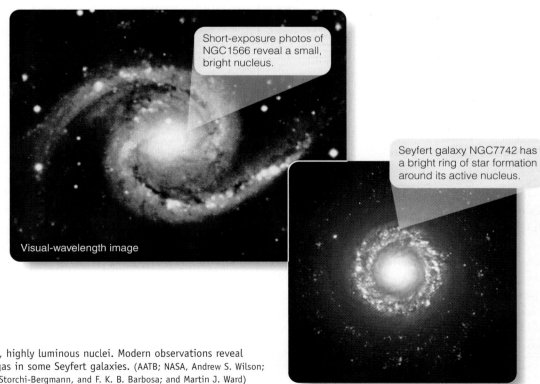

Short-exposure photos of NGC1566 reveal a small, bright nucleus.

Seyfert galaxy NGC7742 has a bright ring of star formation around its active nucleus.

Visual-wavelength image

■ **Figure 10-9**

Seyfert galaxies are spirals with small, highly luminous nuclei. Modern observations reveal bursts of star formation and ejected gas in some Seyfert galaxies. (AATB; NASA, Andrew S. Wilson; Patrick L. Shopbell, Chris Simpson, Thaisa Storchi-Bergmann, and F. K. B. Barbosa; and Martin J. Ward)

False-color visual-wavelength image

Seyfert galaxy NGC7674 is distorted with faint tidal tails.

Tidal tail

Tidal tail

9100 km/s

8620 km/s

8700 km/s

8200 km/s

■ **Figure 10-10**

The Seyfert galaxy NGC7674 appears to be distorted by interacting with its companion galaxies. You can be sure they are a group because they all have similar apparent velocities of recession indicating they are at about the same distance from Earth. (John W. Mackenty, Institute for Astronomy, University of Hawaii)

aligned with the jet axis. Radio observations show that the nucleus must be no more than a light-week in diameter. This evidence is suggestive, but observations of motions around the center are decisive. A high-resolution image shows that the core of M87 is surrounded by a spinning disk, and the jet lies along the axis of the disk. At a distance of 60 ly from the center the gas in the disk is orbiting at 750 km/s, which tells you that the central mass must be roughly 2.4 billion solar masses. Only a supermassive black hole could pack so much mass into such a small region.

seen on smaller scales when matter flows into neutron stars and black holes and forms an accretion disk plus jets.

The accumulated evidence leads modern astronomers to conclude that the core of a Seyfert galaxy contains a supermassive black hole, with a mass as high as a billion solar masses, plus a correspondingly large accretion disk. Gas in the centers of Seyfert galaxies is traveling so fast it would escape from a normal galaxy, and only large central masses could exert enough gravity to hold the gas inside the nuclei. Encounters with other galaxies could throw matter toward the black hole; and, as you learned in Chapter 8, lots of energy can be liberated by matter flowing through an accretion disk into a black hole. In Chapter 9 you learned that the Milky Way Galaxy contains a massive central black hole, but one that seems to be on a starvation diet and is therefore relatively inactive.

Double-Lobed Radio Sources

Beginning in the 1950s, radio astronomers found that some sources of radio energy in the sky consist of pairs of radio-bright regions. When optical telescopes studied the locations of these **double-lobed radio sources,** they revealed galaxies located between the two lobes. The geometry suggests that radio lobes are inflated by jets of excited gas emerging in two directions from the central galaxy. Statistical evidence indicates that jets and radio lobes, like Seyfert nuclei, are associated with interacting galaxies. The AGN jets seem to be related to matter falling into a central black hole via an accretion disk, although the details of this process are not understood.

Consider the giant elliptical galaxy M87 (page 200). It has a very small, bright nucleus, a visible jet of matter 1800 pc long racing out of its core (■ Figure 10-11), and a double-lobed radio source

Quasars

The largest telescopes detect multitudes of faint points of light with peculiar emission spectra, objects called **quasars** (also known as *quasi-stellar objects,* or *QSOs*). Although astronomers now recognize quasars as extreme examples of AGN as well as some of the most distant visible objects in the universe, they were a mystery when they were first identified.

In the early 1960s, photographs of the location of some radio sources that resembled radio galaxies revealed only single starlike points of light. The first of these objects identified was 3C48, and later the source 3C273 was found. They were obviously not normal radio galaxies. Even the most distant photographable galaxies look fuzzy, but these objects looked like stars. Their spectra, however, were totally unlike stellar spectra, so the objects were called quasi-stellar objects.

For a few years, the spectra of quasars were a mystery. A few unidentifiable emission lines were superimposed on a continuous spectrum. In 1963, astronomer Maarten Schmidt calculated that if hydrogen Balmer lines were redshifted by $z = 0.158$ (meaning, each line's change in wavelength divided by its lab wavelength equals 0.158), they would fit the observed lines in 3C273's spectrum (■ Figure 10-12). Other quasar spectra quickly yielded to this type of analysis, revealing even larger redshifts.

To understand the significance of these large redshifts and the large velocities of recession they imply, recall the Hubble law from earlier in this chapter stating that galaxies have apparent velocities of recession proportional to their distances. The large redshifts of the quasars imply that they must be at great distances, some farther away than any known galaxy. Many quasars are evidently so far away that galaxies at those distances are very diffi-

■ Figure 10-11

The jet at the center of giant elliptical galaxy M87 is only a few percent the diameter of the entire galaxy. Images reveal a small, rapidly spinning disk at the center of the galaxy with the jet emerging at nearly half the speed of light along the axis of the disk. (M87: AURA/NOAO/NSF; jet: NASA/STScI; radio image: NRAO and J. Biretta)

cult to detect, yet the quasars are easily photographed. This leads to the conclusion that quasars must be ultraluminous, 10 to 1000 times the luminosity of a large galaxy.

Soon after quasars were discovered, astronomers detected fluctuations in their brightness over time scales of hours or minutes. The rapid fluctuations in quasars showed that they are small objects like AGN, only a few light-minutes or light-hours in diameter. Evidence has accumulated that quasars are the most lu-

minous AGN, located in very distant galaxies. For example, some quasars, like AGN, are at the centers of double radio lobes plus jets (■ Figure 10-13).

Perhaps you have a skeptical question about quasar distances at this point: "How can you be sure quasars really are that far away?" Astronomers faced with explaining how a small object could produce so much energy asked themselves the same question. In the early 1980s, astronomers were able to photograph

■ Figure 10-12

This image of 3C273 shows the bright quasar at the center surrounded by faint fuzz. Note the jet protruding at lower right. (b) The spectrum of 3C273 (top) contains three hydrogen Balmer lines redshifted by 15.8 percent. The drawing shows the unshifted positions of the lines. (Courtesy Maarten Schmidt)

Radio-wavelength image

■ Figure 10-13

This radio image of quasar 3C175 reveals that it is ejecting a jet and is flanked by radio lobes. Presumably you see only one jet because it is directed approximately toward Earth, and the other jet is invisible because it is directed away from Earth. The presence of jets and radio lobes indicates that quasars are distant active galaxies. (NRAO/AUI/NSF)

faint nebulosity surrounding some quasars, which was called quasar fuzz. The spectra of quasar fuzz looked like the spectra of normal but very distant galaxies with the same redshift as the central quasar. In other cases, quasar light shines through the outskirts of a distant galaxy, and the quasar spectrum has extra absorption lines from gas at the redshift of the galaxy, smaller than the quasar's redshift. This means the quasar is farther away than the galaxy containing the gas that absorbed some of the quasar's light on its way to Earth. Both of these observations confirm that galaxy and quasar redshifts indicate distances in a mutually consistent way.

The Search for a Unified Model

Astronomers studying galaxies are now developing a **unified model** of AGN and quasars. A monster black hole is the centerpiece (■ Figure 10-14).

Even a supermassive black hole is quite small compared with a galaxy. A 10-million-solar-mass black hole would be only one-fifth the diameter of Earth's orbit. This means that matter in an accretion disk can get very close to the black hole, orbit very fast, and grow very hot. Theoretical calculations indicate that the disk im-

mediately around the black hole is "puffed up," thick enough to hide the central black hole from some viewing angles. The hot inner disk seems to be the source of the jets often seen coming out of active galaxy cores, but the process by which jets are generated is not understood. The outer part of the disk, according to calculations, is a fat, relatively cold torus (doughnut shape) of dusty gas.

According to the unified model, what you see when you view the core of an AGN or QSO depends on how its accretion disk is tipped with respect to your line of sight (Figure 10-14).

1 If you view the accretion disk from the edge, you cannot see the central zone at all because the thick dusty torus blocks your view. Instead you see radiation emitted by gas lying above and below the central disk that is therefore relatively cool and moving relatively slowly, with small Doppler shifts. Thus, you see narrower spectral lines coming from what is called the narrow-line region.

2 If the accretion disk is tipped slightly, you may be able to see some of the intensely hot gas in the central cavity. This is called the broad-line region because the gas is hot and also orbiting at high velocities. The resulting high Doppler shifts spread out the spectral lines.

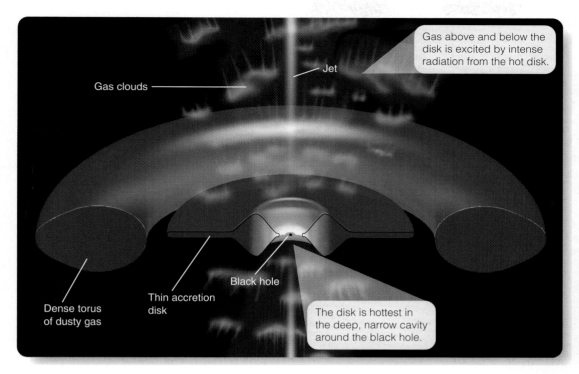

Gas clouds

Jet

Gas above and below the disk is excited by intense radiation from the hot disk.

Dense torus of dusty gas

Thin accretion disk

Black hole

The disk is hottest in the deep, narrow cavity around the black hole.

■ **Figure 10-14**

The features visible in the spectrum of an AGN depend on the angle at which it is viewed. The unified model, shown in cross section, suggests that matter flowing inward passes first through a large, opaque torus; then into a thinner, hotter disk; and finally into a small, hot cavity around the black hole. Telescopes viewing such a disk edge-on would see only narrow spectral lines from cooler gas, but a telescope looking into the central cavity would see broad spectral lines formed by the hot gas. This diagram is not to scale. The central cavity may be only 0.01 pc in radius, while the outer torus may be 1000 pc in radius. (Artist's impression: NASA)

3 If you look directly into the central cavity around the black hole—down the dragon's throat, so to speak—you see the jet emerging perpendicular to the accretion disk and coming straight at you. Model calculations indicate this would result in a very bright and highly variable source with few or no emission lines. This is the appearance of AGN known as **blazars.**

Astronomers are now using this unified model to sort out the different kinds of active galaxies and quasars so they can understand how they are related. For example, about 1 percent of quasars are strong radio sources, and the radio radiation may come from synchrotron radiation (see Chapter 8) produced in the high-energy gas and magnetic fields in the jets. Another example: Using infrared cameras to see through dust, astronomers observed the core of the double-lobed radio galaxy Cygnus A and found an object much like a quasar. Astronomers have begun to refer to such hidden objects as "buried quasars."

The unified model is far from complete. The detailed structure of accretion disks is poorly understood, as is the process by which the disks produce jets. Furthermore, the spiral Seyfert galaxies are clearly different from the giant elliptical galaxies that have double radio lobes. Unification does not explain all of the differences among active galaxies. Rather, it is a model that provides some clues to what is happening in AGN and quasars.

The Origin of Supermassive Black Holes

Naturally you are wondering where these supermassive black holes in the nuclei of galaxies came from, and that question is linked to a second question. What makes a supermassve black hole erupt? Answering those questions will help you understand how galaxies form.

Evidence is accumulating that most galaxies contain a supermassive black hole at their center. Even our own Milky Way Galaxy and the nearby Andromeda Galaxy contain central black holes. Only a few percent of galaxies, however, have obviously active galactic nuclei. That must mean that most of the supermassive black holes are dormant. Presumably they are not being fed large amounts of matter. A slow trickle of matter flowing into the supermassive black hole at the center of our galaxy could explain the relatively mild activity seen there. It would take a larger meal to trigger an eruption such as those seen in active galactic nuclei.

What could trigger a supermassive black hole to erupt? The answer is something that you studied back in Chapter 3—tides. You have seen in a previous section of this chapter how tides twist interacting galaxies and rip matter away into tidal tails. Active galaxies are often distorted: They have evidently been twisted by tidal forces as they interacted or merged with another galaxy. Mathematical models show that those interactions can also throw stars plus clouds of interstellar gas and dust inward toward the galaxies' centers. A sudden flood of matter flowing into a supermassive black hole would trigger it into eruption. ■ Figure 10-15 shows how a passing star would be shredded and partially consumed by a supermassive hole.

Star Falling into a Black Hole

A star, perhaps disturbed by an encounter with another star, drifts toward a supermassive black hole.

As the near side of the star tries to orbit faster than the far side, the star is torn apart by tidal forces.

Most of the mass of the star is flung away from the black hole…

but roughly 1 percent falls into the black hole as an accretion disk forms.

A steady diet of inflowing gas, dust, and an occasional star would keep an AGN powered by a supermassive black hole active.

A few dozen supermassive black holes have measured masses, and their masses are correlated with the masses of the host galaxies' nuclear bulges. In each case, the mass of the black hole is about 0.5 percent the mass of the surrounding nuclear bulge. Apparently, as a galaxy forms its nuclear bulge, a certain fraction of the mass sinks to the center where it forms a supermassive black hole. All of that matter flowing together to form the black hole would release a tremendous amount of energy and trigger a violent eruption. Long ago, when galaxies were actively forming, the birth of the nuclear bulges must have triggered many AGN. Later episodes of AGN activity could be triggered by interactions or mergers with other galaxies.

Quasars through Time

In the next chapter, you will see evidence that the universe began 13.7 billion years ago. Some quasars are over 10 billion light-years away, and because of their large look-back times they appear as they were when the universe was only 10 percent of its present age. The first clouds of gas that formed galaxies would have also made supermassive black holes at the centers of those galaxies' nuclear bulges. The abundance of matter flooding into those early black holes could have triggered outbursts that are seen as quasars.

You should also note that galaxies were closer together when the universe was young and had not expanded very much. Because they were closer together, the forming galaxies collided more often, and you have seen how collisions between galaxies could throw matter into central supermassive black holes and trigger eruptions. Quasars are often located in host galaxies that are distorted as if they were interacting with other galaxies (■ Figure 10-16).

Quasars are most common with redshifts of a little over 2 and less common with redshifts above 2.7. The largest quasar redshifts are over 6, but such high-redshift quasars are quite rare. Evidently, if you looked at quasars with redshifts a bit over 2, you would be looking back to an age when galaxies were actively forming, colliding, and merging. In that era, quasars were about 1000 times more common than they are now, although, even so, only a fraction of galaxies had quasars erupting in their cores at any one time. If you looked back to higher redshifts, you would

■ Figure 10-15

Artist's conception of the cause of an X-ray flare in the galaxy RX J1242-11 detected by X-ray telescopes in orbit around Earth. Equaling the energy of a supernova explosion, the flare was evidently caused when a star wandered too close to the 100-million-solar-mass black hole at the center of the galaxy. When tidal forces ripped the star apart, some of the mass fell into the black hole, and the rest was flung away. That sudden meal for the black hole was enough to trigger an outburst. (ESA) see fewer quasars because you would be looking back to an age when the universe was so young it had not yet begun to form many galaxies and quasars.

Then where are all the dead quasars? There is no way to get rid of supermassive black holes. Astronomers have discovered that nearly all galaxies contain supermassive black holes,

and those black holes may have suffered quasar eruptions when the universe was younger, galaxies were closer together, and infalling gas and dust were more plentiful. Quasar eruptions became less common as galaxies became more stable and as the abundance of gas and dust in the centers of galaxies was exhausted. Our own Milky Way Galaxy is a good example. It could have been a quasar long ago, but today its supermassive black hole is resting. Dormant black holes at the centers of galaxies today can be reawakened to become AGN by galaxy collisions.

■ Figure 10-16

The bright object at the center of each of these images is a quasar. Fainter objects near and around the quasars are galaxies distorted by collisions. Compare the upper left ring-shaped galaxy with the ring galaxy on page 209 and compare the tail in the lower left image with the Antennae Galaxies on page 209. (J. Bahcall, Institute for Advanced Study, Mike Disney, University of Wales, NASA)

Building Scientific Arguments

What evidence can you cite to show that double-lobed radio galaxies and Seyfert galaxies have similar energy sources?

Often a good argument, scientific or otherwise, compares two dissimilar objects and finds the connecting factors. In this case, observations show that double-lobed radio sources are produced by two jets flowing out of a galaxy's core and inflating the radio lobes. The evidence of small size and very-high-speed motions in the core shows that these galaxies must contain large amounts of mass in small regions. Seyfert galaxies also have small nuclei, as shown by their rapid fluctuations, and the high velocities in their cores imply that the cores must contain very large amounts of mass. Consequently, both kinds of objects are hypothesized to contain supermassive black holes at their centers.

This argument has cited evidence and theory to reach a conclusion. Construct a new argument: **Why do astronomers think that galaxy interactions trigger active galactic nuclei?**

■   ■   ■

Summary

10-1 | The Family of Galaxies

What types of galaxies exist?

■ Galaxies can be divided into three classes—elliptical, spiral, and irregular—with subclasses specifying details of the galaxy's shape.

■ Elliptical galaxies contain little gas and dust and cannot make new stars. Consequently, they lack hot, blue stars and have a relatively red color.

■ Spiral galaxies contain more gas and dust and support active star formation, especially along the spiral arms. Some of the newborn stars are massive, hot, and blue, and that gives the galaxies a blue color.

■ The halo and nuclear bulge of a spiral galaxy usually lack gas and dust and contain little star formation. The halos and nuclear bulges have a reddish color because they lack hot, blue stars.

■ Irregular galaxies have no obvious shape but contain gas and dust and support star formation.

10-2 | Measuring the Properties of Galaxies

How do astronomers measure the distances to galaxies, and how does that allow the sizes, luminosities, and masses of galaxies to be determined?

■ Astronomers find the distance to galaxies using distance indicators sometimes called standard candles, objects of known luminosity. The most accurate distance indicators are the Cepheid variable stars. Others include Type Ia supernovae explosions.

■ Astronomers can estimate the distance to a galaxy by dividing its apparent rate of recession (its radial velocity) by the Hubble constant. Apparent rate of recession is related to distance by the Hubble law.

■ Some dwarf ellipticals and dwarf irregular galaxies are only a few percent the size and luminosity of our galaxy, but some giant elliptical galaxies are as much as five times larger in diameter than the Milky Way Galaxy.

■ Astronomers measure the masses of galaxies in several ways. The most precise is the rotation curve method, but it can be applied only to nearby galaxies.

■ Observations of individual galaxies show that galaxies contain 10 to 100 times more dark matter than visible matter. Signs of the dark matter's gravitational influence are more obvious in galaxy outer halos.

■ Stars near the centers of galaxies are orbiting at high velocities, suggesting the presence of supermassive black holes in the centers of most galaxies.

10-3 | The Evolution of Galaxies

Why are there different kinds of galaxies, and how do galaxies evolve?

■ Rich clusters of galaxies contain fewer spirals than do poor clusters of galaxies, and that is evidence that galaxies evolve by collisions and mergers.

■ When galaxies collide, tides twist and distort their shapes, and the resulting compression of gas clouds triggers star formation.

■ Large galaxies can absorb smaller galaxies in what is called galactic cannibalism. There is clear evidence that our own Milky Way Galaxy is devouring some of the small galaxies that orbit nearby and that our galaxy has consumed other small galaxies in the past.

■ Shells of stars, counter-rotating parts of galaxies, streams of stars in the halos of galaxies, and multiple nuclei are evidence that galaxies can merge.

■ The merger of two larger galaxies can scramble star orbits and drive bursts of star formation to use up gas and dust. Most large elliptical galaxies have evidently been produced by past mergers.

■ Spiral galaxies have thin, delicate disks and appear not to have suffered mergers with large galaxies.

■ A galaxy moving through the gas in a cluster of galaxies can be stripped of its own gas and dust and become an S0 galaxy.

■ At great distances and large look-back times, astronomers see that galaxies were smaller, more irregular, and closer together. Also, there were more spirals and fewer ellipticals long ago.

10-4 | Active Galaxies and Quasars

What is the energy source for active galaxies, what can trigger the activity, and what does that reveal about the history of galaxies?

■ Some galaxies have peculiar properties. Seyfert galaxies, for example, are spirals with small, highly luminous cores. Spectra of Seyfert galaxy nuclei show that they contain highly excited, rapidly moving gas.

■ Because AGN can change their brightness quickly, you can conclude they must be small—smaller than our solar system.

■ Double-lobed radio galaxies emit radio energy from areas on either side of the galaxies. These lobes appear to be inflated by jets ejected from the nuclei of the galaxies.

■ Astronomers know that quasars are very far away because they have very high redshifts. To be visible at such great distances, quasars must be ultraluminous.

■ Observations of the spectra of hazy objects near quasars and the spectra of quasar fuzz show that quasars are the active cores of very distant galaxies.

■ Evidence is strong that active galactic nuclei contain supermassive black holes into which matter is flowing through hot accretion disks. This can eject jets in opposite directions that push into the intergalactic medium and inflate radio lobes like balloons.

■ The mass of a central black hole can be found by observing the velocity of stars orbiting the black hole or the rotational speed of its accretion disk. Masses range from millions to billions of solar masses.

■ According to the unified model, what you see depends on the tilt of the accretion disk. If you see into the core, you see broad spectral lines. If you see the disk edge-on, you see only narrow spectral lines. If the jet from the black hole points directly at you, you see a blazar.

■ Many galaxies appear to contain supermassive black holes at their centers, but they are dormant because large amounts of matter are not flowing inward. Only when a supermassive black hole is fed does it erupt.

■ Matter flowing into a hot accretion disk around a black hole can eject high-energy jets of gas and radiation perpendicular to the disk.

■ Interactions between galaxies can throw matter into the center, feed the black hole, and trigger eruptions. This explains why active galaxies are often distorted or have nearby companions.

■ Because quasars lie at great distances, astronomers see them as they were long ago—some over 10 billion years ago—when the universe was young and just forming galaxies.

■ Supermassive black holes powering AGN and quasars cannot be stellar black holes. That is, they cannot have been formed by dying stars but may have formed as the nuclear bulges of their host galaxies began to form and contract.

■ Some quasars may be caused by the formation of the first galaxies when the universe was young.

New Terms

elliptical galaxy (p. 200)

spiral galaxy (p. 200)

barred spiral galaxy (p. 200)

irregular galaxy (p. 201)

Large Magellanic Cloud (p. 201)

Small Magellanic Cloud (p. 201)

megaparsec (Mpc) (p. 202)

standard candle (p. 202)

distance ladder (p. 202)

look-back time (p. 202)

Hubble law (p. 203)

Hubble constant (H) (p. 203)

rotation curve method (p. 203)

rich galaxy cluster (p. 205)

poor galaxy cluster (p. 205)

Local Group (p. 205)

starburst galaxy (p. 207)

tidal tail (p. 208)

galactic cannibalism (p. 208)

ring galaxy (p. 209)

active galaxy (p. 210)

radio galaxy (p. 210)

active galactic nuclei (AGN) (p. 210)

Seyfert galaxy (p. 210)

double-lobed radio source (p. 212)

quasar (p. 212)

unified model (p. 214)

blazar (p. 215)

Review Questions

ThomsonNOW™ Assess your understanding of this chapter's topics with additional quizzing and animations at **www.thomsonedu.com**

1. Why didn't astronomers at the beginning of the 20th century recognize galaxies for what they are?

2. How can a classification system aid a scientist?

3. What is the difference between an E0 galaxy and an E1 galaxy?

4. What is the difference between an Sa and an Sb galaxy? Between an SBb and an Sb galaxy?

5. Why do selection effects make it difficult to decide how common elliptical and spiral galaxies are?

6. Why are Cepheid variable stars good distance indicators? What about supernovae?

7. How is the rotation curve method for determining galaxy masses related to binary stars and Kepler's third law?

8. What evidence is there that galaxies contain dark matter? Supermassive black holes?

9. What evidence is there that galaxies collide and merge?

10. Ring galaxies often have nearby companions. What does that suggest?

11. Propose an explanation for the lack of gas, dust, and young stars in elliptical galaxies.

12. How do deep images by the Hubble Space Telescope support the hypothesis that galaxies have evolved substantially since they formed?

WIYN/NOAO/NSF

13. In the image at right you see two interacting galaxies; one is nearly face-on and the other is nearly edge-on. Discuss the shapes of these galaxies and describe what is happening.

14. What statistical evidence suggests that Seyfert galaxies have suffered recent interactions with other galaxies?

15. Why were quasars first noticed as being peculiar?

16. How do the large redshifts of quasars lead astronomers to conclude they must be very distant?

17. What evidence shows that AGN and quasars must be very small?

18. What evidence is there that quasars occur in distant galaxies?

19. What observations are necessary to identify the presence of a supermassive black hole at the center of a galaxy?

20. What evidence is there that AGN and quasars can be triggered by galaxy collisions and mergers?

21. How does the unified model explain collisions and mergers as triggers for active galaxies and quasars?

22. The image at right combines visual (blue) with radio (red) to show the galaxy radio astronomers call Fornax A. Explain the features of this image. Is it significant that the object is a distorted elliptical galaxy in a cluster?

NRAO/AUI

23. A radio image of quasar 3C334 is shown at the right. Why do you see only one jet? What does it mean that this looks so much like a radio image of a double-lobed radio galaxy?

NRAO/AUI

24. Why are there few quasars at low redshifts and at high redshifts but many at redshifts of a little over 2?

Discussion Questions

1. Should an orbiting infrared telescope find irregular galaxies bright or faint in the far infrared? Why? What about elliptical galaxies?

2. Why do quasars, active galaxies, SS433 (see Chapter 8), and protostars have similar geometries?

3. By custom, astronomers refer to the unified *model* of AGN and not to the unified *hypothesis* or unified *theory*. In your opinion, which of the words seems best?

4. If a quasar is triggered in a galaxy's core, what do you think it would look like to beings living in the outer disk of the galaxy, considering interstellar dust as well as distance? Could life continue in that galaxy?

11 | Cosmology in the 21st Century

LOOK AT YOUR THUMB. The matter in your thumb was present in the fiery beginning of the universe. **Cosmology,** the study of the universe as a whole, can tell you where your matter came from, and it can tell you where your matter is going. | Cosmology is a mind-bendingly weird subject, and you can enjoy it for its strange ideas and theories. It is fun to think about space stretching like a rubber sheet, invisible energy pushing the universe to expand faster and faster, and the origin of vast walls of galaxy clusters. Notice that this is better than speculation—it is all supported by evidence. Cosmology, however strange it may seem, is a serious and logical attempt to understand how the universe works, and it leads to wonderful insights into how you came to be a part of it.

All of the energy and matter in the universe, including the matter in your body, began in the big bang. In this computer model of the evolution of the universe soon after the big bang, matter is being drawn together to form great clouds of galaxies. (MPA and Jörg Colberg)

Guidepost

Looking Back

Since Chapter 1 you have been on an outward journey through the universe. You have studied the appearance of the night sky seen from Earth, the births and deaths of stars, and the interactions of galaxies. Now you have reached the limit of your journey in space and in time, and you can study the universe as a whole.

This Chapter

The ideas in this chapter are the biggest and the most difficult in all of science. Can you imagine a limitless universe or the first instant of time? Perhaps you have heard that the universe began in an event called the big bang. Now it is time to try to imagine what that was like.

As you explore cosmology, you will find answers to four essential questions:

Does the universe have an edge and a center?

How do you know that the universe began with a big bang?

How does the universe expand?

How has the universe evolved, and what will be its fate?

These questions are deceptively simple, but they will lead you to a new understanding of where you are and what you are.

Looking Ahead

Once you have finished this chapter, you will have a modern insight into the nature of the universe, and it will be time to refocus on your local neighborhood in the universe—the subject of the rest of this book.

■ Figure 11-1

The entire sky is filled with galaxies. Some lie in clusters of thousands, and others are isolated in nearly empty voids between the clusters. In this image of a typical spot on the sky, bright objects with spikes caused by diffraction in the telescope are nearby stars. All other objects are galaxies, ranging from the nearby face-on spiral at upper right to the most distant galaxies visible only in the infrared, shown as red in this composite image. (R. Williams, STScI HDF-South Team, NASA)

11-1 Introduction to the Universe

EVERYONE HAS AN IMPRESSION OF THE UNIVERSE as a vast depth filled with galaxies (■ Figure 11-1); but, as you begin exploring the universe, you need to sort out your expectations so they do not mislead you. The first step is to deal with an expectation so obvious that most people, for the sake of a quiet life, don't even think about it.

The Edge–Center Problem

In your daily life, you are accustomed to boundaries. Rooms have walls, athletic fields have boundary lines, countries have borders, oceans have shores. It is natural to think of the universe as having an edge, but that idea can't be right.

If the universe had an edge, imagine going to the edge. What would you find there: A wall? A great empty space? Nothing? Even a child can ask: If there is an edge to space, what's beyond it? A true edge would have to be more than just an end of the distribution of matter, it would have to be an end of space itself; but then, what would happen if you tried to reach past or move past that edge?

An edge to the universe violates common sense, and modern observations (which you will study later in this chapter) indicate that the universe could be infinite and have no edge. Note that you find the centers of things—galaxies, globular clusters, oceans, pizzas—by referring to their edges. If the universe has no edge, then it cannot have a center.

It is a common misconception to imagine that the universe has a center, but, as you have just learned, that is impossible. As you study cosmology, you need to take care to avoid thinking that there is a center of the universe.

The Necessity of a Beginning

Of course you have noticed that the night sky is dark. That is an important observation because reasonable assumptions about the universe can lead to the conclusion that the night sky actually should glow blindingly bright! This conflict between observation and theory is called **Olbers's paradox** after Heinrich Olbers, an Austrian physician and astronomer who publicized the problem in 1826 (*problem* or *question* would be more accurate words than *paradox*). Olbers was not the first to pose the question, but it is named after him because modern cosmologists were not aware of the earlier discussions. You will be able to answer Olbers's question and understand why the night sky is dark by revising your assumptions about the universe.

The point Olber made seems simple. Suppose you assume that the universe is infinite and filled with stars. (The clumping of stars into galaxies can be shown mathematically to make no essential difference.) If you look in any direction, your line of sight must eventually reach the surface of a star. Look at ■ Figure 11-2, which uses the analogy of lines of sight in a forest. When you are deep in a forest, every line of sight ends on a tree trunk, and you cannot see out of the forest. By analogy, every line of sight from Earth out into space should eventually end on the surface of a star, so the entire sky should be as bright as the surface of an average star—like suns crowded "shoulder to shoulder," covering the sky from horizon to horizon. It should not get dark at night.

Today, cosmologists believe they understand why the sky is dark. Olbers's paradox makes an incorrect prediction because it is based on a hidden assumption. The universe may be infinite in size but it is not eternal—that is, not infinitely old. That answer to Olbers's question was suggested by Edgar Allan Poe in 1848. He proposed that the night sky is dark because the universe is not infinitely old but instead began at some time in the past. The more distant stars are so far away that light from them has not yet reached Earth. That is, if you look far enough, the look-back time is greater than the age of the universe. The night sky is dark because the universe had a beginning.

This is a powerful idea because it clearly illustrates the difference between the universe and the observable universe. The universe is everything that exists, but the **observable universe** is the part that you can see. You will learn later that the universe is about 14 billion years old. In that case, the observable universe has a radius of about 14 billion light-years. Do not confuse the

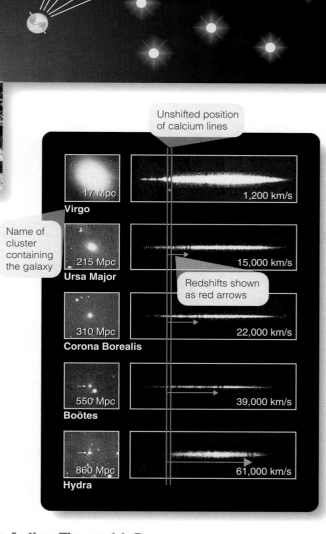

■ Figure 11-2

(a) Every direction you look in a forest eventually reaches a tree trunk, and you cannot see out of the forest. (Photo courtesy Janet Seeds) (b) If the universe is infinite and filled with stars, then any line from Earth should eventually reach the surface of a star. This assumption predicts that the night sky should glow as brightly as the surface of the average star, a puzzle commonly referred to as Olbers's paradox.

observable universe, which is finite, with the universe as a whole, which could be infinite.

Cosmic Expansion

In 1929, Edwin P. Hubble published his discovery that the sizes of galaxy redshifts are proportional to galaxy distances (see Figure 10-4). Nearby galaxies have small redshifts, but more distant galaxies have larger redshifts. These redshifts imply that the galaxies are receding from each other. ■ Figure 11-3 shows spectra of galaxies in clusters at various distances. The Virgo cluster is relatively nearby, and its redshift is small. The Hydra cluster is very distant, and its redshift is so large that the two dark lines formed by ionized calcium are shifted from near-ultraviolet wavelengths well into the visible part of the spectrum.

The expansion of the universe does not imply that Earth is at the center. To see why, look at ■ Figure 11-4, which shows an analogy to baking raisin bread. As the dough rises, it pushes the raisins away from each other uniformly at speeds that are proportional to

■ Active Figure 11-3

These galaxy spectra extend from the near-ultraviolet at left to the blue part of the visible spectrum at right. The two dark absorption lines of once-ionized calcium are prominent in the near-ultraviolet. The redshifts in galaxy spectra are expressed here as apparent velocities of recession. Note that the apparent velocity of recession is proportional to distance. (Caltech)

ThomsonNOW™ Sign in at www.thomsonedu.com and go to ThomsonNOW to see Active Figure "Cosmic Redshift."

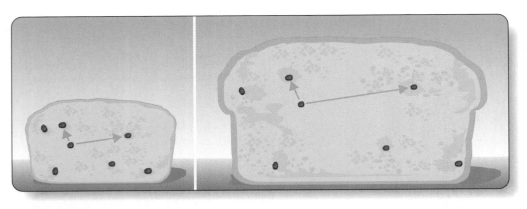

■ **Active Figure 11-4**

A raisin bread analogy for the expansion of the universe. As the dough rises, raisins are pushed apart with speeds proportional to distance. A colony of bacteria living on any raisin will find that the redshifts of the other raisins are proportional to their distances.

ThomsonNOW Sign in at www. thomsonedu.com and go to ThomsonNOW to see Active Figure "Raisin Bread."

their distances from each other. Two raisins that were originally close are pushed apart slowly, but two raisins that were far apart, having more dough between them, are pushed apart faster. If bacterial astronomers lived on a raisin in your raisin bread, they could observe the redshifts of the other raisins and derive a bacterial Hubble law. They would conclude that their universe was expanding uniformly. It does not matter which raisin the bacterial astronomers lived on, they would get the same Hubble law—no raisin has a special viewpoint. Similarly, astronomers in any galaxy will see the same law of expansion—no galaxy has a special viewpoint.

When you look at Figure 11-4, you see the edge of the loaf of raisin bread, and you can identify a center to the loaf of bread. The raisin bread analogy to the expanding universe no longer works when you consider the crust—the edge—of the bread. Remember that the universe cannot have an edge or a center, so there can be no center to the expansion.

11-2 The Big Bang Theory

THE EXPANSION OF THE UNIVERSE LED ASTRONOMERS TO CONCLUDE that the universe must have begun with an event of astounding cosmic intensity.

Necessity of the Big Bang

Imagine that you have a video of the expanding universe, and you run it backward. You would see the galaxies moving toward each other. There is no center to the expansion of the universe, so you would not see galaxies approaching a single spot. Rather, you would see the space between galaxies disappearing, distances between all

galaxies decreasing without the galaxies themselves moving, and eventually galaxies beginning to merge. If you ran your video far enough back, you would see the matter and energy of the universe compressed into a high-density, high-temperature state. You can conclude that the expanding universe began with expansion from that condition of extremely high density and temperature, which modern astronomers call the **big bang.**

How long ago did the universe begin? You can estimate the age of the universe with a simple calculation. If you must drive to a city 100 miles away and you can travel 50 miles per hour, you divide distance by rate of travel and learn the travel time—in this example, 2 hours. To find the age of the universe, you can divide the distance between galaxies by the speed with which they are separating and find out how much time was required for them to have reached their present separation. You will fine-tune your estimate later in this chapter, but for the moment you can conclude that basic observations of the recession of the galaxies require that the universe began with a big bang approximately 14 billion years ago. That estimated span is called the **Hubble time.**

Your instinct is to think of the big bang as an historical event, like the Gettysburg Address—something that happened long ago and can no longer be observed. But the look-back time makes it possible to observe the big bang directly. The look-back time to nearby galaxies is only a few million years, but the look-back time to more distant galaxies is a large fraction of the age of the universe (see Chapter 10). Suppose you were to look between and beyond the distant galaxies, back to the time of the big bang. You should be able to detect the hot gas that filled the universe long ago.

Again, do not think of an edge or a center when you think of the big bang. It is a very common misconception that the big bang was an explosion and that the galaxies are flying away from a center. Although your imagination tries to visualize the big bang as a localized event, you must keep firmly in mind that the big bang did not occur at a single place but filled the entire volume of the universe. You cannot point to any particular place and say, "The big bang occurred over there." At the time of the big bang, all of the galaxies, stars, and atoms in the observable universe were confined to a very small volume. That hot, dense state, the big bang, occurred everywhere, including right where you are now. The matter of which you are made was part of the big bang, so you are inside the remains of that event, and the universe continues to expand around you. In whatever direction astronomers look, at great distances they can see back to the age when the universe was filled with dense hot gas (■ Figure 11-5).

This diagram shows schematically the expansion of a small part of the universe. Although the universe is now filled with galaxies, the look-back time distorts what you see. Nearby you can see galaxies, but at greater distances, the look-back time reveals the universe at earlier stages. At very great distances, the big bang is detectable as infrared and radio energy arriving from the hot gas that filled the universe soon after the big bang.

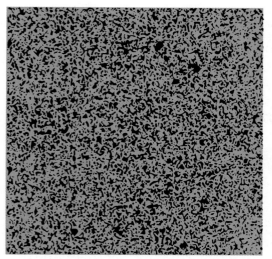

a A region of the universe during the big bang

b A region of the universe now

Milky Way Galaxy

c The present universe as it appears from our galaxy

Evidence for the Big Bang—The Cosmic Microwave Background

The radiation that comes from great distance has a tremendous redshift. The most distant visible objects are faint galaxies and quasars, with redshifts up to about 7 (meaning, change of wavelength divided by original wavelength equals 7). In contrast, the radiation from the hot gas of the big bang is calculated to have a redshift of about 1100. That means the light emitted by the big bang gases arrives at Earth as far-infrared radiation and short-wavelength radio waves. You can't see it with your eyes, but it should be detectable with infrared and radio telescopes. Amazingly, the big bang can still be detected by the radiation it emitted.

In the mid-1960s two Bell Laboratories physicists, Arno Penzias and Robert Wilson, were measuring the radio brightness of the sky (■ Figure 11-6) when they discovered peculiar radio noise coming from all directions. Physicists George Gamow and Ralph Alpher had predicted in the 1940s that the big bang would have emitted blackbody radiation that should now be in the far-infrared and radio parts of the spectrum. Physicist Robert Dicke and his team at Princeton began building a receiver to detect that radiation in the early 1960s. When Penzias and Wilson learned about the preceding work, they realized the radio noise they had detected, now called the **cosmic microwave background (CMB),** is actually radiation from the big bang. They received the 1978 Nobel Prize in physics for their discovery.

The detection of the background radiation was tremendously exciting, but astronomers wanted confirmation. Critical observations in the far-infrared checking whether the CMB really has a blackbody spectrum could not be made from the ground. It was not until January 1990 that satellite measurements confirmed that the CMB is blackbody radiation with an apparent temperature of 2.725 ± 0.002 K—in good agreement with theoretical predictions.

It may seem strange that the hot gas of the big bang appears to have a temperature 2.7 degrees above absolute zero, but recall the tremendous redshift. Observers on Earth see light that has a redshift of about 1100—that is, the wavelengths of the photons are about 1100 times longer than when they were emitted. The gas clouds that emitted the photons had a temperature of about 3000 K, and they emitted blackbody radiation with a λ_{max} of about 1000 nm (see Chapter 5 regarding Wien's law). The expansion of the universe has redshifted the wavelengths about 1100 times longer, so λ_{max} is now about 1 million nm (1 mm). That is why the hot gas of the big bang seems to be 1100 times cooler now, about 2.7 K.

In 1965, Arno Penzias (right) and Robert Wilson first detected the background radiation using the horn antenna behind them in this photograph.

When the CMB was first detected in 1963, technology did not allow measurements at many wavelengths. Not until infrared detectors could be put in orbit was it conclusively shown that the background radiation, as predicted by theory, follows a blackbody curve very precisely. (Photo, AT&T Archives; COBE map, NASA)

Launched in 1989, the COBE satellite showed the background radiation followed the black body curve.

$T = 2.725 \pm 0.002$ K

Intensity

Wavelength (cm)

0.05   0.1   0.5   1

All-sky COBE map of tiny variations in the background radiation.

Particles and Nucleosynthesis— The First Seconds and Minutes

Simple observations of the darkness of the night sky and the redshifts of the galaxies tell you that the universe must have had a beginning. Furthermore, you have seen that the CMB is clear evidence that conditions at the beginning were hot and dense. Theorists can combine these observations with knowledge from physics of how atoms and subatomic particles behave to work out the story of how the big bang occurred.

Cosmologists cannot begin their history of the big bang at time zero, because no one understands the behavior of matter and energy under such extreme conditions, but they can come amazingly close. If you could visit the universe when it was only one 10-millionths of a second old, you would find it filled with high-energy photons having a blackbody temperature well over 1 trillion (10^{12}) K and a density (using Einstein's equation $E = mc^2$ to calculate the mass equivalent to a certain amount

of energy) greater than 5×10^{13} g/cm^3, nearly the density of an atomic nucleus.

If photons have enough energy, two photons can combine and convert their energy into a pair of particles—a particle of normal matter and a particle of **antimatter.** And when an antimatter particle meets its matching particle of normal matter, the two particles annihilate each other and convert their mass back into energy in the form of two gamma rays. In the early universe, photons had enough energy to produce proton–antiproton pairs or neutron–antineutron pairs. When these particles collided with their antiparticles, they converted their mass back into photons. Thus, the early universe was filled with a dynamic soup of energy flickering from photons into particles and back again.

While all this went on, the expansion of the universe cooled the radiation. By the time the universe was 0.0001 second old, its blackbody temperature had fallen to 10^{12} K. At that point the average energy of the gamma-ray photons fell below the energy equivalent to the mass of a proton or a neutron, so the gamma rays could no longer produce such heavy particles, and the creation of protons and neutrons stopped. Those particles combined with their antiparticles and quickly converted most of the mass into photons.

You might guess from this that all of the protons and neutrons would have been annihilated with their antiparticles, and the universe should now consist of nothing but photons. However, for reasons that are poorly understood, a small excess of normal particles apparently existed. For every billion protons annihilated by antiprotons, one survived with no antiparticle to destroy it. Consequently, you live in a world of normal matter, and antimatter is very rare.

Although the gamma rays did not have enough energy to produce protons and neutrons after the universe fell below 10^{12} K, electron–positron pairs, having lower mass, could still be produced. That continued until the universe was about 4 seconds old, at which time the expansion had cooled the gamma rays to the point where they could no longer create even electron–positron pairs. Most of the electrons and positrons combined to form photons, and only one in a billion elections survived. The protons, neutrons, and electrons of which our universe is made were produced during the first 4 seconds of its history.

This soup of hot gas and radiation continued to cool. By the time the universe was about 2 minutes old, protons and neutrons could link to form deuterium, the nucleus of a heavy hydrogen atom, and not be broken apart again. By the end of the next minute, further reactions began converting deuterium into helium, but almost no heavier atoms could be built because there are no stable nuclei with atomic weights of 5 or 8 (in units of the hydrogen atom)—they fall apart as soon as they are created. Cosmic element building during the big bang had to proceed rapidly, step-by-step, like someone hopping up a flight of stairs. The lack of stable nuclei at atomic weights of 5 and 8 meant there were missing steps in the stairway, and the step-by-step reactions had great difficulty jumping over these gaps. Astronomers can calculate that the big bang produced a tiny amount of lithium (atomic weight 7) but no elements heavier than that.

By the time the universe was 3 minutes old, it had become so cool that most nuclear reactions had stopped, and by the time it was 30 minutes old the nuclear reactions had ended completely. At that time about 25 percent of the mass was in the form of helium nuclei, and the rest was in the form of hydrogen nuclei (protons). That is the abundance of hydrogen and helium observed today in the oldest stars. The cosmic abundance of hydrogen and helium were essentially fixed during the first minutes of the universe. The hydrogen nuclei in water molecules in your body have survived unchanged since they formed during the first moments of the big bang. Heavier elements were built by nucleosynthesis inside later generations of massive stars (see Chapters 8 and 9).

Recombination and Reionization—The First Thousands and Millions of Years

At first, the universe was so hot that the gas was totally ionized, and the electrons were not attached to nuclei. Free electrons interact with photons so easily that a photon could not travel very far before it encountered an electron and was deflected (■ Figure 11-7). The radiation and matter interacted continuously with each other and cooled together as the universe expanded.

■ **Figure 11-7**

Photons scatter (bounce) off electrons (blue) easily but hardly at all from the much more massive protons (red). (a) When the universe was very dense and ionized, photons could not travel very far before they scattered off an electron. This means the gas was opaque. (b) As the universe expanded, the electrons were spread further apart, and the photons could travel farther; this made the gas more transparent. (c) After the era of recombination, most electrons were locked to protons to form neutral atoms, and the universe became almost completely transparent.

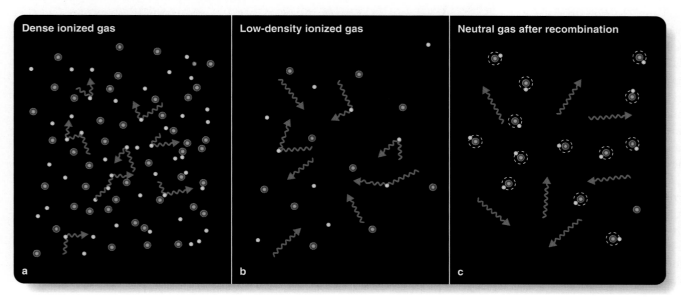

As the young universe expanded, it went through three important changes. First, when the universe reached an age of roughly 50,000 years, the density of the energy present in the form of photons became less than the density of the gas. Before that time, matter could not clump together because the intense sea of photons smoothed the gas out. Once the density of the radiation fell below that of matter, the matter could begin to draw together under the influence of gravity and form the clouds that eventually became galaxy clusters and galaxies.

The expansion of the universe spread the particles of the ionized gas farther and farther apart. As the universe reached the age of about 400,000 years, the second important change began. As the density decreased and the falling temperature of the universe reached 3000 K, protons were able to capture and hold free electrons to form neutral hydrogen, a process called **recombination,** although "combination" (for the first time) would be more accurate. As the free electrons were gobbled up into atoms, they could no longer deflect photons. The photons could travel easily through the gas, so the gas became transparent (Figure 11-7), and the photons retained the blackbody temperature of 3000 K that the gas and photons together had at the time of recombination. Those photons are what are observed today as the CMB, with a large redshift that makes their temperature now about 2.7 K.

Recombination left the gas of the big bang neutral, hot, dense, and transparent. At first the universe was filled with the glow of the hot gas, which would have been partly at visible wavelengths. As the universe expanded and cooled, the glow faded and the universe entered what cosmologists call the **dark age,** a period lasting hundreds of millions of years until the formation of the first stars. During the dark age the universe expanded in darkness.

The dark age ended as the first stars began to form. The gas from which those first stars formed contained almost no metals and was consequently quite transparent. Mathematical models show that the first stars formed from this metal-poor gas would have been very massive, very luminous, and very short lived. That first burst of star formation produced enough ultraviolet light to begin ionizing the gas, and today's astronomers, looking back to the most distant visible quasars and galaxies, can see traces of that **reionization** of the universe (■ Figure 11-8). Reionization marks the end of the dark ages and the beginning of the age of stars and galaxies that continues through today.

Look carefully at ■ Figure 11-9; it summarizes the story of the big bang, from the formation of helium in the first 3 minutes through energy–matter equal-

ity, recombination, and finally reionization of the gas. It may seem amazing that mere humans trapped on Earth can draw such a diagram, but remember that it is based on evidence and on the best understanding of how matter and energy interact.

Building Scientific Arguments

How do you know there was a big bang?

A good scientific argument combines evidence and theory to describe how nature works, and this question calls for a detailed argument. The cosmic microwave background consists of photons emitted by the hot gas of the big bang, so when astronomers detect those photons, they are "seeing" the big bang. Of course, all scientific evidence must be interpreted, so you must understand how the big bang could produce radiation all around the sky before you can accept the background radiation as evidence. First, you must remember that the big bang event filled the entire universe with hot, dense gas. The big bang didn't happen in a single place; it happened everywhere. At recombination, the expansion of the universe reached the

■ **Figure 11-8**

In this artist's conception of reionization, the first stars produce floods of ultraviolet photons that ionized the gas in expanding bubbles. Such a storm of star formation ended an age when the universe had expanded in darkness. Spectra of the most distant quasars reveal that those first galaxies were surrounded by neutral gas that had not yet been fully ionized. Thus the look-back time allows modern astronomers to observe the age of reionization. (K. Lanzetta, SUNY, A. Schaller for STScI, and NASA)

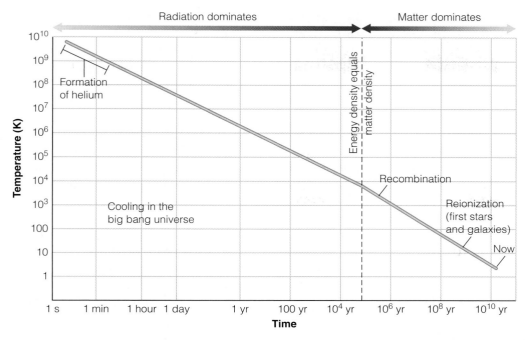

During the first few minutes of the big bang, some hydrogen was fused to produce helium, but the universe quickly became too cool for such fusion reactions to occur. The rate of cooling increased as matter began to dominate over radiation. Recombination freed the radiation from the influence of the gas, and reionization was caused by the birth of the first stars. Note how the exponential scale in time stretches early history and compresses recent history.

stage where the matter became transparent, and the radiation was free to travel through space. Today that radiation from the age of recombination arrives from all over the sky. It is all around you because you are part of the big bang event; and, as you look out into space to great distance, you look back in time and see the hot gas in any direction you look. The radiation isn't seen as visible light because the expansion of the universe has redshifted (lengthened) the wavelengths by a factor of 1100 or so. Astronomers can detect the radiation as photons with far-infrared and short radio wavelengths.

With this interpretation, the cosmic microwave background radiation is powerful evidence that there was a big bang. That tells you how the universe began, but your argument hinges on an important point. **Why do you think the universe cannot have an edge or a center?**

■   ■   ■

11-3) Space and Time

HOW CAN THE BIG BANG HAVE HAPPENED EVERYWHERE? To solve the puzzle, you must put what seem to be reasonable expectations on hold and look carefully at how space and time behave on cosmic scales.

Looking at the Universe

The universe looks about the same whichever way you look. That is called **isotropy.** Of course, there are local differences. If you look toward a galaxy cluster you see more galaxies, but that is only a local variation. On the average, you see similar numbers of galaxies in every direction. Furthermore, the background radiation is also almost perfectly uniform across the sky. The universe is observed to be highly isotropic, almost exactly the same in all directions when viewed from our position.

The universe also seems to be homgenous; **homogeneity** is the property of being the same everywhere. Of course there are local variations. Some regions contain more galaxies and some less. Also, if the universe evolves, then at large look-back times you see galaxies at an earlier stage. If you account for these well-understood variations, then the universe seems to be, on average, the same everywhere. This is harder to check because you can't actually go to the locations of distant galaxies and check in detail that things are about the same there as here, but all astronomical observations indicate that is so.

Isotropy and homogeneity together lead to the **cosmological principle,** that any observer in any galaxy sees that the universe has the same general properties, after accounting for relatively minor local and evolutionary variations. The cosmological principle implies that there are no special places in the universe. What you see from the Milky Way Galaxy is typical of what all intelligent creatures see from their respective home galaxies. Furthermore, the cosmological principle is another way of saying that the universe has no center or edge. Such locations would be special places, and the cosmological principle means there are no special places.

The Cosmic Redshift

Distance is the separation between two points in space; time is the separation between two moments. Einstein's theories of special relativity and general relativity (published respectively in

1905 and 1916) describe how space and time are related and can be considered together as the fabric of the universe, called space-time. You can think of space-time as the canvas on which the universe is painted. Einstein's theories predict that the canvas of space-time can potentially expand (or contract), and, amazingly, that has been confirmed by observations.

The stretching of space-time explains one of the most important observations in cosmology—cosmological redshifts. Modern astronomers understand that, except for small local motions within clusters of galaxies, the galaxies are basically at rest and have kept approximately the same "address" in space since the big bang. The distances between them increase as space-time expands. Furthermore, as space-time expands, it stretches photons traveling through space to longer wavelengths. Photons from distant galaxies spend more time traveling through space and are stretched more than photons from nearby galaxies. That is why redshift depends on distance (■ Figure 11-10). Note that objects like the Milky Way, Earth, and you that are held together by gravity or electromagnetic forces do not expand as the universe expands.

Astronomers often express redshifts as if they were radial velocities, but the redshifts of the galaxies are not Doppler shifts. That is why this book is careful to refer to a galaxy's *apparent* velocity of recession. All a cosmological redshift tells you directly is how much the universe has expanded since the light began its journey to Earth. The formula to calculate the distance a photon has traveled, given its redshift, is complicated, and not all the parameters have been measured precisely. Nevertheless, the Hubble law does apply, and redshifts can be used to estimate the distances to galaxies.

Model Universes

Almost immediately after Einstein published his theory, theorists were able to solve the highly sophisticated mathematics to compute simplified descriptions of the behavior of space-time and matter. Those "model universes" dominated cosmology throughout the 20th century.

The equations allowed three general possibilities. Space-time might be curved in ways that are called an **open universe** or a **closed universe,** or space-time might have no overall curvature at all, a situation that is called a **flat universe.** Most people find the curved space-time models difficult to imagine. Fortunately, modern observations have shown that the flat universe model is almost certainly correct, so you don't have to wrap your brain around the curved models. Note that "flat" does not mean two dimensional. Rather, it means that the familiar rules of geometry you learned in elementary and middle school, for instance that the circumference of a circle is 2π times its radius and that the interior angles of a triangle add up to 180 degrees, are true on the largest scales. Different rules of geometry would apply on large scales if the universe were curved.

A main criterion separating the three models is the average density of the universe, which according to general relativity determines the overall curvature of space-time. If the average density matter and energy in the universe equals what is called the **critical density,** calculated to be about 9×10^{-30} g/cm^3 (depending on the exact value of Hubble constant), space-time will be flat. If the average density is more than the critical density, the universe must be closed; and if it is less, the universe must be open.

The expansions versus time of the three different models are compared in ■ Figure 11-11. The parameter R on the vertical axis is a measure of the extent to which the universe has expanded. You could think of it, essentially, as the average distance between galaxies. In the figure you can see that closed universes may expand and then contract, and open universes may expand forever. Notice also that you can't find the actual age of the universe—the distance on the horizontal axis between the present and the beginning, when R was zero—from the expansion rate alone. You also need to know whether the universe is open, closed, or flat. The Hubble time, discussed earlier in this chapter as a rough estimate of the age of the

■ **Figure 11-10**

A distant galaxy emits a short-wavelength photon toward our galaxy.

The expansion of space-time stretches the photon to longer wavelength as it travels.

The farther the photon has to travel, the more it is stretched.

When the photon arrives at our galaxy, you see it with a longer wavelength — a redshift that is proportional to distance.

The grid shows the expansion of space-time. Like a rubber sheet, space-time stretches, moving the galaxies away from each other and increasing the wavelengths of photons as they travel.

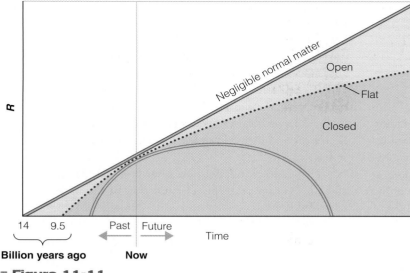

■ Figure 11-11

Open-universe models can expand without end, and the corresponding curves fall in the region shaded orange. Closed models can expand and then contract again (red curve). A flat universe (dotted line) marks the boundary between open and closed universes. The estimated age of the universe for a given value of Hubble's constant depends on the curvature of space-time. Note that these ages are correct for an expansion that always decelerates, but now there is evidence that the expansion first decelerated then later accelerated. That effect makes the ages for universe models older than the values in this plot.

universe, is actually the age the universe would be if it were totally open, which means it would have to contain almost no matter at all, or if it were flat but with an expansion that decelerates then accelerates (discussed in Section 11-4).

Dark Matter in Cosmology

Later in this chapter you will see how observational evidence indicates that the universe is probably flat. For now, you must solve a different problem. If the universe is flat, then its average density must equal the critical density, yet when astronomers added up the matter they could detect, they found only a few percent of the critical density. They wondered if the dark matter made up the rest.

In Chapters 9 and 10, you learned that our galaxy, other galaxies, and galaxy clusters have much stronger gravitational fields than expected based on the amount of visible matter. Even when you add in the nonluminous gas and dust that you expect to find, their gravitational fields are stronger than expected. Galaxies must contain dark matter, in fact, much more dark matter than normal matter (see Figure 9-9).

The protons and neutrons that make up normal matter, including Earth and you, belong to a family of subatomic particles called *baryons*. Modern evidence based on what can be determined about the products of nuclear reactions in the first few minutes after the big bang shows that the dark matter is not baryonic. If there were lots of baryons present during those early moments, they would have (1) collided with and destroyed deuterium nuclei and (2) collided with some of the helium to make lithium.

■ Figure 11-12 shows that the observed amount of deuterium sets a lower limit on the density of the universe, and the observed abundance of lithium-7 sets an upper limit. Those limits indicate that the baryons you and Earth and the stars are made of cannot add up to more than 4 percent of the critical density. Yet, observations show that galaxies and galaxy clusters contain as much as 30 percent of the critical density in the form of dark matter. Only a small amount of the matter in the universe can be baryonic, so the dark matter must be **nonbaryonic matter.**

The true nature of dark matter remains one of the mysteries of astronomy. The most successful models of galaxy formation require that the dark matter be made up of **cold dark matter,** meaning the particles move slowly and can clump into structures with sizes that explain the galaxies and galaxy clusters you see today.

Although the evidence is very strong that dark matter exists, it is not abundant enough by itself to make the universe flat. Dark matter appears to constitute no more than about 30 percent of the critical den-

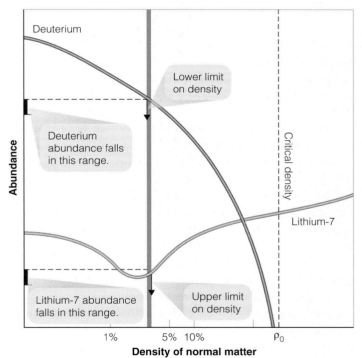

■ Figure 11-12

This diagram compares observation with theory. Theory predicts how much deuterium and lithium-7 you would observe for different densities of normal matter (red and blue curves). The observed density of deuterium falls in a narrow range shown at upper left and sets a lower limit on the possible density of normal matter. The observed density of lithium-7 sets an upper limit. This means the true density of normal matter must fall in a narrow range represented by the green column. Certainly, the density of normal matter is much less than the critical density.

sity. As you will see later in this chapter, there is more to the universe than meets the eye and more even than dark matter.

11-4 21st-Century Cosmology

IF YOU ARE A LITTLE DIZZY FROM THE WEIRDNESS of expanding space-time and dark matter, make sure you are sitting down before you read further. As the 21st century began, cosmologists made a startling discovery, and all around the world astronomers looked at each other and said, "Whoa! What??" To get a running start on these new discoveries, you'll have to go back a couple of decades.

Inflation

By 1980, the big bang model was widely accepted, but it faced two problems that led to the development of an improved theory—a big bang model with an important addition.

One of the problems is called the **flatness problem.** The curvature of space-time seems to be near the transition between an open and a closed universe. That is, the universe seems approximately flat. It seems peculiar that the actual density of the universe is anywhere near the critical density that would make it flat. To be so near critical density now, the density of the universe during its first moments must have been very close, within 1 part in 10^{49}, of the critical density. *So, the flatness problem is: Why was the universe so close to exactly flat, with no space-time curvature, at the time of the big bang?*

The second problem with the original big bang theory is called the **horizon problem.** When astronomers correct for the motion of Earth, they find that the CMB is very isotropic, the same in all directions to a precision of better than 1 part in 1000. Yet, background radiation coming from two points in the sky separated by more than an angle of one degree is from two parts of the big bang far enough apart that they should not have been connected at any previous time. That is, when the CMB photons were released, the universe was not old enough for energy to have traveled at the speed of light from one of those regions to the other—the regions should always have been beyond each other's "horizon" and could not have exchanged heat to make their temperatures equal. *So, the horizon problem is: How did every part of the observable universe get to be so nearly the same temperature by the time of recombination?*

The key to these two problems and to other problems with the simple big bang model may lie with a modified model called the **inflationary big bang** that predicts there was a sudden extra expansion when the universe was very young, even more extreme than that predicted by the original big bang model. According to the inflationary universe model, the universe expanded and cooled until about 10^{-35} seconds after the big bang. Then, the

universe became cool enough that the forces of nature (see Chapter 7), which at earlier extremely high temperatures would have behaved identically, began to differ from each other. Physicists calculate this would have released tremendous amounts of energy and suddenly inflated the universe by a factor of between 10^{20} and 10^{30}. As a result, the part of the universe that is now visible from Earth, the entire observable universe, expanded rapidly from no larger than the volume of an atom to the volume of a cherry pit and then continued a slower expansion to its present state.

That sudden inflation can solve both the flatness problem and the horizon problem. The inflation of the universe would have forced whatever curvature it had toward zero, just as inflating a balloon makes a small spot on its surface flatter. You now live in a universe that has almost perfectly flat space-time geometry because of that sudden inflation long ago. In addition, because the observable part of the universe was no larger in volume than an atom before inflation, it was small enough to have equalized its temperature by then. Now you live in a universe that has the same CMB temperature in all directions.

The inflationary theory predicts that the universe is almost perfectly flat. Observations, however, give evidence that the masses scientists know about (baryonic matter plus dark matter) add up only to about 30 percent of the amount needed to make space-time flat. Can there be more to the universe than baryonic matter and dark matter? What could be weirder than dark matter? Read on.

The Acceleration of the Universe

Both common sense and mathematical models suggest that, as the galaxies recede from each other, the expansion should be slowed by gravity trying to pull the galaxies toward each other. How much the expansion is slowed should depend on the amount of matter in the universe. If the density of matter is less than the critical density, the expansion should be slowed only slightly, and the universe should expand forever. If the density of matter in the universe is greater than the critical density, the expansion should be slowing down dramatically, and the universe should eventually begin contracting.

For decades, astronomers struggled to measure the distance to very distant galaxies directly, compare distances with redshifts, and thereby detect the slowing of the expansion. The rate of slowing would in turn reveal the true curvature of the universe. This was one of the key projects for the Hubble Space Telescope, and two teams of astronomers spent years in competition making the measurements using the same technique. They calibrated type Ia supernovae as standard candles by locating such supernovae occurring in nearby galaxies whose distances were known from Cepheid variables and other reliable distance indicators (see Chapter 9, especially Window on Science 9-1). Once the peak luminosity of type Ia supernovae had been determined, they could be used to find the distances of much more distant galaxies.

Both teams announced their results in 1998. They agreed that the expansion of the universe is not slowing down. Contrary to expectations, it is speeding up! In other words, the expansion of the universe is accelerating (■ Figure 11-13).

The announcement that the expansion of the universe is accelerating was totally unexpected, and astronomers immediately began testing it. The most likely problem was thought to be that the calibration of the supernovae by the original teams might be incorrect. However, that type of problem has been ruled out by measurements of more recently discovered supernovae at very great distances. Other possible problems have been checked and appear to have been eliminated. The universe really does seem to be expanding faster and faster.

Dark Energy and Acceleration

If the expansion of the universe is accelerating, then there must be a force of repulsion in the universe, and astronomers are struggling to understand what it could be. One possibility leads back to Albert Einstein.

When Einstein published his theory of general relativity in 1916, he noticed that his equations describing space-time implied that the universe should contract because of the gravitational attraction of galaxies for each other. In 1916 astronomers did not yet know that the universe was expanding, so Einstein thought he needed to balance the attractive force of gravity by adding a constant to his equations called the **cosmological constant,** representing a force of repulsion that would make the universe hold still. Thirteen years later, in 1929, Edwin Hubble announced that the universe was expanding, and Einstein said

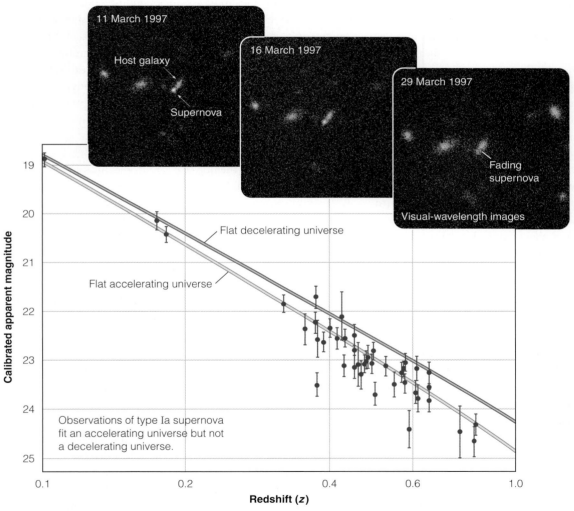

■ Figure 11-13

From the way supernovae fade over time, astronomers can identify those that are type Ia. Once calibrated, those supernovae can be compared with their redshifts, revealing that distant type Ia supernovae are about 25 percent fainter than expected. That must mean they are farther away than expected, given their redshifts. This is strong evidence that the expansion of the universe is accelerating. (ESO) that introducing the cosmological constant was his biggest blunder. Modern astronomers aren't so sure.

One explanation for the acceleration of the universe is that there is a cosmological constant after all, representing a real force that drives a continuing acceleration in the expansion of the universe. The cosmological constant, as its name implies, would be constant in strength over time. Another possibility is a type of energy that is not constant in strength over time. Astronomers have begun referring to this type of energy as **quintessence.** In either case, the observed acceleration is evidence that some form of energy, either a cosmological constant or quintessence, is spread throughout space. Astronomers refer to this as **dark energy,** energy that drives the acceleration of the universe but does not contribute to starlight or the CMB.

You will recall that acceleration and dark energy were first discovered when astronomers found that supernovae a few billion light-years away were slightly *fainter* than expected. Since then even more distant supernovae have been determined to be a bit *brighter* than expected, meaning they are not as far away as the redshifts of their galaxies would seemingly indicate. This means that sometime about 6 billion years ago the universe shifted gears from deceleration to acceleration. The careful calibration of type Ia supernovae allow astronomers to observe this change from deceleration to acceleration. This discovery has the important consequence of increasing previous estimates of the age of the universe by several billion years.

Dark energy can also help you understand the lack of curvature of space-time. The theory of inflation makes the specific prediction that the universe is flat. Dark energy seems to fit with that prediction. As mentioned earlier in this chapter, energy and matter are equivalent, so dark energy is equivalent to mass spread through space. Baryonic matter plus dark matter makes up about a third of the critical density, and dark energy appears to make up two-thirds. That is, when you include dark energy, the total mass-plus-energy density of the universe equals the critical density, making the universe flat.

The Fate of the Universe

For many years, cosmologists have enjoyed saying: "Geometry is destiny." Thinking about models of open, closed, and flat universes, they concluded that the density of a model universe determines its geometry, and its geometry determines its fate. In other words, they were sure that an open universe must expand forever, and a closed universe must eventually begin contracting. That is true, however, only if the universe is ruled by gravity. If the power of dark energy dominates gravity, then geometry is not destiny, and even a closed universe might expand forever.

The ultimate fate of the universe depends on the nature of dark energy. If dark energy is described by the cosmological constant, then the force driving acceleration does not change with time, and our flat universe will expand forever. The galaxies will get farther and farther apart, use up their gas and dust making stars, and the stars will ultimately all die, until each galaxy is isolated, burnt out, dark, and alone. If, however, dark energy is described by quintessence, then its strength could increase with time, and the universe expansion may accelerate faster and faster as space pulls the galaxies away from each other and eventually pulls the galaxies apart, then pulls the stars apart, and finally rips individual atoms apart. This has been called the **big rip.** Don't worry. Even if a big rip is in the future, nothing will be happening for at least 30 billion years.

Maybe there will be no big rip. Critically important observations made by the Chandra X-Ray Observatory have been used to measure the amount of hot gas and dark matter in 25 galaxy clusters. These observations are critical for two reasons. First, the redshifts and distance of these galaxies confirm that the universe expansion initially slowed down, but then shifted gears about 6 billion years ago and is now accelerating. The results are also critical because they are almost good enough to rule out quintessence. If dark energy is described by the cosmological constant and not by quintessence, then there will be no big rip (■ Figure 11-14). Notice that this method does not depend on type Ia supernovae at all, so it is independent confirmation that acceleration is real. When a theory is confirmed by observations of many different types, scientists have much more confidence that it is a true description of nature.

The Origin of Structure

On the largest scales, the universe is isotropic, meaning it looks the same in all directions. On smaller scales there are irregularities. Galaxies are grouped in clusters ranging from a few galaxies to thousands, and those clusters appear to be grouped into **superclusters.** The Local Supercluster, in which we live, is a roughly disk-shaped swarm of galaxy clusters 50 to 75 Mpc in diameter. By measuring the redshifts and positions of thousands of galaxies in great slices across the sky, astronomers have been able to create maps revealing that the superclusters are not scattered at random. They are distributed in long, narrow filaments and thin walls that outline great voids nearly empty of galaxies (■ Figure 11-15). These aggregations and gaps are referred to as **large-scale structure.**

This large-scale structure is a problem because the cosmic microwave background radiation is very uniform, and that means the gas of the big bang must have been extremely uniform at the time of recombination. Yet the look-back time to the farthest known galaxy clusters, galaxies, and quasars is about 93 percent of the way back to the big bang. How did the uniform gas at the time of recombination coagulate so quickly to form galaxy clusters, galaxies, and supermassive black holes in the centers of galaxies, so early in the history of the universe? The answer appears to lie in the characteristics of dark matter.

Baryonic matter is so rare in the universe that it did not have enough gravity to pull itself together quickly after the big bang. As you have read earlier, astronomers propose that dark matter is nonbaryonic and therefore immune to the smoothing effect of the

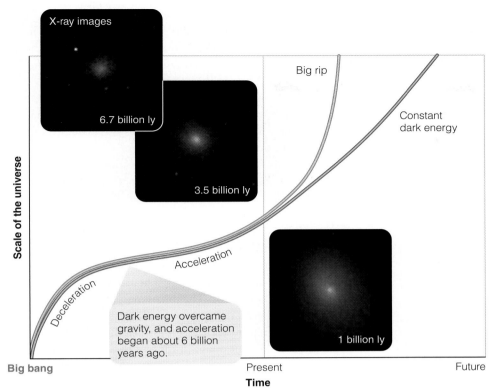

X-ray images

6.7 billion ly

3.5 billion ly

1 billion ly

Scale of the universe

Big rip

Constant
dark energy

Acceleration

Deceleration

Dark energy overcame
gravity, and acceleration
began about 6 billion
years ago.

Big bang

Present

Future

Time

■ **Figure 11-14**

X-ray observations of hot gas in galaxy clusters confirm that in its early history the universe was decelerating because gravity was stronger than the dark energy. As expansion weakened the influence of gravity, dark energy began to produce acceleration. The evidence is not conclusive, but it most directly supports the cosmological constant and weighs against quintessence, which means the universe may not face a big rip. This diagram is only schematic, and the two curves are drawn separated for clarity; at the present time the two curves have not diverged from each other. (NASA/CXC/IoA/S. Allen et al.)

and walls. The structure you see in Figure 11-15 may be the ghostly traces of quantum fluctuations in the infant universe.

CMB Irregularities and the Curvature of Space-Time

Observations of tiny irregularities in the background radiation can also reveal details about inflation and acceleration. In fact, the inflationary theory of the universe makes very specific predictions about the sizes of the irregularities an observer on Earth should see in the CMB. The Wilkinson Microwave Anisotropy Probe (WMAP), a space infrared telescope, has made extensive observations of the type required to test those predictions.

The background radiation is very isotropic—it looks almost exactly the same in all directions. However, when the average

intense radiation that prevented normal matter from contracting. Dark matter was able to collapse into clouds and then pull in the normal matter to begin the formation of galaxies, clusters, and superclusters. Mathematical models have attempted to describe this process, and cold dark matter does seem capable of jump-starting the formation of structure (■ Figure 11-16).

But what started the clumping of the dark matter? Theorists say that space is filled with tiny, random quantum mechanical fluctuations smaller than the smallest atomic particles. At the moment of inflation, those tiny fluctuations would have been stretched to become very large but very subtle variations in gravitational fields that could have later led to the formation of clusters, filaments,

■ **Figure 11-15**

Nearly 70,000 galaxies are plotted in this double slice of the universe, extending outward in the plane of Earth's equator. The nearest galaxies are shown in red and the more distant in green and blue. The galaxies form filaments and walls enclosing empty voids. The Sloan Great Wall is almost 1.4 billion light-years long and is the largest known structure in the universe. The most distant galaxies in this diagram are roughly 3 billion light-years from Earth. (Sloan Digital Sky Survey)

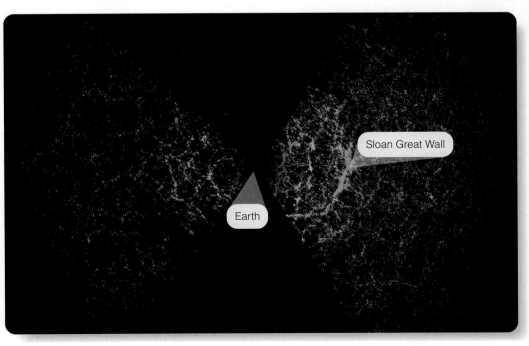

Sloan Great Wall

Earth

Growth of Structure in the Universe

Soon after the big bang, radiation and hot gas are almost uniformly spread through the universe.

Cold dark matter, immune to the influence of light, can contract to form clouds…

which pull in normal gas to form superclusters of galaxies. Gravity continues to pull clusters together.

Statistical tests show the distribution in this model universe resembles the observed distribution of galaxies.

■ Figure 11-16

This computer model traces the formation of structure in the universe from soon after the big bang to the present. Compare with the previous figure's observational data. (Adapted from a model by Kauffmann, Colberg, Diaferio, & White: Max-Planck Institut für Astronomie)

intensity is subtracted from each spot on the sky, small irregularities are evident. That is, some spots on the sky look a tiny bit hotter and brighter than other spots (■ Figure 11-17).

Cosmologists can analyze those irregularities in the intensity of the CMB using sophisticated mathematics to measure how often spots of different sizes occur. The analysis confirms that spots about 1 degree in diameter are the most common, but spots of other sizes occur as well, and it is possible to plot a graph such as ■ Figure 11-18 to show how common irregularities of various sizes are.

Theory predicts that most of the irregularities in the hot gas of the big bang should be about one degree in diameter if the

universe is flat. If the universe were open, the most common irregularities would be smaller. The size of the irregularities in the cosmic background radiation show that the observations fit the flat universe model well, as you can see in Figures 11-17 and 11-18. You learned earlier in this chapter that a flat universe is most easily explained by inflation. Not only is the theory of inflation supported, an exciting result itself, but these data show that the universe is flat, which indirectly confirms the existence of dark energy and the acceleration of the universe.

The results from the WMAP observations make a complicated curve in Figure 11-18, and details of the wiggles tell cosmologists a great deal about the universe. The curve shows that the universe is flat, accelerating, and will expand forever. The age of the universe derived from the data is 13.7 billion years. Furthermore, the smaller peaks in the curve reveal that the universe contains 4 percent baryonic (normal) matter, 23 percent dark matter, and 73 percent dark energy. The Hubble constant is confirmed to be 71 km/s/Mpc. The inflationary theory is confirmed, and the data support the cosmological constant version of dark energy, although quintessence is not ruled out. The dark matter needs to be "cold" (meaning, in this context, slowly-moving) in order to clump together rapidly enough after the big bang to make the first galaxies, quasars, and galaxy clusters.

Please reread the preceding paragraph. It is mind-blastingly amazing. WMAP and other studies of the cosmic microwave background radiation and the distribution of galaxies have revolutionized cosmology. At last, astronomers have accurate observations against which to test theories. The basic constants are known to a precision of a few percent.

On reviewing these results, one cosmologist announced that "Cosmology is solved!"—but that might be premature. Scientists don't understand dark matter or the dark energy that drives the acceleration, so over 95 percent of the universe is not understood. Hearing this, another astronomer suggested that a better phrase is "Cosmology in crisis!" Certainly there are further mysteries to be explored, but cosmologists are growing more confident that they can describe the origin and evolution of the universe.

Building Scientific Arguments

How does inflation theory solve the flatness problem?

This question requires a carefully constructed argument. The flatness problem can be stated as a question: Why is the universe so flat, meaning, why does space-time have little or no curvature? Analyses of WMAP space telescope observations of the cosmic microwave background indicate the universe geometry has close to zero curvature and therefore the combined densities of baryonic matter, dark matter, and dark energy must be close to the critical density. Furthermore, the density must have been extremely close to the critical density when the universe was very young, or it would not be so close

Figure 11-17

You can see the difference yourself. Compare the observations of the irregularities in the background radiation at right with the three simulations at left. The grids represent the curvature of space-time for closed, flat, and open models, respectively. The observed size of the CMB irregularities fits best with cosmological models having flat geometry. Detailed mathematical analysis confirms your visual impression: The universe is flat. (Courtesy of the BOOMERANG Collaboration)

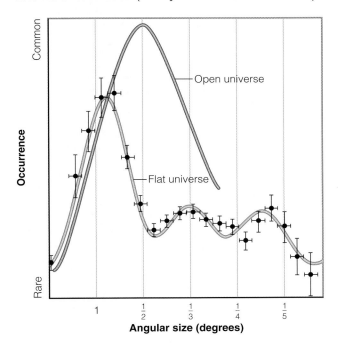

now. The inflationary theory solves this problem by proposing that the universe underwent a moment of rapid inflation when it was a tiny fraction of a second old. That inflation drove the universe toward flatness just as blowing up a balloon makes a spot on the balloon flatter and flatter.

Understanding theory in cosmology is critically important, but science depends on evidence. Now build an argument based on evidence. **What evidence can you cite that the expansion of the universe is accelerating?**

▪ ▪ ▪

Figure 11-18

This graph shows how commonly irregularities of different sizes occur in the cosmic microwave background radiation. Irregularities of about 1 degree in diameter are most common. Models of the universe that are open or closed are ruled out. The data fit a flat model of the universe very well. Crosses on data points show the uncertainty in the measurements.

Summary

11-1 | Introduction to the Universe

Does the universe have an edge and a center?

- Astronomers conclude that it is impossible for the universe to have an edge because it introduces logical inconsistencies. That is, an edge to the universe does not make sense.

- If the universe has no edge, then it cannot have a center.

- The darkness of the night sky leads to the conclusion that the universe is not infinitely old. If it were infinite in extent and age, then every spot on the sky would glow as brightly as the surface of a star. This problem, commonly known as Olbers's paradox, leads to the conclusion that the universe had a beginning.

11-2 | The Big Bang Theory

How do you know that the universe began with a big bang?

- Edwin Hubble's 1929 discovery that the redshift of a galaxy is proportional to its distance is known as the Hubble law and indicates that the universe is expanding. Tracing this expansion backward in time brings you to an initial high-density, high-temperature state commonly called the big bang.

- Although the expanding universe began from the big bang, it has no center. The galaxies do not recede from a single point. They recede from each other as space expands between them.

- The cosmic microwave background (CMB) is blackbody radiation with a temperature of about 2.73 K uniformly spread over the entire sky. It is the light from the big bang released from matter at the time of recombination and now redshifted by a factor of 1100.

- The background radiation is clear evidence that the universe began with a big bang.

- During the first 3 minutes of the big bang, nuclear fusion converted some of the hydrogen into helium but was unable to make many other heavy atoms because no stable nuclei exist with atomic weights of 5 or 8. Today hydrogen and helium are common in the universe, but heavier atoms are rare.

11-3 | Space and Time

How does the universe expand?

- The universe appears isotropic and homogeneous. That is, in its major features, the universe looks the same in all directions and in all locations.

- Isotropy and homogeneity together lead to the cosmological principle, the idea that there are no special places in the universe. Except for local small differences, every place is the same.

- General relativity explains that cosmic redshifts are caused by the stretching of photon wavelengths as they travel through expanding space-time.

- Model universes with flat (uncurved) space-time are infinite and can expand forever. A flat universe would have an average density equal to what is called the critical density. Modern observations indicate that the universe is flat.

- The amounts of deuterium and lithium-7 show that normal baryonic matter can make up only about 4 percent of the critical density. Dark matter must be nonbaryonic and makes up less than 30 percent of the critical density. For the universe to be flat, there must be another major component aside from baryonic matter and dark matter.

11-4 | 21st-Century Cosmology

How has the universe evolved, and what will be its fate?

- The inflationary theory proposes that the universe expanded dramatically a tiny fraction of a second after the big bang.

- Inflation solves the flatness problem because the sudden inflation forced the universe to become flat, just as a spot on an inflating balloon becomes flatter as the balloon inflates.

- Inflation solves the horizon problem because the part of the universe that is now observable was so small before inflation that energy could move and equalize the temperature everywhere.

- Observations of type Ia supernovae reveal that the expansion of the universe is speeding up. This is supposed to be due to energy present in empty space that astronomers have named "dark energy."

- The nature of dark energy is unknown. It may be described by Einstein's cosmological constant, or it may change with time, a property that astronomers label *quintessence*.

- Corrected for inflation and acceleration, the observed value of the Hubble constant implies that the universe is 13.7 billion years old. The future fate of the universe depends on the nature of dark energy. If dark energy increases in strength with time, the universe could end in a big rip.

- The sudden inflation of the universe could have magnified tiny quantum mechanical fluctuations in space-time. These very large-scale but weak differences in gravity worked with dark matter to draw together baryonic (ordinary) matter and create the clusters of galaxies, filaments, and voids that astronomers call large-scale structure.

- Precise observations of irregularities in the CMB and in the large-scale structure of the universe confirm that the universe is flat and contains about 4 percent baryonic matter, 23 percent dark matter, and 73 percent dark energy.

New Terms

cosmology (p. 220)

Olbers's paradox (p. 222)

observable universe (p. 222)

big bang (p. 224)

Hubble time (p. 224)

cosmic microwave background (CMB) (p. 225)

antimatter (p. 226)

recombination (p. 228)

dark age (p. 228)

reionization (p. 228)

isotropy (p. 229)

homogeneity (p. 229)

cosmological principle (p. 229)

open universe (p. 230)

closed universe (p. 230)

flat universe (p. 230)

critical density (p. 230)

nonbaryonic matter (p. 231)

cold dark matter (p. 231)

flatness problem (p. 232)

horizon problem (p. 232)

inflationary big bang (p. 232)

cosmological constant (p. 233)

quintessence (p. 234)

dark energy (p. 234)

big rip (p. 234)

supercluster (p. 234)

large-scale structure (p. 234)

Review Questions

1. How does the darkness of the night sky tell you something important about the universe?

2. How can Earth be located at the center of the observable universe if you accept the Copernican principle?

3. What evidence shows that the universe is expanding? That it began with a big bang?

4. Why couldn't atomic nuclei exist when the universe was less than 2 minutes old?

5. Why are measurements of the present density of the universe important?

6. How does the inflationary universe theory resolve the flatness problem? The horizon problem?

7. Why must the universe have been very uniform during its first 50,000 years?

8. What evidence shows that the expansion of the universe is accelerating?

9. What evidence can you cite that the universe is flat (in other words, that space-time is uncurved)?

10. The image at the right shows irregularities in the background radiation. Why isn't the background radiation perfectly uniform? What does the average size of these irregularities tell you?

BOOMERANG
Collaboration

Discussion Questions

1. Do you think Copernicus would have accepted the cosmological principle? Why or why not?

2. If you reject any model of the universe that has an edge in space because you can't comprehend such a thing, shouldn't you also reject any model of the universe that has a beginning or an ending? Aren't those just edges in time, or is there a difference?

12 | The Origin of the Solar System

What place is this?

Where are we now?

CARL SANDBURG, *GRASS*

THE SOLAR SYSTEM IS YOUR HOME in the universe. Because humans are an intelligent species, we have the right and the responsibility to wonder what we are. Our kind have inhabited this solar system for at least a million years, but only within the last hundred years have we begun to understand what a solar system is.

Guidepost

Looking Back

You have become an expert on the universe. You have studied the appearance, origin, structure, and evolution of stars, galaxies, and the universe itself. But so far, your studies have left out one important class of objects—planets. Now it is time for you to correct that omission; after all, you live on a planet.

This Chapter

In this chapter, you will look back on what you have learned and comprehend your place in the cosmos. What does it mean that your home is Earth? Once you know the evidence for how the solar system formed, you will understand how the processes you have been studying produced your home planet. One important part of that investigation is to understand asteroids, comets, and meteors, objects preserved from the age of planet building. As you explore our solar system in space and time, you will find answers to five essential questions:

What is the theory explaining the origin of the solar system?

What are the observed properties of the solar system that must be explained by the theory of its origin?

What are asteroids and comets and what are their connections to meteors and meteorites?

How do planets form?

What do astronomers know about other solar systems?

Looking Ahead

One reason you should learn about the origin of the solar system is that you live here, but there is another reason. In the next two chapters you will explore each of the planets in more detail. By studying the origin of the solar system before studying individual planets, you have a framework for understanding these fascinating worlds.

12-1 The Great Chain of Origins

YOU ARE LINKED THROUGH A GREAT CHAIN of origins that leads backward through time to the first instant when the universe began 13.7 billion years ago. The gradual discovery of the links in that chain is one of the most exciting adventures of the human intellect. In earlier chapters, you studied some of that story: the origin of the universe in the big bang, the formation of galaxies, the origin of stars, and the production of the chemical elements. Here you will explore further and consider the origin of planets.

The History of the Atoms in Your Body

By the time the universe was 3 minutes old, the protons, neutrons, and electrons in your body had come into existence (see Chapter 11). You are made of very old matter.

Although those particles formed quickly, they were not linked together to form the atoms that are common today. Most of the matter was hydrogen, and about 25 percent of the mass was helium. Very few of the heavier atoms were made in the big bang. Although your body does not contain helium, it does contain many of those ancient hydrogen atoms unchanged since the universe began.

Within a few hundred million years after the big bang, matter began to collect to form galaxies containing billions of stars. You have learned how nuclear reactions inside stars combine low-mass atoms such as hydrogen to make heavier atoms. Generations of stars cooked the original particles, fusing them into atoms such as carbon, nitrogen, and oxygen (see Chapters 8 and 9). Those are common atoms in your body. Even the calcium atoms in your bones were assembled inside stars.

■ Active Figure 12-1

Dust disks around young stars are evident in these Hubble Space Telescope infrared images. The stars are so young that material is still falling inward and being illuminated by the light from the star. Dark bands across the nebulae (arrows) are caused by dense disks of gas and dust from which planets may form. Note that the star at lower left is double. (D. Padgett, IPAC/Caltech; W. Brandner, IPAC; K. Stapelfeldt, JPL; and NASA)

ThomsonNOW Sign in at www.thomsonedu.com and go to ThomsonNOW to see Active Figure "Circumstellar Disk."

Most of the iron in your body was produced by carbon fusion in type Ia supernovae and by the decay of radioactive atoms in the expanding matter ejected by type II supernovae. Atoms heavier than iron, for example iodine, critical in the functioning of your thyroid gland, were created by rapid nuclear reactions that can occur only during supernova explosions (see Chapter 8). Elements uncommon enough to be expensive, such as gold, silver, and platinum in the jewelry that humans wear, also were produced during the violent deaths of rare massive stars.

Our galaxy contains at least 100 billion stars, of which the sun is one. It formed from a cloud of gas and dust about 5 billion years ago, and the atoms in your body were part of that cloud. How the sun took shape, how the cloud gave birth to the planets, how the atoms in your body found their way onto Earth and into you is the story of this chapter. As you explore the origin of the solar system, keep in mind the great chain of origins that created the atoms. As the geologist Preston Cloud remarked, "Stars have died that we might live."

The Origin of the Solar System

Astronomers have a theory for the origin of our solar system that is consistent both with observations of the solar system and with

DG Tau B

Haro 6-5B

500 AU

IRAS 04248+2612

IRAS 04302+2247

Infrared images

observations of star formation, and now they are refining the details.

The **solar nebula hypothesis** supposes that planets form in rotating disks of gas and dust around young stars. You have seen clear evidence that disks of gas and dust are common around young stars. Bipolar flows from protostars (see Chapter 7) were the first evidence of such disks, but modern techniques can image the disks directly (■ Figure 12-1). Our own planetary system formed in such a disk-shaped cloud around the sun. When the sun became luminous enough, the remaining gas and dust were blown away into space, leaving the planets orbiting the sun (■ Figure 12-2).

According to the solar nebula hypothesis, Earth and the other planets of our solar system formed billions of years ago as the sun condensed from the interstellar medium. This hypothesis predicts that most stars should have planets because planet formation is a natural part of star formation, and therefore planets should be very common in the universe, probably more common than stars.

The Solar Nebula Hypothesis

A rotating cloud of gas contracts and flattens...

to form a thin disk of gas and dust around the forming sun at the center.

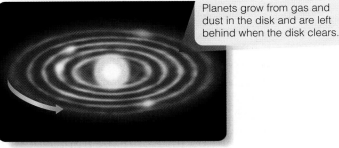
Planets grow from gas and dust in the disk and are left behind when the disk clears.

■ **Figure 12-2**

The solar nebula hypothesis proposes that the planets formed along with the sun.

12-2 A Survey of the Planets

TO TEST HYPOTHESES about the formation of the solar system, astronomers must search the present solar system for evidence of its past. In the next two sections you will survey the solar system and compile a list of its most significant characteristics, potential clues to how it formed.

You should begin with the most general view of the solar system. It is, in fact, almost entirely empty space (see Figure 1-7). Imagine that you reduce the solar system until Earth is the size of a small grain of table salt, about 0.3 mm (0.01 in.) in diameter. The sun is the size of a small plum 4 m (13 ft) from Earth. Jupiter is an apple seed 20 m (66 ft) from the sun. Neptune, at the edge of the solar system, is a large grain of sand located 120 m (400 ft) from the central plum. You can see that planets are tiny specks of matter scattered around the sun—the last significant remains of the solar nebula.

Revolution and Rotation

The planets revolve around the sun in orbits that lie close to a common plane. (Recall from Chapter 2 that the words *revolve* and *rotate* refer to different motions: A planet revolves around the sun but rotates on its axis.) The orbit of Mercury, the closest planet to the sun, is tipped 7° to Earth's orbit. The rest of the planets' orbital planes are inclined by no more than 3.4°. As you can see, the solar system is basically "flat" and disk shaped.

The rotation of the sun and planets on their axes also seems related to the same overall direction of motion. The sun rotates with its equator inclined only 7.25° to Earth's orbit, and most of the other planets' equators are tipped less than 30°. The rotations of Venus and Uranus are peculiar, however. Venus rotates backward compared with the other planets, and Uranus rotates on its "side" (with its equator almost perpendicular to its orbit). You will explore these planets in detail in Chapters 13 and 14, but later in this chapter you will learn a hypothesis about how they may have acquired their peculiar rotations.

Apparently, the preferred direction of motion in the solar system—counterclockwise as seen from the north—is related to the rotation of the disk of material that became the planets. All the planets revolve around the sun in that direction; and, with the exception of Venus and Uranus, they rotate on their axes in that same direction. Furthermore, nearly all of the moons in the solar system, including Earth's moon, orbit around their planets in that direction. With only a few exceptions, most of which astronomers think they understand, revolution and rotation in the solar system both follow a common theme.

Two Kinds of Planets

Perhaps the most striking clue to the origin of the solar system comes from the obvious division of the planets into two categories, the small Earthlike worlds and the giant Jupiter-like worlds. The difference is so dramatic that you are led to say, "Aha, this must mean something!" Study **Terrestrial and Jovian Planets** on pages 246–247 and notice three important points:

1 The two kinds of planets are distinguished by their location. The four inner planets are quite different from the outer four.

2 Almost every solid surface in the solar system is covered with craters.

3 The planets are distinguished by individual properties such as rings, clouds, and moons. Any theory of the origin of the planets needs to explain these properties.

The division of the planets into two families is a clue to how our solar system formed. The present properties of individual planets, however, don't tell everything you need to know about their origins. The planets have all evolved since they formed. For further clues you can look at smaller objects that have remained largely unchanged since the birth of the solar system.

12-3 Space Debris: Planet Building Blocks

THE SUN AND PLANETS are not the only remains of the solar nebula. The solar system is littered with three kinds of space debris: asteroids, comets, and meteoroids. Although these objects represent a tiny fraction of the mass of the system, they are a rich source of information about the origin of the planets.

Asteroids

The **asteroids,** sometimes called minor planets, are small rocky worlds, most but not all of which orbit the sun in a belt between the orbits of Mars and Jupiter. More than 100,000 asteroids have orbits that are charted, of which at least 2000 follow orbits that bring them into the inner solar system where they can occasionally collide with a planet. You will learn in the next section that Earth has been struck many times in its history. Other asteroids share Jupiter's orbit, while some have been found beyond the orbit of Saturn.

About 200 asteroids are more than 100 km (60 mi) in diameter, and tens of thousands are estimated to be more than 10 km (6 mi) in diameter. There are probably a million or more that are larger than 1 km (0.6 mi) and billions that are smaller. Because even the largest are only a few hundred kilometers in diameter, Earth-based telescopes can detect no details on their surfaces, and even the Hubble Space Telescope can image only the largest features.

Photographs returned by robotic spacecraft and space telescopes show that asteroids are generally irregular in shape and battered by impact cratering (■ Figure 12-3a). In fact, some as-

teroids appear to be rubble piles of broken fragments (Figure 12-3b), and a few asteroids are known to be double objects or to have small moons in orbit around them. These are understood by astronomers to be evidence of multiple collisions among the asteroids. A few larger asteroids show signs of volcanic activity on their surfaces that may have happened when those asteroids were young (Figure 12-3c).

Astronomers recognize the asteroids as debris left over by a planet that failed to form at a distance of about 3 AU from the sun. A good theory should explain why a planet didn't form there, but left behind a belt of construction material.

Comets

In contrast to the rocky asteroids, the brightest **comets** are impressively beautiful objects (■ Figure 12-4). Most comets are faint, however, and difficult to locate even at their brightest. A comet may take months to sweep through the inner solar system, during which time it appears as a glowing head with an extended tail of gas and dust.

The beautiful tail of a comet can be longer than an AU, but it is produced by an icy nucleus (sometimes described as a dirty snowball) only a few tens of kilometers in diameter. The nucleus remains frozen and inactive while it is far from the sun. As the nucleus moves along its elliptical orbit into the inner solar system, the sun's heat begins to vaporize the ices, releasing gas and dust. The pressure of sunlight and the solar wind pushes the gas and dust away, forming a long tail. The gas and dust respond differently to the forces acting on them so they sometimes separate into two separate sub-tails (Figure 12-4a). The motion of the nucleus along its orbit, the pressure of sunlight, and the outward flow of the solar wind cause the tails of comets to point always approximately away from the sun (Figure 12-4b).

Comet nuclei contain ices of water and other **volatile** (easily vaporized) compounds such as carbon dioxide, carbon monoxide, methane, ammonia, and so on. These ices are the kinds of compounds that should have condensed from the outer solar nebula, and that makes astronomers think that comets are ancient samples of the gases and dust from which the outer planets formed.

Five spacecraft flew past the nucleus of Comet Halley when it passed through the inner solar system in 1985 and 1986. Since then spacecraft have visited the nuclei of several other comets. Images show that comet nuclei are irregular in shape and very dark, with jets of gas and dust spewing from active regions on the nuclei (■ Figure 12-5). In general, these nuclei are darker than a lump of coal, which suggests they have composition similar to certain dark, water- and carbon-rich meteorites that are described in the next section.

Since 1992, astronomers have discovered roughly a thousand small, dark, icy bodies orbiting in the outer fringes of the solar system beyond Neptune. This collection of objects is called

Figure 12-3

Most asteroids are too small for their gravity to pulls them into a spherical shape. Impacts break them into irregularly shaped fragments. (a) The Near Earth Asteroid Rendezvous (NEAR) spacecraft visited the asteroid Eros in 2000 and found it to be heavily cratered by collisions and covered by a layer of crushed rock and dust. Its density (determined from the gravitational influence of Eros on the NEAR spacecraft) indicates the interior is solid rock. (NEAR/APL) (b) Mathilde has a very low density, meaning it cannot be solid rock. Like many asteroids, Mathilde may be a rubble pile with large empty spaces between fragments. (NEAR/APL) (c) The large asteroid Vesta has a spectrum resembling solidified lava. This image by the Hubble Space Telescope shows that Vesta is not quite spherical and has a huge crater at its south pole. (NASA/STSci) (d) One class of meteorites that are spectroscopically identical to Vesta is made of basalt, a volcanic rock. Vesta seems to have had internal heat and volcanic activity in its past, perhaps due to decay of radioactive minerals. (New England Meteoritical)

1 The distinction between the terrestrial planets and the Jovian planets is dramatic. The inner four planets, Mercury, Venus, Earth, and Mars, are **terrestrial planets,** meaning they are small, dense, rocky worlds with little or no atmosphere. The outer four planets, Jupiter, Saturn, Uranus, and Neptune, are **Jovian planets,** meaning they are large, low-density worlds with thick atmospheres and liquid interiors.

The planets and the sun to scale. Saturn's rings would just reach from Earth to the moon.

Planetary orbits to scale. The terrestrial planets lie quite close to the sun, whereas the Jovian planets are spread far from the sun.

Mercury
Venus
Earth
Mars
Asteroids
Jupiter
Saturn
Uranus
Neptune

1a Of the terrestrial planets, Earth is most massive, but the Jovian planets are much more massive. Jupiter is over 300 Earth masses, and Saturn is nearly 100 Earth masses. Uranus and Neptune are 15 and 17 Earth masses.

Mercury is only 40 percent larger than Earth's moon, and its weak gravity cannot retain a permanent atmosphere. Like the moon, it is covered with craters from meteorite impacts.

Mercury

Earth's moon

2 Craters are common on all of the surfaces in the solar system that are strong enough to retain them. Earth has about 150 impact craters, but many more have been erased by erosion. Besides the planets, the asteroids and nearly all of the moons in the solar system are scarred by craters. Ranging from microscopic to hundreds of kilometers in diameter, these craters have been produced over the ages by meteorite impacts. When astronomers see a rocky or icy surface that contains few craters, they know that the surface is young.

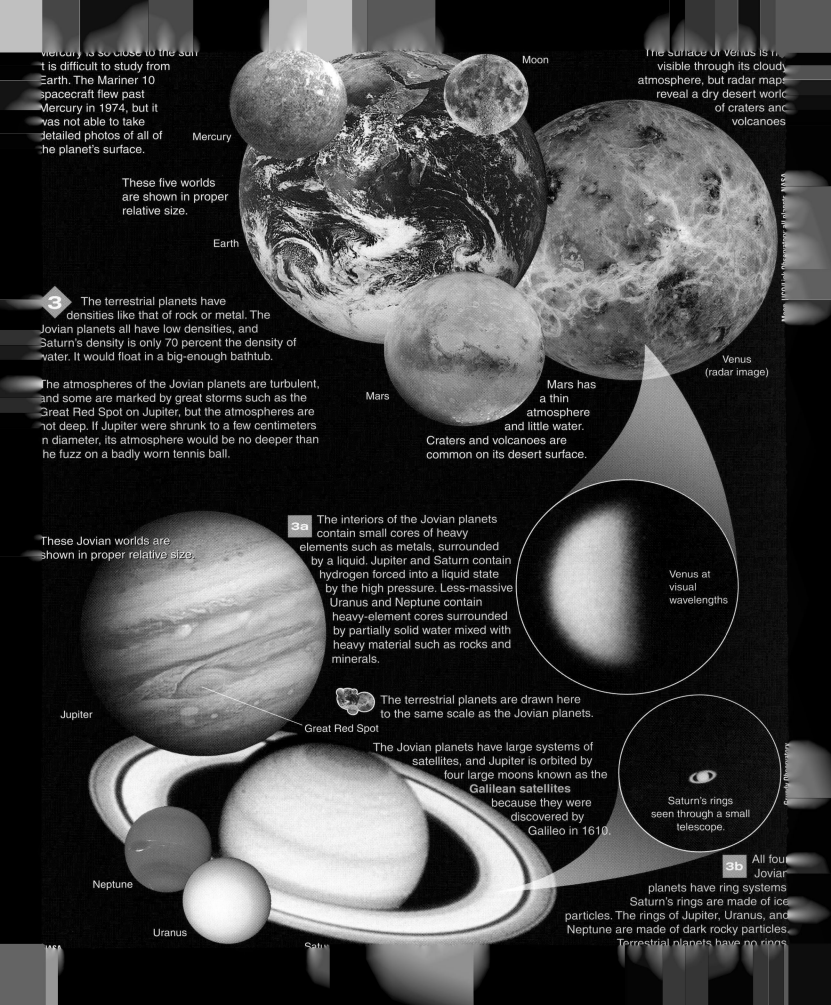

Mercury is so close to the sun it is difficult to study from Earth. The Mariner 10 spacecraft flew past Mercury in 1974, but it was not able to take detailed photos of all of the planet's surface.

Mercury

These five worlds are shown in proper relative size.

Earth

Moon

The surface of Venus is not visible through its cloudy atmosphere, but radar maps reveal a dry desert world of craters and volcanoes

Venus (radar image)

3 The terrestrial planets have densities like that of rock or metal. The Jovian planets all have low densities, and Saturn's density is only 70 percent the density of water. It would float in a big-enough bathtub.

The atmospheres of the Jovian planets are turbulent, and some are marked by great storms such as the Great Red Spot on Jupiter, but the atmospheres are not deep. If Jupiter were shrunk to a few centimeters in diameter, its atmosphere would be no deeper than the fuzz on a badly worn tennis ball.

Mars

Mars has a thin atmosphere and little water. Craters and volcanoes are common on its desert surface.

These Jovian worlds are shown in proper relative size.

3a The interiors of the Jovian planets contain small cores of heavy elements such as metals, surrounded by a liquid. Jupiter and Saturn contain hydrogen forced into a liquid state by the high pressure. Less-massive Uranus and Neptune contain heavy-element cores surrounded by partially solid water mixed with heavy material such as rocks and minerals.

Venus at visual wavelengths

Jupiter

Great Red Spot

The terrestrial planets are drawn here to the same scale as the Jovian planets.

The Jovian planets have large systems of satellites, and Jupiter is orbited by four large moons known as the **Galilean satellites** because they were discovered by Galileo in 1610.

Saturn's rings seen through a small telescope.

Neptune

Uranus

Saturn

3b All four Jovian planets have ring systems. Saturn's rings are made of ice particles. The rings of Jupiter, Uranus, and Neptune are made of dark rocky particles. Terrestrial planets have no rings.

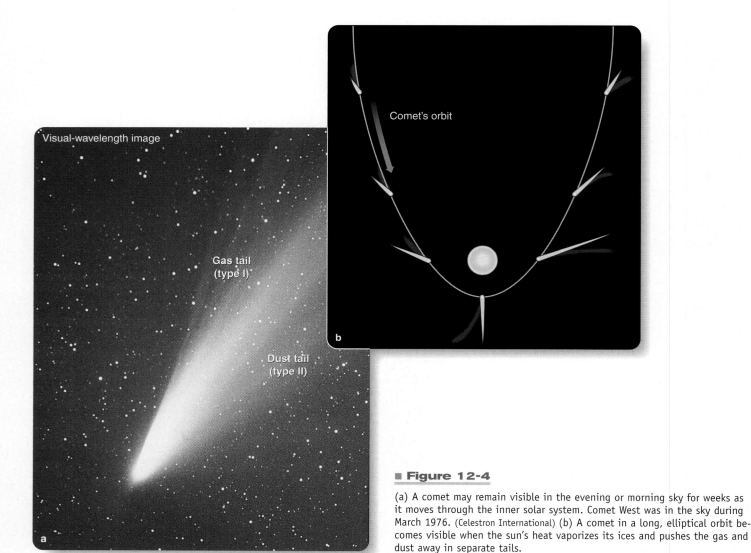

Visual-wavelength image

Gas tail
(type I)

Dust tail
(type II)

a

Comet's orbit

b

■ **Figure 12-4**

(a) A comet may remain visible in the evening or morning sky for weeks as it moves through the inner solar system. Comet West was in the sky during March 1976. (Celestron International) (b) A comet in a long, elliptical orbit becomes visible when the sun's heat vaporizes its ices and pushes the gas and dust away in separate tails.

the **Kuiper belt** after the Dutch-American astronomer Gerard Kuiper (pronounced KIE-per), who predicted their existence in the 1950s. There are probably 100 million bodies larger than 1 km in the Kuiper belt, and any successful theory should explain how they came to be where they are. Astronomers believe that some comets, those that move in the plane of the solar system with relatively short orbital periods, come from the Kuiper belt.

Meteoroids, Meteors, Meteorites

Unlike the stately comets, **meteors** flash across the sky in momentary streaks of light. They are commonly called "shooting stars." Of course, they are not stars but small bits of rock and metal falling into Earth's atmosphere and bursting into incandescent vapor about 80 km (50 mi) above the ground because of friction with the air. This vapor condenses to form dust, which settles slowly to the ground, adding about 40,000 tons per year to Earth's mass.

Technically, the word *meteor* refers to the streak of light in the sky. In space, before its fiery plunge, the object is called a **meteoroid.** Most meteoroids are specks of dust, grains of sand, or tiny pebbles. Almost all the meteors you see in the sky are produced by meteoroids that weigh less than 1 gram. Only rarely is a meteoroid massive and strong enough to survive its plunge, reach Earth's surface, and become what is called a **meteorite.**

Meteorites can be divided into three broad categories. *Iron* meteorites are solid chunks of iron and nickel. *Stony* meteorites are silicate masses that resemble Earth rocks. *Stony-iron* meteorites are mixtures of iron and stone. These types of meteorites are illustrated in ■ Figure 12-6.

One type of stony meteorite called **carbonaceous chondrites** has a chemical composition that resembles a cooled lump of the sun with the hydrogen and helium removed. These meteorites generally contain abundant volatile compounds including significant amounts of carbon and water, and may have similar composition to comet nuclei. Heating would have modified and

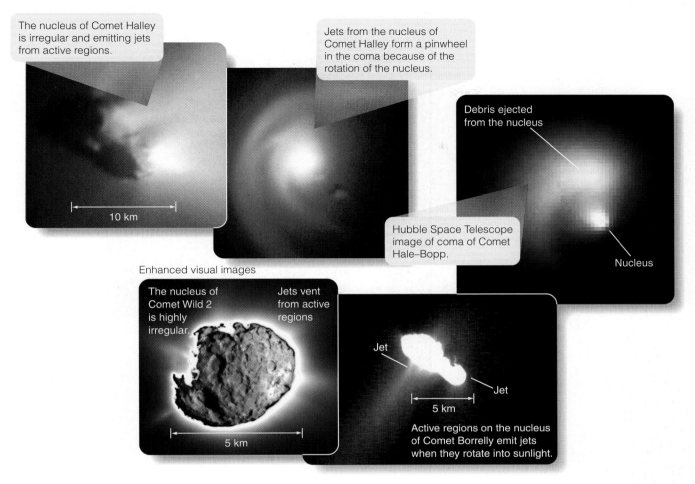

The nucleus of Comet Halley is irregular and emitting jets from active regions.

Jets from the nucleus of Comet Halley form a pinwheel in the coma because of the rotation of the nucleus.

10 km

Debris ejected from the nucleus

Hubble Space Telescope image of coma of Comet Hale–Bopp.

Nucleus

Enhanced visual images

The nucleus of Comet Wild 2 is highly irregular.

Jets vent from active regions

5 km

Jet

Jet

5 km

Active regions on the nucleus of Comet Borrelly emit jets when they rotate into sunlight.

■ **Figure 12-5**

Visual-wavelength images made by spacecraft and by the Hubble Space Telescope show how the nucleus of a comet produces jets of gases from regions where sunlight vaporizes ices. (Halley nucleus: ©1986 Max-Planck Institute; Halley coma: Steven Larson; Comets Borrelly, Hale–Bopp, and Wild2: NASA)

driven off these fragile compounds, so carbonaceous chondrites must not have been heated since they formed. Astronomers conclude that carbonaceous chondrites, unlike the planets, have not evolved and thus give direct information about the early solar system.

You can find evidence of the origin of meteors through one of the most pleasant observations in astronomy by watching a **meteor shower,** a display of meteors that are clearly related by a common origin. For example, the Perseid meteor shower occurs each year in August, and during the height of the shower you might see as many as 40 meteors per hour. The Perseid shower is so named because all its meteors appear to come from a point in the constellation Perseus. Meteor showers are seen when Earth passes near the orbit of a comet (■ Figure 12-7). The meteors in meteor showers must be produced by dust and debris released from the icy head of the comet. In contrast, the orbits of some meteorites have been calculated to lead back into the asteroid belt.

An important reason to mention meteorites here is for one specific clue they can give you concerning the solar nebula: Meteorites can reveal the age of the solar system.

12-4 The Story of Planet Formation

THE CHALLENGE FOR MODERN PLANETARY ASTRONOMERS is to compare the characteristics of the solar system with the solar nebula theory and tell the story of how the planets formed (**Window on Science 12-1**).

The Age of the Solar System

According to the solar nebula theory, the planets should be about the same age as the sun. The most accurate way to find the age of a rocky body is to bring a sample into the laboratory and determine its age by analyzing the radioactive elements it contains.

When a rock solidifies, the process of cooling causes it to incorporate known proportions of the chemical elements. A few of these elements are radioactive and decay into other elements,

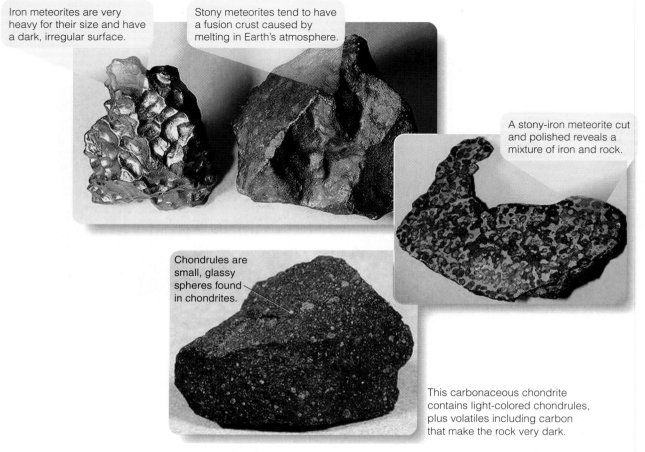

Iron meteorites are very heavy for their size and have a dark, irregular surface.

Stony meteorites tend to have a fusion crust caused by melting in Earth's atmosphere.

A stony-iron meteorite cut and polished reveals a mixture of iron and rock.

Chondrules are small, glassy spheres found in chondrites.

This carbonaceous chondrite contains light-colored chondrules, plus volatiles including carbon that make the rock very dark.

■ **Figure 12-6**

The three main types of meteorites, irons, stones, and stony-irons, have distinctive characteristics. (Lab photos courtesy of Russell Kempton, New England Meteoritical)

called daughter elements or isotopes. The **half-life** of a radioactive element is the time it takes for half of the atoms to decay into the daughter elements. For example, potassium-40 decays into daughter isotopes calcium-40 and argon-40 with a half-life of 1.3 billion years. Another example, uranium-238, decays with a half-life of 4.5 billion years to lead-206 and other isotopes. As time passes, the abundance of a radioactive element in a rock gradually decreases, and the abundances of the daughter elements gradually increase (■ Figure 12-8). You can estimate the original abundances of the elements in the rock from rules of chemistry and observations of rock properties in general. Thus, measuring the present abundances of the parent and daughter elements allows you to find the age of the rock. This works best if you have several radioactive element "clocks" that can be used as independent checks on each other.

Of course, to find a radioactive age, you need a sample in the laboratory, and the only celestial bodies from which scientists have samples are Earth, the moon, Mars, and meteorites. The oldest Earth rocks so far discovered and dated are tiny zircon crystals from Australia that are 4.3 billion years old. The surface of Earth is active, and the crust is continually destroyed and reformed from material welling up from beneath the crust (see Chapter 13). Consequently, the age of these oldest rocks tells you only that Earth is *at least* 4.3 billion years old. Unlike Earth's surface, the moon's surface is not being recycled by constant geologic activity, so you can guess that more of it might have survived unaltered since early in the history of the solar system. The oldest rocks brought back by the Apollo astronauts are 4.48 billion years old. That means the moon must be at least 4.48 billion years old. Although no one has yet been to Mars, over a dozen meteorites found on Earth have been identified by their chemical composition as having come from Mars. The oldest has an age of approximately 4.5 billion years. Mars must be at least that old.

The most important source for determining the age of the solar system is meteorites. As you learned earlier, carbonaceous chondrite meteorites have compositions indicating that they have not been heated much or otherwise altered since they formed. They have a range of ages with a consistent and precise upper limit of 4.56 billion years. This figure is widely accepted as the age of the solar system and is often rounded to 4.6 billion

Reconstructing the Past from Evidence and Hypothesis

Scientists often solve problems in which they must reconstruct the past. Some of these reconstructions are obvious, such as an archaeologist excavating the ruins of a burial tomb, but others are less obvious. In each case, success requires the interplay of hypotheses and evidence to re-create a past that no longer exists.

It is obvious that astronomers reconstruct the past when they use evidence gathered from meteorites to study the origin of the solar system, but a biologist studying a centipede is also reconstructing the past. How did this creature come to have a segmented body with so many legs? How did it evolve the metabolism that allows it to move quickly and hunt prey? Although the problem might

at first seem to be one of mere anatomy, the scientist must reconstruct an environment that no longer exists.

The astronomer's problem is not just to understand what the planets are like but to understand how they got that way. That means planetary astronomers must look at the evidence they can see today and reconstruct a history of the solar system, a past that is quite different from the present. If you had a time machine, it would be a fantastic adventure to go back and watch the planets form. Time machines are impossible, but scientists can use the grand interplay of evidence and theory, the distinguishing characteristic of science, to journey back billions of years and reconstruct a past that no longer exists.

One way science can enrich and inform our lives is by recreating a world that no longer exists.

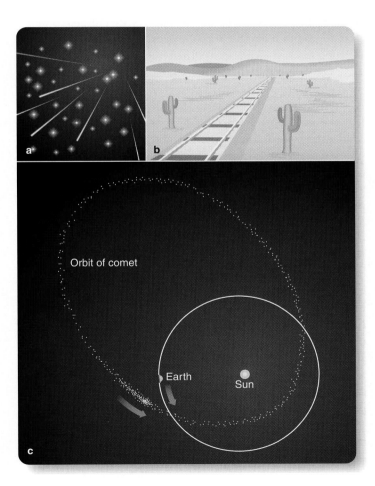

Orbit of comet

Earth

Sun

c

years. That is in agreement with the age of the sun, which is estimated to be 5 billion years plus or minus 1.5 billion years using mathematical models of the sun's interior (see Chapter 7) that are completely independent of meteorite radioactive ages.

Apparently, all the bodies of the solar system formed at about the same time some 4.6 billion years ago. You can add this as the final item to your list of characteristic properties of the solar system that need to be explained by a good theory of its formation (■ Table 12-1).

Chemical Composition of the Solar Nebula

Everything astronomers know about the solar system and star formation suggests that the solar nebula was a fragment of an interstellar gas cloud. Such a cloud would have been mostly hydrogen with some helium and minor traces of the heavier elements.

■ Figure 12-7

(a) Meteors in a meteor shower enter Earth's atmosphere along parallel paths, but perspective makes them appear to diverge from a radiant point in the sky. (b) Similarly, parallel railroad tracks appear to diverge from a point on the horizon. (c) Meteors in a shower are debris left behind as a comet's icy nucleus vaporizes. The rocky and metallic bits of matter spread along the comet's orbit. If Earth passes through such material, you can see a meteor shower.

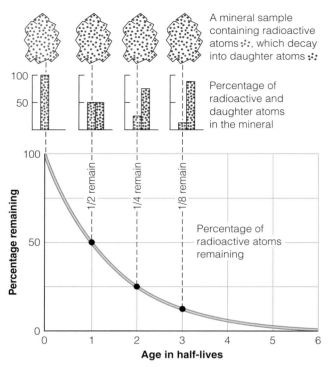

A mineral sample containing radioactive atoms ∴, which decay into daughter atoms ∴

Percentage of radioactive and daughter atoms in the mineral

1/2 remain

1/4 remain

1/8 remain

Percentage of radioactive atoms remaining

■ Active Figure 12-8

The radioactive atoms in a mineral sample (red) decay into daughter atoms (blue). Half the radioactive atoms are left after one half-life, a fourth after two half-lives, an eighth after three half-lives, and so on.

ThomsonNOW™ Sign in at www.thomsonedu.com and go to ThomsonNOW to see Active Figure "Radioactive Decay."

planets are high-density, terrestrial bodies, whereas the outer, Jupiter-like planets are low-density, giant planets. This division is due to the different ways gases condensed into solids in the inner and outer regions of the solar nebula.

Even among the four terrestrial planets, you will find a pattern of slight differences in density. The **uncompressed densities**—the densities the planets would have if their gravity did not compress them, or to put it another way, the densities of their original construction material—can be calculated from the actual densities and masses of each planet. In general, the closer a planet is to the sun, the higher is its uncompressed density.

This density variation is understood to have originated when the solar system first formed solid grains. The kind of matter that condensed in a particular region would depend on the temperature of the gas there. In the inner regions, the temperature seems to have been 1500 K or so. The only materials that can form grains at this temperature are compounds with high melting points, such as metal oxides and pure metals, which are very dense, corresponding to the composition of Mercury. Farther out in the nebula it was cooler, and silicates (rocky material) could condense. These are less dense than metal oxides and metals, corresponding more to the compositions of Venus, Earth, and Mars. Somewhere further from the sun there was a boundary called the **ice line** beyond which the water vapor could freeze to form ice. Not much farther out, compounds such as methane and ammonia could condense to form other ices. Water vapor, methane, and ammonia were abundant in the solar nebula, so beyond the ice line, the nebula was filled with a blizzard of ice particles, and those ices are low-density materials, corresponding to the compositions of Jupiter and the other outer planets.

The sequence in which different materials condensed from the gas at different distances from the sun is called the **condensation sequence** (■ Table 12-2). This is really a hypothesis that says the planets, forming at different distances from the sun,

This is precisely what you see in the composition of the sun (see Table 5-1). Analysis of the solar spectrum shows that the sun is mostly hydrogen, with a quarter of its mass being helium and only about 2 percent being heavier elements. Of course, nuclear reactions have fused some hydrogen into helium, but that happens in the sun's core and has not affected its surface composition. That means the composition of the sun's atmosphere revealed in its spectrum is essentially the composition of the solar nebula gases from which it formed.

You can see that same solar nebula composition reflected in the chemical compositions of the planets. The composition of Jupiter and the other Jovian planets resembles the composition of the sun. Furthermore, if you allowed low-density gases to escape from a blob of sun-stuff, the remaining heavier elements would resemble the composition of Earth and the other terrestrial planets, as well as meteorites.

Condensation of Solids

The key to understanding the process that converted the nebular gas into solid matter is the observed variation in density among solar system objects. You have already noted that the four inner

Table 12-2 | The Condensation Sequence

Temperature (K)	Condensate	Object (Estimated Temperature of Formation; K)
1500	Metal oxides	Mercury (1400)
1300	Metallic iron and nickel	
1200	Silicates	
1000	Feldspars	Venus (900)
680	Troilite (FeS)	Earth (600)
		Mars (450)
175	H_2O ice	Jovian (175)
150	Ammonia–water ice	
120	Methane–water ice	
65	Argon–neon ice	Pluto (65)

would have accumulated from different kinds of materials. The original chemical composition of the solar nebula should have been roughly the same throughout the nebula. The important factor was temperature: The inner nebula was hot, and only metals and rock could condense there, whereas the cold outer nebula could form lots of ices in addition to metals and rock. The ice line seems to have been between Mars and Jupiter, and it separates the formation of the dense terrestrial planets from that of the lower-density Jovian planets.

Formation of Planetesimals

In the development of a planet, three groups of processes operate to collect solid bits of matter—rock, metal, or ices—into larger bodies called **planetesimals** that eventually build the planets. The study of planet building is the study of these three groups of processes: condensation, accretion, and gravitational collapse, which will be described in detail in this section.

According to the solar nebula theory, planetary development in the solar nebula began with the growth of dust grains. These specks of matter, whatever their composition, grew from microscopic size first by condensation, then by accretion.

A particle grows by **condensation** when it adds matter one atom or molecule at a time from a surrounding gas. Snowflakes, for example, grow by condensation in Earth's atmosphere. In the solar nebula, dust grains were continuously bombarded by atoms of gas, some of which stuck to the grains.

The second process is **accretion**, the sticking together of solid particles. You may have seen accretion in action if you have walked through a snowstorm with big, fluffy flakes. If you caught one of those "flakes" on your mitten and looked closely, you saw that it was actually made up of many tiny, individual flakes that had collided as they fell and accreted to form larger particles.

Model calculations indicate that in the solar nebula the dust grains were, on the average, no more than a few centimeters apart, so they collided frequently and accreted into larger particles.

There is no clear distinction between a very large grain and a very small planetesimal, but you can consider an object a planetesimal when its diameter approaches a kilometer or so, like the size of a typical small asteroid or comet. Objects larger than a centimeter were subject to new processes that tended to concentrate them. For example, collisions with the surrounding gas and with each other would have caused growing planetesimals to settle into a thin disk estimated to have been only about 0.01 AU thick in the central plane of the rotating nebula. This concentration of material would have made further planetary growth more rapid. The collapse of the planetesimals into a plane is analogous to the flattening of a forming galaxy. Computer models show that the rotating disk of particles should have been gravitationally unstable and would have been disturbed by spiral density waves much like those found in spiral galaxies. This would have further concentrated the planetesimals and helped them coalesce into objects up to about 100 km (60 mi) in diameter.

Through these processes the nebula became filled with trillions of solid particles ranging in size from pebbles to small planets. As the largest began to exceed 100 km in diameter, new processes began to alter them, and a new stage in planet building began, the formation of protoplanets.

Growth of Protoplanets

The coalescing of planetesimals eventually formed **protoplanets**, massive objects destined to become planets. As these larger bodies grew, new processes began making them grow faster and altered their physical structure.

If planetesimals collided at orbital velocities, it is unlikely that they would have stuck together. The average orbital velocity in the solar system is about 10 km/s (22,000 mph). Head-on collisions at this velocity would have vaporized the material. However, the planetesimals were all moving in the same direction in the nebular plane and didn't collide head-on. Instead, they merely "rubbed shoulders," so to speak, at low relative velocities. Such gentle collisions would have been more likely to combine them than to shatter them.

The largest planetesimals would grow the fastest because they had the strongest gravitational field and could more easily attract additional material. These planetesimals probably grew quickly to protoplanetary dimensions, sweeping up more and more material.

Protoplanets had to begin growing by accumulating solid material because they did not have enough gravity to capture and hold large amounts of gas. In the warm solar nebula, the atoms and molecules of gas were traveling at velocities much larger than the escape velocities of modest-size protoplanets. Because of that, in their early development, the protoplanets could only grow by attracting solid bits of rock, metal, and ice. Once a protoplanet approached a mass of 15 Earth masses or so could it begin to grow by **gravitational collapse,** the rapid accumulation of large amount of infalling gas.

In its simplest form, the theory of terrestrial protoplanet growth supposes that all the planetesimals in each orbital zone had about the same chemical composition. The planetesimals accumulated gradually to form a planet-size ball of material that was of homogeneous composition throughout. Once the planet formed, heat began to accumulate in its interior from the decay of short-lived radioactive elements, and this heat eventually melted the planet and allowed it to differentiate. **Differentiation** is the separation of material according to density. When the planet melted, the heavy metals such as iron and nickel settled to the core, while the lighter silicates floated to the surface to form a low-density crust. The story of planet formation from accretion of planetesimals followed by differentiation is shown in ▪ Figure 12-9.

This process depends on the presence of short-lived radioactive elements whose rapid decay would have released enough heat to melt the interior of planets. Astronomers know such radioactive elements were present because the oldest meteorites contain daughter isotope magnesium-26 that is produced by the decay of aluminum-26 with a half-life of only 0.74 million years. The aluminum-26 and similar short-lived radioactive elements are gone now, but they must have been produced in a supernova explosion that occurred no more than a few million years before the formation of the solar nebula. In fact, many astronomers suspect that this supernova explosion compressed nearby gas and triggered the formation of stars (see Figures 7-12 and 7-13), one of which became the sun. Thus our solar system may exist because of a supernova explosion that occurred about 4.6 billion years ago.

If planets formed and were later melted by radioactive decay, gases released from the planet's interior would have formed an atmosphere. The creation of a planetary atmosphere from a planet's interior is called **outgassing.** Models of the formation of the Earth indicate that the local planetesimals would not have included much water, so some astronomers now think that Earth's water and some of its present atmosphere accumulated late in the formation of the planets as Earth swept up volatile-rich planetesimals forming in the cooling solar nebula. Such icy planetesimals may have formed in the outer parts of the solar nebula and been scattered by encounters with the Jovian planets, creating a solar system–wide bombardment of comets.

According to the solar nebula theory, the Jovian planets began growing by the same processes that built the terrestrial planets, but the outer solar nebula not only contained solid bits of metals and silicates, it also included abundant ices. The Jovian planets grew rapidly and quickly became massive enough to grow by gravitational collapse, drawing in large amounts of gas from the solar nebula. Ices could not condense as solids at the locations of the terrestrial planets, so those planets developed slowly and never became massive enough to grow by gravitational collapse.

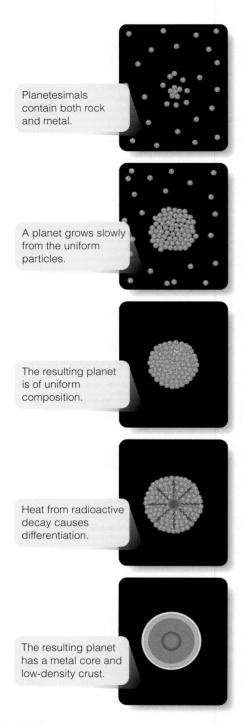

Planetesimals contain both rock and metal.

A planet grows slowly from the uniform particles.

The resulting planet is of uniform composition.

Heat from radioactive decay causes differentiation.

The resulting planet has a metal core and low-density crust.

▪ Figure 12-9

This simple model of accretion assumes planets formed from planetesimals that were of uniform composition, containing both metals and rocky material, and the planets later differentiated, meaning that they melted and separated into layers by density and composition.

The Jovian planets must have grown to their present size in less than about 10 million years, which is when astronomers calculate that the sun become hot and luminous enough to blow away the gas remaining in the solar nebula. The terrestrial planets grew from solids and not from the gas, so they could have

■ Active Figure 12-10

(a) The Barringer Meteorite Crater (near Flagstaff, Arizona) is nearly a mile in diameter and was formed about 50,000 years ago by the impact of an iron meteorite roughly 90 m in diameter. It hit with energy equivalent to that of a 3-megaton hydrogen bomb. Notice the raised and deformed rock strata all around the crater. (M. A. Seeds) (b) Like all larger-impact features, the Barringer Meteorite Crater has a raised rim and scattered ejects. (USGS)

ThomsonNOW™ Sign in at www.thomsonedu.com and go to ThomsonNOW to see the Active Figure called "Impacts on Earth."

continued to grow by accretion from the solid debris left behind when the gas was blown away. Model calculations indicate the process of planet formation was almost completely finished by the time the solar system was about 30 million years old.

Continuing Bombardment of the Planets

Astronomers have good reason to believe that comets and asteroids can hit planets. Meteorites hit Earth every day, and occasionally a large one can form a crater (■ Figure 12-10). Earth is marked by about 150 known meteorite craters. In a sense, this bombardment represents the slow continuation of the accretion of the planets. Earth's moon, Mercury, Venus, Mars, and most of the moons in the solar system are covered with craters. A few of these craters have been formed recently by the steady rain of meteorites that falls on all the planets in the solar system, but most of the craters appear to have been formed more than 4 billion years ago in what is called the **heavy bombardment,** as much of the remaining solid debris in the solar nebula were swept up by the planets. In the next chapter you will learn that the heavy bombardment may have had more than one episode.

Sixty-five million years ago, at the end of the Cretaceous period, over 75 percent of the species on Earth, including the dinosaurs, went extinct. Scientists have found a thin layer of clay all over the world that was laid down at that time, and it is rich in the element iridium—common in meteorites but rare in Earth's crust. This suggests that a large impact altered Earth's climate and caused the worldwide extinction. Mathematical models indicate that a major impact would eject huge amounts of pulverized rock high above the atmosphere. As this material fell back, Earth's atmosphere would be turned into a glowing oven of red-hot meteorites streaming through the air, and the heat would trigger massive forest fires around the world. Soot from such fires has been found in the final Cretaceous clay layers. Once the firestorms cooled, the remaining dust in the atmosphere would block sunlight and produce deep darkness for a year or more, killing off most plant life. Other effects, such as acid rain and enormous tsunamis (tidal waves), are also predicted by the models.

Geologists have located a crater at least 150 km in diameter centered near the village of Chicxulub (pronounced cheek-shoe-lube) in the northern Yucatán region of Mexico (■ Figure 12-11). Although the crater is completely covered by sediments, mineral samples show that it contains shocked quartz typical of impact sites and that it is the right age. The impact of an object 10 to 14 km in diameter formed the crater about 65 million years ago, just when the dinosaurs and many other species died out. Most Earth scientists now believe that this is the scar of the impact that ended the Cretaceous period.

Earthlings watched in awe during six days in the summer of 1994 as 20 or more fragments from the head of comet Shoemaker–Levy 9 slammed into Jupiter and produced impacts

■ **Figure 12-11**

The giant impact scar buried in Earth's crust near the village of Chicxulub in the Yucatán region of Mexico was formed about 65 million years ago by the impact of a large comet or asteroid. This gravity map shows the extent of the crater hidden below limestone deposited long after the impact. (Virgil L. Sharpton, University of Alaska, Fairbanks)

equaling millions of megatons of TNT. Each impact created a fireball of hot gases and left behind dark smudges that remained visible for months afterward (■ Figure 12-12). Such impacts on Jupiter probably occur once every century or two. Major impacts on Earth occur less often because Earth is smaller, but they are inevitable.

We are sitting ducks. All of human civilization is spread out over Earth's surface and exposed to anything that falls out of the sky. Meteorites, asteroids, and comets bombard Earth, producing impacts that vary from dust settling on rooftops to blasts capable of destroying all life. In this case, the scientific evidence is conclusive and highly unwelcome. Statistically you are quite safe. The chance that a major impact will occur during your lifetime is so small that it is hard to estimate. But the consequences of such an impact are so severe that humanity should be preparing. One way to prepare is to find those **NEOs** (**Near Earth Objects**) that could hit this planet, map their orbits in detail, and identify any that are dangerous.

ThomsonNOW™ Sign in at www.thomsonedu.com and go to ThomsonNOW to see the Astronomy Exercise "Cratering."

The Jovian Problem

The solar nebula theory has been very successful in explaining the formation of the solar system. But there are some problems, and the Jovian planets are the troublemakers.

The gas and dust disks around newborn stars don't last long. In Chapter 7 you saw images of dusty gas disks around the young stars in the Orion nebula (page 143). Those disks are being evaporated by the intense ultraviolet radiation from hot stars within

the nebula. Astronomers estimate that most stars form in clusters containing some massive stars, so this evaporation process must happen to most disks. Even if a disk did not evaporate quickly, model calculations predict that the gravitational influence of the crowded stars in a cluster should quickly strip away the outer parts of the disk. This is a troublesome observation because it seems to mean that planet-forming disks around young stars are unlikely to last longer than a few million years, and many must evaporate within 100,000 years or so. That's not long enough to grow a Jovian planet by the processes in the solar nebula theory. Yet, as you will learn in a later section of this chapter, Jovian planets are common in the universe.

A modification to the solar nebula theory has come from mathematical models of the solar nebula. The results show that the rotating gas and dust of the solar nebula could have become unstable and formed outer planets by direct gravitational collapse. That is, massive planets may have been able to form directly from the nebula gas without first accreting a dense core from planetesimals. Jupiters and Saturns can form in these models within a few hundred years. If the Jovian planets formed in this way, they could have formed quickly, long before the solar nebula disappeared.

This new insight into the formation of the outer planets may help explain the formation of Uranus and Neptune. They are so far from the sun that accretion could not have built them rapidly. It is hard to understand how they could have reached their present mass in a region where the material should have been sparse and orbital speeds are slow. Theoretical calculations show that Uranus and Neptune might instead have formed closer to the sun, in the region of Jupiter and Saturn, and then moved outward by gravitational interactions with the other planets or with planetesimals in the Kuiper belt. In any case, the formation of Uranus and Neptune is part of the Jovian problem.

The traditional solar nebula theory proposes that the planets formed by accreting a solid core and then, if they became massive enough, by gravitational collapse of nebula gas. The new theories suggest that some of the outer planets could have skipped the core accretion phase.

■ Figure 12-12

In 1994, fragments of Comet Shoemaker–Levy 9 struck Jupiter. Although these were the first comet impacts ever seen by Earth's inhabitants, such events are probably common occurrences in planetary systems. (Composite and visual images: NASA; IR sequence: Mike Skrutskie; IR image: University of Hawaii)

Explaining the Characteristics of the Solar System

Now you have learned enough to put all the pieces of the puzzle together and explain the distinguishing characteristics of the solar system in Table 12-1.

The disk shape of the solar system is inherited from the solar nebula. The sun and planets revolve and rotate in the same direction because they formed from the same rotating gas cloud. The orbits of the planets lie in the same plane because the rotating solar nebula collapsed into a disk, and the planets formed in that disk.

The solar nebula hypothesis calls on continuing evolutionary processes to gradually build the planets. Scientists call this type of explanation an **evolutionary theory.** In contrast, a **catastrophic theory** invokes special, sudden, even violent, events. For example, as you learned earlier, Uranus rotates on its side and Venus rotates backward. Both these pecularities could have been caused by off-center impacts of massive planetesimals on each planet as they were forming—an explanation of the catastrophic type. On the other hand, computer models suggest that the sun can produce tides in the thick atmosphere of Venus that eventually could have resulted in the reversal of that planet's rotation—an explanation of the evolutionary type.

The second item in Table 12-1, the division of the planets into terrestrial and Jovian worlds, can be understood through the condensation sequence. The terrestrial planets formed in the inner part of the solar nebula, where the temperature was high and only compounds such as metals and silicates could condense to form solid particles. That produced the small, dense terrestrial planets. In contrast, the Jovian planets formed in the outer solar nebula, where the lower temperature allowed the gas to form large amounts of ices, perhaps three times more ices than silicates. That allowed the Jovian planets to grow rapidly and became massive, low-density worlds. Also, Jupiter and Saturn are so massive they eventually were able to grow even larger by drawing in gas directly from the solar nebula. The terrestrial planets could not do this because they never became massive enough.

The **heat of formation** (the energy released by infalling matter) was tremendous for these massive planets. Jupiter must have grown hot enough to glow with a luminosity of about 1 percent that of the present sun. However, because it never got hot enough to start nuclear fusion as a star would, it never generated its own energy. Jupiter is still hot inside. In fact, both Jupiter and Saturn radiate more heat than they absorb from the sun, so they are evidently still cooling.

A glance at the solar system suggests that you should expect to find a planet between Mars and Jupiter at the present location of the asteroid belt. Mathematical models show Jupiter grew into such a massive planet that it was able to gravitationally disturb the motion of nearby planetesimals. The bodies that should have formed a planet between Mars and Jupiter were broken up, thrown into the sun, or ejected from the solar system due to the gravitational influence of massive Jupiter. The asteroids seen today are the last remains of those rocky planetesimals.

The comets, in contrast, are evidently the last of the icy planetesimals. Some may have formed in the outer solar nebula beyond Neptune and Pluto, but many probably formed among the Jovian planets, where ices could condense easily. Mathematical models show that the massive Jovian planets could have ejected some of these icy planetesimals into the far outer solar system, to a region called the **Oort cloud,** from which comets come with very long periods and orbits highly inclined to the plane of the solar system.

The icy Kuiper belt objects, including Pluto, appear to be ancient planetesimals that formed in the outer solar system but were never incorporated into a planet. They orbit slowly far from the light and warmth of the sun and, except for occasional collisions, have not changed much since the solar system was young.

In Table 12-1, point 3 notes that all four Jovian worlds have ring systems, and you can understand this by considering the large mass of these worlds and their remote location in the solar system. A large mass makes it easier for a planet to hold onto orbiting ring particles; and, being farther from the sun, the ring particles are not as easily swept away by the pressure of sunlight and the solar wind. It is hardly surprising, then, that the terrestrial planets, low-mass worlds located near the sun, have no planetary rings.

The last entry in Table 12-1 is the common ages of solar system bodies, and the solar nebula hypothesis has no difficulty explaining that characteristic. If the hypothesis is correct, then the planets formed at the same time as the sun and should have roughly the same age.

Why are there two kinds of planets in our solar system?
This is an opportunity for you to build an argument that closely analyzes the standard solar nebula theory. Planets begin forming from solid bits of matter, not from gas. Consequently, the kind of planet that forms at a given distance from the sun depends on the kind of compounds that can condense out of the gas to form solid particles. In the inner parts of the solar nebula, the temperature was so high that most of the gas could not condense to form solids. Only metals and silicates could form solid grains, and the innermost planets grew from this dense material. Much of the mass of the solar nebula consisted of hydrogen, helium, water vapor, and other gases, and these gases were present in the inner solar nebula but couldn't form solid grains. The small terrestrial planets couldn't grow from these gases, so the terrestrial planets are small and dense.

In the outer solar nebula, the composition of the gas was the same, but it was cold enough to condense ice grains from water vapor. Because hydrogen and oxygen are so abundant, there was lots of ice available. The outer planets grew from solid bits of metal and silicate combined with large amounts of ice. Jupiter and Saturn grew so rapidly that they became massive enough to capture gas directly, and they became the most hydrogen- and helium-rich Jovian worlds.

The condensation sequence combined with the solar nebula hypothesis gives you a way to understand the difference between the terrestrial and Jovian planets. Now expand your argument: **Why do some astronomers argue that the formation of the Jovian planets is a problem that needs further explanation?**

■ ■ ■

12-5 Planets Orbiting Other Stars

ARE THERE OTHER PLANETS? Yes, that is now certain. Are there planets like Earth? The evidence so far makes that seem likely.

Planet-Forming Disks around Other Stars

You have already learned about dense disks of gas and dust around stars that are forming. For example, at least 50 percent of the stars in the Orion nebula are encircled by dense disks of gas and dust with more than enough mass to make planetary systems like ours. The Orion star-forming region is only a few million years old, so it does not seem likely that planets have finished forming there yet. Nevertheless, the important point for astronomers is that disks of gas and dust that could become planetary systems are a common feature of star formation.

Debris Disks

In addition to these planet-forming disks around young stars, infrared astronomers have found cold, low-density dust disks around older stars such as Vega and Beta Pictoris. Although much younger than the sun, these stars are on the main sequence and have completed their formation, so they are clearly in a later stage than the newborn stars in Orion. These low-density disks generally have even lower-density inner zones where planets may have formed (■ Figure 12-13). Such tenuous dust disks are sometimes called **debris disks** because they are understood to be dusty debris released in collisions among small bodies such as comets, asteroids, and Kuiper belt objects. Our own solar system contains such dust, and astronomers have evidence the sun has an extensive debris disk of cold dust extending far beyond the orbits of the planets.

Few planets orbiting stars with debris disks have been detected so far, but the presence of dust particles with lifetimes shorter than the ages of the stars assures you that remnant planetesimals like asteroids and comets are present as the sources of the dust. If planetesimals are there, then you can expect that there are also planets orbiting those stars. Many of the debris disks have details of structure and shape that are probably caused by the gravity of planets orbiting within or at the edges of the debris (Figure 12-13).

Notice the difference between the two kinds of disks that astronomers have found. The low-density dust disks such as the one around Beta Pictoris are produced by dust from collisions among comets, asteroids, and Kuiper belt objects. Such disks are evidence that planetary systems have already formed. The dense disks of gas and dust such as those seen round the stars in Orion are sites where planets could be forming right now.

Extrasolar Planets

A planet orbiting another star is called an **extrasolar planet.** Such a planet would be quite faint and difficult to detect so close to the glare of its star. But there are ways to find these planets. To see how, all you have to do is imagine walking a dog.

■ Figure 12-13

Dust disks have been detected orbiting a number of stars; but, in the visible part of the spectrum, the dust is at least 1000 times fainter than the stars, which must be hidden behind masks. In the far-infrared, the stars are not as bright as the dust. Warps, rings, clumps, and off-center rings in these disks suggest the gravitational influence of planets. (Beta Pic: NASA, Burrows and Krist; Eps Eri: Joint Astronomy Center Jane Greaves)

You will remember that Earth and its moon orbit around their common center of mass, and two stars in a binary system orbit around their center of mass. When a planet orbits a star, the star moves very slightly as it orbits the center of mass of the planet–star system. Think of someone walking a poorly trained dog on a leash; the dog runs around pulling on the leash, and even if it were an invisible dog, you could plot its path by watching how its owner was jerked back and forth. Astronomers can detect a planet orbiting another star by watching how the star moves as the planet tugs on it.

The first planet orbiting a sunlike star detected this way was discovered in 1995. It orbits the star 51 Pegasi. As the planet circles the star, the star wobbles slightly, and this very small motion of the star is detectable as Doppler shifts in the star's spectrum (■ Figure 12-14a). This is the same technique used to study spectroscopic binary stars (see Chapter 6). From the motion of the star and estimates of the star's mass, astronomers can deduce that the planet has half the mass of Jupiter and orbits only 0.05 AU from the star. Half the mass of Jupiter amounts to 160 Earth masses, so this is a large planet, larger than Saturn. Note also that it orbits very close to its star.

Astronomers were not surprised by the announcement that a planet orbited 51 Pegasi; for years they had assumed that many stars had planets. Nevertheless, they greeted the discovery with typical skepticism (**Window on Science 12-2**). That skepticism led to careful tests of the data and further observations that confirmed the discovery. In fact, over 200 planets have been discovered in this way, including at least three planets orbiting the star Upsilon Andromedae—a true planetary system (Figure 12-14b). Over a dozen such multiple-planet systems have been found.

Another way to search for planets is to look for changes in the brightness of the star when the orbiting planet crosses in front of or behind the star. The decrease in light is very small, but it is detectable, and astronomers have used this technique to detect a few planets as they crossed in front of their stars. This is the same technique used to study eclipsing binary stars (see Chapter 6). From the amount of light lost, astronomers can tell that all of these

■ **Figure 12-14**

Just as someone walking a lively dog is tugged around, the star 51 Pegasi is tugged around by the planet that orbits it every 4.2 days. The wobble is detectable in precision observations of its Doppler shift. Someone walking three dogs is pulled about in a more complicated pattern, and you can see something similar in the Doppler shifts of Upsilon Andromedae, which is orbited by at least three planets.

Window on Science 12-2

Scientists: Courteous Skeptics

"Scientists are just a bunch of skeptics who don't believe in anything." That is a common misconception among people who don't understand the methods and goals of science. Yes, scientists are skeptical about new ideas and discoveries, but they do hold strong beliefs about how nature works. Scientists are skeptical not because they want to disprove everything but because they are searching for the truth and want to be sure that a new description of nature is true before it is accepted.

Skeptical scientists question every aspect of a new discovery. They may wonder if another scientist's instruments were properly calibrated or if the scientist's mathematical models are correct. Other scientists will want

to repeat the work themselves using their own instruments to see if they can obtain the same results. Observations are tested, discoveries are confirmed, and only an idea that survives many of these tests begins to be accepted as a scientific truth.

Scientists are prepared for this kind of treatment at the hands of other scientists. In fact, they expect it. Among scientists it is not bad manners to say, "Really, how do you know that?" or "Why do you think that?" or "Show me the evidence!" And it is not just new or surprising claims that are subject to such scrutiny. Astronomers had long expected to discover planets orbiting other stars. But when a planet was finally discovered circling 51 Pegasi, astronomers

were skeptical—not because they thought the observations were flawed but because that is how science works.

The goal of science is to tell stories about nature. Some people use the phrase "telling a story" to describe someone who is telling a fib. But the stories that scientists tell are exactly the opposite; perhaps you could call them antifibs, because they are as true as scientists can make them. Skepticism eliminates stories with logical errors, flawed observations, or misunderstood evidence and eventually leaves only the stories that best describe nature.

Skepticism is not a refusal to hold beliefs. Rather, it is a way for scientists to find those natural principles worthy of belief.

---

planets are roughly the size of Jupiter. The Spitzer Infrared Space Telescope has detected two planets when they passed behind their stars. These planets are hot and emit significant infrared radiation. As they orbit their parent stars, astronomers detect variation in the amount of infrared from the system that further confirms the existence of extrasolar planets and indicates the temperatures and sizes of the planets. That information reveals that these planets have Jupiter-like diameters as well as masses, so astronomers conclude they have Jovian densities and compositions.

The planets discovered so far tend to be massive and have short periods because lower-mass planets or longer-period planets are harder to detect. Low-mass planets don't tug on their stars very much, and present-day spectrographs can't detect the very small velocity changes that these gentle tugs produce. Planets with longer periods are harder to detect because Earth's astronomers have not been making high-precision observations for a long enough time yet. Jupiter takes 11 years to circle the sun once, so it will take years for astronomers to see the longer-period wobbles produced by planets lying farther from their stars. You should not be surprised that the first extrasolar planets discovered are massive and have short periods. This is called a selection effect (see Window on Science 10-1).

The new planets may seem odd for another reason. In our own solar system, the large planets formed farther from the sun where the solar nebula was colder and ices could condense. How could big planets form so near their stars? Model calculations predict that planets that form in an especially dense disk of matter could spiral inward as they sweep up gas or planetesimals.

That means it is possible for a few planets to become the massive, short-period planets that are detected most easily.

The discovery of extrasolar planets gives astronomers added confidence in the solar nebula theory. The theory predicts that planets are common, and astronomers are finding them orbiting many stars.

Building Scientific Arguments

Why are debris disks evidence that planets have already formed?

Sometimes a good scientific argument combines evidence, theory, and an astronomer's past experience, a kind of scientific common sense. Certainly the cold transparent debris disks seen around stars like Vega are not places where planets are forming. They are not hot enough or dense enough to be young disks. Rather, the dust disks must be older, and the dust is being produced by collisions among comets, or asteroids, or the equivalent of objects in our Kuiper belt. Dust would be blown away quickly, so these collisions must be a continuing process. Astronomers know from studying our solar system that comets, asteroids, and Kuiper belt objects are connected to the formation of planets The dusty debris disks seem to be evidence that planets have already formed in such systems.

Now build a new argument. **What direct evidence can you cite that planets orbit other stars?**

■　　■　　■

Summary

12-1 | The Great Chain of Origins

What the is the theory explaining the origin of the solar system?

■ Modern astronomy reveals that the matter in our solar system was formed in the big bang, and the atoms heavier than helium were cooked up in successive generations of stars. The sun and planets evidently formed from a cloud of gas in the interstellar medium.

■ Hot disks of gas and dust have been detected around stars in early stages of formation and are understood to be the kind of disk in which planets could form. Observations show that such disks are common.

■ The solar nebula hypothesis proposes that the planets formed in a disk of gas and dust around the protostar that became the sun.

12-2 | A Survey of the Planets

What are the observed properties of the solar system that must be explained by the theory of its origin?

■ The solar system is disk shaped, with all the orbits lying close to one plane. The orbital revolutions of the planets, as well as most of their axial rotations and most of the orbits of their moons, share a common direction of motion.

■ The planets are divided into two types. The inner four planets are terrestrial—small, rocky, dense Earthlike worlds. The outer four are Jovian planets that are large and low density.

■ All four of the Jovian worlds have ring systems and large families of moons. The terrestrial planets have no rings and few moons.

12-3 | Space Debris: Planet Building Blocks

What are asteroids and comets and what are their connections to meteors and meteorites?

■ Asteroids are irregular in shape and heavily cratered from collisions. Their surfaces are covered by gray, pulverized rock, and some asteroids have such low densities they seem to be fragmented rubble piles.

■ Most asteroids lie in a belt between Mars and Jupiter, although some follow orbits that cross into the inner solar system.

■ Comets are icy bodies that fall into the inner solar system along long elliptical orbits. As the comet approaches the sun, ices vaporize and release dust, and the comet develops a tail or tails that point approximately away from the sun.

■ A comet gas tail is ionized gas carried away by the solar wind. A comet dust tail is solid debris released from the nucleus and blown outward by the pressure of sunlight.

■ Spacecraft flying past comets have revealed that the nucleus of a comet has a very dark, rocky crust and that jets of vapor and dust issue from active regions on the sunlit side.

■ Meteoroids that fall into Earth's atmosphere are vaporized by friction and become visible as meteors. Larger and stronger meteoroids may survive to reach the ground, where they are called meteorites.

■ Stony meteorites that are rich in volatiles and carbon are called carbonaceous chondrites. They are among the least-modified meteorites.

■ Evidence including the orbits of meteorites seen to fall suggests that most meteorites are fragments of asteroids. On the other hand, the vast majority of meteors, including those in meteor showers, appear to be low-density, fragile bits of debris from comets.

12-4 | The Story of Planet Formation

How do planets form?

■ The age of a rocky body can be found by radioactive dating. The oldest rocks from Earth, the moon, and Mars have ages approaching 4.6 billion years. The oldest objects in our solar system are some meteorites that have ages of 4.6 billion years. This is taken to be the age of the solar system.

■ Condensation in the solar nebula converted some of the gas into solid bits of matter, which accreted to form billions of planetesimals.

■ Planets begin growing by accreting solid material. But once a planet approaches about 15 Earth masses, it can begin growing by gravitational collapse as it pulls in gas from the solar nebula.

■ According to the condensation sequence, the inner part of the solar nebula was so hot that only metals and rocky minerals could form solid grains. The dense terrestrial planets grew from those solid particles and did not include many ices or low-density gases.

■ The outer solar nebula was cold enough for large amounts of ices as well as metals and rocky minerals to form solid particles. The Jovian planets grew rapidly and incorporated large amounts of low-density ices and gases.

■ Evidence that the condensation sequence was important in the solar nebula can be found in the densities of the terrestrial planets compared to the Jovian planets.

■ The terrestrial planets may have formed slowly from the accretion of planetesimals of similar composition and then differentiated later when radioactive decay heated the planet's interiors.

■ Earth's first atmosphere was at least partly baked out of Earth's interior ("outgassed") although some of its atmosphere and much of its water may have been carried by impacting comets and asteroids containing volatile substances.

■ Disks of gas and dust may not last long enough to form Jovian planets by core accretion and then by gravitational collapse. Some models suggest Jovian planets can form rapidly by gravitational collapse without core accretion.

■ The asteroids formed as rocky planetesimals between Mars and Jupiter, but Jupiter prevented them from forming a planet. Collisions have broken all but the largest of the asteroids.

■ Comets are believed to have formed as icy planetesimals in the outer solar system, and some were ejected to form the Oort cloud; comets falling in from the Oort cloud become long-period comets.

■ Other icy bodies formed in the outer solar system and now make up the Kuiper belt beyond Neptune that includes Pluto. Objects from the Kuiper belt that fall into the inner solar system can become short-period comets.

■ All of the old surfaces in the solar system were heavily cratered by an early bombardment of the debris that filled the solar system when it was young.

■ An impact 65 million years ago near what is now Chicxulub, Mexico, appears to have triggered the extinction of 75 percent of the species then on Earth, including the dinosaurs.

12-5 | Planets Orbiting Other Stars

What do astronomers know about other solar systems??

■ Cold dust disks, also known as debris disks, appear to be produced by dust released by collisions among objects in other planetary systems resembling our system's comets, asteroids, and Kuiper belt objects. Such disks may be home to planets that have already formed.

■ Planets orbiting other stars have been detected by the way they tug their stars about, creating small Doppler shifts in the stars' spectra. Planets have also been detected as they cross in front of their star and dim the star's light. A few planets have been detected from the variation of infrared radiation astronomers detect as the planets orbit their stars and alternately present their hot and cold sides toward Earth.

■ Nearly all extrasolar planets found so far are massive, Jovian worlds. Lower-mass terrestrial planets are harder to detect but are presumably common.

New Terms

solar nebula hypothesis (p. 243)	condensation sequence (p. 252)
asteroid (p. 244)	planetesimal (p. 253)
comet (p. 244)	condensation (p. 253)
volatile (p. 244)	accretion (p. 253)
terrestrial planet (p. 246)	protoplanet (p. 253)
Jovian planet (p. 246)	gravitational collapse (p. 254)
Galilean satellite (p. 247)	differentiation (p. 254)
Kuiper belt (p. 248)	outgassing (p. 254)
meteor (p. 248)	heavy bombardment (p. 255)
meteoroid (p. 248)	NEO (Near Earth Object) (p. 256)
meteorite (p. 248)	evolutionary theory (p. 257)
carbonaceous chondrite (p. 248)	catastrophic theory (p. 257)
meteor shower (p. 249)	heat of formation (p. 258)
half-life (p. 250)	Oort cloud (p. 258)
uncompressed density (p. 252)	debris disk (p. 259)
ice line (p. 252)	extrasolar planet (p. 259)

Review Questions

ThomsonNOW™ Assess your understanding of this chapter's topics with additional quizzing and animations at
www.thomsonedu.com

1. What produced the helium now present in the sun's atmosphere? In Jupiter's atmosphere? In the sun's core?

2. What produced the iron in Earth's core and the heavier elements like gold and silver in Earth's crust?

3. What evidence can you cite that disks of gas and dust are common around young stars?

4. According to the solar nebula theory, why is the sun's equator nearly in the plane of Earth's orbit?

5. Why does the solar nebula theory predict that planetary systems are common?

6. What evidence indicates that the asteroids are fragmented?

7. Discuss the surface of the asteroid Mathilde. What do you see that tells you something about the history of the asteroids?

Visual NASA

8. What evidence indicates that some asteroids have had active surfaces?

9. What evidence indicates that comet nuclei are rich in ices and dust?

10. What do you see in this image of the nucleus of Comet Borrelly that tells you how comets produce their "heads" (comas) and tails?

Visual NASA

11. Why do astronomers refer to carbonaceous chondrite meteorites as "unmodified"?

12. How do observations of meteor showers reveal one of the sources of meteoroids?

13. How can most meteors be cometary if all meteorites are asteroidal?

14. Why do astronomers think the solar system formed about 4.6 billion years ago?

15. What is the difference between condensation and accretion?

16. How does the solar nebula theory explain the dramatic density difference between the terrestrial and Jovian planets?

17. If you visited some other planetary system in the act of building planets, would you expect to see the condensation sequence at work, or is there reason to believe it was unique to our solar system?

18. Why would you expect to find that planets are differentiated?

19. How does the solar nebula theory help you understand the location of asteroids?

20. How did the bodies in the Kuiper belt and the Oort cloud form?

21. Why do astronomers conclude that the surface of Mercury, shown at right, is old? Why is almost every solid surface in our solar system scarred by craters? When did the majority of these craters form?

NASA

22. What is the difference between the hot disks and cold (debris) disks seen around stars?

23. What evidence can you cite that planets orbit other stars?

Discussion Questions

1. Would you expect that other solar systems have asteroid belts? Should all solar systems show evidence of an age of heavy bombardment?

2. If the solar nebula hypothesis is correct, then there are probably more planets in the universe than stars. Do you agree? Why or why not?

3. If cometary nuclei were heated by internal radioactive decay rather than by solar heat, how would comets differ from what is observed?

4. Do you think the government should spend money to find near-Earth asteroids? How serious is the risk?

5. The human race has intelligence and consequently has both the right and the responsibility to wonder about its origins. Do you agree?

13 | Comparative Planetology of the Terrestrial Planets

That's one small step

for a man . . . one giant leap

for mankind.

NEIL ARMSTRONG
ON THE MOON

Beautiful, beautiful.

Magnificent desolation . . .

EDWIN E. ALDRIN, JR.
ON THE MOON

IF YOU HAD BEEN THE FIRST PERSON to step onto the surface of the moon, what would you have said? Neil Armstrong responded to the historic significance of the first human step on the surface of another world. Buzz Aldrin was second, and he responded to the moon itself. It *is* desolate, and it *is* magnificent. But it is not unusual. Many places in the universe probably look like Earth's moon, and people may someday walk on such worlds and compare them with the moon.   The comparison of one planet with another is called **comparative planetology,** and it is one of the best ways to analyze the worlds in our solar system. You will learn much more by comparing planets than you could by studying them individually.

Guidepost

Looking Back

In the preceding chapter, you learned how our solar system formed as a by-product of the formation of the sun. You also saw how distance from the sun determines the general character of each planet. Now you are ready to visit the planets and get to know them as individuals.

This Chapter

As you study the individual planets, you should seek to understand principles and avoid the pitfall of simply learning facts. One way to focus on principles is to compare the planets with each other and search for similarities and contrasts. Like people, the terrestrial (Earthlike) planets are more alike than they are different, although it is the differences that are most memorable.

As you explore, you will be searching for answers to five essential questions:

How can comparison help you understand the terrestrial planets?

What are the main features of Earth when you view it as a planet?

How does size determine the geological activity and evolution of a planet?

How do distance from the sun and size of a planet affect the properties of its atmosphere?

What is the evidence that surface conditions on Mars were originally more Earthlike than they are at present?

Looking Ahead

Once you are familiar with the family of the terrestrial planets, you will be ready to meet a more peculiar group of characters. The next chapter will introduce you to the worlds of the outer solar system.

13-1 A Travel Guide to the Terrestrial Planets

IN THIS CHAPTER YOU WILL VISIT five Earthlike worlds, and this preliminary section will be your guide to important features and comparisons.

Five Worlds

You are about to visit Earth, Earth's moon, Mercury, Venus, and Mars. It may surprise you that the moon is on your itinerary. It is, after all, just a natural satellite orbiting Earth and isn't one of the planets. But the moon is a fascinating world of its own, a planetlike object two-thirds the size of Mercury, and it makes a striking comparison with the other worlds on your list.

■ Figure 13-1 compares the five worlds you are about to study. The first feature you should notice is diameter. The moon is small, and Mercury is not much bigger. Earth and Venus are large and similar in size to each other, but Mars is a medium-sized world. You will discover that size is a critical factor in determining a world's personality: Small worlds tend to be internally cold and geologically dead, but larger worlds can be geologically active. (Note that the Greek root term *geology*, which originally referred only to Earth, can be used to refer to other worlds as well.)

Core, Mantle, and Crust

Notice that these terrestrial worlds are made up of rock and metal. They are all differentiated, with rocky, low-density crusts, high-density metal cores, and **mantle**s composed of dense rock between the cores and crusts.

■ Figure 13-1

Planets in comparison: Earth and Venus are similar in size, but their atmospheres and surfaces are very different. The moon and Mercury are much smaller, and Mars is intermediate in size. (Moon: UCO/Lick Observatory; All planets: NASA)

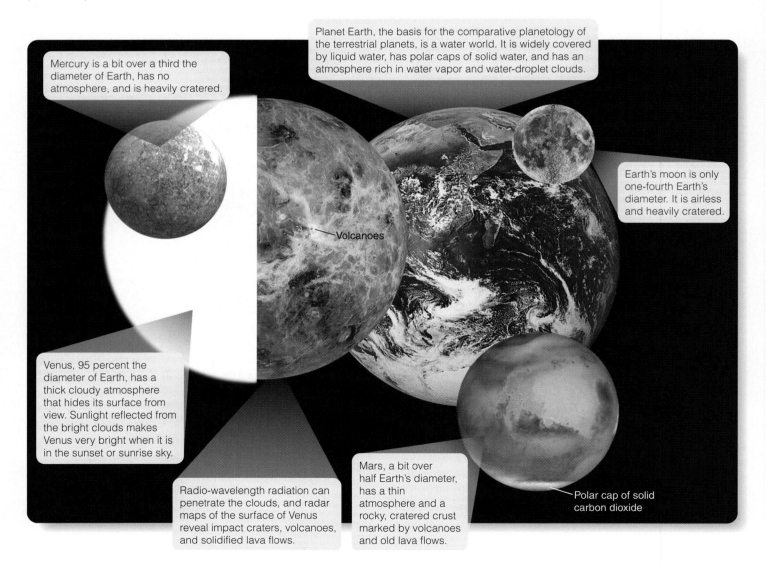

Mercury is a bit over a third the diameter of Earth, has no atmosphere, and is heavily cratered.

Planet Earth, the basis for the comparative planetology of the terrestrial planets, is a water world. It is widely covered by liquid water, has polar caps of solid water, and has an atmosphere rich in water vapor and water-droplet clouds.

Earth's moon is only one-fourth Earth's diameter. It is airless and heavily cratered.

Volcanoes

Venus, 95 percent the diameter of Earth, has a thick cloudy atmosphere that hides its surface from view. Sunlight reflected from the bright clouds makes Venus very bright when it is in the sunset or sunrise sky.

Radio-wavelength radiation can penetrate the clouds, and radar maps of the surface of Venus reveal impact craters, volcanoes, and solidified lava flows.

Mars, a bit over half Earth's diameter, has a thin atmosphere and a rocky, cratered crust marked by volcanoes and old lava flows.

Polar cap of solid carbon dioxide

As you learned in Chapter 12, when the planets formed, their surfaces were subjected to heavy bombardment by leftover planetesimals and fragments; the cratering rate then was as much as 10,000 times what it is at present. You will see lots of craters on the terrestrial worlds, especially on Mercury and the moon. Notice that cratered surfaces are old. For example, if a lava flow covered up some cratered landscape to make a new surface after the end of the heavy bombardment, few craters could be formed afterward on that surface because most of the debris in the solar system was gone. When you see a smooth plain on a planet, you can guess that surface is younger than the cratered areas.

One important way you can study a planet is by following the energy. The heat in the interior of a planet may be left over from the formation of the planet, or it may be heat generated by radioactive decay, but it must flow outward toward the cooler surface where it is radiated into space. In flowing outward, the heat can cause convection currents in the mantle, magnetic fields, plate motions, quakes, faults, volcanism, mountain building, and more. Heat flowing upward through the cooler crust makes a large world like Earth geologically active. In contrast, the moon and Mercury, both small worlds, cooled fast so they have little heat flowing outward now and are relatively inactive.

Atmospheres

When you look at airless Mercury and the moon in Figure 13-1, you can see their craters and plains and mountains, but the surface of Venus is completely hidden by a cloudy atmosphere much thicker than Earth's. Mars, the medium-sized terrestrial planet, has a relatively thin atmosphere.

You might ponder two questions. First, why do some worlds have atmospheres while others do not? You will discover that both size and temperature are important. The second question is more complex. Where did these atmospheres come from? To answer that question, you will have to study the geological history of these worlds.

13-2 Earth: The Active Planet

EARTH IS THE BASIS for your comparative study of the terrestrial planets, so you should pretend to visit it as if you didn't live here. It is an active planet with a molten interior and heat flowing outward that powers volcanism, earthquakes, and an active crust. Almost 75 percent of Earth's surface is covered by liquid water, and the atmosphere contains a significant amount of oxygen.

Earth's Interior

From what you know of the formation of Earth (see Chapter 12), you would expect it to have differentiated, but in science,

evidence rules. What does the evidence reveal about Earth's interior?

Earth's mass divided by its volume tells you its average density, about 5.5 g/cm³ (**Celestial Profile 2**), but the density of Earth's rocky crust is only about half that. Clearly, a large part of Earth's interior must be made of material denser than rock.

Each time an earthquake occurs, seismic waves travel through the interior and register on seismographs all over the world. Analysis of these waves shows that Earth's interior is divided into a metallic core, a dense rocky mantle, and a thin, low-density crust.

The core has a density of 14 g/cm³, denser than lead; models indicate it is composed of iron and nickel at a temperature of roughly 6000 K. The core of Earth is as hot as the gaseous surface of the sun, but high pressure keeps the metal a solid near the middle of the core and a liquid in its outer parts. Two kinds of seismic waves show that the outer core is liquid. The ***P* waves** travel like sound waves, and they can penetrate a liquid; but ***S* waves** travel as a side-to-side vibration that can't travel through a liquid. Scientists can deduce the size of the liquid core by observing where *S* waves get through and where they don't (■ Figure 13-2).

Earth's magnetism gives you further clues about the core. The presence of a magnetic field is evidence that part of Earth's core must be a liquid metal. Convection currents stir the molten liquid, and because it is a very good conductor of electricity and is rotating as Earth rotates it generates a magnetic field through

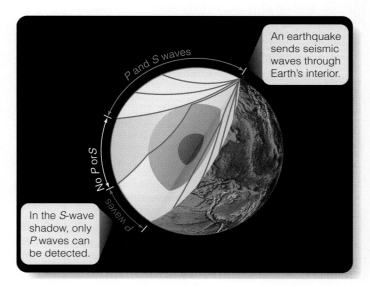

An earthquake sends seismic waves through Earth's interior.

In the *S*-wave shadow, only *P* waves can be detected.

■ **Active Figure 13-2**

P and *S* waves give you clues to Earth's interior. No direct *S* waves reach the far side, which shows that Earth's core is liquid. The size of the *S* wave "shadow" tells you the size of the outer core.

ThomsonNOW™ Sign in at www.thomsonedu.com and go to ThomsonNOW to see the Active Figure "Seismic Waves."

the dynamo effect—a different version of the process that creates the sun's magnetic field (see Chapter 5).

Earth's mantle is a deep layer of dense rock that lies between the molten core and the solid crust. Models indicate the mantle material has the properties of a solid but is capable of flowing slowly, like asphalt used in paving roads, which shatters if struck with a sledgehammer, but bends under the weight of a truck. Just below Earth's crust, where the pressure is less than at greater depths, the mantle flows most easily.

The Earth's rocky crust is made up of low-density rocks. The crust is thickest under the continents, up to 60 km thick, and thinnest under the oceans, where it is only about 10 km thick.

ThomsonNOW™ Sign in at www.thomsonedu.com and go to Thomson-NOW to see Astronomy Exercise "Convection and Magnetic Fields."

Earth's Active Crust

Earth's crust is composed of lower-density rock that floats on the mantle. The image of rock floating may seem odd, but recall that the rock of the mantle is denser than crust rocks. Also, just below the crust, the mantle rock tends to be more fluid, so sections of low-density crust do indeed float on the mantle like great lily pads floating on a pond.

The motion of the crust and the erosive action of water make Earth's crust highly active and changeable. Look at **The Active Earth** on pages 270–271 and notice three important points:

1 The motion of crust plates produces much of the geological activity on Earth. Earthquakes, volcanism, and mountain building are linked to motions in the crust and the location of plate boundaries. While you are thinking about volcanoes, you can correct a common misconception. The molten rock that emerges from volcanoes comes from pockets of melted rock in the upper mantle and lower crust and not from the molten core.

2 The continents on Earth's surface have moved and changed over periods of hundreds of millions of years. A hundred million years is only 0.1 billion years, $\frac{1}{45}$ of the age of Earth, so sections of Earth's crust are in geologically rapid motion.

3 Notice that most of the geological features you know—mountain ranges, the Grand Canyon, and even the outline of the continents—are recent products of Earth's active surface.

Earth's surface is constantly renewed. The oldest Earth material known is small crystals called zircons from western Australia, which are 4.3 billion years old. Most of the crust is much younger than that: The mountains and valleys you see around you are probably no more than a few tens or hundreds of millions of years old.

ThomsonNOW™ Sign in at www.thomsonedu.com and go to ThomsonNOW to see Astronomy Exercise "Convection and Plate Tectonics."

Earth's Atmosphere

When you think about Earth's atmosphere, you should consider three questions: How did it form? How has it evolved? How are humans changing it? Answering these questions will help you understand other planets as well as our own.

Our planet's first atmosphere, its **primary atmosphere,** was once thought to contain gases from the solar nebula, such as hydrogen and methane. Modern studies, however, indicate that the planets formed hot, so gases such as carbon dioxide, nitrogen, and water vapor would have been outgassed from (been cooked out of) the rock and metal (see Chapter 12). In addition, the final stages of planet building may have seen Earth and the other planets accreting planetesimals rich in volatile materials such as water, ammonia, and carbon dioxide. Thus the primary atmosphere must have been rich in carbon dioxide, nitrogen, and water vapor. The atmosphere you breathe today is a **secondary atmosphere** produced later in Earth's history.

Soon after Earth formed, it began to cool; once it cooled enough, oceans began to form, and carbon dioxide began to dissolve in the water. Carbon dioxide is highly soluble in water—which explains the easy manufacture of carbonated beverages. As the oceans removed carbon dioxide from the atmosphere, it reacted with dissolved compounds in the ocean water to form silicon dioxide, limestone, and other mineral sediments. Thus, the oceans transferred the carbon dioxide from the atmosphere to the seafloor and left air richer in other gases, especially nitrogen.

That removal of carbon dioxide is critical to Earth because an atmosphere rich in carbon dioxide can trap heat by a process called the **greenhouse effect.** When visible-wavelength sunlight shines through the glass roof of a greenhouse, it heats the interior. Infrared radiation from the warm interior can't get out through the glass, heat is trapped in the greenhouse, and the temperature climbs until the glass itself grows warm enough to radiate heat away as fast as sunlight enters (■ Figure 13-3a). (Of course, a real greenhouse also retains its heat because the walls prevent the warm air from mixing with the cooler air outside.) This is also called the "parked car effect," for obvious reasons.

Like the glass roof of a greenhouse, a planet's atmosphere can allow sunlight to enter and warm the surface. Carbon dioxide and other greenhouse gases such as water vapor and methane are opaque to infrared radiation, so an atmosphere containing enough of these greenhouse gases can trap heat and raise the temperature of a planet's surface (Figure 13-3b).

It is a common misconception that the greenhouse effect is always bad. Without the greenhouse effect, Earth would be colder by at least 30 K (54 F), with a planetwide average temperature far below freezing. The problem is that human civiliza-

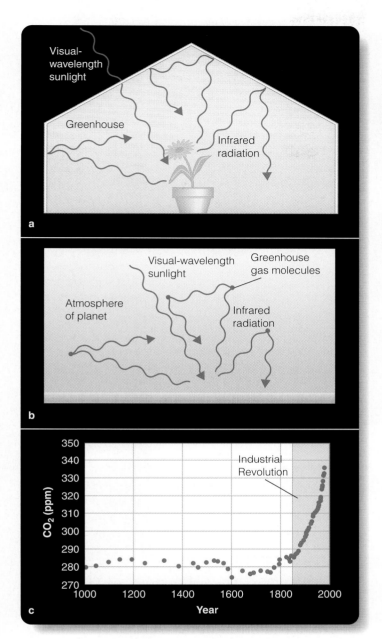

Figure 13-3

The greenhouse effect: (a) Visual-wavelength sunlight can enter a greenhouse and heat its contents, but the longer-wavelength infrared cannot get out. (b) The same process can heat a planet's surface if its atmosphere contains greenhouse gases, such as CO_2. (c) The concentration of CO_2 in Earth's atmosphere, as measured in Antarctic ice cores, remained roughly constant until the beginning of the Industrial Revolution and then began steadily rising. Note that the bottom of the vertical scale is not at zero. (Graph adapted from a figure by Etheridge, Steele, Langenfelds, Francey, Barnola, and Morgan.)

tion is adding greenhouse gases to those that were already in the atmosphere. For 4 billion years, Earth's oceans and plant life have been absorbing carbon dioxide and burying it in the form of carbonates such as limestone and in carbon-rich deposits of coal, oil, and natural gas. In the last century or so, human civilization

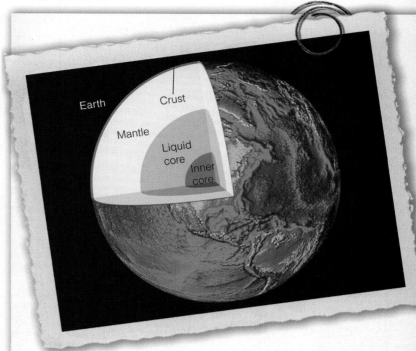

Earth's surface is marked by high continents and low seafloors, but the crust is only 10 to 60 km thick. Below that are a deep mantle and an iron core. (NGDC)

Celestial Profile 2: Earth

Motion:

Average distance from the sun	1.00 AU (1.50×10^8 km)
Eccentricity of orbit	0.017
Inclination of orbit to ecliptic	0° [by definition]
Orbital period	1.000 y (365.25 days)
Period of rotation (with respect to the sun)	24.00 h
Period of rotation (sidereal = with respect to the stars)	23.93 h
Inclination of equator to orbit	23.4°

Characteristics:

Equatorial diameter	1.28×10^4 km
Mass	5.97×10^{24} kg
Average density	5.50 g/cm³ (4.1 g/cm³ uncompressed)
Surface gravity	1.00 Earth gravity
Surface temperature	−90 to 60°C (−130 to 140°F)
Average albedo	0.39
Oblateness	0.0034

The Active Earth

1 Our world is an astonishingly active planet. Not only is it rich in water and therefore subject to rapid erosion, but its crust is divided into moving sections called plates. Where plates spread apart, lava wells up to form new crust; where plates push against each other, they crumple the crust to form mountains. Where one plate slides over another, you see volcanism. This process is called **plate tectonics,** referring to the Greek word for "builder." (An architect is literally an arch builder.)

A **rift valley** forms where continental plates begin to pull apart. The Red Sea has formed where Africa has begun to pull away from the Arabian peninsula.

A typical view of planet Earth

Mountains are common on Earth, but they erode away rapidly because of the abundant water.

William K. Hartmann

Janet Seeds

Midocean rise

Red Sea

Midocean rise

National Geophysical Data Center

1a Evidence of plate tectonics was first found in ocean floors, where plates spread apart and magma rises to form **midocean rises** made of rock called **basalt,** a rock typical of solidified lava. Radioactive dating shows that the basalt is younger near the midocean rise. Also, the ocean floor carries less sediment near the midocean rise. As Earth's magnetic field reverses back and forth, it is recorded in the magnetic fields frozen into the basalt. This produces a magnetic pattern in the basalt that shows that the seafloor is spreading away from the midocean rise.

1b A **subduction zone** is a deep trench where one plate slides under another. Melting releases low-density magma that rises to form volcanoes such as those along the northwest coast of North America, including Mt. St. Helens.

Subduction zone

Midocean rise

Pacific Ocean

Atlantic Ocean

S. America

Africa

Plate motion

Plate motion

Plate motion

Ocean floor

Melting

Mantle

Subduction zone

Hawaiian-Emperor chain

Subduction zone

Hawaii

Midocean rise

Subduction zone

Appalachian Mountains

Midocean rise

Andes mountains

Ural Mountains

Himalaya Mountains

Red Sea

Midocean rise

1c Hot spots caused by rising magma in the mantle can poke through a plate and cause volcanism such as that in Hawaii. As the Pacific plate has moved northwestward, the hot spot has punched through to form a chain of volcanic islands, now mostly worn below sea level. **Folded mountain ranges** can form where plates push against each other. For example, the Ural Mountains lie between Europe and Asia, and the Himalaya Mountains are formed by India pushing north into Asia. The Appalachian Mountains are the remains of a mountain range pushed up when North America collided with Africa.

ThomsonNOW™

Sign in at www.thomsonedu.com and go to ThomsonNOW to see Active Figure "Hot Spot Volcanoes." Notice how the moving plate can produce a chain of volcanic peaks, mostly under water in the case of Earth.

1d The floor of the Pacific Ocean is sliding into subduction zones in many places around its perimeter. This pushes up mountains such as the Andes and triggers earthquakes and active volcanism all around the Pacific in what is called the Ring of Fire. In places such as southern California, the plates slide past each other, causing frequent earthquakes.

Hawaii

Yellow lines on this globe mark plate boundaries. Red dots mark earthquakes since 1980. Earthquakes within the plate, such as those at Hawaii, are related to volcanism over hot spots in the mantle.

Continental Drift

Not long ago, Earth's continents came together to form one continent.

Pangaea

200 million years ago

Pangaea broke into a northern and a southern continent.

Laurasia

Gondwanalan

135 million years ago

Notice India moving north toward Asia.

65 million years ago

The continents are still drifting on the highly plastic upper mantle.

Today

2 The floor of the Atlantic Ocean is not being subducted. It is locked to the continents and is pushing North and South America away from Europe and Africa at about 3 cm per year, a motion called *continental drift*. Radio astronomers can measure this motion by timing pulsars from European and from American radio telescopes. Roughly 200 million years ago, North and South America were joined to Europe and Africa. Evidence of that lies in similar fossils and similar rocks and minerals found in the matching parts of the continents. Notice how North and South America fit against Europe and Africa like a puzzle.

3 Plate tectonics pushes up mountain ranges and causes bulges in the crust, and water erosion wears the rock away. The Colorado River began cutting the Grand Canyon only about 10 million years ago when the Colorado plateau warped upward under the pressure of moving plates. That sounds like a long time ago, but it is only 0.01 billion years. A mile down, at the bottom of the canyon, lie rocks 0.57 billion years old, the roots of an earlier mountain range that stood as high as the Himalayas. It was pushed up, worn away to nothing, and covered with sediment long ago. Many of the geological features we know on Earth have been produced by very recent events.

Formation of Grand Canyon
Age of dinosaurs
Formation of Earth
Breakup of Pangaea
Heavy bombardment
Oldest fossil life
First animals emerge on land
?

4.6   4   3   2   1   Now

Billions of years ago

has begun digging up those fuels, burning them for energy, and releasing the carbon back into the atmosphere as carbon dioxide (Figure 13-3c). This process is steadily increasing the carbon dioxide concentration in the atmosphere and warming Earth's climate in what is called **global warming.**

Global warming is a critical issue not just because it affects agriculture. It changes climate patterns that will warm some areas and cool other areas. It addition, the warming is melting permanently frozen ices in the polar caps, causing sea levels to rise. A rise of just a few feet will flood major land areas. When you visit Venus, you will see a planet dominated by the greenhouse effect.

ThomsonNOW˙ Sign in at www.thomsonedu.com and go to ThomsonNOW to see Astronomy Exercises "Primary Atmospheres" and "The Greenhouse Effect."

Oxygen in Earth's Atmosphere

When Earth was young, its atmosphere had no free oxygen. Oxygen is very reactive and quickly forms oxides in the soil, so plant life is needed to keep a steady supply of oxygen in the atmosphere. Photosynthesis makes energy for the plant by absorbing carbon dioxide and releasing free oxygen. Ocean plants began to manufacture oxygen faster than chemical reactions could remove it about 2 to 2.5 billion years ago. Atmospheric oxygen then increased rapidly.

Because there is oxygen (O_2) in the atmosphere now, there is also a layer of ozone (O_3) at altitudes of 15 to 30 km. Many people hold the common misconception that ozone is bad because they hear it mentioned as part of smog. Breathing ozone is bad for you, but you need the ozone layer in the upper atmosphere to protect you from harmful UV photons. Certain compounds called chlorofluorocarbons (CFCs) used in refrigeration and industry can destroy ozone when they leak into the atmosphere. Since the late 1970s, the ozone concentration has been falling, and the intensity of harmful UV radiation at Earth's surface has been increasing year by year. While this poses an immediate problem for public health on Earth, it is also of interest astronomically. When you visit Mars, you will see the effects of an atmosphere without ozone. Note that ozone depletion is an additional Earth environmental issue that is separate from global warming.

A Short Geological History of Earth

As Earth formed in the inner solar nebula, it passed through three stages that also describe the histories of the other terrestrial planets to varying extents. When you try to tell the story of each planet in our solar system, you pull together all the known facts plus hypotheses and try to make them into a logical history of how the planet got to be the way it is. Of course, your stories will be incomplete because scientists don't yet understand all the factors affecting the history of the planets.

The first stage of planetary evolution is *differentiation,* the separation of each planet's material into layers according to density. Some of that differentiation may have occurred very early as the heat released by infalling matter melted the growing Earth. Some of the differentiation, however, may have occurred later as radioactive decay released more heat and further melted Earth, allowing the denser metals to sink to the core.

The second stage, *cratering and giant basin formation,* could not begin until a solid surface formed. The heavy bombardment of the early solar system cratered Earth just as it did the other terrestrial planets. Some of the largest craters, called basins, were likely big enough to crack through to the upper mantle where rocks are partly molten. That molten rock would have welled up through cracks to fill some of the basins and craters. On planets with liquid water, the basins may also have been the first oceans. As the debris in the solar nebula cleared away, the rate of impacts and crater formation fell to its present low rate.

The third stage, *slow surface evolution,* has continued for at least the past 3.5 billion years. Earth's surface is constantly changing as sections of crust slide over and past each other, push up mountains, and shift continents. In addition, moving air and water erode the surface and wear away geological features. Almost all traces of the earlier stages of differentiation and cratering have been destroyed by the active crust and erosion. Life apparently started on Earth around the beginning of the slow surface evolution stage, and the secondary atmosphere began to replace the primary atmosphere; that may be unique to Earth and may not have happened on the other terrestrial planets.

Terrestrial planets pass through these stages, but differences in masses, temperature, and composition emphasize some stages over others and produce surprisingly different worlds.

Building Scientific Arguments:

What evidence indicates that Earth has a molten metallic core?
A good scientific argument focuses on evidence. In this case, the evidence is indirect because you can never visit Earth's core. Seismic waves from distant earthquakes pass through Earth, but a certain kind of wave, the S waves, do not pass through the core. Because the S waves cannot propagate through a liquid, scientists conclude that Earth's core is partly liquid. Earth's magnetic field gives further evidence of a liquid metallic core. The theory for the generation of magnetic fields, the dynamo effect, requires a convective, conducting liquid (for a planet) or gas (for a star) in the interior. If Earth's core were not partly a liquid metal, it would not be able to generate a magnetic field.

Two different kinds of evidence tell you that our planet has a liquid core. Can you build a new argument? **What evidence can you cite to support the theory of plate tectonics?**

■ ■ ■

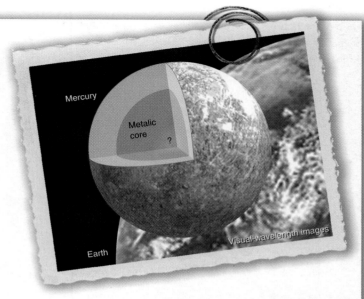

Earth's moon is about one-fourth the diameter of Earth. Its low density indicates that it contains little iron, but the size of its core and the amount of remaining heat are uncertain. (NASA)

Mercury is a bit over one-third the diameter of Earth, but its high density must mean it has a large iron core. The amount of heat it retains is unknown. (NASA)

Celestial Profile 3: The Moon and Mercury

Motion:

	The Moon	Mercury
Average distance from primary	3.84×10^5 km (from Earth)	5.79×10^7 km (0.387 AU) (from Sun)
Eccentricity of orbit	0.055	0.206
Inclination of orbit to ecliptic	5.1°	7.0°
Orbital period	27.3 d	88.0 d (0.241 y)
Period of rotation (sidereal)	27.3 d (synchronous)	58.6 d (direct)
Inclination of equator to orbit	6.7°	0°

Characteristics:

	The Moon	Mercury
Equatorial diameter	3.48×10^3 km (0.272 $D_\oplus$)	4.89×10^3 km (0.382 $D_\oplus$)
Mass	7.35×10^{22} kg (0.0123 $M_\oplus$)	3.31×10^{23} kg (0.0558 $M_\oplus$)
Average density	3.36 g/cm³ (3.3 g/cm³ uncompressed)	5.44 g/cm³ (5.4 g/cm³ uncompressed)
Surface gravity	0.17 Earth gravity	0.38 Earth gravity
Surface temperature	−170 to 130°C (−275 to 265°F)	−170 to 430°C (−275 to 800°F)
Average albedo	0.07	0.1
Oblateness	0	0

13-3 The Moon

YOU WILL NEED A GOOD SPACESUIT to visit the moon. There is no air, and the temperature difference from sunshine to shade is extreme (**Celestial Profile 3**).

Lunar Geology

You could visit two kinds of terrain on the moon. The dark gray areas visible from Earth are the smooth lunar lowlands, which, drawing on the Latin word for *seas,* earlier astronomers named **maria** (plural of **mare,** which is pronounced mah-ray). You could also visit the lighter-colored mountainous lunar highlands. The moon looks quite bright in the night sky seen from Earth. In fact, the **albedo** of the near side of the moon is only 0.07, meaning it reflects only 7 percent of the light that hits it (Celestial Profile 3). The moon looks bright only in contrast to the night sky. It is in reality a dark gray world.

Wherever you go on the moon, you will find craters. The highlands are marked heavily by craters, but the smooth lowlands contain fewer craters. The evidence that craters on the moon were

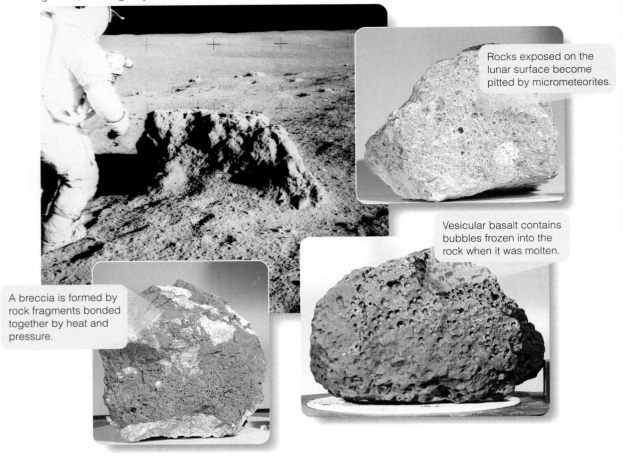

The Apollo astronauts found that all moon rocks are igneous, meaning they solidified from molten rock.

Rocks exposed on the lunar surface become pitted by micrometeorites.

Vesicular basalt contains bubbles frozen into the rock when it was molten.

A breccia is formed by rock fragments bonded together by heat and pressure.

■ Figure 13-4

Rocks returned from the moon show that the moon's surface formed in a molten state, that it was heavily fractured by cratering when it was young, and that it is now affected mainly by micrometeorite erosion. (NASA)

formed by impacts is their distinguishing characteristics such as shape and the material (**ejecta**) they have spread across the moon's surface. Craters range in size from giant basins hundreds of kilometers across to microscopic pits found in moon rocks. Most of the craters on the moon are old; they were formed long ago when the solar system was young. This is clear from comparison of ages of moon rocks collected from the heavily cratered highlands and the somewhat younger, less heavily cratered maria.

Twelve Apollo astronauts visited both the moon's maria and highlands in six expeditions between 1969 and 1972 (■ Figure 13-4). Most of the rocks they found, both in the highlands and the maria, were typical of hardened lava. The maria are actually ancient basalt lava flows. The highlands, in contrast, are composed of low-density rock, for example, **anorthosite,** a light-colored rock that contributes to the highlands' bright contrast with the dark maria and that would be among the first material to solidify and float to the top of molten rock. Many rocks found on the moon are **breccias,** made up of fragments of broken rock cemented together un-

der pressure (Figure 13-4). The breccias show how extensively the moon's surface has been shattered by meteorites, as does the surface layer of powdery dust kicked up by the astronaut's boots.

ThomsonNOW™ Sign in at www.thomsonedu.com and go to ThomsonNOW to see Astronomy Exercise "The Moon's Craters."

The Origin of Earth's Moon

Over the last two centuries, astronomers developed three different hypotheses for the origin of Earth's moon. The *fission hypothesis* proposed that the moon broke from a rapidly spinning proto-Earth. The *condensation hypothesis* suggested that Earth and its moon condensed from the same cloud of matter in the solar nebula. The *capture hypothesis* suggested that the moon formed elsewhere in the solar nebula and was later captured by Earth. Each of these traditional ideas had problems and failed to survive comparison with all the evidence.

In the 1970s, a new hypothesis was proposed that combined some aspects of the three traditional hypotheses. The **large-impact hypothesis** proposes that the moon formed when a planetesimal estimated to have been at least as large as Mars smashed into the proto-Earth. Model calculations indicate that this collision would have ejected a disk of debris into orbit around Earth that would have quickly formed the moon (■ Figure 13-5).

This would explain a number of phenomena. If the collision occurred off-center, it would have spun the Earth–moon system rapidly and would thus explain the present high angular momentum. If the proto-Earth and impactor had already differentiated, the ejected material would have been mostly iron-poor mantle and crust, which would explain the moon's low density and iron-poor composition. Furthermore, the material would have lost its volatile components while it was in space, so the moon also would have formed lacking those materials. Such an impact would have melted the proto-Earth, and the material falling together to form the moon would have been heated hot enough to melt. This fits the evidence that the highland anorthosite in the moon's oldest rocks formed by differentiation of large quantities of molten material. The large-impact hypothesis passes tests of comparison with the known evidence and is now considered likely to be correct.

The History of Earth's Moon

The history of Earth's moon since its formation is dominated by a single fact—the moon is small, only one-quarter the diameter of Earth. The escape velocity is low, so it has been unable to hold any atmosphere, and it cooled rapidly as its internal heat flowed outward into space.

The Apollo moon rocks are the source of information about the timing of events during the moon's history. For example, they show that the moon must have formed in a mostly molten state (Figure 13-4). Planetary geologists now refer to the exterior of the newborn moon as a **magma ocean.** Denser materials sank toward the center, and low-density minerals floated to the top to form a low-density crust. In this way the moon partly differentiated. The radioactive ages of the moon rocks show that the surface solidified about 4.4 billion years ago.

The second stage, cratering and basin formation, began as soon as the crust solidified, and the older highlands show that cratering was intense during the first 0.5 billion years—during the heavy bombardment at the end of the solar system's period of planet building. The moon's crust was shattered, and the largest impacts formed giant **multiringed basins** hundreds of kilometers in diameter (■ Figure 13-6). There is some evidence indicating that roughly 4 billion years ago there was a sudden temporary increase in the cratering rate. Astronomers refer to this as the **late heavy bombardment** and hypothesize that it could have been caused by the final accretion and migration of Uranus and Neptune (see Chapter 12) scattering remnant planetesimals across the solar system to collide with the moon and planets. After about 3.8 billion years ago the cratering rate fell rapidly to the current low rate.

The tremendous impacts that formed the lunar basins cracked the crust as deep as 10 kilometers and led to flooding by lava. Though Earth's moon cooled rapidly after its formation, ra-

The Large-Impact Hypothesis

A protoplanet nearly the size of Earth differentiates to form an iron core.

Another body that has also formed an iron core strikes the larger body and merges, trapping most of the iron inside.

Iron-poor rock from the mantles of the two bodies forms a ring of debris.

Volatiles are lost to space as the particles in the ring begin to accrete larger bodies.

Eventually the moon forms from the iron-poor and volatile-poor matter in the disk.

■ Figure 13-5

Evidence indicates that sometime early in the solar system's history, after Earth differentiated into core, mantle, and crust, a collision produced the moon (the "large-impact hypothesis").

dioactive decay heated the subsurface material, and part of it melted, producing lava that followed the cracks up into the giant basins (Figure 13-6a). Studies of rocks brought back from the moon by Apollo astronauts show that the basins were flooded by successive lava flows of dark basalts, forming the maria, from 3.8 to 3.2 billion years ago. The moon's crust is thinner on the side

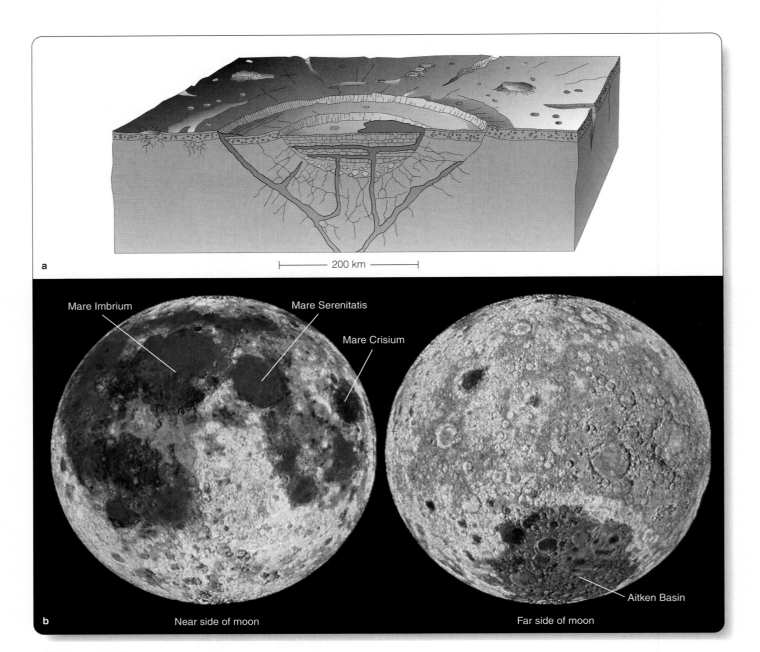

a

|— 200 km —|

Mare Imbrium

Mare Serenitatis

Mare Crisium

Aitken Basin

b   Near side of moon   Far side of moon

■ Figure 13-6

Much of the near side of the moon is marked by great, generally circular lava plains called maria. The crust on the far side is thicker, and there is much less flooding. Even the huge Aitken Basin near the lunar south pole contains little lava flooding. In these maps color marks elevation, with red the highest regions and purple the lowest. (Adapted from a diagram by William Hartmann; NASA/Clementine)

toward Earth, perhaps due to tidal effects. Consequently, while lava flooded the basins on the Earth-facing side, it was unable to rise through the thicker crust to flood the lowlands on the far side.

The third stage, slow surface evolution, was limited both because the moon cooled rapidly and because it lacks water. Flooding on Earth included water, but the moon has never had an atmosphere and thus has never had liquid water. With no air and no water, erosion is limited to the constant bombardment of **mi-**

crometeorites and rare larger impacts. Indeed, a few meteorites found on Earth have been identified as moon rocks ejected from the moon by impacts within the last few million years. As the moon lost its internal heat, volcanism died down, and the moon became geologically dead. Its crust never divided into moving plates—there are no folded mountain ranges—and the moon is now a "one-plate" planet.

Building Scientific Arguments:

Why are the maria nearly free of craters?
Timing is everything, someone has said, and in this case your argument must carefully consider the sequence of events. The moon's crust formed and was heavily cratered before about 4

billion years ago. The impacts of the heavy bombardment cratered the entire lunar surface and created a few very large basins. Later, after the end of that heavy cratering, lava welled up and filled the lowlands of the largest crater basins with basalt to form the mare. Any craters in the basins were covered over, and there were few later impacts to form new craters. Thus, the mare surfaces are nearly free of craters, but the ancient highlands are heavily cratered.

Now build a new argument. **How is timing important in explaining the formation of an iron-poor moon in the large-impact hypothesis?**

■   ■   ■

13-4 Mercury

DON'T RENT A CHEAP SPACESUIT to take to Mercury. The temperature extremes between sunshine and shade could be deadly (see **Celestial Profile 3**). Like Earth's moon, Mercury is small and nearly airless. Mercury cooled too quickly to develop plate tectonics, so you will find it also is a cratered, dead world.

Spacecraft at Mercury

Mercury orbits close to the sun and is difficult to observe from Earth, so little was known about it until 1974–1975 when the Mariner 10 spacecraft flew past Mercury three times and revealed a planet with a heavily cratered surface, much like that of Earth's moon (■ Figure 13-7a). Analysis shows that large areas have been flooded by lava and then cratered. The largest impact feature on Mercury is the Caloris Basin, a multiringed area 1300 km (800 mi) in diameter (Figure 13-7b).

Mercury is quite dense, and models indicate that it must have a large metallic core (Celestial Profile 3). In fact, the metallic core occupies about 70 percent of the radius of the planet. Mercury is essentially a metal planet with a thin rock mantle. Mariner 10 photos revealed long curving ridges up to 3 km (2 mi) high and 500 km (300 mi) long, as shown in Figure 13-7c. The ridges even cut through craters, indicating that they formed after most of the heavy bombardment. Planetary scientists understand these ridges as evidence that the entire crust of Mercury compressed and shrank long ago.

Mariner 10 photographed only half of Mercury during its three visits. The remainder of Mercury will remain unknown until the Messenger spacecraft passes by the planet three times in 2008 and 2009 and goes into orbit in 2011.

A History of Mercury

Like Earth's moon, Mercury is small, and that fact has determined much of its history. Not only is Mercury too small to retain an atmosphere, it has also lost much of its internal heat and thus is not geologically active.

In the first stage, Mercury differentiated to form a metallic core and a rocky mantle. Mariner 10 discovered a magnetic field about 10^{-4} times that of Earth—further evidence of a metallic core. In Chapter 12, you saw that the condensation sequence could explain the abundance of metals in Mercury, but careful model calculations show that Mercury contains even more iron than would be expected from its position in the solar system. Drawing on the large-impact hypothesis for the origin of Earth's moon, scientists have proposed that Mercury suffered a major impact soon after it differentiated, an impact so large it ejected much of the rocky mantle into space. Such catastrophic events are rare in nature, but they do occur, so astronomers must be prepared to consider such hypotheses (see Window on Science 12-1).

In the second stage of planet formation, cratering battered the crust, and lava flows welled up to fill the lowlands, just as they did on the moon. As the small world lost internal heat, its large metal core contracted, and its crust compressed and broke to form the long ridges, much as the peel of a drying apple wrinkles.

Lacking an atmosphere to erode it, Mercury has changed little since the last lava hardened, and it is now a "one-plate" planet like Earth's moon.

13-5 Venus

YOU MIGHT EXPECT VENUS to be much like Earth. It is 95 percent Earth's diameter (**Celestial Profile 4**), has a similar average density, and is only 30 percent closer to the sun. Unfortunately, the surface of Venus is perpetually hidden below thick clouds, and only in the last few decades have planetary scientists discovered that Venus is a hot desert world of volcanoes, lava flows, and impact craters lying at the bottom of a deep hot atmosphere. No spacesuit will allow you to visit the surface of Venus.

The Atmosphere of Venus

In composition, temperature, and density, the atmosphere of Venus is more Hades than Heaven. The air is unbreathable, very hot, and very dense. How do planetary scientists know this? Because U.S. and Soviet space probes have descended into the atmosphere and in some cases landed on the surface.

In composition, the atmosphere of Venus is roughly 96 percent carbon dioxide. The rest is mostly nitrogen, with some argon, sulfur dioxide, and small amounts of sulfuric acid, hydrochloric acid, and hydrofluoric acid. There is only a tiny amount of water vapor. On the whole, the composition is deadly unpleasant, and most certainly smells bad too. Spectra show that the impenetrable clouds that hide the surface are made up of droplets of sulfuric acid and microscopic crystals of sulfur. This unbreathable atmosphere is 90 times denser than Earth's. The air you breathe is 1000 times less dense than water, but on Venus the air is only 10 times less dense than water. If you could survive the

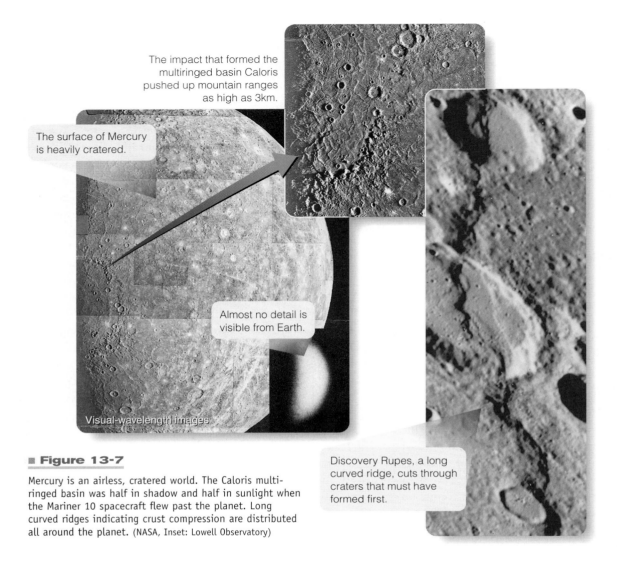

The impact that formed the multiringed basin Caloris pushed up mountain ranges as high as 3km.

The surface of Mercury is heavily cratered.

Almost no detail is visible from Earth.

Visual-wavelength images

■ **Figure 13-7**

Mercury is an airless, cratered world. The Caloris multi-ringed basin was half in shadow and half in sunlight when the Mariner 10 spacecraft flew past the planet. Long curved ridges indicating crust compression are distributed all around the planet. (NASA, Inset: Lowell Observatory)

Discovery Rupes, a long curved ridge, cuts through craters that must have formed first.

unpleasant composition and intense heat, you could strap wings on your arms and fly.

The surface temperature on Venus (Celestial Profile 4) is hot enough to melt lead, and you can understand that because the thick atmosphere creates a severe greenhouse effect. Sunlight filters down through the clouds and warms the surface, but infrared radiation cannot escape because the atmosphere is opaque to infrared. It is the overwhelming abundance of carbon dioxide that makes the greenhouse effect on Venus much more severe than on Earth.

The Surface of Venus

Although the thick clouds and atmosphere on Venus are opaque to visible and infrared light, they are transparent to radio waves. Both U.S. and Soviet spacecraft orbited Venus and mapped its surface by radar, revealing details as small as 100 meters in diameter (■ Figure 13-8).

Radar maps of Venus are reproduced using arbitrary colors. In some maps lowlands are colored blue, but this is in a sense misleading because there are no oceans on Venus. In other maps the scientists have chosen instead to give Venus an overall orange tint because sunlight filtering down through the clouds bathes the landscape in a perpetual sunset glow. Some radar maps are colored gray, the natural color of the rocks.

Radar maps show that Venus is similar to Earth in one way but strangely different in other ways. Nearly 75 percent of Earth is covered by low-lying basaltic sea floors, and 85 percent of Venus is covered by basaltic lowlands. Of course on Venus the lowlands are not seafloors, and the remaining highlands are not the well-defined continents you see on Earth. Whereas Earth is dominated by plate tectonics, something different is happening on Venus.

The highland area Ishtar Terra, named for the Babylonian goddess of love, is about the size of Australia (Figure 13-8). At its eastern edge, the mountain called Maxwell Montes thrusts up 12 km (Everest, the tallest mountain on Earth, is 8.8 km high). The mountains bounding Ishtar Terra, including Maxwell, resemble folded mountain ranges, which suggests that some horizontal motion in the crust as well as volcanism has helped form the highlands.

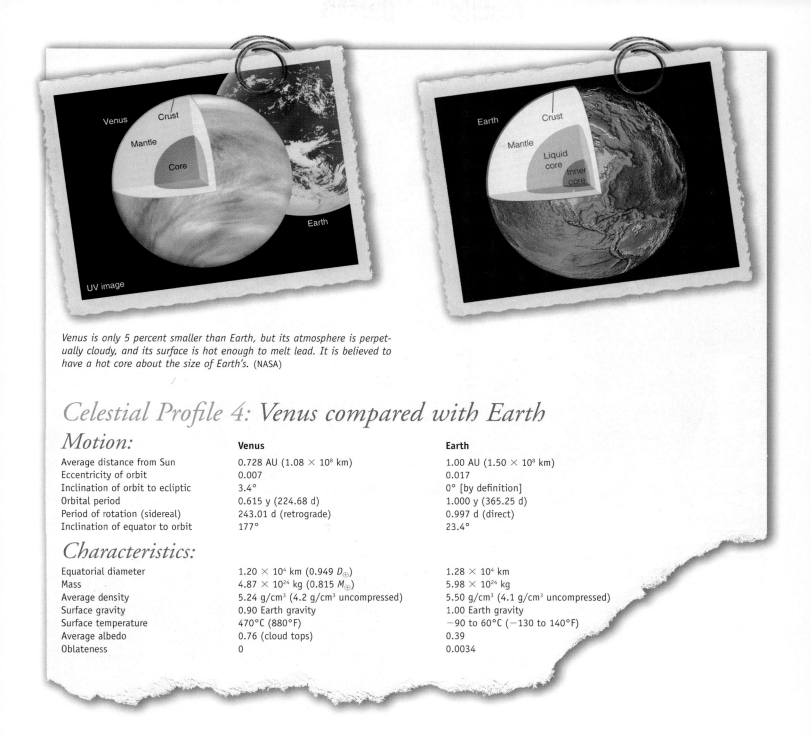

Venus is only 5 percent smaller than Earth, but its atmosphere is perpetually cloudy, and its surface is hot enough to melt lead. It is believed to have a hot core about the size of Earth's. (NASA)

Celestial Profile 4: Venus compared with Earth

Motion:

	Venus	Earth
Average distance from Sun	0.728 AU (1.08×10^8 km)	1.00 AU (1.50×10^8 km)
Eccentricity of orbit	0.007	0.017
Inclination of orbit to ecliptic	3.4°	0° [by definition]
Orbital period	0.615 y (224.68 d)	1.000 y (365.25 d)
Period of rotation (sidereal)	243.01 d (retrograde)	0.997 d (direct)
Inclination of equator to orbit	177°	23.4°

Characteristics:

	Venus	Earth
Equatorial diameter	1.20×10^4 km (0.949 $D_\oplus$)	1.28×10^4 km
Mass	4.87×10^{24} kg (0.815 $M_\oplus$)	5.98×10^{24} kg
Average density	5.24 g/cm^3 (4.2 g/cm^3 uncompressed)	5.50 g/cm^3 (4.1 g/cm^3 uncompressed)
Surface gravity	0.90 Earth gravity	1.00 Earth gravity
Surface temperature	470°C (880°F)	−90 to 60°C (−130 to 140°F)
Average albedo	0.76 (cloud tops)	0.39
Oblateness	0	0.0034

Many features on Venus testify to its volcanic history. Long, narrow lava channels meander for thousands of kilometers. Radar maps reveal many smaller volcanoes, faults, and sunken regions produced when magma below the surface drained away. Other volcanic features include the **coronae,** circular bulges up to 2100 km (1300 mi) in diameter bordered by fractures, volcanoes, and lava flows (■ Figure 13-9). These appear to be produced by rising convection currents of molten magma that push up under the crust. When the magma withdraws, the crust sinks back, but the circular fractures mark the edge of the corona.

Radar images also show that Venus is marked by numerous craters. Meteorites big enough to make craters larger than 3 km in diameter have no trouble penetrating the thick atmosphere but have formed only about 10 percent as many craters on Venus as on the maria of Earth's moon. The number of craters shows that the crust of Venus is not as ancient as the lunar maria, but not as young as Earth's active surface. The average age of the surface of Venus is estimated to be roughly half a billion years. Clearly, geological processes cannot be renewing the surface as rapidly as they do on Earth.

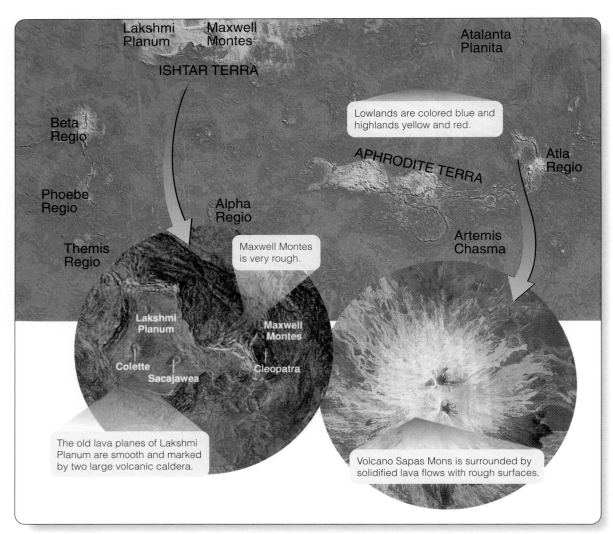

Figure 13-8

Notice how three radar maps show different things. The main radar map, made by the Magellan robot probe, shows elevation over most of the surface of Venus, omitting the polar areas. The detailed map of Maxwell Montes and Lakshmi Planus are colored according to roughness, with orange the roughest terrain. The map of volcano Sapas Mons also shows roughness but is given an orange color to mimic the color of sunlight reaching the surface through the thick atmosphere (Maxwell and Lakshmi Planum map: USGS; Other maps: NASA)

Soviet robot spacecraft landed on the surface of Venus and survived the heat and pressure for a few hours. Some of those spacecraft analyzed the rock and snapped a few photographs (■ Figure 13-10). The surface rocks on Venus are dark gray basalts much like those in Earth's ocean floors. This evidence confirms that volcanism is important.

The History of Venus

To tell the story of Venus you must draw together all the evidence and find hypotheses to explain two things, the thick carbon dioxide atmosphere and the peculiar geology.

Figure 13-9

Radar map of surface features on Venus: Aine Corona, about 200 km in diameter, is marked by faults, lava flows, small volcanic domes, and pancake domes of solidified lava. (NASA)

Calculations show that Venus and Earth have outgassed about the same amount of carbon dioxide, but Earth's oceans have dissolved most of that and converted it to sediments such as limestone. If all of the carbon in Earth's sediments and crust were dug up and converted back to carbon dioxide, our atmosphere would be about as dense as the air on Venus. This suggests that the main difference between Earth and Venus is the lack of water on Venus.

Venus may have had oceans when it was young and much more Earthlike conditions than at present, but, being closer to the sun than Earth, it was warmer, and the carbon dioxide in the atmosphere created a greenhouse effect that made the planet even warmer. That process could have dried up any oceans that did exist and prevented Venus from purging its atmosphere of carbon dioxide. In fact, evidence from the composition of Venus's atmosphere indicates that an ocean's worth of water has been vaporized and lost. As carbon dioxide continued to be outgassed, the greenhouse effect grew even more severe. Thus, planetary scientists conclude that Venus was trapped in a **runaway greenhouse** effect.

The intense heat at the surface may have affected the geology of Venus by making the crust drier and more flexible so that it was unable to break into moving plates as on Earth. There is no sign of plate tectonics on Venus, but there is evidence that convection currents below the crust are deforming the crust to make coronae, push up mountains such as Maxwell, and create some folded mountains like those around Ishtar Terra by minor horizontal crust motions.

The small number of craters on the surface of Venus hints that the entire crust has been replaced within the last half-billion years or so. This may have occurred in a planetwide overturning as the old crust broke up and sank and lava flows created a new crust. Comparing Earth to Venus may eventually reveal more about how our own world's volcanism and tectonics work.

Building Scientific Arguments:

What evidence indicates that Venus does not have plate tectonics?

This argument must cite evidence and use comparison. On Earth, plate tectonics is identifiable by the worldwide network of faults, subduction zones, volcanism, and folded mountain chains that outline the plates. Although some of these features are visible on Venus, they do not occur in a planetwide network that outlines plates. Volcanism is widespread, but folded mountain ranges occur in only a few places. Rather than being dominated by the motion of rigid crust plates, Venus may have a more pliable crust strained by rising plumes of molten rock that produce coronae or break through to form volcanoes and lava flows.

Earth and Venus are twin worlds in size; but they can be no more than half-siblings in terms of their surface processes. Now build an argument to compare atmospheres. **Why isn't Earth's atmosphere like that of Venus?**

■ ■ ■

13-6 Mars

IF YOU EVER VISIT ANOTHER WORLD, Mars may be your best choice (**Celestial Profile 5**). You will need a heated, pressurized spacesuit, but Mars is not as inhospitable as the moon. It is also more interesting, with weather, complex geology, craters, volcanoes, and signs that water once flowed over its surface.

■ **Active Figure 13-10**

The Venera 13 lander touched down on Venus in 1982 and carried a camera that swiveled from side to side to photograph the surface. The orange glow is produced by the thick atmosphere; when that is corrected, as in the lower frame, you can see that the rocks are dark gray. Isotopic analysis suggests they are basalts. An instrument cover and the base of the spacecraft are visible at lower right. (NASA)

ThomsonNOW™ Sign in at www .thomsonedu.com and go to ThomsonNOW to see the Active Figure "Planetary Atmosphere."

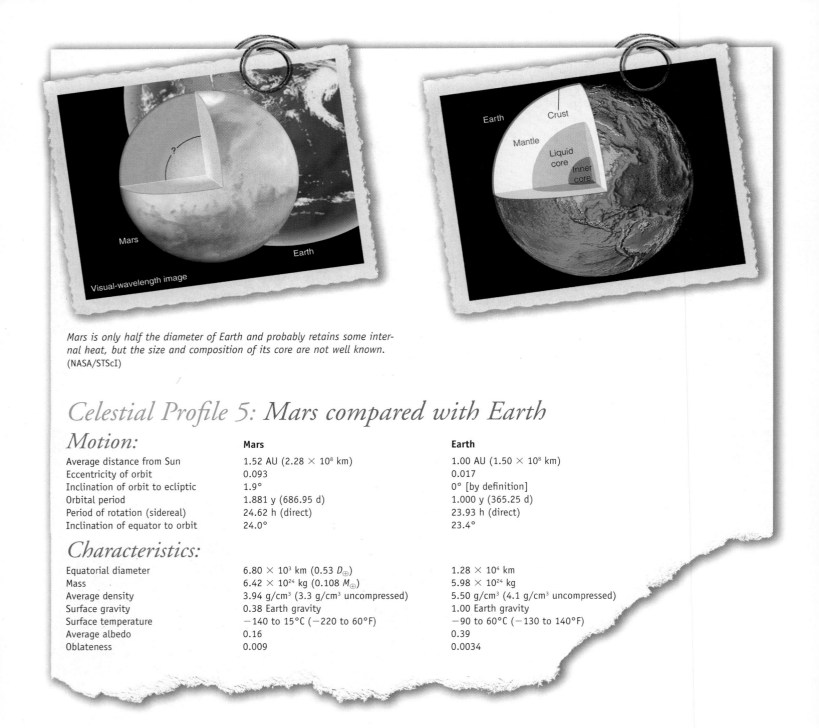

Mars is only half the diameter of Earth and probably retains some internal heat, but the size and composition of its core are not well known. (NASA/STScI)

Celestial Profile 5: Mars compared with Earth

Motion:

	Mars	Earth
Average distance from Sun	1.52 AU (2.28×10^8 km)	1.00 AU (1.50×10^8 km)
Eccentricity of orbit	0.093	0.017
Inclination of orbit to ecliptic	1.9°	0° [by definition]
Orbital period	1.881 y (686.95 d)	1.000 y (365.25 d)
Period of rotation (sidereal)	24.62 h (direct)	23.93 h (direct)
Inclination of equator to orbit	24.0°	23.4°

Characteristics:

	Mars	Earth
Equatorial diameter	6.80×10^3 km (0.53 $D_\oplus$)	1.28×10^4 km
Mass	6.42×10^{24} kg (0.108 $M_\oplus$)	5.98×10^{24} kg
Average density	3.94 g/cm^3 (3.3 g/cm^3 uncompressed)	5.50 g/cm^3 (4.1 g/cm^3 uncompressed)
Surface gravity	0.38 Earth gravity	1.00 Earth gravity
Surface temperature	−140 to 15°C (−220 to 60°F)	−90 to 60°C (−130 to 140°F)
Average albedo	0.16	0.39
Oblateness	0.009	0.0034

The Atmosphere of Mars

The Martian air contains 95 percent carbon dioxide, 3 percent nitrogen, and 2 percent argon. That is similar to the composition of air on Venus, but the Martian atmosphere is very thin, less than 1 percent as dense as Earth's atmosphere and 1/10,000 as dense as Venus's.

There is very little water in the Martian atmosphere. Liquid water cannot survive on the surface of Mars because the air pressure is too low: Any liquid water would immediately boil away. The polar caps appear to be composed of frozen water ice coated over by frozen carbon dioxide (dry ice). Whatever water is present on Mars is frozen either within the polar caps or as **permafrost** in the soil.

Although the present atmosphere of Mars is very thin, you will see evidence that the climate once permitted liquid water to flow over the surface, so Mars must have once had a thicker atmosphere. As a terrestrial planet, it should have outgassed significant amounts of carbon dioxide, nitrogen, and water vapor; but because it was small it could not hold onto its atmosphere. The escape velocity on Mars is only 5 km/s, less than half of Earth's,

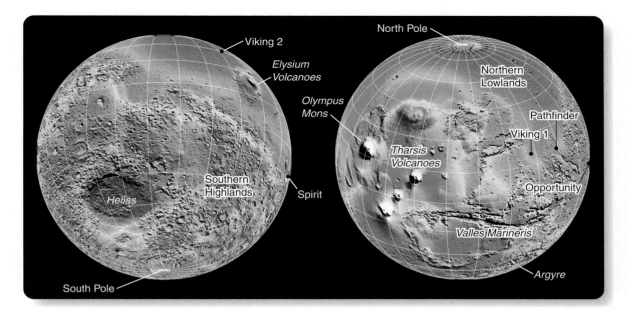

Figure 13-11

These hemisphere maps of Mars are color coded to show elevation. The northern lowlands lie about 4 km (2.5 mi) below the southern highlands. Volcanoes are very high (white), and the giant impact basins, Hellas and Argyre, are low. Note the depth of the great canyon Valles Marineris. (NASA)

so it was easier for rapidly moving gas molecules to escape into space. Another factor is the temperature of a planet. If Mars had been colder, the gas molecules in its atmosphere would have been traveling more slowly and would not have escaped as easily.

Another process that has reduced Mars's atmosphere is the planet's lack of an ozone layer to screen out UV radiation: Sunbathing on Mars would be a fatal mistake. Sunlight UV photons can break atmospheric molecules up into smaller fragments. Water, for example, can be broken by UV into hydrogen and oxygen. The hydrogen then easily escapes into space, and the oxygen easily combines chemically with rocks and soil. Mars may have had a substantial atmosphere when it was young, but it gradually lost much of that both by direct escape and by UV destruction. As a result, with a thin atmosphere that does not provide much greenhouse warming nor permit liquid water to persist on the surface, Mars is now a cold, dry world.

Exploring the Surface of Mars

Data recorded by orbiting satellites show that the southern hemisphere of Mars is a heavily cratered highland region up to 4 billion years old. The northern hemisphere is mostly a much younger lowland plain with few craters (■ Figure 13-11).

Volcanism on Mars is dramatically evident in the Tharsis region, a highland region of volcanoes and lava flows bulging 10 km (6 mi) above the surrounding surface. A similar uplifted volcanic plain, the Elysium region, is more heavily cratered and eroded and appears to be older than the Tharsis bulge. The lack of many impact craters suggests that some volcanoes have been active within the last few hundred million years and maybe even more recently.

All of the volcanoes on Mars are **shield volcanoes,** very broad mountains with gentle slopes that are produced on Earth by hot spots penetrating upward through the crust.

Shield volcanoes are not related to plate tectonics; in fact, the large shield volcanoes on Mars, the largest of which is Olympus Mons, provide evidence that plate tectonics has not been significant on that planet. Olympus Mons is 600 km (almost 400 mi) in diameter at its base and rises 21 km (13 mi) high. In comparison, the largest volcano on Earth, also a shield volcano, is Mauna Loa in Hawaii, rising only 10 km (6 mi) above its base on the seafloor (■ Figure 13-12a). On Earth, volcanoes like those that formed the Hawaiian Islands occur over rising currents of hot material in the mantle. Because the crust plate moves horizontally, a chain of volcanoes is formed instead of a single large feature (Figure 13-12b). A lack of plate motion on Mars has allowed rising currents of magma to heat the crust repeatedly in the same places and build Olympus Mons plus other very large volcanic shields, especially in the Tharsis region (Figure 13-11).

When the crust of a planet is strained, it may break, producing faults and rift valleys. Near the Tharsis region is a great valley, Valles Marineris (Figure 13-11), named after the Mariner spacecraft that first photographed it. The valley is a block of crust that has dropped downward along parallel faults. Erosion and landslides have further modified the valley into a great canyon. It is four times deeper, nearly ten times wider, and over ten times longer than the Grand Canyon. The number of craters in the valley indicates that it is 1 to 2 billion years old, placing its origin sometime before the end of the most active volcanism in the Tharsis region.

Searching for Water on Mars

Before you can tell the story of Mars, you must consider a difficult issue—water. How much water has Mars had, how much has been lost, and how much remains? You cannot expect surface water on Mars now because of its low atmospheric pressure and low temperature. Observations from orbiting spacecraft, however, have revealed landforms that suggest the effects of flowing water on Mars, and rovers on the surface have turned up proof positive of surface water.

In 1976, the two Viking spacecraft reached orbit around Mars and photographed its surface. Those photos revealed two kinds of water-related features. **Outflow channels** appear to have been cut by massive floods carrying as much as 10,000 times the water flowing down the Mississippi River. In a matter of hours or days, such floods swept away geological features and left scarred land such as that shown in ■ Figure 13-13. In contrast, **valley networks** look like meandering riverbeds with sandbars, deltas, and tributaries typical of streams that flowed for extended periods of time. The number of craters on top of these features reveals that they are quite old.

Spacecraft in orbit around Mars have used remote instruments to detect large amounts of water frozen in the soil. A radar study has found frozen water extending at least a kilometer beneath both polar caps. Images made from orbit also show regions of jumbled terrain, suggesting that subsurface ice may have melted and drained away. Gullies leading downslope suggest water seeping from underground sources. Study Figure 13-11, where the lowlands have been color-coded blue, and notice the major outflow channels leading from the highlands into the lowlands northwest of the Viking 1 landing site and southeast of the Pathfinder landing site. The terrain at the edges of the northern lowlands has been compared to shorelines, and some scientists suspect that the northern lowlands were filled with an ocean 3 billion or more years ago.

Time since last eruption (million years)

5   3   1.5   1   0

Kauai   Oahu   Molokai   Maui   Hawaii   Active volcanoes

← Plate motion

Rovers named Spirit and Opportunity landed in January 2004 and carried sophisticated instruments to explore the rocky surface (■ Figure 13-14). Both rovers landed in areas suspected of having had water on their surfaces, and both made exciting discoveries. Using close-up cameras, the rovers found small spherical concretions of the mineral hematite (dubbed "blueberries") that must have formed in water. In other places, they found layers of sediments with ripple marks and crossed layers showing deposits that must have formed in moving water (■ Figure 13-15). Chemical analysis revealed minerals in the soil such as sulfates that are left behind when standing water evaporates.

The History of Mars

Did Mars ever have plate tectonics? Where did the water go? These fundamental questions challenge you to assemble the evidence and hypotheses for Mars and tell the story of its evolution (**Window on Science 13-1**).

The history of Mars is a case of arrested development. The planet began by differentiating into a crust, mantle, and core. Studies of its rotation reveal that it has a dense core but no planetwide magnetic field. The core must have cooled quickly and shut off the dynamo effect that would have produced a field.

The crust of Mars is now quite thick, as shown by the mass of the Olympus Mons volcano (Figure 13-12), but it was probably thinner in the past. Cratering may have broken or at least weakened the crust, triggering lava flows that flooded some basins. Mantle convection may have pushed up the Tharsis and Elysium volcanic regions and broken the crust to form Valles Marineris, but moving crustal plates never dominated Mars. There are no folded mountain ranges on Mars and no sign of plate boundaries. The small planet cooled rapidly, and the crust grew thick and immobile.

For Mars, the last stage of planetary development has been one of slow decline. Volcanoes may still occasionally erupt, but the little planet has lost much of its internal heat, and most volcanism occurred long ago. At some point in the history of Mars, water was abundant enough to flow over the surface in great floods and may have filled an ocean, but the age of liquid water must have ended over 3 billion years ago. The climate on Mars has changed as the atmosphere gradually

■ Figure 13-12

(a) Olympus Mons, a shield volcano on Mars, is much larger than Mauna Loa, a shield volcano that is the largest mountain on Earth. Mauna Loa is so heavy that it has sunk into Earth's crust, producing an undersea moat around its base, but Olympus Mons, two times higher, has not made a moat. Evidently, the crust of Mars is much stronger than Earth's. (b) Volcanoes that make up the Hawaiian Islands have been produced by a hot spot poking upward through the middle of the Pacific plate. In contrast to the situation on Mars, hot spot volcanoes such as these do not grow very large on Earth because the crust plates move horizontally and carry older volcanoes away from the magma source.

Figure 13-13

These visual-wavelength images made by the Viking orbiters and Mars Global Surveyor show some of the features that suggest liquid water on Mars. Outflow channels and runoff channels are old, but some gullies may be quite recent. (Malin Space Science Systems and NASA)

became thinner. Atmospheric gases and water were lost to space, and the volcanic activity that could have replaced them had nearly stopped.

The water remaining on Mars today is frozen in the polar caps or in the soil. However, note that water is the first necessity of life, so its presence long ago on Mars is exciting. Someday an astronaut may scramble down an ancient Martian streambed, turn over a rock, and find a fossil.

The Moons of Mars

Unlike Mercury or Venus, Mars has moons. Small and irregular in shape, Phobos (28 × 23 × 20 km in diameter) (■ Figure 13-16a) and Deimos (16 × 12 × 10 km) (Figure 13-16b) are almost certainly captured asteroids (see Chapter 12). Phobos and Deimos are not just small; they are tiny. An athletic astronaut who could jump 2 m (6 ft) high on Earth could jump almost 3 km (2 mi) on Phobos. However they formed, these moons are so small that any interior heat would have leaked away very quickly, and there is no evidence of any internally driven geologic activity on either object.

Some futurists suggest that the first human missions to Mars may not land on the planet's surface but instead could build a colony on Phobos or Deimos. These plans involve speculation that these moon and other asteroids may have water in deep interior rocks that colonists could use.

Visual-wavelength image

■ Figure 13-14

Mars is a red desert planet, as shown in this true-color photograph by the Spirit rover. Rock layers are fractured by meteorite impacts, and dust is blown around the planet by winds. Dust suspended in the atmosphere colors the sky red. Data from orbiters, landers, and rovers show that billions of years ago Mars was less of a desert and had water on its surface. (NASA)

■ Figure 13-15

Rover Opportunity photographed hematite concretions weathered from rock. The spheres ("blueberries") appear to have grown as minerals collected around small crystals in the presence of water. Similar concretions are found on Earth. (b) The layers in this rock were deposited as sand and silt in rapidly flowing water. From the way the layers curve and cross each other, geologists can estimate that the water was at least 10 centimeters (4 in.) deep. A few "blueberries" are also visible in this image. (NASA/JPL/Cornell/USGS)

The Present Is the Key to the Past

Geologists are fond of saying, "The present is the key to the past." By that they mean that you can learn about the history of Earth by looking at the present condition of Earth's surface. The position and composition of various rock layers in the Grand Canyon, for example, tell you that the western United States was once at the floor of an ocean. This principle of geology was first formulated in the late 1700s, and it continues to be relevant today as you try to understand other worlds such as Venus and Mars.

In the late 18th century, naturalists first recognized that the present gave them clues to the history of Earth. This was astonishing because most people assumed that Earth had no history. That is, they assumed either that Earth had been created in its present state as described in the Old Testament or that Earth was eternal. In either case, people commonly assumed that the hills and mountains they saw around them had always existed more or less as they were. The 18th-century naturalists began to see evidence that the hills and mountains were not eternal but were the result of past processes and were slowly changing. That gave birth to the idea that Earth had a history.

As the naturalists of the 18th century made the first attempts to thoughtfully and logically explain the nature of Earth by looking at the evidence, they were inventing modern geology as a way of understanding Earth. What Copernicus, Kepler, and Newton did for the heavens in the 1500s and 1600s, the first geologists did for Earth beginning in the late 1700s. Of course, the invention of geology as the study of Earth led directly to the modern attempts to understand the geology of other worlds.

Geologists and astronomers share a common goal: They are attempting to reconstruct the past (see Window on Science 12-1). Whether you study Earth, Venus, or Mars, you are looking at the present evidence and trying to reconstruct the past history of the planet by drawing on observations and logic to test each step in the story. How did Venus get to be covered with lava? How did Mars lose its atmosphere? The final goal of planetary astronomy is to draw together all of the available evidence (the present) to tell the story (the past) of how the planet got to be the way it is. Those first geologists of the late 1700s would be fascinated by the stories planetary astronomers tell today.

Layers of rock in the Martian crater Terby hint at a time when the crater was filled with a lake. (NASA/JPL/Malin Space Science Systems)

Visual-wavelength images

a

b

■ Figure 13-16

The moons of Mars are too small to pull themselves into spherical shape. The two irregular moons, to scale here showing that the smallest dimension of Phobos is about the same size as the largest dimension of Deimos, were named for the mythical attendants of Mars, the god of war. Phobos was the god of fear, and Deimos was the god of dread. (Phobos: Damon Simonelli and Joseph Ververka, Cornell University/NASA; Deimos: NASA)

Summary

13-1 | A Travel Guide to the Terrestrial Planets

How can comparison help you understand the terrestrial planets?
- The terrestrial worlds differ mainly in size, but they all have low-density crusts, mantles of dense rock, and metallic cores.
- Comparative planetology alerts you to expect that cratered surfaces are old, that heat flowing out of a planet drives geological activity, and that the nature of a planet's atmosphere depends on both the size of the planet and its temperature.
- Earth's moon is included in this study because it is a complex world and illustrates important principles such as cratering and flooding of basins by lava.

13-2 | Earth: The Active Planet

What are the main features of Earth when you view it as a planet?
- Earth has passed through three stages as it has evolved: (1) differentiation; (2) cratering and basin formation; and (3) continued slow surface evolution. These stages also apply in varying degrees to the other terrestrial planets.
- Seismology shows that Earth has differentiated to form a liquid metallic core, which generates a magnetic field.
- Earth is dominated by plate tectonics that breaks the crust into moving plates. This process is driven by heat flowing upward from the interior.
- Earth's primary atmosphere was probably mostly carbon dioxide, but that gas dissolves in seawater, and plant life has created oxygen, so the primary atmosphere has been replaced by a secondary atmosphere.

13-3 | The Moon

How does size determine the geological activity and evolution of a planetary body?
- Earth's moon formed in a molten state and partly differentiated, but it contains little metal and has a low density.
- Because it is a small world, Earth's moon has lost most of its internal heat and is no longer geologically active. Its old highlands are heavily cratered, but the lowland maria are filled by smooth lava flows that formed soon after the end of the heavy bombardment.
- The large-impact hypothesis suggests the moon formed when an impact between Earth and a very large planetesimal ejected debris that formed a disk around Earth. The moon formed from that disk.

13-4 | Mercury

- Mercury is smaller than Earth but larger than Earth's moon. It is airless and has an old, heavily cratered surface.
- Mercury has a much higher density than Earth's moon and must have a large metallic core. It may have suffered a major impact when it was young that drove off some of the lower-density crust and mantle rock and left Mercury with a metallic core that is larger than the condensation sequence would imply.
- Mercury has long curving ridges formed by compression of the crust when its large metallic core solidified and contracted.

13-5 | Venus

How do distance from the sun and size of a planet affect the properties of its atmosphere?
- Although Venus is almost as big as Earth, it has a thick, cloudy atmosphere of carbon dioxide that hides the surface from sight. The surface can be studied by radar mapping.
- The carbon dioxide atmosphere drives an intense greenhouse effect and makes Venus's surface hot enough to melt lead.
- Venus is slightly closer to the sun than Earth and was too warm for liquid water oceans to persist and dissolve carbon dioxide from the atmosphere, so Venus suffered a runaway greenhouse effect.
- The hot crust of Venus is dominated by volcanism but not by plate tectonics.

13-6 | Mars

What is the evidence that surface conditions on Mars were originally more Earthlike than they are at present?
- Mars is about half the diameter of Earth; it has a thin atmosphere and also has lost much of its internal heat.
- The loss of atmospheric gases depends on the size of a planet and its temperature. Mars is cold, but it is small and has a low escape velocity, and many of its lighter gases have leaked away.
- Some water may have leaked away because UV radiation from the sun broke it into hydrogen and oxygen, but some water is frozen in the polar caps and in the soil.
- Outflow channels and valley networks visible from orbit seem to have been cut by sudden floods or by longer-term drainage, but water cannot exist as a liquid on Mars now because of its low temperature and low atmospheric pressure.
- The southern hemisphere of Mars is old cratered terrain, but some large volcanoes lie in the north. The sizes of these volcanoes indicate that the crust does not have horizontal motions and plate tectonics.
- Some volcanism may still occur on Mars, but because the planet is small it has cooled and is not very active geologically.
- Orbiters have measured large amounts of water frozen below the surface.
- Rovers have found clear signs that liquid water flowed over the surface in at least some places and therefore evidence that the Martian climate was different in the past. The northern lowlands may even have held an ocean once.
- The two moons of Mars are probably captured asteroids.

New Terms

comparative planetology (p. 264)

mantle (p. 266)

P wave (p. 267)

S wave (p. 267)

primary atmosphere (p. 268)

secondary atmosphere (p. 268)

greenhouse effect (p. 268)

plate tectonics (p. 270)

rift valley (p. 270)

midocean rise (p. 270)

basalt (p. 270)

subduction zone (p. 270)

folded mountain range (p. 270)

global warming (p. 272)

mare, maria (p. 273)

albedo (p. 273)

ejecta (p. 274)

anorthosite (p. 274)

breccia (p. 274)

large-impact hypothesis (p. 274)

magma ocean (p. 275)

multiringed basin (p. 275)

late heavy bombardment (p. 275)

micrometeorite (p. 276)

coronae (p. 279)

runaway greenhouse (p. 281)

permafrost (p. 282)

shield volcano (p. 283)

outflow channel (p. 284)

valley network (p. 284)

Review Questions

ThomsonNOW™ Assess your understanding of this chapter's topics with additional quizzing and animations at
www.thomsonedu.com

1. What are the stages in common in the development of terrestrial planets?

2. Why would you expect planets to have differentiated?

3. How does plate tectonics create and destroy Earth's crust?

4. Why do planetary scientists suspect that Earth's primeval atmosphere was rich in carbon dioxide?

5. Why doesn't Earth have as many craters as its moon or Venus?

6. What kind of erosion is now active on Earth's moon?

7. Discuss the hypothesis and evidence concerning the origin of Earth's moon.

8. How did Earth avoid the greenhouse effect that made Venus so hot?

9. What evidence indicates that plate tectonics does not occur on Venus? On Mars?

10. What evidence suggests that Venus has been resurfaced within the last half-billion years?

11. Volcano Sif Mons on Venus is shown in this radar image. What kind of volcano is it, and why is it orange in this image? What color would the rock be if you could see it with your own eyes?

12. Why is the atmosphere of Venus rich in carbon dioxide? Why is the atmosphere of Mars rich in carbon dioxide?

13. Why do planetary scientists conclude that the crust on Mars must be thick?

14. What evidence indicates that there has been liquid water on Mars?

15. Olympus Mons on Mars is a very large volcano. In this image you can see superimposed caldera (volcanic craters) at the top. What do those multiple caldera and the volcano's immense size indicate about the geology of Mars?

Discussion Questions

1. If you visited a planet in another solar system and discovered oxygen in its atmosphere, what might you guess about its surface conditions and history?

2. If liquid water is rare on the surface of planets, then most terrestrial planets in other solar systems must have CO_2-rich atmospheres. Why?

14 | Comparative Planetology of the Outer Planets

There wasn't a breath

in that land of death . . .

ROBERT SERVICE
THE CREMATION OF SAM MCGEE

THE WORLDS OF THE OUTER SOLAR SYSTEM CAN BE STUDIED from Earth, but much of what scientists know has been radioed back to Earth from robot spacecraft. Voyager 2 flew past each of the outer planets in the 1970s and 1980s, the Galileo spacecraft circled Jupiter dozens of times in the late 1990s, and the Cassini/Huygens orbiter and probe arrived at Saturn in 2004. Throughout this discussion, you will find images and data returned by those robotic explorers.

Jupiter is mostly a liquid hydrogen planet with a small core of heavy elements, not much bigger than Earth. (NASA/JPL/University of Arizona)

Density, oblateness, and gravity measurements made by planetary probes allow planetary astronomers to model Saturn's interior. (NASA/STScI)

Celestial Profile 6: *Jupiter and Saturn*

Motion:

	Jupiter	Saturn
Average distance from the sun	5.20 AU	9.54 AU
Eccentricity of orbit	0.0484	0.0560
Inclination of orbit to ecliptic	1.3°	2.5°
Orbital period	11.9 y	29.5 y
Period of rotation	9.92 h	10.66 h
Inclination of equator to orbit	3.1°	26.4°

Characteristics:

	Jupiter	Saturn
Equatorial diameter	1.43×10^5 km (11.2 $D_\oplus$)	1.30×10^5 km (9.4 $D_\oplus$)
Mass	1.90×10^{27} kg (318 $M_\oplus$)	5.69×10^{26} kg (95 $M_\oplus$)
Average density	1.34 g/cm³	0.69 g/cm³
Gravity at base of clouds	2.5 Earth gravities	1.2 Earth gravities
Temperature at cloud tops	140 K (−200°F)	95 K (−290°F)
Albedo	0.51	0.61
Oblateness	0.064	0.102

Jupiter's Family of Moons

Jupiter has four large moons and at least 25 smaller moons. Larger telescopes and modern techniques are rapidly finding more small moons orbiting each of the Jovian planets. Most of the small moons are probably captured asteroids. In contrast, the four largest moons (■ Figure 14-7), called the Galilean moons after their discoverer, Galileo, are clearly related to each other and probably formed with Jupiter.

The outermost Galilean moons, Ganymede and Callisto, are about the size of Mercury, one and a half times the size of Earth's moon. In fact, Ganymede is the largest moon in the solar system.

Ganymede and Callisto have low densities of only 1.9 and 1.8 g/cm³, respectively, meaning they must consist roughly of half rock and half ice. Observations of their gravitational fields by the Galileo spacecraft reveal that both moons have rocky or metallic cores and lower-density icy exteriors, so they have both differentiated. Both moons interact with Jupiter's magnetic field in a way that shows they probably have mineral-rich layers of liquid water 100 km or more below their icy crusts.

Callisto's surface and most of Ganymede's surface appear old because they are heavily cratered and very dark. The continuous blast of micrometeorites evaporates surface ice, leaving behind

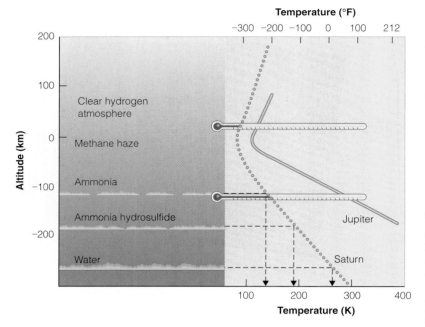

Temperature (°F)

| −300 | −200 | −100 | 0 | 100 | 212 |

Clear hydrogen atmosphere

Methane haze

Ammonia

Ammonia hydrosulfide

Water

Jupiter

Saturn

Temperature (K)

cal orbit caused by its interactions with the other moons. Jupiter's gravitational field flexes Io with tides, and the resulting friction heats its interior. That heat flowing outward causes the volcanism. Europa is not as active as Io, but it too must have a heat source, presumably tidal heating. Ganymede is no longer active, but when it was younger it must have had internal heat to break the crust and produce the grooved terrain.

A History of Jupiter

Can you put all of the evidence together and tell the story of Jupiter? Creating such a logical argument from evidence and hypotheses is the ultimate goal of planetary astronomy.

Jupiter formed far enough from the sun to incorporate large numbers of icy planetesimals, and it must have grown rapidly. Once it was about 10 to 15 times more massive than Earth, it could grow by gravitational collapse (see Chapter 12), capturing gas directly from the solar nebula. Thus, it grew rich in hydrogen and helium from the solar nebula. Its present composition resembles the composition of the solar nebula and is also quite sunlike. Jupiter's gravity is strong enough to hold on to all of its gases, even hydrogen.

The large family of moons may be mostly captured asteroids, and Jupiter may still encounter a wandering asteroid or comet now and then. Some of these are deflected, some captured into orbit, and some, like a comet that struck Jupiter in 1994 (see Figure 12-12), actually fall into the planet. Dust blasted off of the inner moons by micrometeorites settles into the equatorial plane to form Jupiter's rings.

The four Galilean moons are large and seem to have formed like a mini-solar system in a disk of gas and dust around the forming planet. The innermost, Io, is densest, and the densities of the others decrease as you move away from Jupiter, similar to the way the densities of the planets decrease with distance from the sun. Perhaps the inner moons incorporated less ice because they formed closer to the heat of the growing planet. You can recognize that tidal heating also has been important, and the intense warming of the inner moons could have driven off much of their ices. Thus two processes together are probably responsible for the differences in compositions of the Galilean moons.

embedded minerals to form a dark skin like the grimy crust on an old snowbank, so surfaces get darker with age. More recent impacts dig up cleaner ice and leave bright craters, as you can see on Callisto in Figure 14-7. Ganymede has some younger, brighter grooved terrain believed to be systems of faults in the brittle crust. Some sets of grooves overlap other sets of grooves, suggesting extended episodes of geological activity.

The density of the next moon inward, Europa, is 3 g/cm³, high enough to mean that the moon is mostly rock with a thin icy crust. The visible surface is very clean ice, contains very few craters, has long cracks in the icy crust, and has complicated terrain that resembles blocks of ice in Earth's Arctic Ocean (■ Figure 14-8). The pattern of folds on its surface suggests that the icy crust breaks as the moon is flexed by tides (see Chapter 3). Europa's gravitational influence on the Galileo spacecraft reveals that a liquid-water ocean perhaps 200 km deep lies below the 10- to 100-km-thick crust. The lack of craters tells you that Europa is an active world where craters are quickly erased.

Images from spacecraft reveal that Io, the innermost of the four Galilean moons, has over 100 volcanic vents on its surface (■ Figure 14-9). The active volcanoes throw sulfur-rich gas and ash high above the surface; the ash falls back to bury the surface at a rate of a few millimeters a year. This explains why you see no impact craters on Io—they are covered up as fast as they form. Io's density is 3.6 g/cm³, showing that it is not ice but rather rock and metal. Its gravitational influence on the passing Galileo spacecraft revealed that it is differentiated into a large metallic core, a rocky mantle, and a low-density crust.

The activity you see in the Galilean moons must be driven by energy flowing outward, yet these objects are too small to have remained hot from the time of their formation. Io's volcanism seems to be driven by **tidal heating.** Io follows a slightly ellipti-

Zones are brighter than belts because rising gas forms clouds high in the atmosphere, where sunlight is strong.

Altitude

Belt

Zone

North

Equator

a

b

■ Figure 14-4

(a) On Earth, the temperature difference between poles and equator drives a wave-shaped high-speed wind (the "jet stream") that organizes high- and low-pressure areas into circulation patterns familiar from weather maps. The poles and equator on Jupiter are about the same temperature, perhaps because of heat rising from the interior. Consequently, instead of wave-shaped winds, the planet's rapid rotation stretches high- and low-pressure areas into belts and zones that circle the planet. (b) Zones are brighter than belts because rising gas forms clouds high in the atmosphere where sunlight is stronger and reflected back to outside viewers.

14-3 Saturn

THE ROMAN GOD SATURN, protector of the sowing of seed, was celebrated in a weeklong Saturnalia at the time of the winter solstice in late December. Early Christians took over the holiday to celebrate Christmas.

Saturn is most famous for its beautiful rings, which are easily visible through the telescopes of modern amateur astronomers. Large Earth-based telescopes have explored the planet's atmosphere, rings, and moons. The two Voyager spacecraft flew past Saturn in 1979, and the Cassini spacecraft went into orbit around Saturn in 2004 on an extended exploration of the planet, its rings, and its moons (**Window on Science 14-1**).

Saturn the Planet

As you can see in Figure 14-1, Saturn has only faint belt–zone circulation, but Voyager, Hubble Space Telescope, and Cassini images show that belts and zones are present and that the associated winds blow up to three times faster than on Jupiter. Belts and zones on Saturn are less visible than on Jupiter because they occur deeper in the cold atmosphere, below a layer of methane haze (Figure 14-3).

Saturn is less dense than water (it would float!), suggesting that it is, like Jupiter, rich in hydrogen and helium. As photographs show, Saturn is the most oblate of the planets, and that evidence tells you that its interior is mostly liquid with a small core of heavy elements (Celestial Profile 6). Because its internal pressure is lower, Saturn has less liquid metallic hydrogen than Jupiter. Perhaps that is why Saturn's magnetic field is 20 times weaker than Jupiter's. Like Jupiter, Saturn radiates more energy than it receives from the sun, and models predict that it has a very hot interior.

Saturn's Rings

Study **The Ice Rings of Saturn** on pages 300–301 and notice three things:

1 The rings are made up of billions of ice particles, each in its own orbit around the planet. The ring particles you observe now can't be as old as Saturn. The rings must be replenished now and then by impacts on Saturn's moons or other processes. The same is true of the rings around the other Jovian planets.

2 The gravitational effects of small moons can confine some rings in narrow strands or keep the edges of rings sharp. Moons can also produce waves in the rings that are visible as tightly wound ringlets.

3 The ring particles are confined in a thin plane spread among small moons and confined by gravitational interactions with larger moons. The rings of Saturn, and the rings of the other Jovian worlds, are created by and controlled by the planet's moons. Without the moons, there would be no rings.

Jupiter's atmosphere is a very thin layer of turbulent gas above the liquid interior. It makes up only about 1 percent of the radius of the planet.

■ **Figure 14-5**

The Great Red Spot is a giant circulating storm embedded in one of the southern hemisphere zones in Jupiter's atmosphere. It has lasted at least 300 years since astronomers noticed it after the invention of the telescope. Other smaller spots are also circulating storms. (NASA/JPL)

■ **Active Figure 14-6**

The main ring of Jupiter, illuminated from behind, glows brightly in this visual-wavelength image made by the Galileo spacecraft from within Jupiter's shadow. (NASA)

ThomsonNOW™ Sign in at www.thomsonedu.com and go to ThomsonNOW to see the Active Figure "Roche Limit."

Visual-wavelength images

Size of Earth's moon

■ **Figure 14-7**

The Galilean moons of Jupiter, from left to right, are Io, Europa, Ganymede, and Callisto. The circle shows the size of Earth's moon. (NASA)

■ **Figure 14-8**

Like icebergs on an arctic ocean, blocks of crust on Europa appear to have floated apart. The blue icy surface is stained brown by mineral-rich water venting from below the crust. White areas are ejecta from the impact that formed nearby crater Pwyll. (NASA)

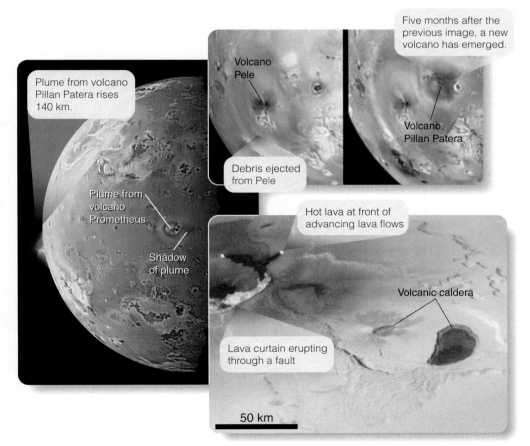

Five months after the previous image, a new volcano has emerged.

Plume from volcano Pillan Patera rises 140 km.

Volcano Pele

Volcano Pillan Patera

Plume from volcano Prometheus

Debris ejected from Pele

Shadow of plume

Hot lava at front of advancing lava flows

Volcanic caldera

Lava curtain erupting through a fault

50 km

■ **Figure 14-9**

These color images of volcanic features on Io were produced by combining visual and near-infrared images and then digitally enhancing the color. To human eyes, most of Io would look pale yellow and light orange. (NASA)

The Ice Rings of Saturn

1 The brilliant rings of Saturn are made up of billions of ice particles ranging from microscopic specks to chunks bigger than a house. Each particle orbits Saturn in its own circular orbit. Much of what astronomers know about the rings was learned when the Voyager 1 spacecraft flew past Saturn in 1980, followed by the Voyager 2 spacecraft in 1981. The Cassini Spacecraft reached orbit around Saturn in 2004. From Earth, astronomers see three rings labeled A, B, and C. Voyager and Cassini images reveal over a thousand ringlets within the rings.

Saturn's rings can't be leftover material from the formation of Saturn. The rings are made of ice particles, and the planet would have been so hot when it formed that it would have vaporized and driven away any icy material. Rather, the rings must be debris from collisions between passing comets and Saturn's icy moons. Such impacts should occur every 10 million years or so, and they would scatter ice throughout Saturn's system of moons. The ice would quickly settle into the equatorial plane, and some would become trapped in rings. Although the ice may waste away due to meteorite impacts and damage from radiation in Saturn's magnetosphere, new impacts could replenish the rings with fresh ice. The bright, beautiful rings you see today may be only a temporary enhancement caused by an impact that occurred since the extinction of the dinosaurs.

Encke's division

Cassini's division

A ring

B ring

C ring
The Crepe ring

Earth to scale

Visual-wavelength image

NASA

As in the case of Jupiter's ring, Saturn's rings lie inside the planet's Roch´e limit where the ring particles cannot pull themselves together to form a moon.

Because it is so dark, the C ring has been called the Crepe ring, referring to the black, semitransparent cloth associated with funerals.

1a An astronaut could swim through the rings. Although the particles orbit Saturn at high velocity, all particles at the same distance from the planet orbit at about the same speed, so they collide gently at low velocities. If you could visit the rings, you could push your way from one icy particle to the next. This artwork is based on a model of particle sizes in the A ring.

The C ring contains boulder-size chunks of ice, whereas most particles in the A and B ring are more like golf balls, down to dust-size ice crystals. Further, C ring particles are less than half as bright as particles in the A and B rings. Cassini observations show that the C ring particles contain less ice and more minerals.

2 Because of collisions among ring particles, planetary rings should spread outward. The sharp outer edge of the A ring and the narrow F ring are confined by **shepherd satellites** that gravitationally usher straying particles back into the rings.

Some gaps in the rings, such as Cassini's Division, are caused by resonances with moons. A particle in Cassini's Division orbits Saturn twice for each orbit of the moon Mimas and three times for each orbit of Enceladus. On every other orbit, the particle feels a gravitational tug from Mimas and, on every third orbit, a tug from Enceladus. These tugs always occur at the same places in the orbit and force the orbit to become slightly elliptical. Such an orbit crosses the orbits of other particles, which results in collisions, and that removes the particle from the gap.

Pandora

Visual-wavelength image

F ring close up

Prometheus

The F ring is clumpy and braided because of two shepherd satellites.

F ring

This image was recorded by the Cassini spacecraft looking up at the rings as they were illuminated by sunlight from above. Saturn's shadow falls across the upper side of the rings.

Visual-wavelength images

Waves in the A ring

Encke's Division

Encke's Division is not empty. Note the ripples at the inner edge. A small moon orbits inside the division.

Saturn does not have enough moons to produce all of its ringlets by resonances. Many are produced by tightly wound waves, much like the spiral arms found in disk galaxies.

Cassini's Division

A ring

Encke's Division

This combination of UV images has been given false color to show the ratio of mineral material to pure ice. Blue regions such as the A ring are the purest ice, and red regions such as Cassini's division are the dirtiest ice. How the particles become sorted by composition is unknown.

3 How do moons happen to be at just the right places to confine the rings? That puts the cosmic cart before the horse. The ring particles get caught in the most stable orbits among Saturn's innermost moons. The rings push against the inner moons, but those moons are locked in place by resonances with larger, outer moons. Without the moons, the rings would spread and dissipate.

Saturn's rings are a very thin layer of particles and nearly vanish when the rings turn edge-on to Earth. Although ripples in the rings caused by waves may be hundreds of meters high, the sheet of particles may be only 10 meters thick.

Ultraviolet image

Who Pays for Science?

Science is an expensive enterprise, and that raises the question of payment. Some science has direct technological applications with immediate practical use, so the business community funds much of this type of research. For example, auto manufacturers need inexpensive, durable, quick-drying paint for their cars, and they find it worth the cost to hire chemists to study the way paint dries. Many industries have large research budgets, and some industries, such as pharmaceutical manufacturers, depend exclusively on scientific research to discover and develop new products.

If a field of research has immediate potential to help society, it is likely that government will supply funds. Research on public health is funded mostly by government institutions such as the National Institutes of Health and the National Science Foundation. The practical benefit of finding new ways to prevent disease, for example, is well worth the tax dollars.

Basic science, however, has no immediate practical use. That doesn't mean it is useless, but it does mean that the practical-minded stockholders of a company will not approve major investments in such research. Digging up dinosaur bones, for instance, is very poorly funded because no industry can make a profit from the discovery of a new dinosaur. Astronomy is another field of science that has few direct applications, and thus very little astronomical research is funded by industry.

The value of basic research is twofold. Discoveries that have no known practical use today may be critically important years from now. Thus, society needs to continue basic research to protect its own future. But basic research, such as studying Jupiter's atmosphere or Saturn's rings, is also of cultural value because it tells us something about Earth and indirectly about what we are. Each of us benefits in intangible ways from such research, and thus society needs to fund basic research for the same reason it funds art galleries and national parks—to make our lives richer and more fulfilling.

Because there is no immediate financial return from this kind of research, it falls to government institutions and private foundations to pay the bill. The Keck Foundation has built two giant telescopes with no expectation of financial return, and the National Science Foundation has funded thousands of astronomy research projects for the benefit of society. Debates rage as to how much money is

Sending the Cassini spacecraft to Saturn cost each American $0.56 per year over the life of the project. (NASA)

enough and how much is too much, but, ultimately, funding basic scientific research is a public responsibility that society must balance against other needs. There isn't anyone else to pick up the tab.

The Moons of Saturn

Saturn has nearly 50 known moons, many of which are small and all of which contain mixtures of ice and rock. Many are probably captured objects.

The largest of Saturn's moons is Titan, a bit larger than the planet Mercury. Its density suggests that it must contain a rocky core under a thick mantle of ices. Titan is so cold that its gas molecules do not travel fast enough to escape, so it has an atmosphere composed mostly of nitrogen with traces of argon and methane. Sunlight converts some of the methane into complex carbon-rich molecules that collect into small particles filling the atmosphere with orange smog (■ Figure 14-10a). These particles settle slowly downward to coat the surface with what has been described as dark organic goo, meaning it is composed of carbon-rich molecules.

Titan's surface is mainly composed of ices of water and methane at −180°C (−290°F). The Cassini spacecraft dropped the Huygens probe into the atmosphere of Titan, and it photographed dark drainage channels suggesting that liquid methane falls as rain, washes the dark goo off the higher terrain, and drains into the low-lands (Figure 14-10b). Such methane downpours may be rare, however. No direct evidence of liquid methane was detected as the probe descended, but later radar images made by the Cassini orbiter have detected what appear to be lakes presumably containing liquid methane. Infrared images suggest the presence of methane volcanoes that replenish the methane in the atmosphere, so Titan must have some internal heat source to power the activity.

Most of the remaining moons of Saturn are small and icy, have no atmospheres, and are heavily cratered. Most have dark, ancient surfaces. The moon Enceladus, however, shows signs of recent geological activity. Some parts of its surface contain 1000 times fewer craters than other regions, and infrared observations show that its south polar region is unusually warm and venting water and ice geysers (■ Figure 14-11). Evidently, a reservoir of liquid water lies only tens of meters below the surface. At some point in its history, this moon must have been caught in a resonance with another moon and was warmed by tidal heating. Ice particles flying into space from geysers on Enceladus appear to maintain Saturn's large, low-density E ring, which extends far beyond the visible rings.

Visual-wavelength images

Infrared images

■ **Figure 14-10**

The Huygens probe descended by parachute through the hazy atmosphere of Titan and photographed the surface. From an altitude of 8 km (5 mi), the surface showed clear signs that liquid methane had drained over the surface and into the lowlands. Once Huygens landed on the surface, it radioed back photos showing a level plain and chunks of ice rounded by a once-moving liquid. (ESA/NASA/JPL/University of Arizona)

14-4 Uranus

NOW THAT YOU ARE FAMILIAR with the gas giants in our solar system, you will be able to appreciate how weird the ice giants, Uranus and Neptune, are. Uranus, especially, seems to have forgotten how to behave like a planet.

Uranus was discovered in 1781 by the scientist William Herschel, a German expatriate living in England. He named it *Georgium Sidus,* George's Star, after the English King George III. European astronomers, especially the French, refused to accept a planet named after an English king. They called it Herschel. Years later, German astronomer J. E. Bode suggested Uranus, the oldest of the Greek gods.

Planet Uranus

Uranus is only one-third the diameter of Jupiter and only one-twentieth as massive, and, being four times farther from the sun, its atmosphere is almost 100°C colder than Jupiter's (**Celestial Profile 7**).

Uranus never grew massive enough to capture large amounts of gas from the nebula as Jupiter and Saturn did, so it has much less hydrogen and helium. Its internal pressure is enough lower than Jupiter's that it should not contain any liquid metallic hy-drogen. Models of Uranus based in part on its density and oblateness suggest that it has a small core of heavy elements and a deep mantle of partly solid water. Although referred to as ice, this material would not be anything like ice on Earth, given the temperatures and pressures inside Uranus. The mantle also contains rocky-composition material and dissolved ammonia and methane. Circulation in this electrically conducting mantle may generate the planet's peculiar magnetic field, which is highly inclined to its axis of rotation. Above this mantle lies a deep hydrogen and helium atmosphere.

Uranus rotates on its side, with its equator inclined about 98° to its orbit. With an orbital period of 84 years, each of the four seasons lasts 21 years, and the winter–summer contrast is extreme. During a season when one of its poles is pointed nearly at the sun (a solstice), a citizen of Uranus would never see the sun rise or set. Compare this with seasons on Earth discussed in Chapter 2. Uranus's odd rotation may have been produced when it was struck by a very large planetesimal late in its formation, or by tidal interactions with the other giant planets as it migrated outward early in the history of the solar system (see Chapter 12).

Voyager 2 photos show a nearly featureless ball (■ Figure 14-12a). The atmosphere is mostly hydrogen and helium, but traces of methane absorb red light and thus make the atmosphere look green-blue. There is no belt–zone circulation visible in the

Among Saturn's moons, a few stand out. Enceladus is venting water and ice from geysers at its south pole. Iapetus is dark on its leading side and lighter at its poles and trailing side and has a peculiar equatorial ridge up to 13 km (8 mi) high. (NASA/JPL/Space Science Institute)

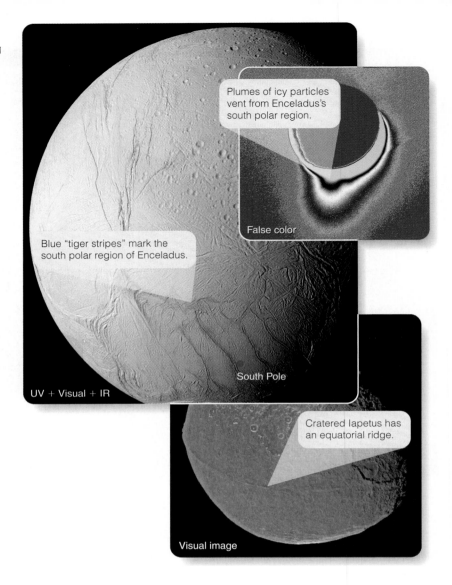

Plumes of icy particles vent from Enceladus's south polar region.

False color

Blue "tiger stripes" mark the south polar region of Enceladus.

South Pole

UV + Visual + IR

Cratered Iapetus has an equatorial ridge.

Visual image

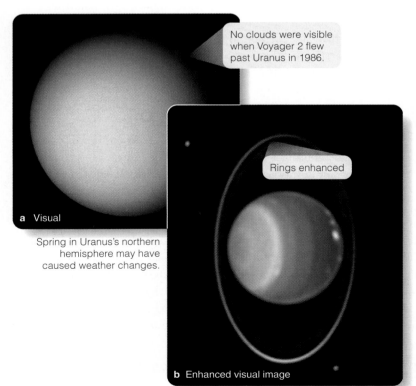

No clouds were visible when Voyager 2 flew past Uranus in 1986.

Rings enhanced

a Visual

Spring in Uranus's northern hemisphere may have caused weather changes.

b Enhanced visual image

■ **Figure 14-12**

If you visited Uranus, it would probably look like a bland green-blue ball, but observations at certain wavelengths would reveal traces of belt–zone circulation deep in the atmosphere. (NASA and Erich Karkoschka, University of Arizona)

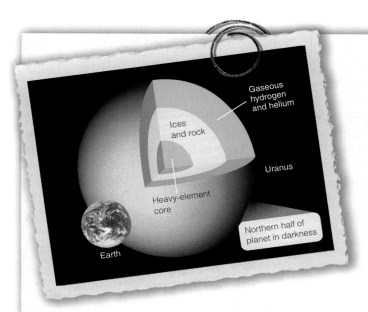

Uranus rotates on its side, and, when Voyager 2 flew past in 1986, the planet's south pole was pointed almost directly at the sun. (NASA)

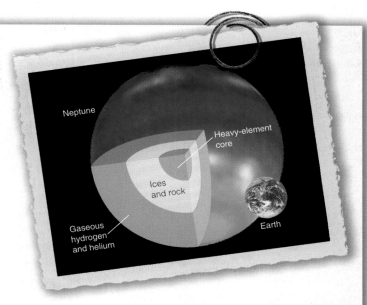

Neptune's axis was tipped slightly away from the sun when the Hubble Space Telescope recorded this image. The interior is much like Uranus's, but there is more heat flowing outward from Neptune. (NASA)

Celestial Profile 7: Uranus and Neptune

Motion:

	Uranus	Neptune
Average distance from the sun	19.2 AU	30.1 AU
Eccentricity of orbit	0.0461	0.0100
Inclination of orbit to ecliptic	0.8°	1.8°
Orbital period	84.0 y	164.8 y
Period of rotation	17.23 h	16.05 h
Inclination of equator to orbit	97.9°	28.8°

Characteristics:

	Uranus	Neptune
Equatorial diameter	5.11×10^4 km (4.01 $D_\oplus$)	4.95×10^4 km (3.93 $D_\oplus$)
Mass	8.69×10^{25} kg (14.5 $M_\oplus$)	1.03×10^{26} kg (17.2 $M_\oplus$)
Average density	1.29 g/cm³	1.66 g/cm³
Gravity at cloud tops	0.9 Earth gravities	1.2 Earth gravities
Temperature above cloud tops	55 K (−360°F)	55 K (−360°F)
Albedo	0.35	0.35
Oblateness	0.023	0.017

Voyager photographs, although extreme computer enhancement revealed a few clouds and bands around the south pole. In the decades since Voyager 2 flew past Uranus, spring has come to the northern hemisphere of Uranus and autumn to the southern hemisphere. Images made by the Hubble Space Telescope and modern Earth-based telescopes reveal changing clouds and cloud bands in both hemispheres (Figure 14-12b).

Infrared measurements show that Uranus is radiating about the same amount of energy that it receives from the sun, meaning it has much less heat flowing out of its interior than Jupiter or Saturn (or Neptune). This may account for its limited atmo-

spheric activity. Astronomers are not sure why Uranus differs in this respect from the other Jovian worlds.

The Uranian Moons

Until recently, astronomers could see only five moons orbiting Uranus. Voyager 2 discovered 10 more small moons in 1986, and yet more have been found in images recorded by new, giant telescopes on Earth.

The five major moons of Uranus are smaller than Earth's moon and have old, dark, cratered surfaces. A few have deep cracks, produced, perhaps, when the interior froze and expanded. In some cases, liquid water "lava" appears to have erupted and smoothed over some regions. Miranda, the innermost moon, is only 14 percent the diameter of Earth's moon, but its surface is marked by grooves called **ovoids** (■ Figure 14-13). These may have been caused by internal heat driving convection in the icy mantle. By counting craters on the ovoids, astronomers conclude that the entire surface is old, and the moon is no longer active.

The Uranian Rings

The rings of Uranus (Figure 14-12) are dark and faint, contain little dust, are confined by shepherd satellites, and must be continuously resupplied with material from the moons. They are not easily visible from Earth: The first hint that Uranus had rings came from **occultations,** the passage of the planet in front of a star. Most of what astronomers know about these rings comes from the observations of the Voyager 2 spacecraft. Their composition appears to be water ice mixed with methane that has been darkened by exposure to radiation.

In 2006, astronomers found two new, very faint rings orbiting far outside the previously known rings of Uranus. The newly discovered satellite Mab appears to be the source of particles for the larger ring, and the smaller of the new rings is confined between the orbits of the moons Portia and Rosalind. Note that the International Astronomical Union (IAU) has declared that the moons of Uranus are to be named after characters in Shakespeare's plays.

Visual-wavelength image

■ **Figure 14-13**

Geological activity on Uranus's moon Miranda. The face of Miranda is marked by ovoids, which are believed to have formed when internal heating caused slow convection in the ice of the moon's mantle. Note the 5-km-high cliff at the lower edge of the moon. (NASA)

Building Scientific Arguments

Why are the rings of Uranus so narrow?
Unlike the rings of Jupiter and Saturn, the rings of Uranus are quite narrow, like hoops of wire. You would expect collisions among ring particles to gradually spread the rings out into thin sheets, so something must be confining the narrow rings. In fact, two small moons have been found orbiting just inside and outside the epsilon ring. If a ring particle drifts away from the ring, the corresponding moon's gravity will nudge it back into the ring. Moons too small to detect are thought to shepherd the other rings. Thus, the rings of Uranus resemble Saturn's narrow F ring.

Now expand your argument. **How do moons happen to be in the right places to keep the rings narrow?**

■ ■ ■

14-5 Neptune

A BRITISH AND A FRENCH ASTRONOMER independently calculated the existence and location of Neptune from irregularities in the motion of Uranus. British astronomers were too slow; Neptune was discovered in 1846, and the French astronomer got the credit. Neptune looks like a tiny blue dot with no visible cloud features. Because of its blue color, astronomers named it after the god of the sea. In 1989, Voyager 2 flew past and revealed some of Neptune's secrets.

Planet Neptune

Almost exactly the same size as Uranus, Neptune has a similar interior. A small core of heavy elements lies within a slushy mantle of water, ices, and minerals (rock) below a hydrogen-rich atmosphere (Celestial Profile 7). Yet Neptune looks quite different; it is dramatically blue and has active cloud formations. Neptune has a dark blue tint because its atmosphere contains one and a half times more methane than Uranus. Methane absorbs red photons better than blue and scatters blue photons better than red, giving Neptune a blue color and Uranus a green-blue color.

Atmospheric circulation on Neptune is much more dramatic than on Uranus. When Voyager 2 flew by Neptune in 1989, the largest feature was the Great Dark Spot (■ Figure 14-14a). Roughly the size of Earth, the spot seemed to be an atmospheric circulation similar to Jupiter's Great Red Spot. Smaller spots were visible in Neptune's atmosphere, and photos showed they were circulating like hurricanes. More recently, the Hubble Space Telescope has photographed Neptune and found that the Great Dark Spot is gone, and new cloud formations have appeared, as shown in Figure 14-14b. Evidently, the weather on Neptune is changeable.

The atmospheric activity on Neptune is apparently driven by heat flowing from the interior plus some contribution from dim sunlight 30 AU from the sun. Neptune may have more at-

■ Figure 14-14

Neptune's equator is inclined almost 29 degrees in its orbit, and it experiences seasons that each last about 40 years. Since 1989, spring has come to the southern hemisphere, and the weather has clearly changed, which is surprising because sunlight at Neptune is 900 times dimmer than at Earth. (NASA, L. Sromovsky, and P. Fry, University of Wisconsin–Madison)

mospheric activity than Uranus because it has more heat flowing out of its interior, for reasons that are unclear.

Like Uranus, Neptune has a highly inclined magnetic field that must be linked to circulation in the interior. In both cases, astronomers suspect that ammonia dissolved in the liquid water mantle makes the mantle a good electrical conductor and that convection in the water, coupled with the rotation of the planet, drives the dynamo effect and generates the magnetic field.

The Neptunian Moons

Neptune has two moons that were discovered from Earth before Voyager 2 flew past in 1989. The passing spacecraft discovered six more very small moons. Since then a few more small moons have been found by astronomers using large Earth-based telescopes.

The two largest moons have peculiar orbits. Nereid, about one-tenth the size of Earth's moon, follows a large, elliptical orbit, taking nearly an Earth year to circle Neptune once. Triton, nearly 80 percent the size of Earth's moon, orbits Neptune backward—clockwise as seen from the north. These odd orbits suggest that the system was disturbed long ago in an interaction with some other body, such as a massive planetesimal.

■ Figure 14-15

Visible-wavelength image of Triton. Triton's southern polar cap is formed of nitrogen frost. Note the absence of craters and the dark smudges caused by nitrogen geysers. (NASA)

With a temperature of 37 K (−393°F), Triton has an atmosphere of nitrogen and methane about 10^5 times less dense than Earth's. A significant part of Triton is ice, and deposits of nitrogen frost are visible at the southern pole (■ Figure 14-15). Many features on Triton suggest it has had an active past. It has few craters on its surface, but it does have long faults that appear to

■ Figure 14-16

The bright disk of Neptune is hidden behind the black bar in this Voyager 2 image. Two narrow rings are visible, and a wider, fainter ring lies closer to the planet. More ring material is visible between the two narrow rings. The rings are bright in forward-scattered light, indicating that the rings contain significant amounts of very small dust particles with short lifetimes. Like the rings of Uranus, the rings of Neptune are dark (reflect very little light) and probably contain methane-rich ice darkened by radiation. (NASA)

have formed when the icy crust broke, plus large basins that seem to have been flooded repeatedly by liquids from the interior. Most exciting of all are the dark smudges visible in the southern polar cap that are interpreted as sunlight-darkened deposits of methane erupted out of liquid nitrogen geysers.

The Neptunian Rings

Neptune's rings are faint and very hard to detect from Earth, but they illustrate some interesting processes of comparative planetology (■ Figure 14-16). Neptune's rings are similar to those of Uranus but contain more dust. One of Neptune's moons is producing short arcs in the outermost ring. Neptune's ring system, like the others, is apparently resupplied by impacts on moons scattering debris that fall into the most stable places among the orbits of the moons.

Building Scientific Arguments

Why is Neptune blue?

In building this argument, you must be careful not to be misled by the words you use. The light Earth receives from Neptune is sunlight that is scattered from various layers of Neptune and journeys to your eyes. Sunlight entering Neptune's atmosphere must pass through hydrogen gas that contains a small amount of methane, which is a good absorber of longer wavelengths. As a result, red photons are more likely to be absorbed than blue photons, and that makes the light bluer. Furthermore, when the light is scattered in deeper layers, the shorter-wavelength photons are most likely to be scattered, and thus the light that finally emerges from the atmosphere and reaches your telescope lacks photons with longer wavelengths. It looks blue.

This argument shows how a careful, step-by-step analysis of a natural process can help you better understand how nature works. Now expand your argument. **Why do the clouds on Neptune look white?**

■ ■ ■

14-6) Pluto: Planet No More

OUT ON THE EDGE OF THE SOLAR SYSTEM orbits a family of small, icy worlds. Pluto was the first to be discovered, in 1930, but modern telescopes have found more.

You may have learned in school that there are nine planets in our solar system, but in 2006 the International Astronomical Union voted to remove Pluto from the list of planets. Pluto is a very small, icy world: It isn't Jovian, and it isn't terrestrial. Its orbit is highly inclined and so elliptical that Pluto actually comes closer to the sun that Neptune at times. To understand Pluto's status, you must use comparative planetology to analyze Pluto and then compare it with its neighbors.

Pluto is very difficult to observe from Earth. It has only 65 percent the diameter of Earth's moon. In Earth-based telescopes it never looks like more than a faint point of light and even in Space Telescope images it shows little detail. Orbiting so far from the sun, Pluto is cold enough to freeze most compounds you think of as gases, and spectroscopic observations have found evidence of nitrogen ice. It has a thin atmosphere of nitrogen and carbon monoxide with small amounts of methane.

Pluto has three moons. Two, named Nix and Hydra, are quite small, but Charon is relatively large, with half of Pluto's diameter. Charon orbits Pluto with a period of 6.4 days in an orbit highly inclined to the ecliptic (■ Figure 14-17). Pluto and Charon are tidally locked to face each other, so Pluto's rotation is also highly inclined.

Charon's orbit size and period plus Kepler's third law reveal that the mass of the system is only about 0.002 Earth mass. Most of that mass is Pluto, which has about 12 times the mass of Charon. Knowing the diameters and masses of Pluto and Charon allows astronomers to calculate that their densities are both about 2 g/cm³. This indicates that Pluto and Charon must each contain about 35 percent ice and 65 percent rock.

The best photos by the Hubble Space Telescope reveal almost no surface detail, but you know enough about icy moons to guess that Pluto has craters and probably shows signs of tidal heating caused by interaction with its large moon Charon. The New Horizons spacecraft will fly past Pluto in July 2015, and the images radioed back to Earth will certainly show that Pluto is an interesting world.

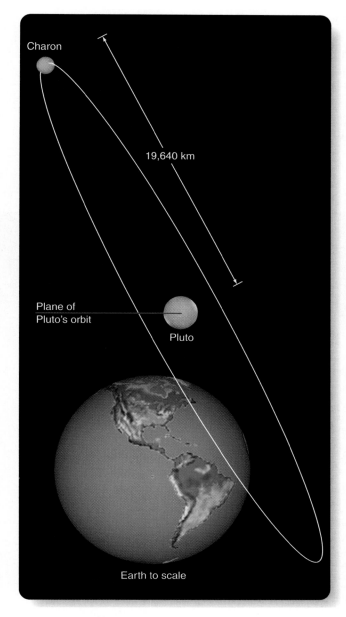

■ Figure 14-17

The circular orbit of Pluto's moon Charon is seen here at an angle. The orbit is only a few times bigger than Earth and is tipped 118° to the plane of Pluto's orbit around the sun.

What Defines a Planet?

To understand why Pluto is no longer considered a planet, you should recall the Kuiper belt (see Chapter 12). Since 1992 astronomers have discovered roughly a thousand icy bodies orbiting beyond Neptune. There may be as many as 100 million objects in the Kuiper belt larger than 1 km in diameter. They appear to be icy bodies left over from the outer solar nebula. Some of the Kuiper belt objects are quite large, and one, named Eris, is 5 percent larger in diameter than Pluto. Three other Kuiper belt

objects found so far, Sedna, 2004DW, and Quaoar (pronounced kwah-o-wahr), are half the size of Pluto or larger. Some of these objects have moons of their own. In that way, they resemble Pluto and its three moons.

A bit of comparative planetology shows that Pluto is not related to the Jovian or terrestrial planets; it is obviously a member of a newfound family of icy worlds that orbit beyond Neptune. These bodies must have formed at about the same time as the eight classical planets of the solar system, but they did not grow massive to clear their orbital zones of remnant planetesimals and remain embedded among a swarm of other objects in the Kuiper belt.

One of the IAU's criteria for planet status is that an object must be large enough to dominate and gravitationally clear its orbital region of most or all other objects. Eris and Pluto, the largest objects found so far in the Kuiper belt, and Ceres, the largest object in the asteroid belt, do not meet that standard. On the other hand all three are large enough for their gravities to have pulled them into spherical shapes, so they are the prototypes of a new class of objects defined by the IAU as **dwarf planets.**

Pluto and the Plutinos

No, this section is not about a 1950s rock and roll band. It is about the history of the dwarf planets, and it will take you back 4.6 billion years to watch the outer planets form.

Over a hundred Kuiper belt objects are known that are caught with Pluto in a 3:2 resonance with Neptune. That is, they orbit the sun twice while Neptune orbits three times. These Kuiper belt objects have been named **plutinos.** The plutinos formed in the outer solar nebula, but how did they get caught in resonances with Neptune? You have seen (Chapter 12) that models of the formation of the planets suggest that Uranus and Neptune may have formed closer to the sun and that sometime later, gravitational interactions among the Jovian planets could have gradually shifted Uranus and Neptune outward. As Neptune migrated outward, its orbital resonances could have swept up small objects like a strange kind of snowplow. The Plutinos are caught in the 3:2 resonance, and other Kuiper belt objects are caught in other resonances. This appears to support the models that predict that Uranus and Neptune migrated outward.

The migration of the outer planets would have dramatically upset the motion of some of these Kuiper belt objects, and some could have been thrown inward where they could interact with the Jovian planets. Some of those objects may have been captured as moons, and astronomers wonder if moons such as Neptune's Triton could have started life as Kuiper belt objects. Other objects may have impacted bodies in the inner solar system and caused the late heavy bombardment episode especially evident on the surface of Earth's moon. The small frozen worlds on the fringes of the solar system may hold clues to the formation of the planets 4.6 billion years ago.

What evidence indicates that cataclysmic impacts have occurred in our solar system?

To build this argument, you must cite lots of evidence. The peculiar orbits of Neptune's moons Triton and Nereid, the peculiar rotation of Uranus, and Pluto's large satellite Charon all hint that impacts and encounters with large planetesimals or comet nuclei have been important in the history of these worlds. Furthermore, the existence of planetary rings suggests that impacts have scattered small particles and replenished the ring systems. Even in the inner solar system, the backward rotation of Venus, the high density of Mercury, and the formation of Earth's moon are possible consequences of major impacts when the solar system was young.

Now assemble evidence in a new argument. **How does the origin of the plutinos give a clue about one possible source of impacting bodies in the solar system?**

■   ■   ■

Study and Review

Summary

14-1 | A Travel Guide to the Outer Planets

What are the properties of the Jovian (Jupiter-like) planets?

■ The Jovian planets, Jupiter, Saturn, Uranus, and Neptune, are large, low-density worlds rich in hydrogen and helium.

■ The atmospheres of the Jovian planets are marked by cloud belts parallel to their equators.

■ Jupiter and Saturn, usually called "gas giants," are composed mostly of liquid hydrogen and might instead be called "liquid giants." Uranus and Neptune contain water in liquid and solid form and therefore are sometimes called "ice giants."

■ All of the Jovian worlds have large systems of satellites and rings that have had complex histories.

14-2 | Jupiter

■ Jupiter's atmosphere contains three layers of clouds formed of hydrogen-rich molecules such as ammonia and water.

■ The clouds are in bands parallel to the equator called zones and belts. Zones are high-pressure regions of rising gas, and belts are lower-pressure areas of sinking gas.

■ Spots in Jupiter's atmosphere, including the Great Red Spot, are circulating weather patterns.

■ Models indicate that Jupiter has a core of heavy elements and a deep mantle of liquid metallic hydrogen in which the planet's magnetic field is generated.

■ The magnetic field around Jupiter traps high-energy particles from the sun to form intense radiation belts.

■ Jupiter must be very hot inside because heat is flowing out of it at a high rate.

■ Jupiter's ring is composed of dark particles that strongly forward-scatter light, which means the particles are very small. They are probably composed of dust resulting from meteorite impacts on moons.

■ Jupiter's ring, like all of the rings in the solar system, lies inside the planet's Roche limit, where moons would be torn apart or unable to form.

How do you know that some moons have been geologically active?

■ Grooves on Ganymede, smooth ice and cracks on Europa, and active volcanoes on Io show that tidal heating has made these moons active.

14-3 | Saturn

■ Saturn is less dense than water and contains a small core and less metallic hydrogen than Jupiter.

■ The cloud layers on Saturn occur at the same temperature as those on Jupiter, but, because Saturn is further from the sun and colder, the cloud layers are deeper in the hydrogen atmosphere below a layer of methane haze.

How are planetary rings formed and maintained?

■ Saturn's rings are composed of icy particles ranging in size from boulders to dust. In some regions the ice is purer than in other regions.

■ Grooves and other features in the rings can be produced by resonances with moons or by waves that propagate through the rings.

■ Narrow rings and sharp ring edges can be confined by shepherd satellites.

■ The rings are short lived. They cannot be as old as Saturn and must be replenished now and then with material from meteorites and comets colliding with moons. The same is true for the other Jovian planet rings.

■ Saturn's largest moon, Titan, has a cold, cloudy nitrogen and methane atmosphere. Sunlight entering Titan's atmosphere can convert methane into complex carbon-rich molecules to form haze and particles that settle out to coat the surface with dark organic goo. Methane lakes may have been detected on Titan's surface in radar images made by the Cassini probe.

■ Enceladus has a light surface with some uncratered regions. Geysers of water and ice spray from the south polar region and provide ice particles to the E ring.

14-4 | Uranus

How do the two smaller Jovian worlds—Uranus and Neptune—differ from their more massive siblings?

■ Uranus is much less massive than Jupiter, and its internal pressure cannot produce liquid hydrogen. It has a heavy-element core and a mantle of slushy or solid ice and rock below a hydrogen-rich atmosphere.

- The atmosphere of Uranus is almost featureless at visual wavelengths with a pale green-blue color caused by traces of methane, which absorbs red light.
- The larger moons of Uranus are icy and heavily cratered, with signs on some of the moons showing past geological activity.
- The rings of Uranus are narrow hoops confined by shepherd satellites. The particles appear to be ice with traces of methane darkened by radiation.
- Uranus rotates "on its side," with its axis nearly in the plane of its orbit, perhaps because of a major impact during its early history.

14-5 | Neptune

- Neptune is an ice giant like Uranus with little or no liquid hydrogen in its interior.
- The atmosphere of Neptune, marked by faint patterns of belt–zone circulation, is rich in hydrogen and colored blue by traces of methane.
- Neptune's satellite system is odd in that distant Nereid follows an extremely elliptical orbit and Triton orbits backwards.
- Triton is icy with a thin atmosphere and frosty polar caps. Smooth areas suggest past geological activity, and dark smudges mark the location of active nitrogen geysers.
- The rings of Neptune are made of dark icy particles in narrow hoops, like the rings of Uranus. Neptune's rings contain arcs produced by the gravitational influence of one or more moons.

14-6 | Pluto

- Pluto is a small, icy world with three moons, one of which is relatively large. The moons are in orbits highly inclined to Pluto's orbit around the sun. Pluto has a thin atmosphere.

Why did astronomers redefine Pluto as a "dwarf planet"?

- Pluto is evidently related to other icy objects in the Kuiper belt orbiting beyond Neptune. At least one of those objects is a bit larger than Pluto, and some, called plutinos, follow similar orbits.
- In 2006 the International Astronomical Union defined a planet as an object orbiting the sun that is large enough for its gravity to force it into a spherical shape and to have mostly or completely cleared its orbital zone of smaller objects. Pluto is spherical, but it is just one of many objects in the Kuiper belt, so it qualifies as a "dwarf planet" rather than a planet.

New Terms

oblateness (p. 293)

liquid metallic hydrogen (p. 293)

magnetosphere (p. 293)

belt–zone circulation (p. 293)

forward scattering (p. 294)

Roche limit (p. 294)

tidal heating (p. 296)

shepherd satellite (p. 299)

ovoid (p. 306)

occultation (p. 306)

dwarf planet (p. 309)

plutino (p. 309)

Review Questions

ThomsonNOW™ Assess your understanding of this chapter's topics with additional quizzing and animations at
www.thomsonedu.com

1. Why is Jupiter so much richer in hydrogen and helium than Earth?
2. How can Jupiter have a liquid interior and not have a liquid surface?
3. How does the dynamo effect account for the magnetic fields of Jupiter and the other Jovian planets?
4. Why are the belts and zones on Saturn less distinct than those on Jupiter?
5. Why do astronomers conclude that neither Jupiter's ring nor the rings of the other Jovian planets can be left over from the formation of the planets?
6. Explain why geological activity on Jupiter's moons varies with distance from the planet.
7. How can a moon produce a gap in a planetary ring system?
8. What makes Saturn's F ring and the rings of Uranus and Neptune so narrow?
9. What is the evidence that Enceladus is geologically active?
10. Why is the atmospheric activity of Uranus unlike that of Jupiter, Saturn, and Neptune?
11. What are the seasons on Uranus like?
12. Why are Uranus and Neptune greenish-blue and blue, respectively?
13. What evidence is there that Neptune's moon Triton has been geologically active recently?
14. What evidence indicates that catastrophic impacts have occurred in the solar system's past?
15. This photo shows a segment of the surface of Jupiter's moon Callisto. Why is the surface dark? Why are some craters dark and some white? What does this image tell you about the history of Callisto?

NASA/JPL

16. The Cassini spacecraft recorded this photo of Saturn's A ring and Encke's division. What do you see in this photo that tells you about processes that confine and shape planetary rings?

NASA/JPL/Space Science Institute

17. Two images of Uranus show it as it would look to the eye and through a red filter that enhances methane clouds in the northern hemisphere. Why didn't Voyager 2 photograph the northern hemisphere? What do the visible features tell you about atmospheric circulation on Uranus?

NASA and Heidi Hammel

Discussion Questions

1. Some astronomers argue that Jupiter and Saturn are unusual, while other astronomers argue that all planetary systems should contain one or two such giant planets. What do you think? Support your argument with evidence.
2. Why don't the terrestrial planets have rings? If you were to search for a ring among the terrestrial planets, where would you look first?

15 | Life on Other Worlds

Did I solicit thee from

darkness to promote me?

Adam, to God,
in JOHN MILTON's
PARADISE LOST

AS A LIVING THING, you have been promoted from darkness. The atoms of carbon, oxygen, and other heavy elements that are necessary components of your body did not exist at the beginning of the universe but were created by successive generations of stars.

The elements from which you are made are common everywhere in the observable universe, so it is possible that life began on other worlds and evolved to intelligence there as well. If so, perhaps those other civilizations will be detected from Earth. Future astronomers may discover distant alien races completely different from any life on Earth. Your goal in this chapter is to try to understand the most intriguing of puzzles—the origin and evolution of life on Earth and possibly on other worlds (Window on Science 15-1).

Every life form on Earth has evolved to survive in some ecological niche. The Wekiu bug lives with the astronomers at an altitude of 13,800 feet atop the Hawaiian volcano Mauna Kea. It inhabits spaces between the icy cinders and eats insects carried up by ocean breezes. (Kris Koenig/Coast Learning Systems)

Guidepost

Looking Back

This chapter is either unnecessary or vital, depending on your point of view. If you believe that astronomy is the study of the physical universe above the clouds, then you are done; the previous 14 chapters completed your study of astronomy. But if you believe that astronomy is the study of your position in the universe, not just your physical location but also your role as a living being in the evolution of the universe, then everything you have learned so far from this book was preparation for this chapter.

This Chapter

You will begin by tackling a question that has troubled scientists and philosophers for centuries: What is life? You won't get more than the beginning of an answer to that question here, but thinking about it will illuminate the problem and prepare you to search for life on other worlds. As you read this chapter, you will ask four related questions:

What is life?

How did life originate on Earth?

Could life begin on other worlds?

Can Earthlings communicate with civilizations on other worlds?

These are difficult questions, but often in science asking a question is more important than getting an immediate answer.

Looking Ahead

You have explored the universe from the phases of the moon to the big bang, from the origin of Earth to the death of the sun. Astronomy is important, not because it is about stars and galaxies, but because it is about you: It can tell you what you are. Now that you know astronomy, you see yourself and your world in a different way. Astronomy has changed you.

Window on Science | 15-1

The Nature of Scientific Explanation

Science is a way of understanding the world around you, and at the heart of that understanding are explanations that science gives you for natural phenomena. Whether you call these explanations stories, histories, hypotheses, or theories, they are attempts to describe how nature works based on evidence and intellectual honesty. While you may take these explanations as factual truth, you should understand that they are not the only explanations that describe the natural world.

A separate class of explanations involves religion. The Old Testament description of the creation of the world does not fit with scientific observations, but it is a way of understanding the world nonetheless. Religious explanations are based partly on faith rather than on strict rules of logic and evidence, and it is wrong to demand that they follow the same rules as scientific explanations. In the same way, it is wrong to demand that scien-

tific explanations take into account religious beliefs. The so-called conflict between science and religion arises when people fail to recognize that science and religion are different ways of knowing about the world.

Scientific explanations are very convincing because science has been very successful at producing technological innovations that have changed the world you live in. From new vaccines to digital music players to telescopes that can observe the most distant galaxies, the products of the scientific process are all around you. Scientific explanations have provided tremendous insights into the workings of nature, and it is easy to forget that there are other explanations.

Science and religion both have a lot to offer by their differing ways of explaining the world, but the two ways follow separate rules and cannot be judged by each other's standards. The trial of Galileo (see Chapter 3) can be understood as a conflict between these two ways of knowing.

Galileo's telescope gave him a new way to know about the world.

15-1 The Nature of Life

WHAT IS LIFE? Philosophers have struggled with that question for thousands of years, and it is not possible to answer it in a single chapter. An attempt at a general definition of what living things do, distinguishing them from nonliving things, might be: Life is a *process* by which an organism extracts energy from the surroundings, maintains itself, and modifies the surroundings to promote its own survival and reproduction.

One very important observation is that all living things on Earth, no matter how apparently different, share certain characteristics in how they perform the process of life.

The Physical Basis of Life

The physical basis of life on Earth is the element carbon (■ Figure 15-1). Because of the way carbon atoms bond to each other and to other atoms, they can form long, complex, stable chains that are capable of storing and transmitting information. A large amount of information of some sort is necessary to maintain the forms and control the functions of living things.

It is possible that life on other worlds could use silicon instead of carbon, but that seems unlikely because silicon chains are harder to assemble and disassemble than their carbon counterparts and can't be as lengthy. Science fiction has proposed even

stranger life forms based on electromagnetic fields and ionized gas, and none of these possibilities can be ruled out. These hypothetical life forms make for fascinating speculation, but they can't be studied as systematically as life on Earth. This chapter is concerned with the origin and evolution of life as it is on Earth, based on carbon, not because of lack of imagination, but because it is the only form of life about which we know anything.

Even carbon-based life has its mysteries. What makes a lump of carbon-based molecules a living thing? An important part of the answer lies in the transmission of information from one molecule to another.

Information Storage and Duplication

Almost every action performed by a living cell is carried out by chemicals it manufactures. Cells must store recipes for all these chemicals, use them when they need them, and pass them on to their offspring.

Study **DNA: The Code of Life** on pages 316–317 and notice three important points:

1 The chemical recipes of life are stored as templates on DNA molecules. DNA templates automatically guide specific chemical reactions within the cell.

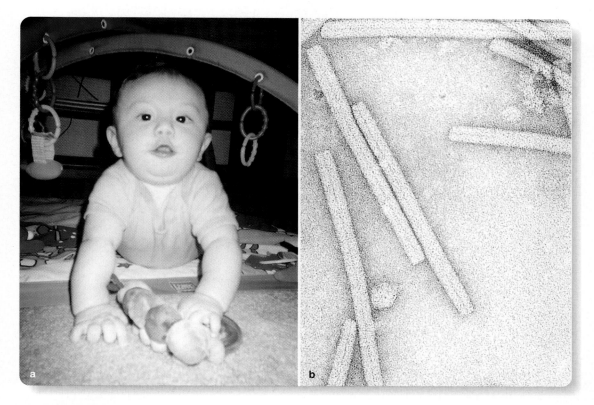

■ Figure 15-1

All living things on Earth are based on carbon chemistry. Even the long molecules that carry genetic information, DNA and RNA, have a framework defined by chains of carbon atoms. (a) Jason, a complex mammal, contains about 30 AU of DNA. (Jamie Backman) (b) Each rod of the tobacco mosaic virus contains a single spiral strand of RNA about 0.01 mm long. (L. D. Simon)

2 The instructions stored in DNA are genetic information handed down to offspring.

3 The DNA molecule reproduces itself when a cell divides so that each new cell contains a copy of the original information.

To produce viable offspring, a cell must be able to make copies of its DNA. Surprisingly, it is important for the continued existence of all life that not all the copies be exact duplicates.

Modifying the Information

Earth's environment changes continuously. To survive, species must change as their food supply, climate, or home terrain changes. If the information stored in DNA could not change, then life would quickly go extinct. The process by which life adjusts itself to its changing environment is called **biological evolution.**

When an organism reproduces, its offspring receive a copy of its DNA. Sometimes external effects like radiation alter the DNA during the parent organism's lifetime, and sometimes mistakes are made in the copying process, so that occasionally the copy is slightly different from the original. Offspring born with random alterations to their DNA are called **mutants.** Most mutations make no difference, but some mutations are fatal, killing the afflicted organisms before they can reproduce. In rare but vitally important cases, a mutation can actually help an organism survive.

These changes produce variation among the members of a species. All of the squirrels in the park may look the same, but they carry a range of genetic variation. Some may have slightly longer tails or faster-growing claws. These variations make almost no difference until the environment changes. For example, if the environment becomes colder, a squirrel with a heavier coat of fur will, on average, survive longer and produce more offspring than its normal contemporaries. Likewise, the offspring that inherit this beneficial variation will also live longer and have more offspring of their own. These differing rates of survival and reproduction are examples of **natural selection.** Over time, the beneficial variation increases in frequency, and a species can evolve until the entire population shares the trait. In this way, natural selection adapts species to their changing environments by selecting, from the huge array of random variations, those that would most benefit the survival of the species.

DNA: The Code of Life

1 The key to understanding life is information — the information that guides all of the processes in an organism. In most living things on Earth, that information is stored on a long spiral molecule called **DNA (deoxyribonucleic acid).**

1a The DNA molecule looks like a spiral ladder with rails made of phosphates and sugars. The rungs of the ladder are made of four chemical bases arranged in pairs. The bases always pair the same way. That is, base A always pairs with base T, and base G always pairs with base C.

1b Information is coded on the DNA molecule by the order in which the base pairs occur. To read that code, molecular biologists have to "sequence the DNA." That is, they must determine the order in which the base pairs occur along the DNA ladder.

The Four Bases

A — Adenine

C — Cytosine

G — Guanine

T — Thymine

2 DNA automatically combines raw materials to form important chemical compounds. The building blocks of these compounds are relatively simple **amino acids.** Segments of DNA act as templates that guide the amino acids to join together in the correct order to build specific **proteins,** chemical compounds important to the structure and function of organisms. Some proteins called **enzymes** regulate other processes. In this way, DNA recipes regulate the production of the compounds of life.

Jamie Backman

The traits you inherit from your parents, the chemical processes that animate you, and the structure of your body are all encoded in your DNA. When people say "you have your mother's eyes" they are talking about DNA codes.

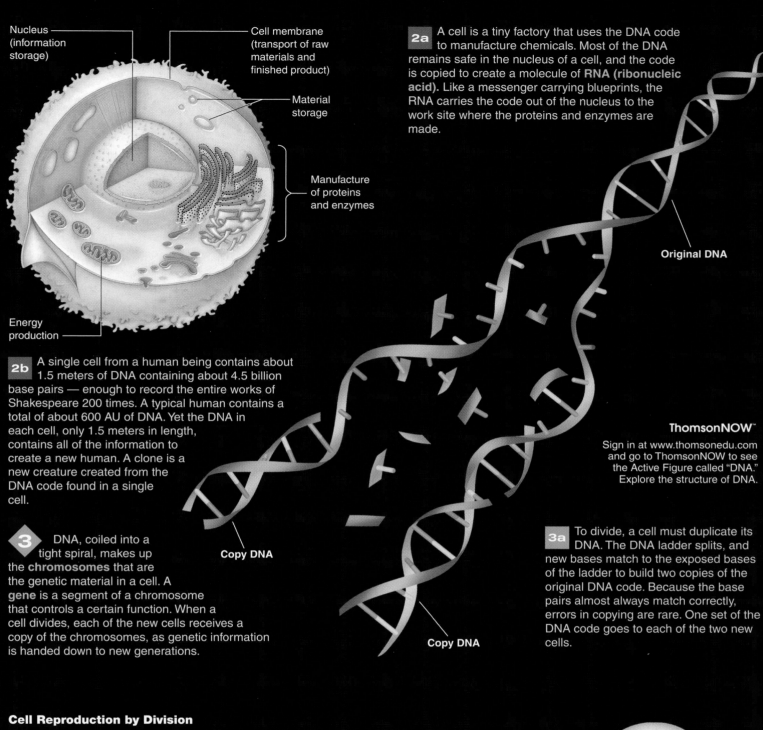

Nucleus
(information
storage)

Cell membrane
(transport of raw
materials and
finished product)

Material
storage

Manufacture
of proteins
and enzymes

Energy
production

2a A cell is a tiny factory that uses the DNA code to manufacture chemicals. Most of the DNA remains safe in the nucleus of a cell, and the code is copied to create a molecule of **RNA (ribonucleic acid).** Like a messenger carrying blueprints, the RNA carries the code out of the nucleus to the work site where the proteins and enzymes are made.

Original DNA

2b A single cell from a human being contains about 1.5 meters of DNA containing about 4.5 billion base pairs — enough to record the entire works of Shakespeare 200 times. A typical human contains a total of about 600 AU of DNA. Yet the DNA in each cell, only 1.5 meters in length, contains all of the information to create a new human. A clone is a new creature created from the DNA code found in a single cell.

ThomsonNOW™

Sign in at www.thomsonedu.com and go to ThomsonNOW to see the Active Figure called "DNA." Explore the structure of DNA.

3 DNA, coiled into a tight spiral, makes up the **chromosomes** that are the genetic material in a cell. A **gene** is a segment of a chromosome that controls a certain function. When a cell divides, each of the new cells receives a copy of the chromosomes, as genetic information is handed down to new generations.

Copy DNA

Copy DNA

3a To divide, a cell must duplicate its DNA. The DNA ladder splits, and new bases match to the exposed bases of the ladder to build two copies of the original DNA code. Because the base pairs almost always match correctly, errors in copying are rare. One set of the DNA code goes to each of the two new cells.

Cell Reproduction by Division

As a cell begins to divide, its DNA duplicates itself.

The duplicated chromosomes move to the middle.

The two sets of chromosomes separate, and . . .

the cell divides to produce . . .

two cells, each containing a full set of the DNA code.

It is commonly believed that evolution is random, but that is not true. The underlying variation within species is random, but natural selection is not random because progressive changes in a species are directed by changes in the environment.

Building Scientific Arguments

Why isn't the information in DNA always copied perfectly?
Sometimes the most valuable scientific arguments are those that challenge common misconceptions. It seems obvious that mistakes shouldn't be made in copying DNA, but variation is necessary for long-term survival. For example, the DNA in a starfish contains all the information the starfish needs to grow, develop, survive, and reproduce. The information must be passed on to the starfish's offspring for them to survive. But that information must change if the environment changes. A change in the ocean's temperature may kill the specific shellfish that the starfish eat. If none of the starfish are able to digest another kind of food—if all the starfish have exactly the same DNA—they will all die. But if a few starfish are born with the ability to make enzymes that can digest a different kind of shellfish, the species may be able to carry on.

Variations in DNA are caused both by external factors such as radiation and by occasional mistakes in the copying process. The survival of life depends on this delicate balance between reliable reproduction and the introduction of small variations in DNA. Now build a new argument. **How does DNA make copies of itself?**

■   ■   ■

15-2 The Origin of Life

IT IS OBVIOUS that the 4.5 billion chemical bases that make up human DNA did not just come together in the right order by chance. The key to understanding the origin of life lies in the processes of evolution. The complex interplay of environmental factors with the DNA of generation after generation of organisms drove some life forms to become more sophisticated over time, until they became the unique and specialized creatures on Earth today.

This means that life on Earth could have begun very simply, even as simple a form as carbon chain molecules able to copy themselves. Of course, this is a hypothesis for which you can seek evidence. What evidence exists regarding the origin of life on Earth?

The Origin of Life on Earth

The oldest fossils are all the remains of sea creatures, and this indicates that life began in the sea. Identifying the oldest fossils is not easy, however. Rocks from western Australia that are more

■ Figure 15-2

A fossil stromatolite from western Australia that is more than 3 billion years old, constituting some of the oldest evidence of life on Earth. Stromatolites are formed, layer upon layer, by mats of bacteria living in shallow water where they are covered repeatedly by sediments. (Chip Clark, National Museum of Natural History © 2006 Smithsonian Institution)

than 3 billion years old contain features that experts identify as **stromatolites,** fossilized remains of colonies of single-celled organisms (■ Figure 15-2). Fossils this old are difficult to recognize because the earliest living things contained no easily preserved hard parts like bones or shells, and because the individual organisms were microscopic. The evidence, though scarce, indicates that simple organisms lived in Earth's oceans 3.4 billion or more years ago, less than 1.1 billion years after Earth formed. Where did these simple organisms come from?

An important experiment performed by Stanley Miller and Harold Urey in 1952 sought to recreate the conditions in which life on Earth began. The **Miller experiment** consisted of a sterile, sealed glass container holding water, hydrogen, ammonia, and methane. An electric arc inside the apparatus created sparks to simulate the effects of lightning in Earth's early atmosphere (■ Figure 15-3).

Miller and Urey let the experiment run for a week and then analyzed the material inside. They found that the interaction between the electric arc and the simulated atmosphere had produced many organic molecules from the raw material of the experiment, including such important building blocks of life as amino acids. When the experiment was run again using different energy sources such as hot silica to represent molten lava spilling into the ocean, similar molecules were produced. Even the amount of UV radiation present in sunlight was sufficient to produce complex organic molecules.

According to updated models of the formation of the solar system and Earth (see Chapters 12 and 13), Earth's early atmosphere probably consisted mostly of carbon dioxide, nitrogen, and water vapor instead of the mix of hydrogen, ammonia, and methane assumed by Miller and Urey. When gases corresponding to the newer understanding of the early Earth atmosphere are processed in a Miller apparatus, lesser but still significant numbers of organic molecules are created.

Figure 15-3

(a) The Miller experiment circulated gases through water in the presence of an electric arc. This simulation of primitive conditions on Earth produced many complex organic molecules, including amino acids, the building blocks of proteins. (b) Stanley Miller with a Miller apparatus. (Stanley Miller)

The Miller experiment is important because it shows that complex organic molecules form naturally in a wide variety of circumstances. Lightning, sunlight, and hot lava pouring into the oceans are just some of the energy sources that naturally rearrange simple common molecules into the complex molecules that make life possible. If you could travel back in time, you would expect to find Earth's first oceans filled with a rich mixture of organic compounds called the **primordial soup.**

Many of these organic compounds would have been able to link up to form larger molecules. Amino acids, for example, can link together to form proteins by joining ends and releasing a water molecule (■ Figure 15-4). It was initially thought that this must have occurred in sun-warmed tidal pools where organic molecules were concentrated by evaporation. However, violent episodes of volcanism and catastrophic meteorite impacts would probably have destroyed any evolving life forms at the surface, so

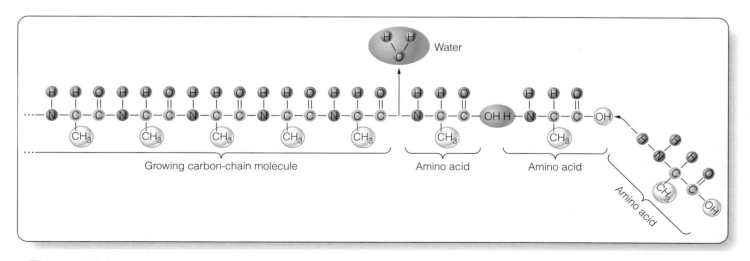

Figure 15-4

Amino acids can link together via the release of a water molecule to form long carbon-chain protein molecules. The amino acid in this hypothetical example is alanine, one of the simplest.

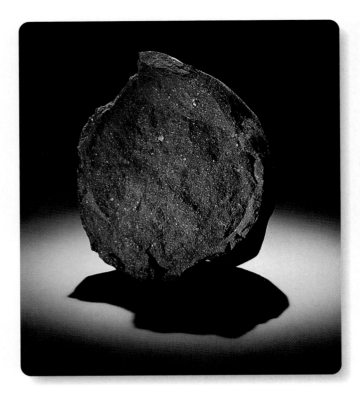

■ Figure 15-5

A piece of the Murchison meteorite, a carbonaceous chondrite (see Chapter 12) that fell in 1969 near Murchison, Australia. Analysis of the interior of the meteorite revealed the presence of amino acids. Whether the first building blocks of life originated in space is unknown, but the amino acids found in meteorites illustrate how commonly amino acids and other complex organic molecules occur in the universe, even in the absence of living things. (Chip Clark, National Museum of Natural History)

■ Figure 15-6

Single amino acids can be assembled into long proteinlike molecules. When such material cools in water, it can form microspheres, microscopic spheres with double-layered boundaries similar to cell membranes. Microspheres may have been an intermediate stage in the evolution of life from complex molecules to living cells holding molecules reproducing genetic information. (Sidney Fox and Randall Grubbs)

it is now believed that the early linkage of complex molecules more likely took place on the ocean floor perhaps near the hot springs at midocean rises (see page 270).

These complex organic molecules were still not living things. Even though some proteins may have contained hundreds of amino acids, they did not reproduce but rather linked and broke apart at random. Because some molecules are more stable than others, and some bond together more easily than others, scientists hypothesize that a process of **chemical evolution** eventually concentrated the various smaller molecules into the most stable larger forms. Eventually, according to the hypothesis, somewhere in the oceans, after sufficient time, a molecule formed that could copy itself. At that point, the chemical evolution of molecules became the biological evolution of living things.

An alternate theory for the origin of life holds that reproducing molecules may have arrived here from space. Radio astronomers have found a wide variety of organic molecules in the interstellar medium, and similar compounds have been found inside meteorites (■ Figure 15-5). The Miller experiment showed how easy it is to create organic molecules, so it is not surprising to find them in space. Although speculation is fun, the hypothesis that life arrived on Earth from space is presently more difficult to test than the hypothesis that life on Earth originated on Earth.

Whether the first reproducing molecules formed here on Earth or in space, the important thing is that they could have formed by natural processes. Scientists know enough about these processes to feel confident about them, even though some of the steps remain unknown.

The details of the evolution of the first cells are unknown, but the first reproducing molecule to surround itself with a protective membrane must have gained an important survival advantage. Experiments have shown that microscopic spheres the size of cells containing organic molecules form relatively easily in water (■ Figure 15-6), so the evolution of the cell-like structures is not surprising.

The first cells must have been simple single-celled organisms much like modern bacteria. As you learned earlier, these kinds of cells are preserved in stromatolites (Figure 15-2), mineral formations formed by layers of photosynthetic bacteria and shallow ocean sediments. Stromatolites formed in rocks with radioactive ages of 3.4 billion years, and they also still form in some places today. Stromatolites and other photosynthetic organisms would have begun adding oxygen, a product of photosynthesis, to

Artist's conception of a scene on the young Earth, 3.4 billion years ago, with stromatolite bacterial mats growing near the shores of an ocean. (Mural by Peter Sawyer/National Museum of Natural History © 2006 Smithsonian Institution)

ThomsonNOW™ Sign in at www.thomsonedu.com and go to ThomsonNOW to see the Active Figure called "Future of the Sun."

Earth's early atmosphere (■ Figure 15-7). An oxygen abundance of only 0.1 percent would have created an ozone screen, protecting organisms from the sun's ultraviolet radiation and later allowing life to colonize the land.

Over the course of eons, the natural processes of evolution gave rise to stunningly complex **multicellular** life forms with their own widely differing ways of life. It is a common misconception to imagine that life is too complex to have evolved from such simple beginnings. It is possible because small variations can accumulate, but that accumulation requires huge amounts of time.

Geologic Time

Life has existed on Earth for at least 3.4 billion years, but there is no evidence of anything more than simple organisms until about 600 million years ago, when life suddenly branched into a wide variety of complex forms like the trilobites (■ Figure 15-8).

This sudden increase in complexity is known as the **Cambrian explosion,** and it marks the beginning of the Cambrian period.

If you represented the entire history of Earth on a scale diagram, the Cambrian explosion would be near the top of the column, as shown at the left of ■ Figure 15-9. The emergence of most animals familiar to you today, including fishes, amphibians, reptiles, birds, and mammals, would be crammed into the topmost part of the chart, above the Cambrian explosion.

If you magnify this portion of the diagram, as shown on the right side of Figure 15-9, you can get a better idea of when these events occurred in the history of life. Humanoid creatures have walked the Earth for about 3 million years. This is a long time by the standard of a human lifetime, but it makes only a narrow red line at the top of the diagram. All of recorded history would be a microscopically thin line at the very top of the column.

To understand just how thin that line is, imagine that the entire 4.5-billion-year history of the Earth has been compressed onto

■ Figure 15-8

Trilobites made their first appearance in the Cambrian oceans. The smallest were almost microscopic, and the largest were bigger than dinner plates. This example, about the size of a human hand, lived 400 million years ago in an ocean floor that is now a limestone deposit in Pennsylvania. (Grundy Observatory)

a yearlong video and that you began watching this video on January 1. You would not see any signs of life until March or early April, and the slow evolution of the first simple forms would take up the next 6 or 7 months. Suddenly, in mid-November, you would see the trilobites and other complex organisms of the Cambrian explosion.

You would see no life of any kind on land until November 28, but once it appeared it would diversify quickly, and by December 12 you would see dinosaurs walking the continents. By the evening of Christmas Day they would be gone, and mammals and birds would be on the rise.

If you watched closely, you might see the first humanoid forms by suppertime on New Year's Eve, and by late evening you could see humans making the first stone tools. The Stone Age would last until 11:45 PM, and the first towns and cities would appear at about 11:54. Suddenly things would begin to happen

at lighting speed. Babylon would flourish, the Pyramids would rise, and Troy would fall. The Christian era would begin 14 seconds before the New Year. Rome would fall, the Middle Ages and the Renaissance would flicker past. The American and French revolutions would occur one second before the end of the video.

By imagining the history of Earth as a yearlong video, you have gained some perspective on the rise of life. Tremendous amounts of time were needed for the first simple living things to evolve in the oceans; but, as life became more complex, new forms arose more and more quickly as the hardest problems—how to reproduce, how to take energy efficiently from the environment, how to move around—were solved. The easier problems, like what to eat, where to live, and how to raise their young, were solved in different ways by different organisms, leading to the diversity that is seen today.

Even intelligence—that which appears to set humans apart from other animals—may be a unique solution to an evolutionary problem posed to humanity's ancient ancestors. A smart animal is better able to escape predators, outwit its prey, and feed and shelter itself and its offspring, so under certain conditions evolution is likely to select for intelligence. Could intelligent life arise on other worlds? To try to answer this question, you can estimate the chances of any type of life arising on other worlds, then assess the likelihood of that life developing intelligence.

Life in Our Solar System

Could there be carbon-based life elsewhere in our solar system? Liquid water seems to be a requirement of carbon-based life, necessary both for vital chemical reactions and as a medium to transport nutrients and wastes. It is not surprising that life developed in Earth's oceans and stayed there for billions of years before it was able to colonize the land.

Scientists are in agreement that any world harboring living things must have significant quantities of some type of liquid. Water is a cosmically abundant substance with properties such as high heat capacity that set it apart from other common molecules that are liquid at the temperatures of planetary surfaces.

Many worlds in the solar system can be eliminated immediately as hosts for life because liquid water is not possible there. The moon and Mercury are airless, and water would boil away into space immediately. Venus has traces of water vapor in its atmosphere, but it is too hot for liquid water to survive on the surface. The Jovian planets have deep atmospheres, and at a certain level it is likely that water condenses into liquid droplets. However, it seems unlikely that life could have originated there; the Jovian planets do not have solid surfaces (see Chapter 14). Isolated water droplets could never mingle to mimic the rich primordial oceans of Earth, where organic molecules grew and interacted. Additionally, powerful

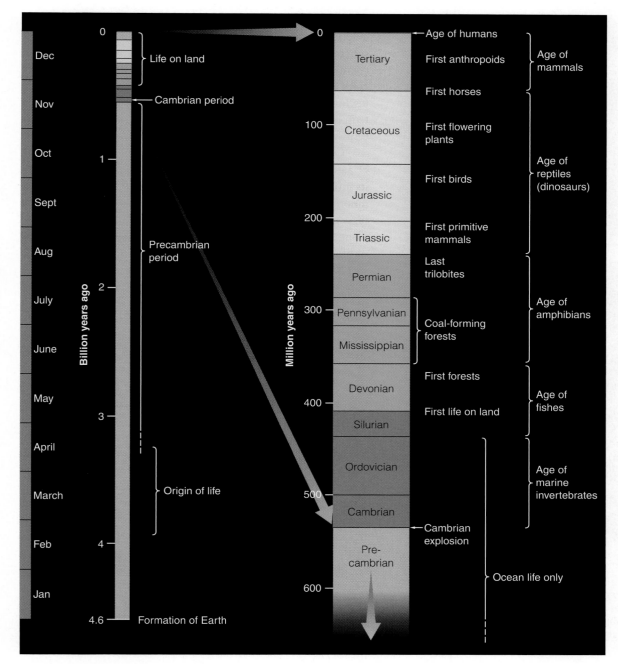

■ **Active Figure 15-9**

Complex life has developed on Earth only recently. If the entire history of Earth were represented in a time line (left), you would have to magnify the end of the line to see details such as life leaving the oceans and dinosaurs appearing. The age of humans would still be only a thin line at the top of your diagram. If the history of Earth were a yearlong videotape, humans would not appear until the last hours of December 31.

ThomsonNOW™ Sign in at www.thomsonedu.com and go to ThomsonNOW to see the Active Figure called "Earth Calendar."

Saturn's moon Titan is rich in organic molecules. Chapter 14 described how sunlight converts the methane in Titan's atmosphere into organic smog particles that settle to the surface. The chemistry of any life that could have evolved from those molecules and survived in Titan's lakes of methane is unknown. It is fascinating to consider possibilities, but Titan's extremely low temperature of $-180°C$ ($-290°F$) could make chemical reactions slow to the point where life processes seem unlikely.

Observations of water venting from the south-polar region of Saturn's moon Enceladus show that it has liquid water below its crust. It is possible that life could exist in that water, but the moon is very small and has been warmed by tidal heating that may operate only occasionally. Enceladus may not have had plentiful liquid water for the extended time necessary for the rise of life.

Mars is the most likely place for life to exist in the solar system because, as you learned in Chapter 13, there is a great deal of evidence that liquid water once flowed on its surface. Even so, results from searches for signs of life on Mars are not encouraging. The robotic spacecraft Viking 1 and Viking 2 landed on Mars in 1976 and tested soil samples for living organisms. Some of the tests had puzzling semipositive results that scientists hypothesize were caused by nonbiological chemical reactions in the soil. There is no clear evidence of life or even of organic molecules in the Martian soil.

In 1996 there were prominent news stories regarding chemical and physical traces of life on Mars discovered inside a Martian meteorite found in Antarctica (■ Figure 15-10). Scientists were excited by the announcement, but they immediately began testing the evidence. Their results suggest that the unusual chemical signatures in the rock may have formed by processes that did not involve life. Tiny features in the rock that were originally thought to be fossils of ancient Martian microorganisms could be nonbiological mineral formations. This is the only direct evidence yet found regarding potential life on Mars, but it remains highly controversial. Conclusive evidence of life on Mars may have to wait until a geologist from Earth can scramble down dry Martian streambeds and crack open rocks looking for fossils.

There is no strong evidence for the existence of life elsewhere in the solar system. Now your search will take you to distant planetary systems.

Life in Other Planetary Systems

Could life exist in other planetary systems? You already know that there are many different kinds of stars and that many of these stars have planetary systems. As a first step toward answering this question, you can try to identify the kinds of stars that seem most likely to have stable planetary systems where life could evolve.

■ **Figure 15-10**

(a) Meteorite ALH84001 is one of a dozen meteorites known to have originated on Mars. A research group claimed that the meteorite contained chemical and physical traces of ancient life on Mars, including (b) what appear to be fossils of microscopic organisms. That evidence has not been confirmed, and the claim continues to be debated. (NASA)

currents in the gas giants' atmospheres would quickly carry any reproducing molecules that did form there down into inhospitably hot regions of the atmosphere.

As you learned in Chapter 14, at least one of the Jovian satellites could potentially support life. Jupiter's moon Europa appears to have a liquid-water ocean below its icy crust, and minerals dissolved in the water could provide a source of raw material for chemical evolution. Europa's ocean is kept warm and liquid now by tidal heating. There also may be liquid water layers under the surfaces of Ganymede and Callisto. That can change as the orbits of the moons interact; Europa, Ganymede and Callisto may have been frozen solid at other times in their histories..

If a planet is to be a suitable home for living things, it must be in a stable orbit around its sun. That is simple in a planetary system like our own, but most planet orbits in binary star systems are unstable unless the component stars are very close together or very far apart. Astronomers can calculate that in binary systems with stars separated by distances of a few AU, the planets should eventually be swallowed up by one of the stars or ejected from the system. Half the stars in the galaxy are members of binary systems, and many of them are unlikely to support life on planets.

Moreover, just because a star is single does not necessarily make it a good candidate for sustaining life. Earth required between 0.5 and 1 billion years to produce the first cells and 4.6 billion years for intelligence to emerge. Massive stars that live only a few million years will not do. If the history of life on Earth is at all representative, then stars more massive and luminous than about spectral type F5 are too short lived for complex life to develop. Main-sequence stars of types G and K, and possibly some of the M stars, are the best candidates.

The temperature of a planet is also important, and that depends on the type of star it orbits and its distance from the star. Astronomers have defined a **habitable zone** around a star as a region within which planets have temperatures that permit the existence of liquid water. The sun's habitable zone extends from around the orbit of Venus to the orbit of Mars, with Earth right in the middle. A low-luminosity star has a small habitable zone, and a high-luminosity star has a large one.

Scientists on Earth are finding life in places previously judged inhospitable, such as the bottoms of ice-covered lakes in Antarctica and far underground inside solid rock. Life has also been found in boiling hot springs with highly acidic water. As a result, it is difficult for scientists to pin down a range of environments and state with certainty that life cannot exist outside those conditions. You should also note that three of the environments listed as possible havens for life, Europa, Titan, and Enceladus, are in the outer solar system and lie far outside the sun's conventional habitable zone. Stable planets inside the habitable zones of long-lived stars are the places where life seems most likely, but, given the tenacity and resilience of Earth's life forms, there might be other, seemingly inhospitable, places in the universe where life exists.

Building Scientific Arguments

What evidence indicates that life is at least possible on other worlds?

A good scientific argument involves careful analysis of evidence. Fossils on Earth show that life originated in the oceans at least 3.4 billion years ago, and biologists have proposed relatively simple chemical processes that could have yielded the first reproducing molecules. Fossils show that life developed slowly at first. The pace of evolution quickened about half a billion years ago, when life took on complex forms. Later, when life emerged onto the land, it evolved rapidly into diverse forms. Intelligence is a relatively recent development: It is only a few million years old.

If this process occurred on Earth, it seems reasonable that it could have occurred on other worlds as well, because Earth-like worlds may be plentiful in the universe. Life may begin and eventually evolve to intelligence on any world where conditions are right. **What are the conditions you should expect on other worlds that host life?**

■   ■   ■

15-3 Communication with Distant Civilizations

VISITING EXTRASOLAR PLANETS is, for now, impossible. Nevertheless, if other civilizations exist, it is possible humans can communicate with them. Nature puts restrictions on such conversations, but the main problem lies in the unknown life expectancy of civilizations.

Travel between the Stars

The distances between stars are almost beyond comprehension. The space shuttle would take about 150,000 years to reach the nearest star. The obvious way to overcome these huge distances is with tremendously fast spaceships, but even the closest stars are many light-years away.

Nothing can exceed the speed of light, and accelerating a spaceship close to the speed of light takes huge amounts of energy. Even if you travel more slowly, your rocket would require massive amounts of fuel. If you were piloting a spaceship with the mass of a yacht to the nearest star, 4 light years away, and you wanted to travel at half the speed of light so as to arrive in 8 years, the trip would require 40,000 times as much energy as the entire United States consumes in a year.

These limitations not only make it difficult for humans to leave the solar system, but they would also make it difficult for aliens to visit Earth. Reputable scientists have studied "unidentified flying objects" (UFOs) and have never found any evidence that Earth is being visited or has ever been visited by aliens (**Window on Science 15-2**). Humans are unlikely ever to meet aliens face to face. However, communication by radio across interstellar distances takes relatively little energy.

Radio Communication

Nature puts restrictions on travel through space, and it also restricts astronomers' ability to communicate with distant civilizations by radio. One restriction is based on simple physics. Radio

UFOs and Space Aliens

If you conclude that there is likely to be life on other worlds, then you might be tempted to use UFO sightings as evidence to test your hypothesis. Scientists don't do this for two reasons.

First, the reputation of UFO sightings and alien encounters does not inspire confidence that these data are reliable. Most people hear of such events in grocery store tabloids, daytime talk shows, or sensational "specials" on viewer-hungry cable networks. You should take note of the low reputation of the media that report UFOs and space aliens. Most of these reports, like the reports that Elvis is alive and well, are simply made up for the sake of sensation, and you cannot use them as reliable evidence.

Second, the few UFO sightings that are not made up do not survive careful examination.

Most are mistakes and unintentional misinterpretations of natural events or human-made objects, committed by honest people. Over many decades, experts have studied these incidents and found none that are convincing. In short, despite false claims to the contrary on TV shows, there is no dependable evidence that Earth has ever been visited by aliens.

That's too bad. A confirmed visit by intelligent creatures from beyond our solar system would answer many questions. It would be exciting, enlightening, and, like any real adventure, a bit scary. But there is not yet any direct evidence of life on other worlds.

UFOs from space are fun to think about, but there is no evidence that they are real.

signals are electromagnetic waves and travel at the speed of light. Due to the distances between the stars, the speed of radio waves would severely limit astronomers' ability to carry on normal conversations with distant civilizations. Decades could elapse between asking a question and getting an answer.

So, rather than try to begin a conversation, one group of astronomers decided in 1974 to broadcast a simple message of greeting toward the globular cluster M13, 26,000 light years away, using the Arecibo radio telescope (see Figure 4-12a). When the signal arrives 26,000 years in the future, alien astronomers may be able to decode it.

The Arecibo beacon is an anticoded message, meaning that it is intended to be decoded by beings about whom we know nothing except that they build radio telescopes. The message is a string of 1679 pulses and gaps. Pulses represent 1s, and gaps represent 0s. The string can be arranged in only two possible ways: as 23 rows of 73 or as 73 rows of 23. The second arrangement forms a picture containing information about life on Earth (■ Figure 15-11).

Although the 1974 Arecibo beacon was the only powerful signal sent purposely from Earth to other solar systems, Earth is sending out many other signals. Short-wave radio signals, such as TV and FM, have been leaking into space for the last 50 years or so. Any civilization within 50 light years could already have detected Earth's civilization. That works both ways: Alien signals, whether intentional messages of friendship or the blather of their

equivalent to daytime TV, could be arriving at Earth now. Astronomers all over the world are pointing radio telescopes at the most likely stars and listening for alien civilizations.

Which channels should astronomers monitor? Wavelengths longer than 30 cm would get lost in the background noise of the Milky Way Galaxy, while wavelengths shorter than about 1 cm are absorbed in Earth's atmosphere. Between those wavelengths is a radio window that is open for communication. Even this restricted window contains millions of possible radio-frequency bands and is too wide to monitor easily, but astronomers may have found a way to narrow the search. Within this window lie the 21-cm spectral line of neutral hydrogen and the 18-cm line of OH (■ Figure 15-12). The interval between those lines has low background interference and is named the **water hole** because H plus OH yields water. Any civilizations sophisticated enough to do radio astronomy research must know of these lines and might appreciate their significance in the same way as do Earthlings.

A number of searches for extraterrestrial radio signals have been made, and some are now under way. This field of study is known as **SETI,** Search for Extra-Terrestrial Intelligence, and it has generated heated debate among astronomers, philosophers, theologians, and politicians. Congress funded a NASA search for a short time but ended support in the early 1990s because it feared public reaction. In fact, the annual cost of a major search is only about as much as a single Air Force attack helicopter, but

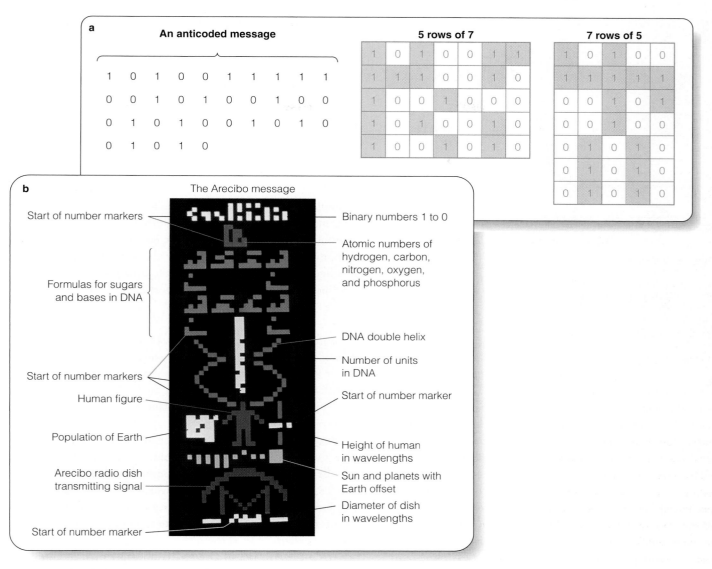

■ Active Figure 15-11

(a) An anticoded message is designed for easy decoding. Here a string of 35 radio pulses, represented as 1s and 0s, can be arranged in only two ways, as 5 rows of 7 or 7 rows of 5. The second way produces a friendly message. (b) The Arecibo message describes life on Earth (color added for clarity). Binary numbers give the height of the human figure (1110) and the diameter of the telescope dish (100101111110) in units of the wavelength of the signal, 12.3 cm. (NASA)

ThomsonNOW Sign in at www.thomsonedu.com and go to ThomsonNOW to see the Active Figure called "Interstellar Communication."

much of the reluctance to fund searches probably stems from issues other than cost. One point to keep in mind is that the discovery of alien intelligence would cause a huge change in humans' worldview, akin to Galileo's discovery that the moons of Jupiter do not go around the Earth. Some turmoil would inevitably result.

In spite of the controversy, the search continues. The NASA SETI project canceled by Congress was renamed Project Phoenix and completed using private funds. The SETI Institute, founded in 1984, managed Project Phoenix plus several other important searches and is currently building a new radio telescope array in northern California, collaborating with the University of California, Berkeley, and partly funded by Paul Allen of Microsoft.

There is even a way for you to help with searches. The Berkeley SETI team (note: they are separate from the SETI In-

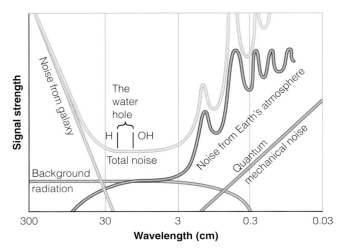

Figure 15-12

Radio noise from various astronomical sources makes it difficult to detect distant signals at wavelengths longer than 30 cm or shorter than 1 cm. In this range, radio emission lines from H atoms and from OH molecules mark a small wavelength range named the water hole that may be a likely channel for communication.

stitute), with the support of the Planetary Society, has recruited about 4 million owners of personal computers that are connected to the Internet. Participants download a screen saver that searches data files from the Arecibo radio telescope for signals whenever the owner is not using the computer. For information, locate the seti@home project at http://setiathome.ssl.berkeley.edu/.

The search continues, but radio astronomers struggle to hear anything against the worsening babble of noise from human civilization. Wider and wider sections of the electromagnetic spectrum are being used for earthly communication, and this, combined with stray electromagnetic noise from electronic devices including everything from computers to refrigerators, makes hearing faint radio signals difficult. It would be ironic if humans fail to detect faint signals from another world because our own world has become too noisy. Ultimately, the chances

of success depend on the number of inhabited worlds in the galaxy.

How Many Inhabited Worlds?

Given enough time, the searches will find other worlds with civilizations, assuming that there are at least a few out there. If intelligence is common, scientists should find signals relatively soon—within the next few decades—but if intelligence is rare, it may take much longer.

Simple arithmetic can give you an estimate of the number of technological civilizations in the Milky Way Galaxy with which you might communicate, N_c. The formula proposed for discussions about N_c is named the **Drake equation** after the radio astronomer Frank Drake, a pioneer in the search for extraterrestrial intelligence. The version of the Drake equation presented here is modified slightly from its original form:

$$N_c = N_* \cdot f_P \cdot n_{HZ} \cdot f_L \cdot f_I \cdot F_S$$

N_* is the number of stars in our galaxy, and f_P represents the fraction of stars that have planets. If all single stars have planets, f_P is about 0.5. The factor n_{HZ} is the average number of planets in each solar system suitably placed in the habitable zone. This factor actually means the number of planets possessing liquid water. Europa and Enceladus in our solar system show that liquid water can exist due to tidal heating outside the conventional habitable zone that in our system contains Earth's orbit. Thus, n_{HZ} may be larger than had been previously thought. The factor f_L is the fraction of suitable planets on which life begins, and f_I is the fraction of those planets where life evolved to intelligence. These factors can be roughly estimated, but the final factor, F_S, is much more uncertain.

F_S is the fraction of a star's life during which the life form is communicative. If a society survives at a technological level for only 100 years, the chances of communicating with it are small. But a society that stabilizes and remains technological

Table 15-1 | The Number of Technological Civilizations per Galaxy

	Variables		Estimates	
			Pessimistic	Optimistic
N_*	Number of stars per galaxy		2×10^{11}	2×10^{11}
f_P	Fraction of stars with planets		0.01	0.5
n_{LZ}	Number of planets per star that lie in habitable zone for longer than 4 billion years		0.01	1
f_L	Fraction of suitable planets on which life begins		0.01	1
f_I	Fraction of planets where life forms evolve to intelligence		0.01	1
F_S	Fraction of star's life during which a technological society survives		10^{-8}	10^{-4}
N_c	Number of communicative civilizations per galaxy		<1	1×10^7

for a long time is much more likely to be detected. For a star with a life span of 10 billion years, F_S can range from 10^{-8} for extremely short-lived societies to 10^{-4} for societies that survive for a million years. ■ Table 15-1 summarizes what many scientists consider a reasonable range of values for F_S and the other factors.

If the optimistic estimates are true, there could be a communicative civilization within a few dozen light-years of Earth. On the other hand, if the pessimistic estimates are true, Earth may be the only planet in our galaxy capable of communication.

What Are We?

There are over four thousand religions around the world, and nearly all hold that humans have a dual nature. We are physical objects made of atoms, but we are also spiritual beings. Science is unable to examine the spiritual side of existence, but it can tell us about our physical nature.

You are made of very old atoms. The matter you are made of appeared in the big bang and was cooked into a wide range of elements inside stars. Your atoms may have been inside at least two or three generations of stars. Eventually, your atoms became part of a nebula that contracted to form the sun and the planets of the solar system.

Your atoms have been part of Earth for the last 4.6 billion years. They have been recycled many times through dinosaurs, stromatolites, fish, bacteria, grass, birds, worms, and other living things. Now you are using your atoms, but when you are done with them, they will go back to Earth and be used again and again.

When the sun swells into a red giant star and dies in a few billion years, Earth's atmosphere and oceans will be driven away, and at least the outer few kilometers of Earth's crust will be vaporized and blown outward to become part of the nebula around the white-dwarf remains of the sun. Your atoms are destined to return to the interstellar medium and will become part of future generations of stars and planets.

The message of astronomy is that humans are not just observers: We are participants, we are part of the universe. Among all of the galaxies, stars, planets, planetesimals, and bits of matter, humans are objects that can think, and that means we can understand what we are.

Is the human race the only thinking species? If so, we bear the sole responsibility to understand and admire the universe. The detection of signals from another civilization would demonstrate that we are not alone, and such communication would end the self-centered isolation of humanity and stimulate a reevaluation of the meaning of human existence.

Building Scientific Arguments

Why does the number of civilizations that could be detected depend on how long civilizations survive at a technological level?
This scientific argument depends on the timing of events. If you turned a radio telescope to the sky and scanned millions of frequency bands for many stars, you would be taking a snapshot of the universe at a particular time. Other civilizations must be broadcasting at this time if they are to be detected. If most civilizations survive for a long time, there is a much greater chance that you will detect one of them in your snapshot than if civilizations tend to disappear quickly, for instance due to nuclear war or environmental collapse. If most civilizations last only a short time, there may be none capable of transmitting during the short interval when Earthlings are capable of building radio telescopes to listen for them.

The speed at which astronomers can search for signals is limited because computers must search many frequency intervals, but not all frequencies inside Earth's radio window are subject to intensive search. Build a new argument to explain: **Why is the water hole an especially good frequency band to listen?**

■   ■   ■

Summary

15-1 | The Nature of Life

What is life?
- The process of life extracts energy from the environment, maintains the organism, and modifies the surroundings to promote the organism's survival.
- Living things have a physical basis—the arrangement of matter and energy that makes life possible. Life on Earth is based on carbon chemistry.
- Living things must also have controlling units of information, which can be passed to each new generation.
- Genetic information for life on Earth is stored in long carbon-chain molecules such as DNA.
- The DNA molecule stores information in the form of chemical bases linked together like the rungs of a ladder. Copied by the RNA molecule, the patterns of bases act as recipes for the manufacture of proteins and enzymes that are respectively the main structural and control components of the life process.
- When a cell divides, the DNA molecules split lengthwise and duplicate themselves so that each of the new cells can receive a copy of the genetic information.
- Errors in duplication or damage to the DNA molecule can produce mutants, organisms that contain new DNA information and have new properties. Variation in genetic codes can become widespread among individuals in a species.
- Natural selection determines which of these variations are best suited to survive, and the species evolves to fit its environment.

15-2 | The Origin of Life

How did life originate on Earth?
- The oldest fossils on Earth are at least 3.4 billion years old. Those fossils provide evidence that life began in the oceans.
- Fossil evidence indicates that life began on Earth as simple organisms like bacteria and evolved into more complex creatures.
- The Miller experiment shows that the chemical building blocks of life form naturally under a wide range of circumstances.
- Biologists hypothesize that chemical evolution concentrated simple molecules into a diversity of larger stable molecules, but those molecules did not reproduce copies of themselves.
- Biological evolution began when molecules started reproducing.
- Not until about 0.5 billion years ago did life forms become large and complex, during what is called the Cambrian explosion.
- Life emerged from the oceans only about 0.4 billion years ago, and human intelligence developed over the last 3 million (0.003 billion) years.

Could life begin on other worlds?
- Life as it is known on Earth requires liquid water and thus a specific range of temperatures.
- No other planet in our solar system appears to harbor life at present. Most are too hot or too cold, although life might have begun on Mars before it became too cold and dry.
- Liquid water exists, and therefore Earthlike life is at least possible, under the surfaces of Jupiter's moon Europa, Saturn's moon Enceladus, and possibly Jupiter's moons Ganymede and Callisto. Saturn's moon Titan has abundant organic compounds but does not have liquid water.
- Because the origin of life and its evolution into intelligent creatures took so long on Earth, scientists do not consider short-lived stars such as middle- and upper-main-sequence stars as likely hosts for life-bearing planets.
- Main-sequence G and K stars are thought to be likely candidates to host planets with life. The fainter M stars are also possibilities.
- The habitable zone around a star may be larger than scientists had expected, given the wide variety of living things now found in extreme environments on Earth and the possibility of tidal heating of moons orbiting large planets.

15-3 | Communication with Distant Civilizations

Can Earthlings communicate with civilizations on other worlds?
- Because of distance, speed, and fuel, travel between the stars seems almost impossible for humans or for aliens who might visit Earth.
- Radio communication may be possible, but a real conversation would be difficult because of very long travel times for radio signals.
- Broadcasting a radio beacon of pulses would distinguish the signal from naturally occurring radio emission and identify the source as a technological civilization. The signal can be anticoded in hope it would be easy for another civilization to decode.
- The best part of the radio spectrum for communication is called the water hole, the wavelength range from the 21-cm spectral line of hydrogen to the 18-cm line of OH. Even so, millions of radio wavelengths need to be tested to fully survey the water hole for a given target star.
- Sophisticated searches are now underway to detect radio transmissions from civilizations on other worlds, but such SETI programs are hampered by limited computer power and radio noise pollution from human civilization.
- The number of civilizations in our galaxy that are at a technological level and able to communicate while humans are listening is limited by the lifetimes of their and our civilizations.

New Terms

biological evolution (p. 315)	stromatolite (p. 318)
mutant (p. 315)	Miller experiment (p. 318)
natural selection (p. 315)	primordial soup (p. 319)
DNA (deoxyribonucleic acid) (p. 316)	chemical evolution (p. 320)
amino acid (p. 316)	multicellular (p. 321)
protein (p. 316)	Cambrian explosion (p. 321)
enzyme (p. 316)	habitable zone (p. 325)
RNA (ribonucleic acid) (p. 317)	water hole (p. 326)
chromosome (p. 317)	SETI (p. 326)
gene (p. 317)	Drake equation (p. 328)

Review Questions

1. If life is based on information, what is that information?

2. What would happen to a life form if the information handed down to off-spring was always the same? How would that endanger the future of the life form?

3. How does the DNA molecule produce a copy of itself?

4. Give an example of natural selection acting on new DNA patterns to select the most advantageous characteristics.

5. What evidence do scientists have that life on Earth began in the sea?

6. Why do scientists think that liquid water is necessary for the origin of life?

7. What is the difference between chemical evolution and biological evolution?

8. What was the significance of the Miller experiment?

9. How does intelligence make a creature more likely to survive?

10. Why are upper-main-sequence (high-luminosity) stars unlikely sites for intelligent civilizations?

11. Why is it reasonable to suspect that travel between stars is nearly impossible?

12. How does the stability of technological civilizations affect the probability that Earth can communicate with them?

13. What is the water hole, and why would it be a good "place" to look for other civilizations?

14. The star cluster NGC 2264, shown here, contains cool red giants and main-sequence stars from hot blue stars all the way down to red dwarfs. Discuss the likelihood that planets orbiting any of these stars might be home to life. Don't neglect to estimate the age of the cluster.

Visual

T. A. Rector, B. A. Wolpa, NRAO/ NOAO/AURA/NSF

15. If you could search for life in the galaxy shown to the right, would you look among disk stars or halo stars? Discuss the factors that influence your decision.

Visual

ESO

Discussion Questions

1. Do you expect hypothetical alien recipients of the Arecibo message will be able to decode it? Why or why not?

2. What do you think it would mean if decades of more careful searches for radio signals for extraterrestrial intelligence turn up nothing?

The aggregate of all our joys and sufferings, thousands of confident religions, ideologies and economic doctrines, every hunter and forager, every hero and coward, every creator and destroyer of civilizations, every king and peasant, every young couple in love, every hopeful child, every mother and father, every inventor and explorer, every teacher of morals, every corrupt politician, every superstar, every supreme leader, every saint and sinner in the history of our species, lived there on a mote of dust, suspended in a sunbeam.

CARL SAGAN (1934–1996)

Earth photographed by Voyager 1 from the edge of the solar system. (NASA)

OUR JOURNEY TOGETHER IS OVER, but before we part company, let's ponder one final time the primary theme of this book—humanity's place in the universe. Astronomy gives us some comprehension of the workings of stars, galaxies, and planets, but its greatest value lies in what it teaches us about ourselves. Now that you have surveyed astronomical knowledge, you can better understand your own position in nature.

To some, the word *nature* conjures up visions of furry rabbits hopping about in a forest glade. To others, nature is the blue-green ocean depths, and still others think of nature as windswept mountaintops. As diverse as these images are, they are all Earthbound. Having studied astronomy, you can see nature as a beautiful mechanism composed of matter and energy interacting according to simple rules to form galaxies, stars, planets, mountaintops, ocean depths, forest glades, and people.

Perhaps the most important astronomical lesson is that humanity is a small but important part of the universe. Most of the universe is probably lifeless. The vast reaches between the galaxies appear to be empty of all but the thinnest gas, and stars are much too hot to preserve the chemical bonds that seem necessary for life to survive and develop. It seems that only on the surfaces of a few planets, where temperatures are moderate, can atoms link together in special ways to form living matter.

If life is special, then intelligence is very precious. The universe must contain many planets devoid of life, planets where the wind has blown unfelt for billions of years. There may also exist planets where life has developed but has not become complex, planets on which the wind stirs wide plains of grass and rustles through dark forests. On some planets, equivalents of Earth insects, fish, birds, and animals may watch the passing days at most dimly aware of their own existence. It is intelligence, human or alien, that gives meaning to the landscape.

Science is the process by which Earth intelligence has tried to understand the physical universe. Science is not the invention of new devices or processes. It does not create home computers, cure the mumps, or manufacture plastic spoons—those are engineering and technology, the adaptation of scientific understanding for practical purposes. Science is the understanding of nature, and astronomy is that understanding on the grandest scale. Astronomy is the science by which the universe, through its intelligent lumps of matter, tries to understand its own existence.

As the primary intelligent species on this planet, we are the custodians of a priceless gift—a planet filled with living things. This is especially true if life is rare in the universe. In fact, if Earth is the only inhabited planet, our responsibility is overwhelming. We are the only creatures who can take action to preserve the existence of life on Earth, and, ironically, our own actions are the most serious hazards.

The future of humanity is not secure. We are trapped on a tiny planet with limited resources and a population growing faster than our ability to produce food. In our efforts to survive, we have already driven some creatures to extinction and now threaten others. If our civilization collapses because of starvation, or if our race destroys itself somehow, the only bright spot is that the rest of the creatures on Earth may be better off for our absence.

But even if we control our population and conserve and recycle our resources, life on Earth is doomed. In 5 billion years, the sun will leave the main sequence and swell into a red giant, incinerating Earth. However, Earth will be lifeless long before that. Within the next few billion years the growing luminosity of the sun will first alter Earth's climate and then vaporize its oceans. Earth, like everything else in the universe, is only temporary.

To survive, humanity must eventually leave Earth and search for other planets. Colonizing the moon and other planets of our solar system will not save us because they will face the same fate as Earth when the sun dies. On the other hand, travel to other stars is tremendously difficult and may be impossible with the limited resources we have in our small solar system. We, and all of the living things on Earth, may be trapped.

This is a sad prospect, but a few factors are comforting. First, everything else in the universe is also temporary. Stars die, galaxies die; perhaps the entire universe will someday end. Our distant future is limited, and this assures us that we are a part of a much larger whole. Second, we have a few billion years to prepare, and a billion years is a very long time. Only a few million years ago, our ancestors were starting to walk upright and communicate. A billion years ago, our ancestors were microscopic organisms living in the oceans. To suppose that a billion years hence there will be beings who resemble today's humans, or that humans will still be the dominant intelligence on Earth, or that humans will even exist, are ultimately conceits.

Our responsibility is not to save our race for all eternity but to behave as dependable custodians of our planet, preserving it, admiring it, and trying to understand it. That calls for drastic changes in our behavior toward other living things and a revolution in our attitude toward our planet's resources. Whether we can change our ways is debatable—humanity is far from perfect in its understanding, abilities, or intentions. However, you must not imagine that we, and our civilization, are less than precious. We have the gift of intelligence, and that is the most wonderful thing this planet has ever produced.

> We shall not cease from exploration
> And the end of all our exploring
> Will be to arrive where we started
> And know the place for the first time.
> —*T. S. Eliot, "Little Gidding"*

Appendix A

Units and Astronomical Data

The Metric System and SI Units

A SYSTEM OF MEASUREMENT is based on the three fundamental units for length, mass, and time. By international agreement, there is a preferred subset of metric units known as the *Système International d'Unités* (SI units) that is based on the meter, kilogram, and second.

Density should be expressed in SI units as kilograms per cubic meter, but no human hand can enclose a cubic meter, so that unit does not help you easily grasp the significance of a given density. Instead, this book refers to density in grams per cubic centimeter. A gram is roughly the mass of a paperclip, a cubic centimeter is the size of a small sugar cube, and a density of 1 g/cm^3 equals the density of water, all tangible units.

Americans generally use the British system of units. In that system (now common only in the United States, Tonga, and Southern Yemen, but, ironically, no longer in Great Britain), the fundamental unit of length is the foot, composed of 12 inches. For conceptual purposes this book expresses some quantities in both metric and English units. Instead of saying that the average person would weigh 133 N on the moon, it might help some readers to see that weight described as 30 lb. Consequently, this text usually gives quantities in metric form followed by the British form in parentheses. For example the radius of the moon is 1738 km (1080 mi).

Log in at www.thomsonedu.com/login and click the *Perspectives on Astronomy* companion website link to view **Supplemental Tables** including a list of the fundamental SI units and their abbreviations in ■ **Table S-1.** The metric prefixes and their abbreviations, used to describe multiples or fractions of the fundamental units, are given in ■ **Table S-2.** Conversion factors between English and metric units are given in ■ **Table S-3.** These and other tables supporting this book that are on the book's companion website are listed in ■ Table A-1.

Temperature Scales

IN ASTRONOMY, AS IN MOST OTHER SCIENCES, temperatures are expressed on the Kelvin scale, although the centigrade scale is also used. The centigrade scale is also called the Celsius scale

> ■ **Table A-1** | **List of Supplemental Tables Available on Book Website at www.thomsonedu.com/astronomy/perspectives1**
>
> | | |
> |---|---|
> | S-1 | SI *(Système International)* Metric Units |
> | S-2 | Metric Prefixes |
> | S-3 | Conversion Factors between British and Metric Units |
> | S-4 | Temperature Scales and Conversion Formulas |
> | S-5 | The Greek Alphabet |
> | S-6 | The Periodic Table and List of the Chemical Elements |

after its inventor, the Swedish astronomer Anders Celsius (1701–1744). Temperatures on the centigrade scale are measured from the freezing point of water (0°F). Temperatures on the Kelvin scale are measured from absolute zero, the temperature of an object that contains no extractable heat (0 K). In practice, no object can be as cold as absolute zero, although laboratory apparatuses have reached temperatures less than 10^{-6} K. The Kelvin scale is named after the Scottish mathematical physicist William Thomson, Lord Kelvin (1824–1907). Centigrade and Kelvin degree steps are the same size.

The Fahrenheit scale fixes the freezing point of water at 32°F and the boiling point at 212°F. Named after the German physicist Gabriel Daniel Fahrenheit (1686–1736), who made the first successful mercury thermometer in 1720, the Fahrenheit scale is used routinely only in the United States.

Log in at www.thomsonedu.com/login and click the *Perspectives on Astronomy* companion website link to view **Supplemental Tables** including a list of numerical values and formulas for conversion of temperatures from one scale to another in ■ **Table S-4.**

Astronomy Units and Constants

ASTRONOMY, AND SCIENCE IN GENERAL, is a way of learning about nature and understanding the universe. To test hypotheses about how nature works, scientists use observations of nature.

The tables that follow contain some of the basic observations that support science's best understanding of the astronomical universe. Of course, these data are expressed in the form of numbers, not because science reduces all understanding to mere numbers, but because the struggle to understand nature is so demanding that science must use every good means available. Quantitative thinking—reasoning mathematically—is one of the most powerful techniques ever invented by the human brain. Thus these tables are not nature reduced to mere numbers but numbers that support humanity's growing understanding of the natural world around us.

■ **Table A-2** defines units of measure commonly used by astronomers. ■ **Table A-3** gives physical constants such as the speed of light, plus basic astronomical data such as the mass and luminosity of the sun. ■ **Table A-4** presents the characteristics of different spectral types of main-sequence stars. ■ Tables **A-5** and **A-6**, respectively, list the 15 nearest and the 15 brightest stars with their individual properties. The fact that these two star lists have almost no overlap is an important truth about stars that you can read more about in Chapter 6. ■ **Table A-7** gives the physical and orbital properties of the major planets in our solar system, consolidating information presented within the Celestial Profiles in Chapters 13 and 14. ■ **Table A-8** lists the larger moons of each planet and their characteristics. The outer planets have many smaller moons that are not listed here, with more being discovered each year.

Log in at www.thomsonedu.com/login and click the *Perspective on Astronomy* website link to view **Supplemental Tables** including ■ **Table S-5** containing the Greek alphabet used in naming the brighter stars. You will also find ■ **Table S-6,** the periodic chart and list of the chemical elements. One of the primary themes of this book is the history of the atoms in your body: the story of how hydrogen and helium were produced in the big bang and how the rest of the stable elements (up to uranium, atomic number 92) were produced inside stars, either slowly during giant star phases or rapidly during supernova explosions.

* * *

■ **Table A-3 I Astronomical Constants**

Velocity of light (c)	$= 3.00 \times 10^8$ m/s
Gravitational constant (G)	$= 6.67 \times 10^{-11}$ m³/s²kg
Mass of H atom	$= 1.67 \times 10^{-27}$ kg
Mass of Earth ($M_\oplus$)	$= 5.98 \times 10^{24}$ kg
Earth equatorial radius ($R_\oplus$)	$= 6.38 \times 10^6$ m
Mass of sun ($M_\odot$)	$= 1.99 \times 10^{30}$ kg
Radius of sun ($R_\odot$)	$= 6.96 \times 10^8$ m
Solar luminosity ($L_\odot$)	$= 3.83 \times 10^{26}$ J/s
Mass of moon	$= 7.35 \times 10^{22}$ kg
Radius of moon	$= 1.74 \times 10^3$ kg

■ **Table A-2 I Units Used in Astronomy**

1 angstrom (Å)	$= 10^{-8}$ cm
	$= 10^{-10}$ m
	$= 10$ nm
1 astronomical unit (AU)	$= 1.50 \times 10^{11}$ m
	$= 93.0 \times 10^6$ miles
1 light-year (ly)	$= 6.32 \times 10^4$ AU
	$= 9.46 \times 10^{15}$ m
	$= 5.88 \times 10^{12}$ miles
1 parsec (pc)	$= 2.06 \times 10^5$ AU
	$= 3.09 \times 10^{16}$ m
	$= 3.26$ ly
1 kiloparsec (kpc)	$= 1000$ pc
1 megaparsec (Mpc)	$= 1,000,000$ pc

Spectral Type	Absolute Visual Magnitude (M_v)	Luminosity*	Temp. (K)	λ_{max} (nm)	Mass*	Radius*	Average Density (g/cm³)
O5	−5.8	501,000	40,000	72	40	17.8	0.01
B0	−4.1	20,000	28,000	100	18	7.4	0.1
B5	−1.1	790	15,000	190	6.4	3.8	0.2
A0	+0.7	79	9900	290	3.2	2.5	0.3
A5	+2.0	20	8500	340	2.1	1.7	0.6
F0	+2.6	6.3	7400	390	1.7	1.4	1.0
F5	+3.4	2.5	6600	440	1.3	1.2	1.1
G0	+4.4	1.3	6000	480	1.1	1.0	1.4
G5	+5.1	0.8	5500	520	0.9	0.9	1.6
K0	+5.9	0.4	4900	590	0.8	0.8	1.8
K5	+7.3	0.2	4100	700	0.7	0.7	2.4
M0	+9.0	0.1	3500	830	0.5	0.6	2.5
M5	+11.8	0.01	2800	1000	0.2	0.3	10.0
M8	+16	0.001	2400	1200	0.1	0.1	63

*Luminosity, mass, and radius are given in terms of the sun's luminosity, mass, and radius.

■ Table A-5 | The 15 Nearest Stars

Name	Absolute Magnitude (M_v)	Distance (ly)	Spectral Type	Apparent Visual Magnitude (m_v)
Sun	4.83		G2	−26.8
Proxima Cen	15.45	4.28	M5	11.05
α Cen A	4.38	4.3	G2	0.1
α Cen B	5.76	4.3	K5	1.5
Barnard's Star	13.21	5.9	M5	9.5
Wolf 359	16.80	7.6	M6	13.5
Lalande 21185	10.42	8.1	M2	7.5
Sirius A	1.41	8.6	A1	−1.5
Sirius B	11.54	8.6	white dwarf	7.2
Luyten 726-8A	15.27	8.9	M5	12.5
Luyten 726-8B (UV Cet)	15.8	8.9	M6	13.0
Ross 154	13.3	9.4	M5	10.6
Ross 248	14.8	10.3	M6	12.2
ε Eri	6.13	10.7	K2	3.7
Luyten 789-6	14.6	10.8	M7	12.2

■ Table A-6 | The 15 Brightest Stars

Star	Name	Apparent Visual Magnitude (m_v)	Spectral Type	Absolute Visual Magnitude (M_v)	Distance (ly)
α CMa A	Sirius	−1.47	A1	1.4	8.7
α Car	Canopus	−0.72	F0	−3.1	98
α Cen	Rigil Kentaurus	−0.01	G2	4.4	4.3
α Boo	Arcturus	−0.06	K2	−0.3	36
α Lyr	Vega	0.04	A0	0.5	26.5
α Aur	Capella	0.05	G8	−0.6	45
β Ori A	Rigel	0.14	B8	−7.1	900
α CMi A	Procyon	0.37	F5	2.7	11.3
α Ori	Betelgeuse	0.41	M2	−5.6	520
α Eri	Achernar	0.51	B3	−2.3	118
β Cen AB	Hadar	0.63	B1	−5.2	490
α Aql	Altair	0.77	A7	2.2	16.5
α Tau A	Aldebaran	0.86	K5	−0.7	68
α Cru	Acrux	0.90	B2	3.5	260
α Vir	Spica	0.91	B1	−3.3	220

PHYSICAL PROPERTIES (EARTH = ⊕)

Planet	Equatorial Radius (km)	Equatorial Radius (⊕ = 1)	Mass (⊕ = 1)	Average Density (g/cm³)	Surface Gravity (⊕ = 1)	Escape Velocity (km/s)	Sidereal Period of Rotation	Inclination of Equator to Orbit
Mercury	2439	0.38	0.056	5.44	0.38	4.3	58.65^d	0°
Venus	6052	0.95	0.815	5.24	0.90	10.3	243.02^d	177°
Earth	6378	1.00	1.000	5.50	1.00	11.2	23.93^h	23°.5
Mars	3396	0.53	0.108	3.94	0.38	5.0	24.62^h	25°.3
Jupiter	71,494	11.20	317.8	1.34	2.54	61	9.92^h	3°.1
Saturn	60,330	9.42	95.2	0.69	1.16	36	10.23^h	26°.4
Uranus	25,559	4.01	14.5	1.19	0.92	22	17.23^h	97°.9
Neptune	24,750	3.93	17.2	1.66	1.19	25	16.05^h	28°.8

ORBITAL PROPERTIES

Planet	Semimajor Axis (a) (AU)	Semimajor Axis (a) (10⁶ km)	Orbital Period (P) (y)	Orbital Period (P) (days)	Average Orbital Velocity (km/s)	Orbital Eccentricity	Inclination to Ecliptic
Mercury	0.39	57.9	0.24	87.97	47.9	0.206	7°.0
Venus	0.72	108.2	0.62	224.68	35.0	0.007	3°.0
Earth	1.00	149.6	1.00	365.26	29.8	0.017	0° (by defn)
Mars	1.52	227.9	1.88	686.95	24.1	0.093	1°.9
Jupiter	5.20	778.3	11.87	4334.3	13.1	0.048	1°.3
Saturn	9.54	1427.0	29.46	10,760	9.6	0.056	2°.5
Uranus	19.18	2869.0	84.01	30,685	6.8	0.046	0°.8
Neptune	30.06	4497.1	164.79	60,189	5.4	0.010	1°.8″

Planet	Satellite	Radius (km)	Average Distance from Planet (10^3 km)	Orbital Period (days)	Orbital Eccentricity	Orbital Inclination
Earth	Moon	1738	384.4	27.32	0.055	5°.1
Mars	Phobos	14 × 12 × 10	9.4	0.32	0.018	1°.0
	Deimos	8 × 6 × 5	23.5	1.26	0.002	2°.8
Jupiter	Amalthea	135 × 100 × 78	182	0.50	0.003	0°.4
	Io	1820	422	1.77	0.000	0°.3
	Europa	1565	671	3.55	0.000	0°.5
	Ganymede	2640	1071	7.16	0.002	0°.2
	Callisto	2420	1884	16.69	0.008	0°.2
	Himalia	~85	11,470	250.6	0.158	27°.6
Saturn	Janus	110 × 80 × 100	151.5	0.70	0.007	0°.1
	Mimas	196	185.5	0.94	0.020	1°.5
	Enceladus	250	238.0	1.37	0.004	0°.0
	Tethys	530	294.7	1.89	0.000	1°.1
	Dione	560	377	2.74	0.002	0°.0
	Rhea	765	527	4.52	0.001	0°.4
	Titan	2575	1222	15.94	0.029	0°.3
	Hyperion	205 × 130 × 110	1484	21.28	0.104	~0°.5
	Iapetus	720	3562	79.33	0.028	14°.7
	Phoebe	110	12,930	550.4	0.163	150°
Uranus	Miranda	242	129.9	1.41	0.017	3°.4
	Ariel	580	190.9	2.52	0.003	0°
	Umbriel	595	266.0	4.14	0.003	0°
	Titania	805	436.3	8.71	0.002	0°
	Oberon	775	583.4	13.46	0.001	0°
Neptune	Proteus	205	117.6	1.12	~0	~0°
	Triton	1352	354.59	5.88	0.00	160°
	Nereid	170	5588.6	360.12	0.76	27.7°

~ signifies *approximately*.

Appendix B

Observing the Sky

OBSERVING THE SKY WITH THE NAKED EYE is as important to modern astronomy as picking up pretty pebbles is to modern geology. The sky is a natural wonder unimaginably bigger than the Grand Canyon, the Rocky Mountains, or any other site that tourists visit every year. To neglect the beauty of the sky is equivalent to geologists neglecting the beauty of the minerals they study. This supplement is meant to act as a tourist's guide to the sky. You analyzed the universe in the textbook's chapters; here you can admire it.

The brighter stars in the sky are visible even from the centers of cities with their air and light pollution. But in the countryside, only a few miles beyond the cities, the night sky is a velvety blackness strewn with thousands of glittering stars. From a wilderness location, far from the city's glare, and especially from high mountains, the night sky is spectacular.

Using Star Charts

THE CONSTELLATIONS ARE A FASCINATING CULTURAL HERITAGE of our planet, but they are sometimes a bit difficult to learn because of Earth's motion. The constellations above the horizon change with the time of night and the seasons.

Because Earth rotates eastward, the sky appears to rotate westward around Earth. A constellation visible in the overhead soon after sunset will appear to move westward, and in a few hours it will disappear below the horizon. Other constellations will rise in the east, so the sky changes gradually through the night.

In addition, Earth's orbital motion makes the sun appear to move eastward among the stars. Each day the sun moves about twice its own diameter, about one degree, eastward along the ecliptic. Consequently, each night at sunset, the constellations are about one degree farther toward the west.

Orion, for instance, is visible in the evening sky in January; but, as the days pass, the sun moves closer to Orion. By March, Orion is difficult to see in the western sky soon after sunset. By June, the sun is so close to Orion it sets with the sun and is in-

visible. Not until late July is the sun far enough past Orion for the constellation to become visible rising in the eastern sky just before dawn.

Because of the rotation and orbital motion of Earth, you need more than one star chart to map the sky. Which chart you select depends on the month and the time of night. The charts printed inside the covers of this book show the evening sky for each season, as viewed from the northern hemisphere at a latitude typical of the United States or central Europe.

To use the charts, select the appropriate chart and hold it overhead as shown in ■ Figure B-1. If you face south, turn the chart until the words *southern horizon* are at the bottom of the chart. If you face other directions, turn the chart appropriately. Note that hours are in Standard Time; for Daylight Savings Time (in late spring, summer, and early fall) add one hour.

■ **Figure B-1**

To use the star charts in this book, select the appropriate chart for the season and time. Hold it overhead and turn it until the direction at the bottom of the chart is the same as the direction you are facing.

Glossary

Pronunciation Guide

ā pay	ē be	ī pie	ō so	ŭ cut
ă hat	ĕ pet	ĭ pit	ŏ pot	
ä father		î pier	ô paw	
			o͞o food	

Numbers in parentheses refer to the page where the term is first discussed in the text.

absolute visual magnitude *(M$_v$)* (106) Intrinsic brightness of a star. The apparent visual magnitude the star would have if it were 10 pc away.

absolute zero (79) The theoretical lowest possible temperature at which a material contains no extractable heat energy. Zero on the Kelvin temperature scale.

absorption line (86) A dark line in a spectrum produced by the absence of photons absorbed by atoms or molecules.

absorption spectrum (dark-line spectrum) (86) A spectrum that contains absorption lines.

accretion (253) The sticking together of solid particles to produce a larger particle.

accretion disk (156) The whirling disk of gas that forms around a compact object such as a white dwarf, neutron star, or black hole as matter is drawn in.

active galactic nuclei (AGN) (210) The centers of active galaxies that are emitting large amounts of excess energy. *See also* active galaxy.

active galaxy (210) A galaxy whose center emits large amounts of excess energy, often in the form of radio emission. Active galaxies are suspected of having massive black holes in their centers into which matter is flowing.

active region (93) A magnetic region on the solar surface that includes sunspots, prominences, flares, and the like.

adaptive optics (63) A computer-controlled optical system in an astronomical telescope used to partially correct for seeing.

albedo (273) The ratio of the light reflected from an object divided by the light that hits the object. Albedo equals 0 for perfectly black and 1 for perfectly white.

amino acid (ŭ·mē′nō) (316) Carbon-chain molecule that is the building block of protein.

angstrom (Å) (ăng′strŭm) (58) A unit of distance commonly used to measure the wavelength of light. 1 Å = 10^{-10} m.

angular diameter (17) The angle formed by lines extending from the observer to opposite sides of an object.

angular distance (17) The angle formed by lines extending from the observer to two locations.

angular momentum (155) A measure of the tendency of a rotating body to continue rotating. Mathematically, the product of mass, velocity, and radius.

annular eclipse (27) A solar eclipse in which the solar photosphere appears around the edge of the moon in a bright ring, or annulus. The corona, chromosphere, and prominences cannot be seen.

anorthosite (ăn·ôr′thŭ·sĭt) (274) Rock of aluminum and calcium silicates found in the lunar highlands.

antimatter (226) Matter composed of antiparticles, which upon colliding with a matching particle of normal matter annihilate and convert the mass of both particles into energy. The antiproton is the antiparticle of the proton, and the positron is the antiparticle of the electron.

aphelion (ŭ·fē′le·ŭn) (21) The orbital point of greatest distance from the sun.

apparent visual magnitude *(m$_v$)* (14) The brightness of a star as seen by human eyes on Earth.

arc minute (17) $\frac{1}{60}$th of a degree.

arc second (17) $\frac{1}{60}$th of a minute of arc.

array detector (67) Device for collecting and recording electromagnetic radiation using multiple individual detectors arrayed on the surface of a chip, for example a CCD electronic camera.

asterism (12) A named grouping of stars that is not one of the recognized constellations. Examples are the Big Dipper and the Pleiades.

asteroid (244) Small, rocky world. Most asteroids lie between Mars and Jupiter in the asteroid belt.

Astronomical Unit (AU) (5) Average distance from Earth to the sun; 1.53108 km, or 933106 mi.

atmospheric window (60) Wavelength region in which our atmosphere is transparent—at visual, infrared, and radio wavelengths.

atom (79) The smallest unit of a chemical element, consisting of a nucleus containing protons and neutrons plus a surrounding cloud of electrons.

aurora (ô·rôr′ŭ) (97) The glowing light display that results when a planet's magnetic field guides charged particles toward the north and south magnetic poles, where they strike the upper atmosphere and excite atoms to emit photons.

autumnal equinox (20) The point on the celestial sphere where the sun crosses the celestial equator going southward. Also, the time when the sun reaches this point and autumn begins in the Northern Hemisphere—about September 22.

Babcock model (94) A model of the sun's magnetic cycle in which the differential rotation of the sun winds up and tangles the solar magnetic field in a 22-year cycle. This is thought to be responsible for the 11-year sunspot cycle.

Balmer series (87) A series of spectral lines produced by hydrogen in the near-ultraviolet and visible parts of the spectrum. The three longest-wavelength Balmer lines are visible to the human eye.

barred spiral galaxy (200) A spiral galaxy with an elongated nucleus resembling a bar from which the arms originate.

basalt (270) Dark igneous rock, characteristic of solidified lava.

belt–zone circulation (293) The atmospheric circulation typical of Jovian planets in which dark belts and bright zones encircle the planet parallel to its equator.

big bang (224) The high-density, high-temperature state from which the expanding universe of galaxies began.

big rip (234) The fate of the universe if dark energy increases with time and galaxies, stars, and even atoms are eventually ripped apart by the accelerating expansion of the universe.

binary stars (113) Pairs of stars that orbit around their common center of mass.

binding energy (82) The energy needed to pull an electron away from its atom.

biological evolution (315) The process of mutation, variation, and natural selection by which life adjusts itself to its changing environment.

blackbody radiation (79) Radiation emitted by a hypothetical perfect radiator. The spectrum is continuous, and the wavelength of maximum emission depends on the body's temperature.

black dwarf (133) The end state of a white dwarf that has cooled to low temperature.

black hole (167) A mass that has collapsed to such a small volume that its gravity prevents the escape of all radiation. Also, the volume of space from which radiation may not escape.

blazars *See* BL Lac objects.

BL Lac objects (215) Objects that resemble quasars, thought to be highly luminous cores of distant active galaxies.

blueshift (84) A Doppler shift toward shorter wavelengths caused by a velocity of approach.

Bok globule (140) Small, dark cloud only about 1 ly in diameter that contains 10 to 1000 solar masses of gas and dust, believed to be related to star formation.

breccia (brĕch′ē·ŭ) (274) Rock composed of fragments of earlier rocks bonded together.

calibration (177) The establishment of the relationship between a parameter that is easily determined and a parameter that is more difficult to determine. For example, the periods of Cepheid variables have been calibrated to reveal absolute magnitudes, which can then be used to find distance. Thus astronomers say Cepheids have been calibrated as distance indicators.

Cambrian explosion (kăm′brē·ŭn) (321) A geological period 0.6 to 0.5 billion years ago during which life on Earth became diverse and complex. Cambrian rocks contain the oldest easily identifiable fossils.

carbonaceous chondrite (kär·bŭ·nā′·shŭs kŏn′-drĭt) (248) Stony meteorite that contains both chondrules and volatiles. These chondrites may be the least-altered remains of the solar nebula still present in the solar system.

carbon–nitrogen–oxygen (CNO) cycle (131) A series of nuclear reactions that use carbon as a catalyst to combine four hydrogen atoms to make one helium atom plus energy, effective in stars more massive than the sun.

Cassegrain focus (kăs′ŭ·grān) (65) The optical design in which the secondary mirror reflects light back down the tube through a hole in the center of the objective mirror.

catastrophic theory (257) An explanation of a phenomenon involving special, sudden, perhaps violent events.

CCD *See* charge-coupled device.

celestial equator (16) The imaginary line around the sky directly above Earth's equator.

celestial sphere (14) An imaginary sphere of very large radius surrounding Earth to which the planets, stars, sun, and moon seem to be attached.

center of mass (51) The balance point of a body or system of masses. The point about which a body or system of masses rotates in the absence of external forces.

Cepheid variable star (sĕ·fē′ĭd) (177) Variable star with a period of 60 days whose period of variation is related to luminosity.

Chandrasekhar limit (shăn′drä·sā′·kär) (154) The maximum mass of a white dwarf, about 1.4 solar masses. A white dwarf of greater mass cannot support itself and will collapse.

charge-coupled device (CCD) (67) An electronic device consisting of a large array of light-sensitive elements used to record very faint images.

chemical evolution (320) The chemical process that led to the growth of complex molecules on primitive Earth. This did not involve the reproduction of molecules.

chromatic aberration (krō·măt′ĭk) (61) A distortion found in refracting telescopes because lenses focus different colors at slightly different distances. Images are consequently surrounded by color fringes.

chromosome (317) A body within a living cell that contains genetic information responsible for the determination and transmission of hereditary traits.

chromosphere (krō′mŭ·sfĭr) (88) Bright gases just above the photosphere of the sun.

circular velocity (50) The velocity an object needs to stay in orbit around another object.

circumpolar constellation (17) A constellation so close to one of the celestial poles that it never sets or never rises as seen from a particular latitude.

closed orbit (51) Either a circular orbit or an elliptical orbit that returns to the same starting point over and over..

closed universe (230) A model universe in which the average density is great enough to stop the expansion and make the universe contract.

CNO cycle *See* carbon–nitrogen–oxygen cycle.

cold dark matter (231) Mass in the universe, as yet undetected except for its gravitational influence, which is made up of slow-moving particles.

comet (244) One of the small, icy bodies that orbit the sun and produce tails of gas and dust when they approach the sun.

compact object (154) A stellar remnant that generates no nuclear energy and is much smaller and denser than a star.

comparative planetology (264) The study of planets in relation to one another.

condensation (253) The growth of a particle by addition of material from surrounding gas, atom by atom.

condensation sequence (252) The sequence in which different materials condense from the solar nebula as we move outward from the sun.

conservation of energy (126) One of the basic laws of stellar structure: The amount of energy flowing out of the top of a shell must equal the amount coming in at the bottom plus whatever energy is generated within the shell.

conservation of mass (126) One of the basic laws of stellar structure: The total mass of the star must equal the sum of the masses of the shells, and the mass must be distributed smoothly through the star.

constellation (12) One of the stellar patterns identified by name, usually of mythological gods, people, animals, or objects. Also, the region of the sky containing that star pattern.

continuous spectrum (86) A spectrum in which there are no absorption or emission lines.

convection (81) Circulation in a fluid driven by heat. Hot material rises and cool material sinks.

convective zone (94) The region inside a star where energy is carried outward as rising hot gas and sinking cool gas.

corona (88) On the sun, the faint outer atmosphere composed of low-density, high-temperature gas.

coronae (279) On Venus, large, round geological faults in the crust caused by the intrusion of magma below the crust.

coronagraph (89) A telescope designed to photograph the inner corona of the sun.

coronal hole (97) An area of the solar surface that is dark at X-ray wavelengths, thought to be associated with divergent magnetic fields and the source of the solar wind.

coronal mass ejection (CME) (97) Matter ejected from the sun's corona in powerful surges guided by magnetic fields.

cosmic microwave background radiation (225) Radiation from the hot clouds of the big bang explosion. The large red shift makes it appear to come from a body whose temperature is only 2.7 K.

cosmological constant (233) A constant in Einstein's equations of space and time that represents a force of repulsion.

cosmological principle (229) The assumption that any observer in any galaxy sees the same general features of the universe.

cosmology (220) The study of the nature, origin, and evolution of the universe.

Coulomb barrier (koo·lôm) (130) The electrostatic force of repulsion between bodies of like charge, commonly applied to atomic nuclei.

Coulomb force (82) The electrostatic force of repulsion or attraction between charged bodies.

critical density (230) The average density of the universe needed to make its curvature flat.

dark age (228) The period of time after the glow of the big bang faded into the infrared and before the birth of the first stars, during which the universe expanded in darkness.

dark energy (234) The energy believed to fill empty spaces and drive the acceleration of the expanding universe.

dark halo (182) The low-density extension of the halo of our galaxy, believed to be composed of dark matter.

dark-line spectrum *See* absorption spectrum.

dark matter (182) Nonluminous matter that is detected only by its gravitational influence.

dark nebula (137) A cloud of gas and dust seen silhouetted against a brighter nebula.

debris disk (259) A disk of dust found by infrared observations around some stars. The dust is debris from collisions among asteroids, comets, and Kuiper belt objects.

degenerate matter (151) Extremely high-density matter in which pressure no longer depends on temperature due to quantum mechanical effects.

density (78) Mass per volume.

density wave theory (184) Theory proposed to account for spiral arms as compressions of the interstellar medium in the disk of the galaxy.

deoxyribonucleic acid (316) The long carbon-chain molecule that records information to govern the biological activity of the organism. DNA carries the genetic data passed to offspring.

deuterium (130) An isotope of hydrogen in which the nucleus contains a proton and a neutron.

differential rotation (94) The rotation of a body in which different parts of the body have different periods of rotation. This is true of the sun, the Jovian planets, and the disk of the galaxy.

differentiation (254) The separation of planetary material according to density.

diffraction fringe (62) Blurred fringe surrounding any image, caused by the wave properties of light. Because of this, no image detail smaller than the fringe can be seen.

digitized (67) Information that is converted to numerical data that can be read directly into a computer memory for later analysis.

disk component (179) All material confined to the plane of the galaxy.

distance ladder (202) The calibration used to build a distance scale reaching from the size of Earth to the most distant visible galaxies.

DNA *See* deoxyribonucleic acid.

Doppler effect (84) The change in the wavelength of radiation due to relative radial motion of source and observer.

double-lobed radio source (212) A galaxy that emits radio energy from two regions (lobes) located on opposite sides of the galaxy.

Drake equation (328) The equation that estimates the total number of communicative civilizations in our galaxy.

dwarf planet (309) A body that orbits the sun, is not a satellite of a planet, is massive enough to pull itself into a spherical shape but not massive enough to clear out other bodies in and near its orbit. For example, Pluto, Eris, and Ceres.

dynamo effect (94) The process by which a rotating, convecting body of conducting matter, such as Earth's core, can generate a magnetic field.

east point (16) One of the four cardinal directions, the point on the horizon directly east.

eccentricity *(e)* (42) A number between 1 and 0 that describes the shape of an ellipse, the distance from one focus to the center of the ellipse divided by the semimajor axis.

eclipsing binary system (115) A binary star system in which the stars eclipse each other.

ecliptic (19) The apparent path of the sun around the sky.

ejecta (274) Pulverized rock scattered by meteorite impacts on a planetary surface.

electromagnetic radiation (58) Changing electric and magnetic fields that travel through space and transfer energy from one place to another; examples are light or radio waves.

electron (79) Low-mass atomic particle carrying a negative charge.

ellipse (41) A closed curve around two points, called the foci, such that the total distance from one focus to the curve and back to the other focus remains constant.

elliptical galaxy (200) A galaxy that is round or elliptical in outline and contains little gas and dust, no disk or spiral arms, and few hot, bright stars.

emission line (85) A bright line in a spectrum caused by the emission of photons from atoms.

emission nebula (136) A cloud of glowing gas excited by ultraviolet radiation from hot stars.

empirical (43) Description of a phenomenon without explaining why it occurs.

energy level (82) One of a number of states an electron may occupy in an atom, depending on its binding energy.

energy transport (126) Flow of energy from hot regions to cooler regions by one of three methods: conduction, convection, or radiation.

enzyme (316) Special protein that controls processes in an organism.

epicycle (ĕp′ŭ-sī-kŭl) (37) The small circle followed by a planet in the Ptolemaic theory. The center of the epicycle follows a larger circle (the deferent) around Earth.

escape velocity (51) The initial velocity an object needs to escape from the surface of a celestial body.

evening star (19) Any planet visible in the sky just after sunset.

event horizon (167) The boundary of the region of a black hole from which no radiation may escape. No event that occurs within the event horizon is visible to a distant observer.

evolutionary theory (257) An explanation of a phenomenon involving slow, steady processes of the sort seen happening in the present day.

excited atom (82) An atom in which an electron has moved from a lower to a higher energy level.

extrasolar planet (259) A planet orbiting a star other than the sun.

eyepiece (60) A short-focal-length lens used to enlarge the image in a telescope. The lens nearest the eye.

false-color image (67) A representation of graphical data with added or enhanced color to reveal detail.

field of view (4) The area visible in an image. Usually given as the diameter of the region.

filament (89) A solar prominence, seen from above, silhouetted against the bright photosphere.

filtergram (89) A photograph (usually of the sun) taken in the light of a specific region of the spectrum—for example, an H$_\alpha$ filtergram.

first principle (34) A principle that seems obviously true to everyone and supposedly needs no further examination.

flare (97) A violent eruption on the sun's surface.

flatness problem (232) In cosmology, the peculiar circumstance that the early universe must have contained almost exactly the right amount of matter to make space-time flat.

flat universe (230) A model of the universe in which space-time is not curved.

flux (14, 105) A measure of the flow of energy out of a surface. Usually applied to light.

focal length (60) The focal length of a lens is the distance from the lens to the point where it focuses parallel rays of light.

folded mountain range (270) A long range of mountains formed by the compression of a planet's crust.

forward scattering (294) The optical property of finely divided particles to preferentially direct light in the original direction of the light's travel.

galactic cannibalism (208) The theory that large galaxies absorb smaller galaxies.

galaxy (7) A large system of stars, star clusters, gas, dust, and nebulae orbiting a common center of mass.

Galilean satellites (găl-ŭ-lē′ŭn) (247) The four largest satellites of Jupiter, named after their discoverer, Galileo.

gamma ray (59) The shortest-wavelength electromagnetic waves.

gamma-ray burst (171) A sudden, powerful burst of gamma rays.

gene (317) A unit of DNA—or sometimes RNA—information responsible for controlling an inherited physiological trait.

general theory of relativity (164) A theory that describes gravity as due to curvature of space-time.

geocentric universe (34) A model universe with Earth at the center, such as the Ptolemaic universe.

geosynchronous satellite (50) A satellite that orbits eastward around Earth with a period of 24 hours and remains above the same spot on Earth's surface.

giant (110, 148) Large, cool, highly luminous star in the upper right of the H–R diagram, typically 10 to 100 times the diameter of the sun.

global warming (272) The gradual increase in the surface temperature of Earth caused by human modifications to Earth's atmosphere.

globular cluster (152) A star cluster containing 100,000 to 1 million stars in a sphere about 75 ly in diameter, generally old, metal-poor, and found in the spherical component of the galaxy.

granulation (81) The fine structure of bright grains covering the sun's surface.

grating (68) A piece of material in which numerous microscopic parallel lines are scribed. Light encountering a grating is dispersed to form a spectrum.

gravitational collapse (254) The process by which a forming body such as a planet gravitationally captures gas from its surroundings.

gravitational redshift (168) The lengthening of the wavelength of a photon due to its escape from a gravitational field.

greenhouse effect (268) The process by which a carbon dioxide atmosphere traps heat and raises the temperature of a planetary surface.

ground state (83) The lowest permitted electron energy level in an atom.

habitable zone (325) A region around a star within which planets have temperatures that permit the existence of liquid water.

half-life (250) The time required for half of the atoms in a radioactive sample to decay.

halo (179) The spherical region of a spiral galaxy, containing a thin scattering of stars, star clusters, and small amounts of gas.

heat (79) Energy stored in a material as agitation among its particles.

heat of formation (258) In planetology, the heat released by infalling matter during the formation of a planetary body.

heavy bombardment (255) The intense cratering during the first 0.5 billion years in the history of the solar system.

heliocentric universe (35) A model of the universe with the sun at the center, such as the Copernican universe.

helioseismology (91) The study of the interior of the sun by the analysis of its modes of vibration.

Herbig–Haro object (140) A small nebula that varies irregularly in brightness, believed to be associated with star formation.

Hertzsprung–Russell (H–R) diagram (hĕrt′sprŭng·rŭs·ŭl) (110) A plot of the intrinsic brightness versus the surface temperature of stars. It separates the effects of temperature and surface area on stellar luminosity and is commonly plotted as absolute magnitude versus spectral type but also as luminosity versus surface temperature or color.

homogeneity (229) The assumption that, on the large scale, matter is uniformly spread through the universe.

horizon (16) The circular boundary between the sky and Earth.

horizon problem (232) In cosmology, the circumstance that the primordial background radiation seems much more isotropic than can be explained by the standard big bang theory.

horizontal branch (149) In the H–R diagram, stars fusing helium in a shell and evolving back toward the red giant region.

H–R diagram *See* Hertzsprung–Russell diagram.

HII region (136) A region of ionized hydrogen around a hot star.

Hubble constant *(H)* (203) A measure of the rate of expansion of the universe, the average value of velocity of recession divided by distance, presently believed to be about 70 km/s/Mpc.

Hubble law (203) The linear relation between the distances to galaxies and their velocity of recession.

Hubble time (224) The age of the universe, equivalent to 1 divided by the Hubble constant. The Hubble time is the age of the universe if it has expanded since the big bang at a constant rate.

hydrostatic equilibrium (126) The balance between the weight of the material pressing downward on a layer in a star and the pressure in that layer.

hypernova (171) Produced when a very massive star collapses into a black hole. Thought to be a possible source of gamma-ray bursts.

hypothesis (42) A conjecture, subject to further tests, that accounts for a set of facts.

ice line (252) Boundary beyond which water vapor could freeze to form ice.

inflationary big bang (232) A version of the big bang theory, derived from grand unified theories, that includes a rapid expansion when the universe was very young.

infrared (IR) (59) The portion of the electromagnetic spectrum with wavelengths longer than red light, ranging from 700 nm to about 1 mm, between visible light and radio waves.

instability strip (177) The region of the H–R diagram in which stars are unstable to pulsation. A star passing through this strip becomes a variable star.

interferometer (62) Separated telescopes combined to produce a virtual telescope with the resolution of a much larger-diameter telescope.

interstellar dust (134) Microscopic solid grains in the interstellar medium.

interstellar medium (134) The gas and dust distributed between the stars.

interstellar reddening (135) The process in which dust scatters blue light out of starlight and makes the stars look redder.

intrinsic brightness (105) A measure of the amount of energy a star produces.

inverse square relation (48) A rule that the strength of an effect (such as gravity) decreases in proportion as the distance squared increases.

ion (82) An atom that has lost or gained one or more electrons.

ionization (82) The process in which atoms lose or gain electrons.

irregular galaxy (201) A galaxy with a chaotic appearance, large clouds of gas and dust, and both population I and II stars, but without spiral arms.

isotopes (82) Atoms that have the same number of protons but a different number of neutrons.

isotropy (ī·sŏt′rŭ·pē) (229) The assumption that, in its general properties, the universe looks the same in every direction.

Jovian planet (246) Jupiter-like planet with a large diameter and low density.

Kelvin temperature scale (79) A temperature scale using Celsius degrees and based on zero at absolute zero.

kiloparsec (kpc) (179) A unit of distance equal to 1000 pc or 3260 ly.

Kirchhoff's laws (86) A set of laws that describe the absorption and emission of light by matter.

Kuiper belt (248) The collection of icy planetesimals believed to orbit in a region from just beyond Neptune out to 100 AU or more.

large-impact hypothesis (274) The theory that the moon formed from debris ejected during a collision between Earth and a large planetesimal.

Large Magellanic Cloud (201) An irregular galaxy, visible in the southern sky, that is a satellite of our Milky Way Galaxy.

large-scale structure (234) The distribution of clusters and superclusters of galaxies in filaments and walls enclosing voids.

late heavy bombardment (275) The sudden temporary increase in the cratering rate in our solar system that occurred about 4 billion years ago.

light curve (115) A graph of brightness versus time commonly used in analyzing variable stars and eclipsing binaries.

light-gathering power (62) The ability of a telescope to collect light, proportional to the area of the telescope's objective lens or mirror.

lighthouse model (163) The explanation of a pulsar as a spinning neutron star sweeping beams of electromagnetic radiation around the sky.

light pollution (64) The illumination of the night sky by waste light from cities and outdoor lighting, which prevents the observation of faint objects.

light-year (ly) (6) The unit of distance light travels in 1 year.

liquid metallic hydrogen (293) A form of liquid hydrogen that is a good electrical conductor, found in the interiors of Jupiter and Saturn.

Local Group (205) The small cluster of a few dozen galaxies that contains our Milky Way Galaxy.

look-back time (202) The amount by which you look into the past when you look at a distant galaxy, a time equal to the distance to the galaxy in light-years.

luminosity *(L)* (106) The total amount of energy a star radiates in 1 second.

luminosity class (112) A category of stars of similar luminosity, determined by the widths of lines in their spectra.

lunar eclipse (27) The darkening of the moon when it moves through Earth's shadow.

Lyman series (lī′měn) (87) Spectral lines in the ultraviolet spectrum of hydrogen produced by transitions whose lowest energy level is the ground state.

magma ocean (275) The exterior of the newborn moon, a shell of molten rock hundreds of kilometers deep.

magnetic carpet (89) The network of small magnetic loops that covers the solar surface.

magnetosphere (măg·nē′tō·sfîr) (293) The volume of space around a planet within which the motion of charged particles is dominated by the planetary magnetic field rather than the solar wind.

magnifying power (63) The ability of a telescope to make an image larger.

magnitude scale (13) The astronomical brightness scale. The larger the number, the fainter the star.

main sequence (110) The region of the H–R diagram running from upper left to lower right, which includes roughly 90 percent of all stars.

mantle (266) The layer of dense rock and metal oxides that lies between the molten core and Earth's surface or a similar layer in another planet.

mare (mä′rā) (plural: **maria**) (273) One of the lunar lowlands filled by successive flows of dark lava, from the Latin for "sea."

mass (47) A measure of the amount of matter making up an object.

mass–luminosity relation (117) The more massive a star is, the more luminous it is.

Maunder butterfly diagram (môn′dŭr) (92) A graph showing the latitude of sunspots versus time, first plotted by W. W. Maunder in 1904.

Maunder minimum (môn′dŭr) (93) A period between 1645 and 1715 of less numerous sunspots and other solar activity.

megaparsec (Mpc) (202) A unit of distance equal to 1,000,000 pc.

metals (187) In astronomical usage, all atoms heavier than helium.

meteor (248) A small bit of matter heated by friction to incandescent vapor as it falls into Earth's atmosphere.

meteorite (248) A meteor that survives its passage through the atmosphere and strikes the ground.

meteoroid (248) A meteor in space before it enters Earth's atmosphere.

meteor shower (249) A multitude of meteors that appear to come from the same region of the sky, believed to be caused by comet debris.

micrometeorite (276) Meteorite of microscopic size.

midocean rise (270) One of the undersea mountain ranges that push up from the seafloor in the center of the oceans.

Milky Way (7) The hazy band of light that circles our sky, produced by the glow of our galaxy.

Milky Way Galaxy (7) The spiral galaxy containing our sun, visible in the night sky as the Milky Way.

Miller experiment (318) An experiment that reproduced the conditions under which life began on Earth and manufactured amino acids and other organic compounds.

millisecond pulsar (164) A pulsar with a pulse period of only a few milliseconds.

model *See* scientific model.

molecular cloud (138) A dense interstellar gas cloud in which atoms are able to link together to form molecules such as H_2 and CO.

molecule (79) Two or more atoms bonded together.

monolithic collapse model (188) An early hypothesis that says that the galaxy formed from a single large cloud of turbulent gas over 13 billion years ago.

morning star (19) Any planet visible in the sky just before sunrise.

multicellular (321) An organism composed of more than one cell.

multiringed basin (275) Large impact feature (crater) containing two or more concentric rims formed by fracturing of the planetary crust.

mutant (315) Offspring born with altered DNA.

nadir (16) The point on the celestial sphere directly below the observer; the opposite of the zenith.

nanometer (nm) (58) A unit of distance equaling one-billionth of a meter (10^{-9} m).

natural law (42) A theory that is almost universally accepted as true.

natural selection (315) The process by which the best traits are passed on, allowing the most able to survive.

neap tide (49) Ocean tide of low amplitude occurring at first- and third-quarter moon.

nebula (134) A glowing cloud of gas or a cloud of dust reflecting the light of nearby stars.

NEO (Near Earth Object) (256) A small solar system body (asteroid or comet) with an orbit near enough to Earth that it poses some threat of eventual collision.

neutrino (131) A neutral, massless atomic particle that travels at or nearly at the speed of light.

neutron (79) An atomic particle with no charge and about the same mass as a proton.

neutron star (161) A small, highly dense star, with radius about 10 km, composed almost entirely of tightly packed neutrons.

Newtonian focus (65) The optical design in which a diagonal mirror reflects light out the side of the telescope tube for easier access.

nonbaryonic matter (231) Proposed dark matter made up of particles other than protons and neutrons (baryons).

north celestial pole (16) The point on the celestial sphere directly above Earth's North Pole.

north point (16) One of the four cardinal directions, the point on the horizon directly north.

nova (146) From the Latin, meaning "new," a sudden brightening of a star making it appear as a new star in the sky, believed to be associated with eruptions on white dwarfs in binary systems.

nuclear bulge (179) The spherical cloud of stars that lies at the center of spiral galaxies.

nuclear fission (129) Reactions that break the nuclei of atoms into fragments.

nuclear fusion (130) Reactions that join the nuclei of atoms to form more massive nuclei.

nucleus (of an atom) (79) The central core of an atom containing protons and neutrons that carries a net positive charge.

oblateness (293) The flattening of a spherical body, usually caused by rotation.

observable universe (222) The part of the universe that you can see from your location in space and in time.

occultation (306) The passage of a larger body in front of a smaller body.

Olbers's paradox (ôl′bŭrs) (222) The conflict between observation and theory about why the night sky should or should not be dark.

Oort cloud (ōrt) (258) The hypothetical source of comets, a swarm of icy bodies believed to lie in a spherical shell extending to 100,000 AU from the sun.

opacity (127) The resistance of a gas to the passage of radiation.

open cluster (152) A cluster of 100 to 1000 stars with an open, transparent appearance, usually relatively young and located in the disk of the galaxy. The stars are not tightly grouped.

open orbit (51) An orbit that carries an object away, never to return to its starting point.

open universe (230) A model of the universe in which the average density is less than the critical density needed to halt the expansion.

optical telescope (61) Telescopes that gather visible light.

outflow channel (284) Geological feature on Mars produced by the rapid motion of floodwaters.

outgassing (254) The release of gases from a planet's interior.

ovoid (306) The oval features found on Miranda, a satellite of Uranus.

paradigm (39) A commonly accepted set of scientific ideas and assumptions.

parallax (36) The apparent change in position of an object due to a change in the location of the observer. Astronomical parallax is measured in seconds of arc.

parsec (pc) (pär′sĕk) (105) The distance to a hypothetical star whose parallax is 1 second of arc. 1 pc = 206,265 AU = 3.26 ly.

Paschen series (pä′shŭn) (87) Spectral lines in the infrared spectrum of hydrogen produced by transitions whose lowest energy level is the third.

penumbra (pǐ·nŭm′brŭ) (26) The portion of a shadow that is only partially shaded.

perihelion (pĕr·ŭ·hē′lē·ŭn) (21) The orbital point of closest approach to the sun.

period–luminosity relation (177) The relation between period of pulsation and intrinsic brightness among Cepheid variable stars.

permafrost (282) Permanently frozen soil.

permitted orbit (82) One of the energy levels that an electron may occupy in an atom.

photographic plate (67) The first image-recording device used with telescopes; it measures the brightness of objects, but with only moderate precision.

photometer (67) Sensitive astronomical instrument that measures the brightness of individual objects very precisely.

photon (59) A quantum of electromagnetic energy that carries an amount of energy that depends inversely on its wavelength.

photosphere (81) The bright visible surface of the sun.

planet (5) A small, nonluminous body formed by accretion in a disk around a protostar.

planetary nebula (151) An expanding shell of gas ejected from a star during the latter stages of its evolution.

planetesimal (plăn·ŭ·tĕs′ŭ·mŭl) (253) One of the small bodies that formed from the solar nebula and eventually grew into protoplanets.

plate tectonics (270) The constant destruction and renewal of Earth's surface by the motion of sections of crust.

plutino (309) One of the icy Kuiper belt objects that, like Pluto, is caught in a 3:2 orbital resonance with Neptune.

poor galaxy cluster (205) An irregularly shaped cluster that contains fewer than 1000 galaxies, many of which are spiral, and no giant ellipticals.

population I (187) Stars rich in atoms heavier than helium, nearly always relatively young stars found in the disk of the galaxy.

population II (187) Stars poor in atoms heavier than helium, nearly always relatively old stars found in the halo, globular clusters, or the nuclear bulge.

precession (14) The slow change in the direction of Earth's axis of rotation. One cycle takes nearly 26,000 years.

primary atmosphere (268) A planet's first atmosphere.

primary lens (60) In a refracting telescope, the largest lens.

primary mirror (60) In a reflecting telescope, the largest mirror.

prime focus (65) The point at which the objective mirror forms an image in a reflecting telescope.

primordial soup (319) The rich solution of organic molecules in Earth's first oceans.

prominence (96) Eruption on the solar surface. Visible during total solar eclipses.

proper motion (177) The rate at which a star moves across the sky, measured in seconds of arc per year.

protein (316) Complex molecule composed of amino acid units.

proton (79) A positively charged atomic particle contained in the nucleus of an atom. The nucleus of a hydrogen atom.

proton–proton chain (130) A series of three nuclear reactions that builds a helium atom by adding together protons. The main energy source in the sun.

protoplanet (253) Massive object, destined to become a planet, resulting from the coalescence of planetesimals in the solar nebula.

protostar (140) A collapsing cloud of gas and dust destined to become a star.

pulsar (162) A source of short, precisely timed radio bursts, believed to be spinning neutron stars.

***P* wave** (267) A pressure wave. a type of seismic wave produced in Earth by the compression of the material.

quantum leap (83) Jumps of electrons from one orbit or energy state to another.

quantum mechanics (82) The study of the behavior of atoms and atomic particles.

quasar (quasi-stellar object, or QSO) (kwā′zär) (212) Small, powerful source of energy believed to be the active core of a very distant galaxy.

quintessence (234) The postulated energy that fills empty space and drives the acceleration of the universe.

radial velocity (V_r) (85) That component of an object's velocity directed away from or toward Earth.

radio galaxy (210) A galaxy that is a strong source of radio signals.

radio telescope (61) Telescopes that gather radio radiation.

rays (380) Ejecta from meteorite impacts forming white streamers radiating from some lunar craters.

recombination (228) The stage, within 300,000 years of the big bang, when the gas became transparent to radiation.

reconnection (97) On the sun, the merging of magnetic fields to release energy in the form of flares.

red dwarf (112) A faint, cool, low-mass, main-sequence star.

redshift (84) A Doppler shift toward longer wavelengths caused by a velocity of recession.

reflection nebula (136) A nebula produced by starlight reflecting off dust particles in the interstellar medium.

refracting telescope (60) A telescope that forms images by bending (refracting) light with a lens.

reionization (228) The stage in the early history of the universe when ultraviolet photons from the first stars ionized the gas filling space.

resolving power (62) The ability of a telescope to reveal fine detail. Depends on the diameter of the telescope objective.

retrograde motion (36) The apparent backward (westward) motion of planets as seen against the background of stars.

revolution (18) Orbital motion about a point located outside the orbiting body. *See also* rotation.

ribonucleic acid (RNA) (317) Long carbon-chain molecules that use the information stored in DNA to manufacture complex molecules necessary to the organism.

rich galaxy cluster (205) A cluster containing over 1000 galaxies, mostly elliptical, scattered over a volume about 3 Mpc in diameter.

rift valley (270) A long, straight, deep valley produced by the separation of crustal plates.

ring galaxy (209) A galaxy that resembles a ring around a bright nucleus, believed to be the result of a head-on collision of two galaxies.

RNA *See* ribonucleic acid.

Roche limit (rōsh) (294) The minimum distance between a planet and a satellite that holds itself together by its own gravity. If a satellite's orbit brings it within its planet's Roche limit, tidal forces will pull the satellite apart.

Roche lobe (155) The volume of space a star controls gravitationally within a binary system.

rotation (18) Motion around an axis passing through the rotating body. *See also* revolution.

rotation curve (182) A graph of orbital velocity versus radius in the disk of a galaxy.

rotation curve method (203) A method of determining a galaxy's mass by observing the orbital velocity and radius of stars in the galaxy.

runaway greenhouse (281) A situation in which the greenhouse effect grows more severe.

Sagittarius A* (190) The powerful radio source located at the core of the Milky Way Galaxy.

Saros cycle (sĕ′rōs) (29) An 18-year, 11⅓-day period after which the pattern of lunar and solar eclipses repeats.

Schmidt-Cassegrain focus (65) The optical design that uses a thin corrector plate at the entrance to the telescope tube. A popular design for small telescopes.

Schwarzschild radius (R_s) (schwôrts′ shēld) (167) The radius of the event horizon around a black hole.

scientific model (14) A tentative description of a phenomenon for use as an aid to understanding.

scientific notation (5) The system of recording very large or very small numbers by using powers of 10.

secondary atmosphere (268) The gases outgassed from a planet's interior; rich in carbon dioxide.

secondary mirror (65) In a reflecting telescope, a mirror that directs the light from the primary mirror to a focal position.

seeing (62) Atmospheric conditions on a given night. When the atmosphere is unsteady, producing blurred images, the seeing is said to be poor.

self-sustaining star formation (186) The process by which the birth of stars compresses the surrounding gas clouds and triggers the formation of more stars, proposed to explain spiral arms.

semi-major axis (*a*) (42) Half of the longest diameter of an ellipse.

SETI (326) The Search for Extra-Terrestrial Intelligence.

Seyfert galaxy (sē′fûrt) (210) An otherwise normal spiral galaxy with an unusually bright, small core that fluctuates in brightness, believed to indicate that the core is erupting.

shepherd satellite (299) A satellite that, by its gravitational field, confines particles to a planetary ring.

shield volcano (283) Wide, low-profile volcanic cone produced by highly liquid lava.

shock wave (139) A sudden change in pressure that travels as an intense sound wave.

sidereal tracking (sī·dîr′e·ŭl) (64) The continuous movement of a telescope to keep it pointed at a star.

singularity (167) The object of zero radius into which the matter in a black hole is believed to fall.

Small Magellanic Cloud (201) An irregular galaxy, visible in the southern sky, that is a satellite of our Milky Way Galaxy.

solar eclipse (23) The event that occurs when the moon passes directly between Earth and the sun, blocking your view of the sun.

solar nebula hypothesis (243) The hypothesis that the planets formed from the same cloud of gas and dust that formed the sun.

solar system (5) The sun and its planets, asteroids, comets, and so on.

solar wind (89) Rapidly moving atoms and ions that escape from the solar corona and blow outward through the solar system.

south celestial pole (16) The point on the celestial sphere directly above Earth's South Pole.

south point (16) One of the four cardinal directions, the point on the horizon directly south.

spectral class (108) A star's position in the temperature classification system O, B, A, F, G, K, M, based on the appearance of the star's spectrum.

spectral line (76) A line in a spectrum at a specific wavelength produced by the absorption or emission of light by certain atoms.

spectral sequence (108) The arrangement of spectral classes (O, B, A, F, G, K, M) ranging from hot to cool.

spectrograph (68) A device that separates light by wavelengths to produce a spectrum.

spectroscopic binary system (115) A star system in which the stars are too close together to be visible separately. We see a single point of light, and only by taking a spectrum can we determine that there are two stars.

spectroscopic parallax (112) The method of determining a star's distance by comparing its apparent magnitude with its absolute magnitude as estimated from its spectrum.

spectrum (68) A range of electromagnetic radiation spread into its component wavelengths (colors), for example a rainbow; also, representation of a spectrum as a graph showing intensity of radiation as a function of wavelength or frequency.

spherical component (179) The part of the galaxy including all matter in a spherical distribution around the center (the halo and nuclear bulge).

spicule (spĭk′yo͞ol) (89) A small, flamelike projection in the chromosphere of the sun.

spiral arm (7, 179) Long spiral pattern of bright stars, star clusters, gas, and dust. Spiral arms extend from the center to the edge of the disk of spiral galaxies.

spiral galaxy (200) A galaxy with an obvious disk component containing gas; dust; hot, bright stars; and spiral arms.

spiral tracer (183) Object used to map the spiral arms—for example, O and B associations, open clusters, clouds of ionized hydrogen, and some types of variable stars.

spring tide (49) Ocean tide of high amplitude that occurs at full and new moon.

standard candle (202) Object of known brightness that astronomers use to find distance—for example, Cepheid variable stars and supernovae.

star (5) A globe of gas held together by its own gravity and supported by the internal pressure of its hot gases, which generate energy by nuclear fusion.

star cluster (139) A group of stars that formed together and orbit a common center of mass.

starburst galaxy (207) A galaxy undergoing a rapid burst of star formation.

Stefan-Boltzmann law (80) A law stating that hotter objects emit more energy than cooler objects of the same size, in proportion to the fourth power of temperature.

stellar association (139) A group of stars that are not gravitationally bound to one another.

stellar model (128) A table of numbers representing the conditions in various layers within a star.

stellar parallax (*p*) (104) A measure of stellar distance. *See also* parallax.

stromatolite (318) A layered fossil formation caused by ancient mats of algae or bacteria, which built up mineral deposits season after season.

subduction zone (270) A region of a planetary crust where a tectonic plate slides downward.

summer solstice (20) The point on the celestial sphere where the sun is at its most northerly point. Also, the time when the sun passes this point, about June 22, and summer begins in the Northern Hemisphere.

sunspot (81) Relatively dark spot on the sun that contains intense magnetic fields.

supercluster (8, 234) A cluster of galaxy clusters.

supergiant (112, 148) Exceptionally luminous star whose diameter is 10 to 1000 times that of the sun.

supernova (146) The explosion of a star in which it increases its brightness by a factor of about a million.

supernova remnant (160) The expanding shell of gas marking the site of a supernova explosion.

***S* wave** (267) A shear wave. a type of seismic wave produced in Earth by the lateral motion of the material.

synchrotron radiation (sĭn′krū·trŏn) (159) Radiation emitted when high-speed electrons move through a magnetic field.

temperature (79) A measure of the agitation among the atoms and molecules of a material, the intensity of heat.

terrestrial planet (246) An Earthlike planet—small, dense, rocky.

theory (42) A system of assumptions and principles applicable to a wide range of phenomena that has been repeatedly verified.

tidal heating (296) The heating of a planet or satellite because of friction caused by tides.

tidal tail (208) A long streamer of stars, gas, and dust torn from a galaxy during its close interaction with another passing galaxy.

time dilation (168) The slowing of moving clocks or clocks in strong gravitational fields.

transition (87) The movement of an electron from one atomic energy level to another.

transit of Venus (78) Rare occasions when Venus can be seen as a tiny dot directly between Earth and the sun.

T Tauri star (tôrē) (140) A young star surrounded by gas and dust, believed to be contracting toward the main sequence.

turnoff point (152) The point in an H–R diagram at which a cluster's stars turn off of the main sequence and move toward the red-giant region, revealing the approximate age of the cluster.

type I supernova (158) A supernova explosion caused by the collapse of a white dwarf.

type II supernova (158) A supernova explosion caused by the collapse of a massive star.

ultraviolet (UV) (59) The portion of the electromagnetic spectrum with wavelengths shorter than violet light, between visible light and X rays.

umbra (ŭm′brū) (23) The region of a shadow that is totally shaded.

uncompressed density (252) The density a planet would have if its gravity did not compress it.

unified model (214) An attempt to explain the different types of active galactic nuclei using a single model viewed from different directions.

uniform circular motion (36) The classical belief that the perfect heavens could move only by the combination of uniform motion along circular orbits.

valley network (284) A system of dry drainage channels on Mars that resembles the beds of rivers and tributary streams on Earth.

vernal equinox (20) The place on the celestial sphere where the sun crosses the celestial equator moving northward. Also, the time of year when the sun crosses this point, about March 21, and spring begins in the Northern Hemisphere.

visual binary system (114) A binary star system in which the two stars are separately visible in the telescope.

volatile (244) Easily vaporized.

water hole (326) The interval of the radio spectrum between the 21-cm hydrogen radiation and the 18-cm OH radiation; these are likely wavelengths to use in the search for extraterrestrial life.

wavelength (58) The distance between successive peaks or troughs of a wave, usually represented by *l*.

wavelength of maximum intensity (λ_{max}) (80) The wavelength at which a perfect radiator emits the maximum amount of energy. Depends only on the object's temperature.

weight (47) The force that gravity exerts on an object.

west point (16) One of the four cardinal directions, the point on the horizon directly west.

white dwarf (112) Dying star at the lower left of the H–R diagram that has collapsed to the size of Earth and is slowly cooling off.

Wien's law (80) A law stating that the hotter an object is, the shorter will be the wavelength of its most intense output.

winter solstice (20) The point on the celestial sphere where the sun is farthest south. Also, the time of year when the sun passes this point, about December 22, and winter begins in the Northern Hemisphere.

X ray (59) Electromagnetic waves with wavelengths shorter than ultraviolet light.

Zeeman effect (93) The splitting of spectral lines into multiple components when the atoms are in a magnetic field.

zenith (16) The point on the sky directly above the observer.

zero-age main sequence (ZAMS) (133) The location in the H–R diagram where stars first reach stability as hydrogen-burning stars.

zodiac (19) A band centered on the ecliptic and encircling the sky.

Index

Boldface page numbers indicate definitions of key terms.

A, **58**
A020-00, 171
Absolute visual magnitude (M_v), **106**
Absolute zero, **79**
Absorption line, **86**
Absorption spectrum, **86**
Acceleration of universe, 232–234
Accretion, **253**
Accretion disk, **156**
Active galactic nuclei (AGN), **210**
Active galaxies, **210**
Active region, **93**
Adams, Douglas, 220
Adaptive optics, **63**
AGN, **210**
Al Magisti, 37
Albedo, **273**
Aldebaran, 13, 148
Aldrin, Edwin E., Jr., 264
Almagest, 37
Alpha Centauri, 105
Alpher, Ralph, 225
AM 0644-741, 209
Amino acids, **316,** 319
Ancient universe, 36–37
Andromeda Alpheratz, 12
Andromeda Galaxy, 202
Angström, Anders, 58
Angstrom (A), **58**
Angular distance, **17**
Angular momentum, **155–156**
Annual eclipse, **27**
Anorthosite, **274**
Antennae Galaxies, 210
Antimatter, **226**
Aphelion, **21**
Apparent brightness, 106
Apparent visual magnitudes (m_v), **14**
Arc minute, **17**
Arc second, **17**
Arecibo beacon, 326
Arecibo Observatory, 67
Aristotle, 34
Aristotle's universe, 34, 36–37
Armstrong, Neil, 264
Array detectors, **67**
Asterism, **12**
Asteroid, **244,** 245
Astrology, 22
Astronomer's triangulation method, 104–105

Astronomia Nova (Kepler), 41
Astronomical telescopes and instruments, 56–75
  electromagnetic spectrum, 58–60
  imaging systems, 67
  modern telescopes, 65
  observatories in space, 69–73
  observatories on earth, 64–67
  photometers, 67
  radiation, 58–60
  spectrographs, 68
  telescopes, 60–64
  telescopes in space, 71
Astronomical unit (AU), **5**
Atmospheric windows, **60**
Atom, **79**
Atomic spectra, 86–87
AU, **5**
Aurora, **97**
Autumnal equinox, **20**

Baade, Walter, 161
Babcock model, **94–95**
Balmer lines of hydrogen, 88
Balmer series, **87**
Barberini, Cardinal, 44, 45
Baronius, Cardinal, 46
Barred spiral galaxies, **200**
Barringer Meteorite Crater, 255
Baryon, 231
Basalt, **270**
Baseline, 104
Basic science, 302
Bell, Jocelyn, 162
Bellarmine, Cardinal, 44, 45
Belt-zone circulation, **293**
Betelgeuse, 13, 80, 112, 148
Big bang, **224**
Big Dipper, 12, 13
Big rip, **234**
Binary pulsars, 164
Binary stars, **113–117**
Binary systems, 114–117, 155–156
Binding energy, **82**
Biological evolution, **315**
Black Eye Galaxy, 207
Black hole, 166–171, **167,** 205, 215–216
Blackbody radiation, **79, 80**
Blazar, **215**
Blueberries, 284, 286
Blueshift, **84**
Bode, J. E., 303
Bok globule, **140**
Bombardment of planets, 255–256
Brahe, Tycho, 40

Breccia, **274**
Brecht, Bertolt, 46
Bright-line spectra, **86**
Brown dwarfs, **133**
Bruno, Giordano, 45

Calibration, **177,** 178
Callisto, 295–296, 301
Cambrian explosion, **321**
Canis Major, 13
Canis Majoris, 13
Cannon, Annie J., 108
Capture hypothesis, 274
Carbon-chain molecules, 319
Carbonaceous chondrites, **248,** 249
Cassegrain focus, **65**
Cassini spacecraft, 302
Cassini's Division, 299
Catastrophic theory, **257**
Cat's Eye, 147
Cause and effect, 52
CCD, **67**
Celestial equator, **16**
Celestial pole, **16**
Celestial sphere, **14,** 16–17
Center of mass, **51**
Cepheid variable stars, **177**
CFCs, 272
Chains of inference, 115
Chandra X-Ray Observatory, 73, 234
Chandrasekhar, Subrahmanyan, 73
Chandrasekhar limit, **154**
Charge-coupled device (CCD), **67**
Charon, 308
Chemical evolution, **320**
Chicxulub crater, 256
Chlorofluorocarbons (CFCs), 272
Chromatic aberration, **61**
Chromosome, **317**
Chromosphere, **88**
Circular velocity, **50**
Circumpolar constellation, **17**
Closed orbit, **51**
Closed universe, **230,** 231
CMB, **225**
CMB irregularities, 235–236
CME, **97**
CNO (carbon-nitrogen-oxygen) cycle, **131**
Cold dark matter, **231**
Colliding galaxies, 207
Coma cluster of galaxies, 205, 206
Comet, **244,** 248
Comet Halley, 244, 249

Communication with distant civilizations, 325–329
Compact object, **154,** 161, 170
Comparative planetology, **264**–311
  atmosphere, 267, 292–293
  core, mantle, crust, 266–267
  Earth, 267–272. *See also* Earth
  interiors, 266–267, 292–293
  Jupiter, 293–296. *See also* Jupiter
  Mars, 281–287. *See also* Mars
  Mercury, 273, 277
  moon, 273–276. *See also* Moon
  Neptune, 305, 306–308
  outer planets, 290–311
  Pluto, 308–309
  Saturn. *See* Saturn
  terrestrial planets, 264–289
  Uranus, 303–306
  Venus, 277–281
Compton Gamma Ray Observatory, 72, 171
Condensation, **253**
Condensation hypothesis, 274
Condensation sequence, **252,** 253
Confidence, 98
Confirmation, 98
Confucius, 2
Conservation of energy, **126**
Conservation of mass, **126**
Consolidation, 98
Constellation, **12**
Continuous spectrum, **86**
Convection, **81**
Convective zone, **94**
Copernican hypothesis, 39
Copernican model, 35–39
Copernicus, Nicolaus, 34–39
Corona, **88, 279**
Coronagraphs, **89**
Coronal hole, **97**
Coronal mass ejection (CME), **97**
Cosmic expansion, 223–224
Cosmic microwave background (CMB), **225**
Cosmic redshift, 229–230
Cosmological constant, **233**
Cosmological principle, **229**
Cosmology, **220**–239
  acceleration of universe, 232–234
  big bang theory, 224–228
  CMB, 225
  CMB irregularities, 235–236
  cosmic expansion, 223–224
  cosmic redshift, 229–230
  cosmological principle, 229

Cosmology *(continued)*
  dark energy, 234
  dark matter, 231
  edge-center problem, 222
  fate of universe, 234
  inflationary big bang, 232
  model universes, 230
  observable universe, 222
  Olbers's paradox, 222
  origin of structure, 234–235
  space-time, 230, 235–236
Coulomb, Charles-Augustin de, 82
Coulomb barrier, **130,** 131
Coulomb force, **82**
Crab Nebula, 159, 163
Cratering and giant basin formation, 272
Critical density, **230**
Cygnus X-1, 169, 171

Dark age, **228**
Dark energy, 234, **234**
Dark halo, **182**
Dark-line spectra, **86**
Dark matter, **182,** 204–205, 231
Dark nebula, **137**
De Revolutionibus Orbium
  Coelestium (Copernicus),
  35–39
Debris disks, **259**
Deep images, 199
Degenerate matter, **151**
Degree, **17**
Deimos, 285, 287
Density, **78**
Density wave theory, **184**
Deuterium, **130**
Dialogo Dei Due Massimi Sistemi
  (Galileo), 45
Dicke, Robert, 225
Differential rotation, **94,** 181
Differentiation, **254,** 272
Diffraction fringe, **62**
Digitized, **67**
Disk component, **179**
Disk population stars, 187
Distance ladder, **202**
DNA (deoxyribonucleic acid), **316**
DNA ladder, 316
Doppler effect, **84–85**
Doppler shift, 84–85
Double-lobed radio sources, **212**
Drake equation, **328**
Dwarf
  brown, 133
  red, 150
  white, 151, 154
Dwarf elliptical galaxies, 203
Dwarf planets, **309**
Dynamo effect, **94**

Eagle Nebula, 103
Earth, 267–272
  atmosphere, 268–272
  characteristics, 269
  crust, 268
  end of, 154–155
  geological history, 272
  interior, 267–268
  Mars, compared, 282
  oxygen, 272
  Venus, compared, 278
East point, **16**
Eccentricity, **42**
Eclipse
  lunar, 27–29
  solar, 23–27
Eclipsing binary system, **115**–116
Ecliptic, **19**
Edge-center problem, 222
Einstein, Albert, 164, 167,
  229–230, 233
Ejecta, **274**
Electromagnetic radiation, **58,** 60
Electromagnetic spectrum, 58–60
Electron-neutrinos, 131–132
Electron shells, 82
Electrons, **79**
Element-building cycle, 187–188
Ellipse, **41**
Elliptical galaxies, **200**
Emission nebula, **136**
Emission spectrum, **86**
Empirical, **43**
Enceladus, 302, 304
Encke's Division, **299**
Energy levels, **82**
Energy transport, **126**–128
Epicycle, **37**
Eris, 309
Eros, 245
Escape velocity, **51**
Europa, 301
Evening star, **19**
Event horizon, **167**
Evolution of thought. *See* Historical
  overview
Evolutionary theory, **257**
Excitation of atoms, 82–84
Excited atom, **82**
Extrasolar planet, **259**–260
Extraterrestrial intelligence,
  325–329
Extraterrestrial radio signals, 326
Eyepiece, **60**

False-color images, **67**
Far-infrared range, **69**
Field of view, **4**
51 Pegasi, 260
Filament, **89**
Filtergram, **89**

First principles, **34**
Fission hypothesis, 274
Flare, **97**
Flat universe, **230,** 231
Flatness problem, **232**
Flux, **14, 105**
Focal length, **60**
Forerunner of Dissertations on the
  Universe, Containing the Mys-
  tery of the Universe, The (Kepler), 40
Forward scattering, **294**
4U 1543-47, 171
Frost, Robert, 56
Funding science, 302

Galactic cannibalism, **208**
Galaxy, **7,** 196–219
  active, 210
  barred spiral, 200
  classification, 198, 200–201
  clusters, 205
  collisions, 207
  dark matter, 204–205
  distance, 202, 203
  double-lobed/radio sources, 212
  dwarf elliptical, 203
  elliptical, 200
  evolution of, 205–210
  Hubble law, 203
  interacting, 208–209
  irregular, 201
  look-back time, 202
  luminosities, 203
  mass, 203–204
  Milky Way. *See* Milky Way
    Galaxy
  origin of, 207–210
  quasar, 212–214, 216–217
  radio, 210
  ring, 209
  Seyfert, 210–212
  spiral, 200
  starburst, 207
  supermassive black holes, 205,
    215–216
  time machines, 202
  unified model, 214–215
Galilean moons, 295–296, 301
Galilean satellites, **247**
Galileo, 91
Galileo spacecraft, 290
Gamma-ray burst, 170–**171**
Gamma rays, **59**
Gamow, George, 225
Ganymede, 295–296, 301
Gemini telescopes, 64
Gene, **317**
General theory of relativity, **164**
Geocentric, **34**
Geologic time, 321–322, 323

Geosynchronous satellite, **50**
Giant star, **110, 148**–150
Global warming, **272**
Globular cluster, **152**
GOODS, 198
Granulation, **81**
Grating, **68**
Gravitational collapse, **254**
Gravitational radiation, **164**
Gravitational redshift, **168**
Gravity, **48**
Great chain of origins, 242–243
Great Dark Spot, 306, 307
Great Nebula of Orion, 141,
  142–143
Great Red Spot, 300
Great Square of Pegasus, 12
Greenhouse effect, **268,** 269
Ground state, **83**

H-alpha Balmer line, 89
H-R diagram, **110,** 111
Habitable zone, **325**
Hale-Bopp, 70
Half-life, **250**
Halo, **179**
Halo population stars, 187
Harmonice Mundi (Kepler), 41
Hawking, Stephen W., 167
Heat, **79**
Heat of formation, **258**
Heavy bombardment, **255**
Heinlein, Robert, 146
Heliocentric, **35**
Helium fusion (giant start),
  148–149
Heloseismology, **91,** 94
Herbig-Haro objects, **140**
Hercules X-1, 164
Herschel, Caroline, 176
Herschel, William, 59, 60, 176,
  303
Hertzsprung-Russell (H-R) diagram, **110,** 111
Hewish, Anthony, 162
HII region, **136**
Hipparchus, 13, 14
Hipparcos, 105
Historical overview, 32–55
  ancient universe, 36–37
  Aristotle's universe, 34, 36–37
  Brahe, 40
  Copernicus, 34–39
  Galileo, 43–46
  Kepler, 40–43
  Newton, 46–49
Homogeneity, **229**
Horizon, **16**
Horizon problem, **232**
Horizontal branch, **149**
Horsehead Nebula, 137

Hubble, Edwin, 181, 196, 198, 203, 223, 233
Hubble constant, **203**
Hubble law, **203**
Hubble Space Telescope, 72, 202
Hubble time, **224**
Hulse, Russell, 164
Huygens probe, 302, 303
Hyades, 152
Hydra, 308
Hydra cluster, 223
Hydrogen Balmer absorption lines, 107
Hydrogen fusion, 130–131
Hydrostatic equilibrium, **126**
Hypernova, **171**
Hypothesis, **42**

Iapetus, 304
Ice line, **252**
Imaging systems, 67
Inflationary big bang, **232**
Infrared Astronomical Satellite (IRAS), 71
Infrared (IR), **59**
Instability strip, **177**
Interacting galaxies, 207, 208–209
Interferometer, **62**
International Ultraviolet Explorer (IUE), 71
Interstellar dust, **134**
Interstellar medium (ISM), **134**
Interstellar reddening, **135**
Intrinsic brightness, **105**, 106
Inverse square relation, **48,** 106
Io, 301
Ion, **82**
Ionize, **82**
IR, **59**
IRAS, 71
Iron meteorites, 248, 250
Irregular galaxies, **201**
ISM, **134**
Isotope, **82**
Isotropy, **229**
IUE, 71

J1655-40, 171
James Webb Space Telescope, 73
Jets, 170
John Paul II, Pope, 46
Jovian planets, **246**–247. *See also* Comparative planetology
Jovian problem, 256
Jupiter, 293–296
  atmosphere, 293–294, 300
  characteristics, 295
  history of, 296
  interior, 293
  moons, 295–296, 301
  ring, 294, 300

KAO, 69
Kelvin temperature scale, **79**
Kepler, Johannes, 40–43
Keplerian motion, 182
Kepler's laws of planetary motion, 42–43
Kepler's first law, 42–43
Kepler's second law, 43
Kepler's third law, 43
Kerr, Roy P., 167
Kiloparsec, **179**
Kirchhoff, Gustav, 86
Kirchhoff's laws, **86**
Kitt Peak National Observatory, 66
Kuhn, Thomas, 39
Kuiper, Gerard, 248
Kuiper Airborne Observatory (KAO), 69
Kuiper belt, **248**

Large Binocular Telescope (LBT), 64, 66
Large-impact hypothesis, **274**
Large Magellanic Cloud, **201**
Large-scale structure, **234**
Late heavy bombardment, **275**
LBT, 64, 66
Leavitt, Henrietta, 177
Life
  DNA, 316–317
  geologic time, 321–322, 323
  Miller experiment, 318, 319
  natural selection, 315, 318
  nature of, 314–318
  origin of, 318–321
  other planetary systems, in, 324–325
  physical basis of, 314
  solar system, in, 322–324
Light, matter and motion, 81–85
  Doppler effect, 84–85
  electron shells, 82
  excitation of atoms, 82–84
Light curve, **115**
Light-gathering power, **62**
Light pollution, **64**
Light-year (ly), **6**
Lighthouse model, **163**
Liquid metallic hydrogen, **293**
LMC X-3, 171
Local Group, **205,** 206
Look-back time, **202**
Luminosity classes, **112**
Luminosity, **106**–107, 110–113
Luminosity spectral classification, 112
Lunar eclipse, **27**
ly, **6**
Lyman series, **87**
Lyra, 151

m_v, **14**
M64, 207, 209
M87, 213
Magma ocean, **275**
Magnetic carpet, **89**
Magnetic solar phenomena, 96–97
Magnetosphere, **293**
Magnifying power, **63**
Magnitude scale, **13**
Main sequence star, **110,** 132–134
Mantle, **266**
Maria (mare), **273**
Mars, 281–287
  atmosphere, 282–283
  characteristics, 282
  Earth, compared, 282
  history of, 284–285
  moons, 285, 287
  surface, 283
  water, 283–284
Mass, **47**
Mass-luminosity relation, **117,** 118, 132–133
Massive stars, 156–161
Mathematical models, 128
Mathematical Syntaxis (Ptolemy), 37
Mathilde, 245
Mauna Loa, 283, 284
Maunder, E. Walter, 92
Maunder butterfly diagram, **92**
Maunder minimum, **93**
Measurement, power/accuracy, 63
Medium-mass (sunlike) stars, 150
Megaparsec (Mpc), 202
Mercury, 273, 277
Merope, 137
Metals, **187**
Meteor, 248–249
Meteor shower, **249,** 251
Meteorite, 248–249, 250
Meteoroid, **248**–249
Micrometeorites, **276**
Midocean rise, **270**
Milky Way Galaxy, **7,** 174–195
  age of, 187
  center of galaxy, 177–179, 190–191
  discovery of, 176
  disk component, 179
  element-building cycle, 187–188
  history, 188–189
  mass, 181–182
  monolithic collapse hypothesis, 189, 190
  nucleus, 190–191
  origin of, 187
  size, 176–177
  spherical component, 179–181
  spiral arms, 183–186
  star clusters, 177–179

star formation, 184–186
stellar populations, 187, 188
Miller, Stanley, 318, 319
Miller experiment, **318,** 319
Millisecond pulsars, **164**
Milton, John, 312
Minor planets, 244
Minute of arc, **17**
Miranda, 306
Model universes, 230
Molecular clouds, **138**
Molecules, **79**
Monolithic collapse hypothesis, 189, 190
Monolithic collapse method, **188**
Moon, 273–276
  cycles of, 22–23
  history, 275–276
  lunar geology, 273–274
  motion of, 23
  origin of, 274–275
  phases, 23, 24–25
Moonrise, 25
Moonset, 25
Morning star, **19**
Mpc, **202**
Multicellular, **321**
Multiringed basins, **275**
Murchison meteorite, 320
Mutant, **315**
Mysterium Cosmographicum (Kepler), 41

Nadir, **16**
Nanometer (nm), **58**
Natural law, **42**
Natural selection, **315,** 318
NCG4039, 209
NCG7251, 209
Neap tides, **49**
Near-infrared, 69
NEAR spacecraft, 245
Nebula, **134,** 136–137
NEO (Near Earth Object), **256**
Neptune, 305, 306–308
Nereid, 307
Neutrino, **131**
Neutron stars, **161**–166
Neutrons, **79,** 161
Newton, Isaac, 46–49
Newtonian focus, **65**
NGC1569, 207
NGC4038, 209
NGC5128, 208
NGC7674, 212
Nix, 308
nm, **58**
Nonbaryonic matter, **231**
North celestial pole, **16**
North point, **16**
Nova, **146**

Nuclear binding energy, 129–130
Nuclear bulge, **179,** 181
Nuclear fission, **129**
Nuclear force, **129**
Nuclear fusion, 129–132, **130**
Nuclear fusion (massive stars),
    156–158
Nucleus, **79**

Oblateness, **293**
Observable universe, **222**
Observatories, 64–67, 69–73
Occultation, **306**
Olbers, Heinrich, 222
Olbers's paradox, **222**
Olympus Mons, 283, 284
Oort cloud, **258**
Opacity, **127**
Open cluster, **152**
Open orbit, **51**
Open universe, **230,** 231
Opportunity, 284, 286
Optical telescopes, **61**
Orbital motion, 48–49
Origin of structure, 234–235
Orion Nebula, 141, 142–143
Orion nebula, 258
Other worlds, 324–329
Outer planets. *See* Comparative
    planetology
Outflow channels, **284**
Outgassing, **254**
Ovoid, **306**
Ozone, 272

P waves, **267**
Paradigm, **39**
Parallax, **36,** 78
Parked car effect, 268
parsec (pc), **105**
Partial eclipse, 26, 29
Paschen series, **87**
Paying for basic science, 302
Payne, Cecilia, 85, 88, 107
pc, **105**
Pegasus, 12
Penumbra, **26**
Penzias, Arno, 225, 226
Perihelion, **21**
Period-luminosity relation, **177**
Permafrost, **282**
Permitted orbits, **82**
Perseid meteor shower, 249
Phobos, 285, 287
Photographic plate, **67**
Photometer, **67**
Photon, **59**
Photosphere, **81**
Pickering, James S., 174
Pixel, 63

Planet, **5.** *See also* Comparative
    planetology; Solar system
Planet formation, 249–258
Planetary nebula, 150–**151,** 154
Planetesimal, **253**
Plate tectonics, **270**
Pleiades, 137
Plutino, **309**
Pluto, 308–309
Poe, Edgar Allan, 222
Polaris, 148
Poor galaxy clusters, **205**
Population I stars, **187**
Population II stars, **187**
Precession, **14,** 15
Predictions, 53
Present as key to past, 287
Pressure-temperature thermostat,
    132
Primary atmosphere, **268**
Primary lens, **60**
Primary mirror, **60**
Prime focus, **65**
Primordial soup, **319**
Principia (Newton), 46
Project Phoenix, 327
Prominence, **96**
Proof, 165
Proper motion, **177**
Protein, **316**
Proton-proton chain, **130,** 131
Protons, **79**
Protoplanet, **253**
Protostar, **140**
Proxima Centauri, 6
Pseudoscience, 22
PSR1257+12, 165–166
PSR1913+16, 164
Ptolemy, Claudius, 34, 37
Pulsar, **162**–165
Pulsar planets, 165–166

QSO, 212
Quantum leap, **83**
Quantum mechanics, **82**
Quasar, **212**–214, 216–217
Quasi-stellar objects (QSO), 212
Quintessence, **234**
QZ Vul, 171

Radial velocity, **85**
Radiation, 58–60
Radio dish, **64**
Radio galaxies, **210**
Radio telescopes, **61**
Recombination, **228**
Reconnections, **97**
Reconstructing the past, 251
Red dwarfs, **112,** 150
Redshift, **84**
Reflecting telescope, **60,** 61

Reflection nebula, **136**
Refracting telescope, **60**–61
Reionization, **228**
Resolving power, **62**
Retrograde motion, **36**
Revolution, **18**
Rich galaxy clusters, **205**
Riegel, 80
Rift valley, **270**
Ring galaxy, **209**
Ring Nebula, 151
RNA (ribonucleic acid), **317**
Roche limit, **294**
Roche lobe, **155**
Rotation, **18**
Rotation curve, **182**
Rotation curve method, **203**–204
Rubin, Vera, 204
Runaway greenhouse, **281**

S0 galaxies, 210
S waves, **267**
Sagittarius A* (Sgr A*), **190,**
    192–193
San Andreas fault, 115
Sandburg, Carl, 240
Saros cycle, **29**
Saturn, 297–304
    characteristics, 295
    moons, 302–304
    rings, 297–299
Schmidt, Maarten, 212
Schmidt-Cassegrain focus, **65**
Schwarzschild, Karl, 167
Schwarzschild black hole, 167–168
Schwarzschild radius, **167**
Scientific explanation, 314
Scientific model, **14,** 15
Scientific notation, **5**
Scientific predictions, 53
Scientific revolutions, 39
Seasons, 19–22
Second of arc, **17**
Secondary atmosphere, **268**
Secondary mirror, **65**
Seeing, **62,** 63
Selection effects, 199, 261
Self-sustaining star formation, **186**
Semi-major axis, **42**
Service, Robert, 290
SETI, **326,** 327
Seyfert, Carl K., 210
Seyfert galaxies, **210**–212
Sgr A*, **190,** 192–193
Shakespeare, William, 102
Shapley, Harlow, 177–179
Shepherd satellites, **299**
Shield volcano, **283**
Shock wave, **139**
Shoemaker-Levy 9, 257
Shooting stars, 248

Sidereal tracking, **64**
Sidereus Nuncius (Galileo), 44
SIM, 71
Singularity, **167**
Sirius, 151
Skepticism, 261
Sky and its motion, 14–18
Sloan Digital Sky Survey, 119
Slow surface evolution, 272
Small Magellanic Cloud, 177
Small Megallanic Cloud, **201**
SN1987A, 159–160
SOFIA, 69, 71
Solar eclipse, **23**–27
Solar nebula hypothesis, **243**
Solar system, **5,** 240–311. *See also*
    Comparative planetology
    age, 249–251
    asteroids, 244, 245
    bombardment of planets,
        255–256
    catastrophic theory, 257
    characteristic properties, 252
    chemical composition of solar
        nebula, 251–252
    comets, 244, 248
    condensation of solids, 252–253
    debris disks, 259
    evolutionary theory, 257
    extrasolar planets, 259–260
    great chain of origins, 242–243
    Jovian planets, 246–247
    Jovian problem, 256
    meteoroids/meteorites, 248–249
    other planets?, 258–261
    planet formation, 249–258
    planetesimals, 253
    protoplanets, 253–255
    revolution, 243
    rotation, 243
    solar nebula hypothesis, 243
    space debris, 244–249
    terrestrial planets, 246–247
Solar wind, **89**
South celestial pole, **16**
South point, **16**
Space aliens, 326
Space debris, 244–249
Space Interferometry Mission
    (SIM), 71
Space-time, 230, 235–236
Spectral classes, **108**
Spectral lines, **76,** 85
Spectral sequence, **109**
Spectrograph, **68**
Spectroscopic binary system,
    115–116
Spectroscopic parallax, **112**
Spectrum, **68**
Spherical component, **179**
Spicule, **89**

Spiral arm, 7, **179**
Spiral galaxies, **200**
Spiral tracers, **183**
Spirit, 284
Spitzer Space Telescope, 73
Spring tides, **49**
SS433, 170
St. Augustine, 46
Standard candles, **202**
Star, **5**, 102–173
  absolute visual magnitude, 106
  binary stars, 113–117
  birth of, 134–143, 184–186
  black hole, 166–171
  brightness, 13–14, 105–106
  brown dwarfs, 133
  clusters, 149–150, 152–153
  constellations, 12
  distances, 104–105
  family of stars, 120–121
  giant, 148–150
  hydrogen fusion, 130–131
  instellar medium, 134–138
  luminosity, 106–107, 110–113
  main-sequence, 132–134
  mass-density-luminosity relation, 117–118
  massive, 156–161
  names, 13
  neutron, 161–166
  nova, 156
  nuclear fusion, 129–132
  planetary nebulae, 150–151
  population I, 187
  population II, 187
  protostar, 139–140
  pulsar, 162–165
  red dwarfs, 150
  size, 109–113
  spiral arms, 184–186
  stellar structure, 126–129
  sunlike, 150
  supernova, 158–160
  surveying the, 118–119
  T Tauri, 140
  temperature, 107–109
  white dwarfs, 151, 154
  young, 141
Star cluster, **139**, 149–150, 152–153, 177
Star formation, 134–143
Star temperatures, 107–109
Starburst galaxies, **207**
Statistical evidence, 211
Stefan-Boltzmann Law, **80**
Steinbeck, John, 32
Stellar association, **139**

Stellar model, **128,** 129
Stellar parallax, **104**
Stellar populations, 187, 188
Stellar spectra, 109
Stellar structure, 126–129
Stoney meteorites, 248, 250
Stony-iron meteorites, 248, 250
Stratospheric Observatory for In-frared Astronomy (SOFIA), 69, 71
Stromatolites, **318**
Subduction zone, **270**
Sudbury Neutrino Observatory, 132
Summer solstice, **20**
Sun
  annual motion, 18–19
  atmosphere, 85–89
  Babcock model, 94–95
  basic characteristics, 78, 79
  chemical composition, 85–88
  chromosphere, 88–89
  chromospheric/coronal activity, 98
  corona, 89
  cycle of, 18–22
  density, 78
  distance, 78
  elements, 88
  fate of, 154–155
  heat flow, 81
  inside the interior, 91
  magnetic cycle, 91, 94
  magnetic solar phenomena, 96–97
  mass, 78
  nuclear fusion, 129–130
  origin of sunlight, 78–80
  photosphere, 81
  size, 78
  spectra, 85, 86–87
  sunspots/sunspot cycle, 91, 92–93
  surface, 81
Sunlight, 78–80
Sunlike stars, 150
Sunrise, 25
Sunset, 24
Sunspots/sunspot cycle, **81,** 91, 92–93
Supercluster, **8, 234**
Supergiant star, **112,** 148
Supermassive black holes, 205, 215–216
Supernova, **146,** 158–160
Supernova explosion, 139
Supernova remnant, **160**

Surveyor's triangulation method, 104
Synchrotron radiation, **159**

T Tauri star, **140**
Tarantular Nebula, 201
Taurus the Bull, 159
Taylor, Joseph, 164
Technological civilizations per galaxy, 328
Telescope. *See* Astronomical tele-scopes and instruments
Temperature, **79**
Temperature spectral classification, 107–109
Terby, 287
Terrestrial planets, **246**–247. *See also* Comparative planetology
Testing a theory by prediction, 53
Tharsis region, 283
Theory, **42**
3C48, 212
3C175, 214
3C219, 212
3C273, 212, 213
Tidal healing, **296**
Tides, 49
Time dilation, **168**
Titan, 302, 303
Total eclipse, 26
Total lunar eclipse, 28
Total solar eclipse, **26**
Transit of Venus, **78**
Transition, **87**
Triangulation method, 104–105
Trilobites, 321
Triton, 307
Turnoff point, **152**
Twain, Mark, 124
Two Micron All Sky Survey (2MASS), 119
Type I supernova, **158**
Type Ia supernova, 158
Type Ib supernova, 158
Type II supernova, **158**

UFO sightings, 326
Ultraviolet (UV), **59**
Umbra, **23**
Uncompressed densities, **252**
Unified model, **214**–215
Uniform circular motion, **36**
Universal gravitation, 48
Universe. *See* Cosmology
Upsilon Andromedae, 260
Uranus, 303–306

Urban VIII, 45
Urey, Harold, 318
UV, **59**

V404 Cygni, 171
V4641 Sgr, 171
Valles Marineris, 283
Valley networks, **284**
Venus, 277–281
Vernal equinox, **20**
Very Large Array (VLA), 66, 67
Very Large Telescope (VLT), 64, 66
Very Long Baseline Array (VLBA), 66
Vesta, 245
Virgo cluster, 223
Visual binary system, **114**
VLA, 66, 67
VLBA, 66
VLT, 64, 66
Volatile, **244**
Voyager 2, 290, 303, 306

Waning crescent, 25
Waning gibbous, 25
Water hole, **326**
Wavelength, **58,** 59
Wavelength of maximum intensity, **80**
Waxing crescent, 23, 24
Waxing gibbous, 23, 24
Weight, **47**
Wekiu bug, 313
West point, **16**
White dwarfs, **112,** 151, 154
Wien's law, **80**
Wilkinson Microwave Anisotropy Probe (WMAP), 235
Wilson, Robert, 225, 226
Winter solstice, **20**
WMAP, 235

X ray, **59**
X-ray binaries, 164, 169
XTEJ1118+480, 171

Yellow giants, 149
Young stars, 141

ZAMS, **133**
Zeeman effect, **93**
Zenith, **16**
Zero-age main sequence (ZAMS), **133**
Zodiac, **19**
Zwicky, Fritz, 161, 162